Safe Science

Workplace Hazardous Materials Information System (WHMIS)

- compressed gas
- dangerously reactive material
- flammable and combustible material
- biohazardous infectious material
- oxidizing material
- poisonous and infectious material causing immediate and serious toxic effects
- corrosive material
- poisonous and infectious material causing other toxic effects

Hazardous Household Product Symbols (HHPS)

Symbol	Danger
	Explosive This container can explode if it is heated or punctured.
	Corrosive This product will burn skin or eyes on contact, or throat and stomach if swallowed.
	Flammable This product, or its fumes, will catch fire easily if exposed to heat, flames, or sparks.
	Poisonous Licking, eating, drinking, or sometimes smelling, this product is likely to cause illness or death.

Practise Safe Science in the Classroom

Be science ready

- Come prepared with your textbook, notebook, pencil, and anything else you need.
- Tell your teacher about any allergies or medical problems.
- Keep yourself and your work area tidy and clean. Keep aisles clear.
- Wear closed shoes (not sandals).
- Do not wear contact lenses while doing investigations.
- Read all written instructions carefully before you start an activity or investigation.

Follow instructions

- Do not enter a laboratory unless a teacher is present or you have permission to do so.
- Listen to your teacher's directions. Read written instructions. Follow them carefully.
- Ask your teacher for directions if you are not sure what to do.
- Wear eye protection or other safety equipment when instructed by your teacher.
- Never change anything, or start an activity or investigation on your own, without your teacher's approval.
- Get your teacher's approval before you start an investigation that you have designed yourself.

Act responsibly

- Pay attention to your own safety and the safety of others.
- Know the location of MSDS (Material Safety Data Sheet) information, exits, and all safety equipment, such as the first-aid kit, fire blanket, fire extinguisher, and eyewash station.
- Alert your teacher immediately if you see a safety hazard such as broken glass, a spill, or unsafe behaviour.
- Stand while handling equipment and materials.
- Avoid sudden or rapid motion in the laboratory, especially near chemicals or sharp instruments.
- Never eat, drink, or chew gum in the laboratory.
- Do not taste, touch, or smell any substance in the laboratory unless your teacher asks you.
- Clean up and put away any equipment when finished.
- Wash your hands with soap and water when done.

NELSON

SCIENCE CONNECTIONS 9

Grade 9 Author Team

Lucille Davies, B.A., M.Sc., B.Ed.
Head of Science, Frontenac Secondary School, Limestone District School Board

Maurice DiGiuseppe, Ph.D.
University of Ontario Institute of Technology (UOIT)

Douglas Fraser
District School Board Ontario North East

Michelle Kane, B.Sc. (Hons.), B.Ed., M.Sc.
Head of Science, Aurora High School, York Region District School Board

Barry LeDrew
Curriculum and Educational Resources Consulting Ltd.

Milan Sanader, B.Sc., B.Ed., M.Ed.
Formerly of Dufferin-Peel Catholic District School Board

Henri M. van Bemmel, B.Sc. (Hons.), B.Ed.
Marc Garneau Collegiate Institute, Toronto District School Board

Rob Vucic, B.Sc. (Hons.), B.Ed.
Head of Science, T.L. Kennedy Secondary School, Peel District School Board

Program Consultants

Maurice DiGiuseppe, Ph.D.
University of Ontario Institute of Technology (UOIT)

Douglas Fraser
District School Board Ontario North East

NELSON EDUCATION

NELSON EDUCATION

Nelson Science Connections 9

Program Consultants
Maurice DiGiuseppe
Douglas Fraser

Authors
Lucille Davies
Maurice DiGiuseppe
Douglas Fraser
Michelle Kane
Barry LeDrew
Milan Sanader
Henri van Bemmel
Rob Vucic

Contributing Authors
Andrew Fazekas
Michael Stubitsch

Vice President, Publishing
Janice Schoening

Acting Publisher, Science
David Spiegel

Managing Editor, Development
Susan Ball

National Director of Research and Teacher In-Service
Jennette MacKenzie

Product Manager
Lorraine Lue

Secondary Sales Specialist
Rhonda Sharp

Program Manager
Carmen Yu

Contributing Editor
Jennifer Hounsell

Developmental Editors
Christina D'Alimonte
Sarah Tanzini

Project Editor
Ananda Fulton

Editorial Assistants
Vytas Mockus
Amy Rotman
Wally Zeisig

Director, Content and Media Production
Linh Vu

Senior Content Production Manager
Debbie Davies-Wright

Content Production Editors
Jane High
Rebecca Rosenblum
Cheryl Tiongson

Copyeditor
Susan McNish

Proofreader
Linda Szostak

Indexer
Noeline Bridge

Production Manager
Helen Jager Locsin

Senior Production Coordinators
Sharon Latta Paterson
Kathrine Pummell

Design Director
Ken Phipps

Interior Design
Glenn Toddun

Feature Pages Design
Jarrel Breckon
Courtney Hellam
Eugene Lo

Series Cover Design
Bill Smith Studio

Cover Design
Courtney Hellam

Cover Image
© Josh Westrich/Corbis

Asset Coordinator
Suzanne Peden

Art Tracking
Franca Amore
Hedy Sellers

Illustrators
Steve Corrigan
Deborah Crowle
Steven Hall
Joel Harris
Sharon Harris
Stephen Hutching
Sam Laterza
Mapping Specialists
Dave Mazierski
Dave McKay
Allan Moon
Nesbitt Graphics, Inc.
Precision Graphics
Jan-John Rivera
Theresa Sakno
Ann Sanderson
Thomson Digital
Bart Vallecoccia
Wilkinson Studios

Compositors
Nelson Gonzalez
Nelson Media Production Services
Nesbitt Graphics, Inc.

Photo Shoot Coordinator
Lynn McLeod

Photo/Permissions Researchers
Natalie Barrington
Daniela Glass
Natalie Russell

Cover Research
Natalie Russell

Printer
Transcontinental Printing, Ltd.

ISBN-13: 978-0-17-624128-5
ISBN-10: 0-17-624128-0

Printed and bound in Canada
1 2 3 13 12 11 10

For more information contact Nelson Education Ltd., 1120 Birchmount Road, Toronto, Ontario M1K 5G4. Or you can visit our Internet site at http://www.nelson.com.

Reviewers

Accuracy Reviewers

Sean Clark
Science Department Head, Sacred Heart High School, Ottawa Catholic DSB

Dr. Paul Ragogna, B.Sc., Ph.D.
Associate Professor of Chemistry, The University of Western Ontario

Assessment Consultants

Aaron Barry, M.B.A., B.Sc., B.Ed.
Sudbury Catholic DSB

Mike Sipos, B.Ph.Ed., B.Ed.
Sudbury Catholic DSB

Catholicity Reviewer

Ted Laxton
Faculty of Education, St. Jerome's University/Wilfrid Laurier University

Environmental Education Consultant

Mark Stabb

ESL Consultants

Paula Markus, B.A., M.A.
ESL/ELD Program Coordinator, Toronto DSB

Lyle Pearson
Formerly of Waterloo Region DSB

Literacy Consultants

Phil Davison
Literacy Coordinator, Halton DSB

Tanya Korostil, B.A. (Hons.), B.Ed., M.Ed.
Peel DSB

Numeracy Consultant

Justin DeWeerdt
Mathematics Teacher/Mathematics Coach, Trillium Lakelands DSB

Safety Consultant

Jim Agban
Past Chair, Science Teachers' Association of Ontario (STAO) Safety Committee

STSE Consultants

Joanne Nazir
Ontario Institute for Studies in Education (OISE), University of Toronto

Ron M. Ricci, B.E.Sc., B.Ed.
Greater Essex County DSB

Technology/ICT Consultant

Luciano Lista, B.A., B.Ed., M.A.
Academic Information Communication Technology Consultant
Online Learning Continuing Education Principal, Toronto Catholic DSB

Advisory Panel and Teacher Reviewers

Anca Bogorin
Avon Maitland DSB

Mark Carlin
Waterloo Region DSB

Charles Cohen
Community Hebrew Academy of Toronto

Carolyn Craven
Hamilton-Wentworth DSB

Greg Dick
Waterloo Region DSB

Matt DiFiore
Dufferin-Peel Catholic DSB

Ed Donato
Simcoe Muskoka Catholic DSB

Daniel Gajewski
Ottawa Catholic DSB

Theresa H. George
Dufferin-Peel Catholic DSB

Sharon Gillies
London District Catholic School Board

Nicole Hagley
Queen's University

Robert Hammond
Upper Canada DSB

Susan Hewitt
Ottawa-Carleton DSB

Ted Hill
Durham DSB

Shawna Hopkins
Niagara DSB

Jan Johns
Upper Canada DSB

Michelle Kane
York Region DSB

Roche Kelly
Durham DSB

Katherine Lalonde
Ottawa-Carleton DSB

Nadine Morrison
Hamilton-Wentworth DSB

Kerry Odlum
York Region DSB

Dermot O'Hara
Toronto Catholic DSB

Robert Pickett
Catholic DSB of Eastern Ontario

Z. Peter Reiter
Toronto Catholic DSB

Adrian Renzetti
Upper Grand DSB

Ron Ricci
Greater Essex County DSB

Athanasios Seliotis
York Region DSB

David Spencer
Upper Canada DSB

Charles Stewart
Peel DSB

Carl Twiddy
Formerly of York Region DSB

Vicki Wraight
Kawartha Pine Ridge Board of Education

Contents

Chapter 4

Canada

Chapter 9

Chapter 10

Discover Your Textbook

This textbook will be your guide to the exciting world of science. The following pages introduce the important features you will find inside. GET READY includes all of the features at the beginning of each unit and chapter. GET INTO IT shows you all the features within each chapter. Finally, WRAP IT UP shows you the features at the end of each chapter and unit.

Get Ready

Focus on STSE
These articles bring the topics you will be learning about into the real world.

Unit Opener
Each unit has a letter and a title. The photo will help you predict what you might learn in the unit.

Unit Preview
The Unit Preview describes what you will learn about in the unit.

Big Ideas
The Big Ideas are the concepts you need to remember after you complete the unit.

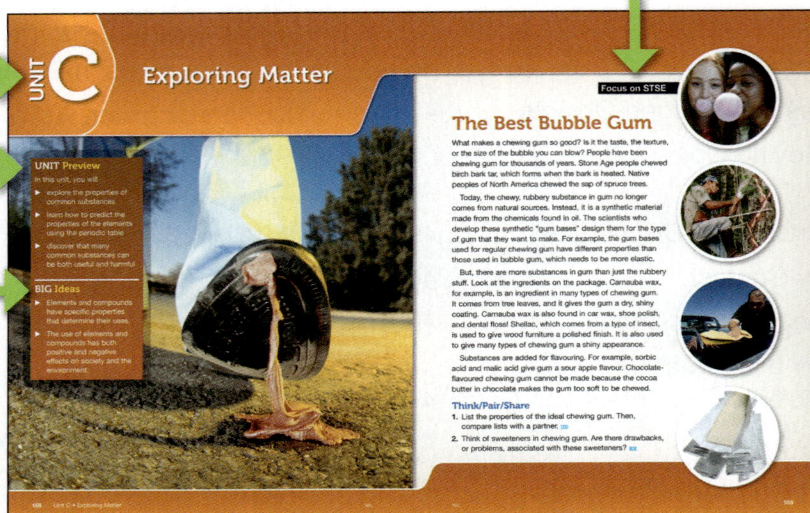

Concept Map
The Concept Map is a description of the topics, connected to picture clues, to help you predict what you will be learning in the unit.

Getting Started
This section gives a quick review of concepts and skills you developed in previous grades. Use the questions to see what you might already know before you start the unit.

Unit Task Preview
Find out about the Unit Task that you will complete at the end of each unit.

Unit Task Assessment
The Assessment Box tells you how you will demonstrate what you have learned by the end of the unit.

Unit Task Bookmark
When you see the Unit Task Bookmark, think about how the section relates to the Unit Task.

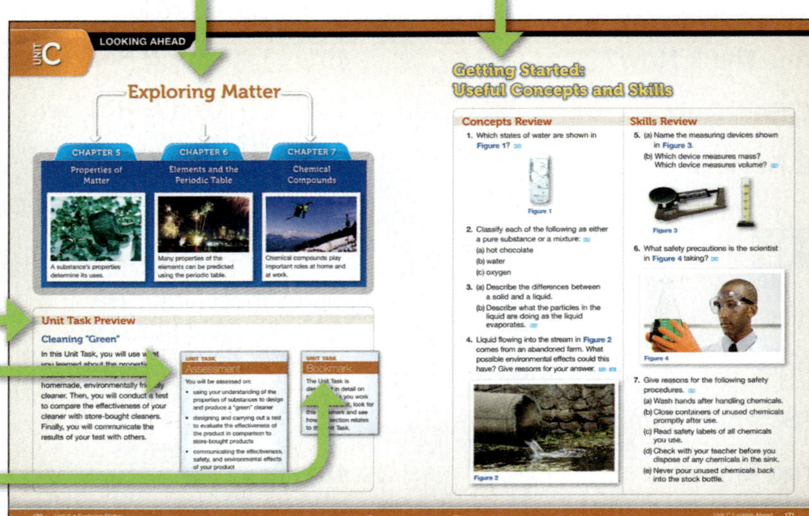

Chapter Opener
Each chapter has a number, title, and a Key Question. You should be able to answer the question by the end of the chapter.

Key Concepts
The Key Concepts are the main ideas discussed in the chapter.

Engage in Science
These articles connect a real-world topic to the content in the chapter.

What Do You Think?
In this section, you will use what you already know to agree or disagree with statements that connect to ideas introduced in the chapter.

Writing Tip
Writing Tips are useful strategies to help you improve your writing skills. Look for these tips throughout the chapter.

Focus on Reading/ Focus on Writing
These reading and writing strategies help you learn science concepts and develop reading and writing skills for the OSSLT.

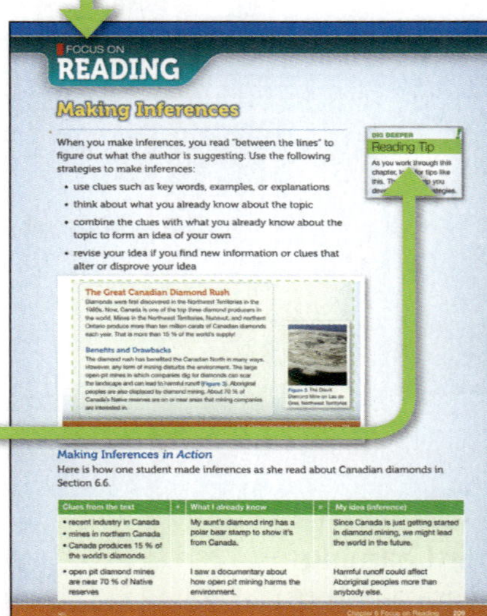

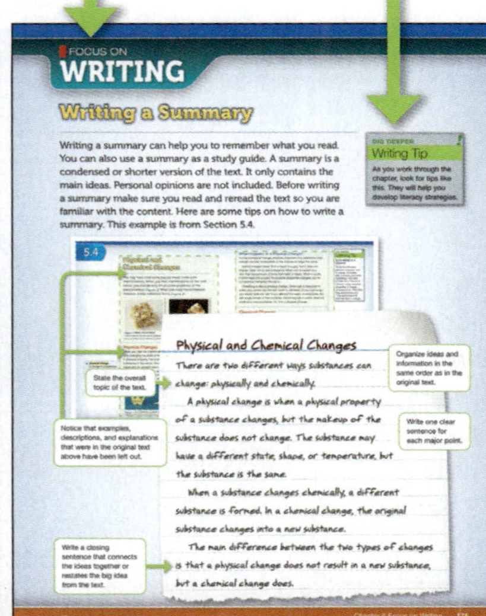

Reading Tip
Reading Tips suggest strategies to help you understand the science concepts covered in the text. Look for these tips throughout the chapter.

Get Into It

Learning Tip
Learning Tips are useful strategies to help you learn new ideas and make sense of what you are reading.

Vocabulary
You will learn many new terms as you work through the chapter. These key terms are highlighted and in bold print. Their definitions are in the margins, as well as the Glossary at the back of the book.

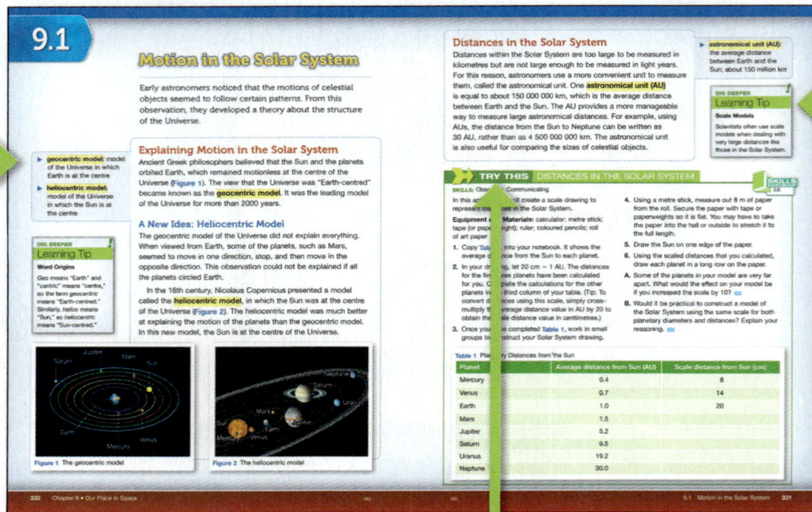

Try This
These are quick, fun activities designed to help you understand concepts and improve your science skills.

Did You Know?
Read interesting facts about real-world events that relate to the topics you are learning.

Citizen Action
These activities encourage you to be a good citizen and a steward of the environment by taking action at home and in your community.

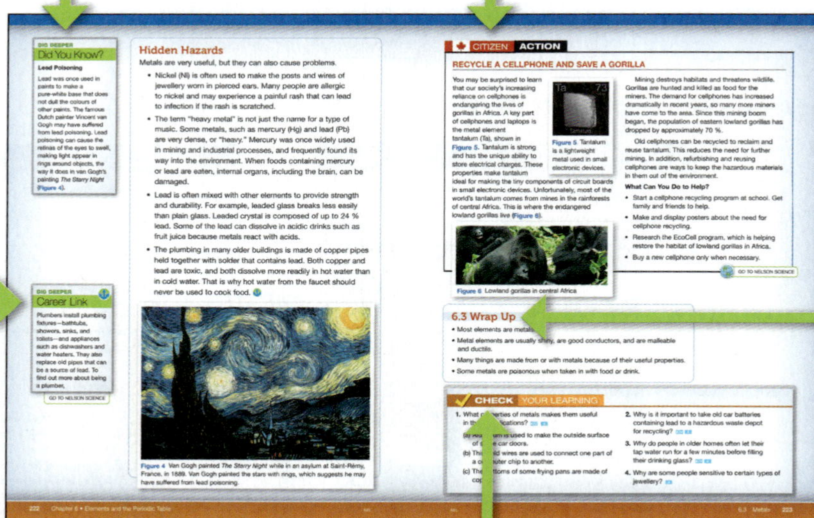

Career Link
The Career icon lets you know that you can learn about science-related careers by visiting the Nelson Science website.

Check Your Learning
Complete these questions at the end of each content section to make sure you understand the concepts you have just learned.

Wrap Up
At the end of each content section, this quick summary of the main ideas will help you review what you learned.

Sample Problems

This feature shows you how to solve problems that appear in the chapter. Be sure to complete the Practice Problems to test your understanding.

Research This

These research-based activities will help you relate science and technology to the world around you and improve your critical thinking and decision-making skills.

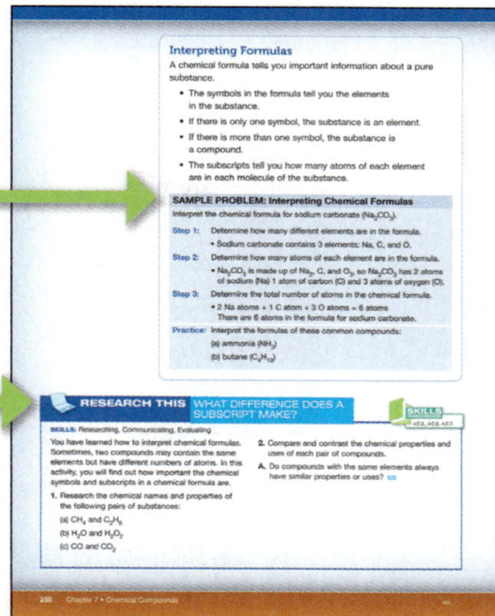

Focus on Math

These sections focus on mathematical skills you will need to better understand the concepts in the chapter.

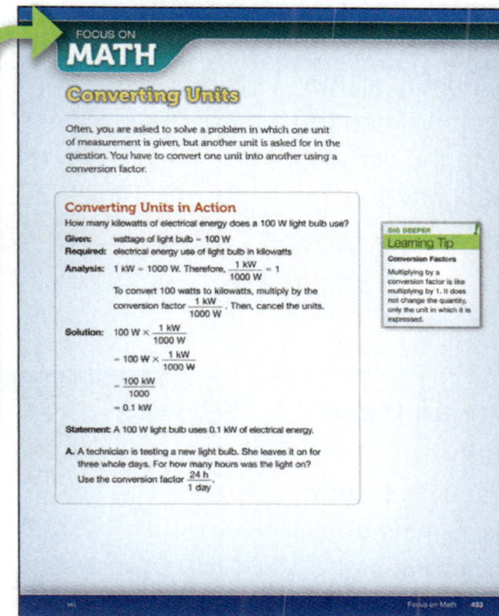

Magazine Features

Look for these special features to learn more about exciting developments in science, cool new technology, career links, or how science relates to your everyday life.

OSSLT Icon

This icon lets you know that the material will help you develop literacy skills in preparation for the OSSLT.

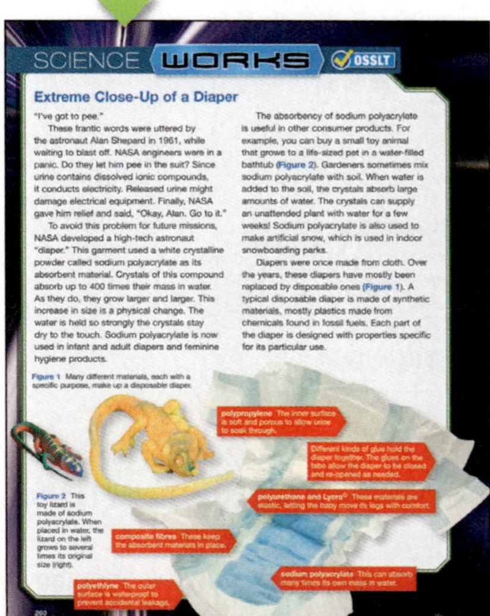

Perform an Activity

These hands-on activities allow you to observe the science that you are learning.

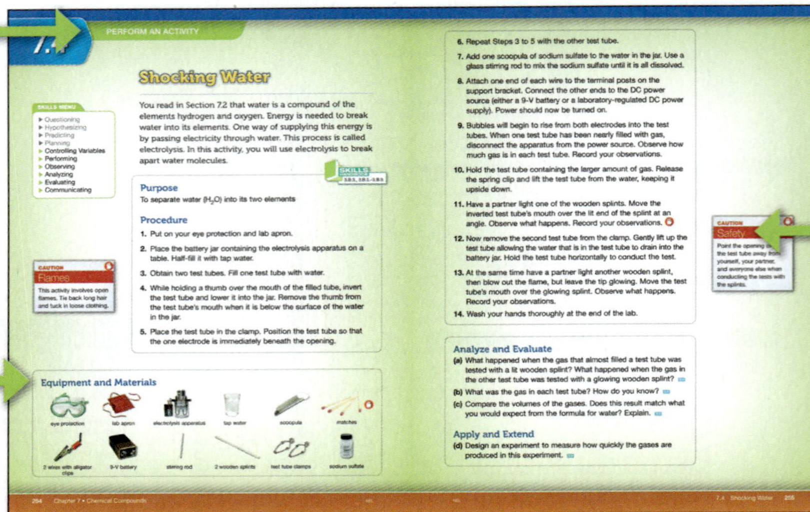

Equipment and Materials

These photos help you to identify all the items you will need to complete the activity or investigation.

Safety Warnings

Look for these warnings about possible dangers in investigations and activities. They are in bright red and have a safety icon.

Explore an Issue Critically

These activities allow you to examine social and environmental issues related to the unit. They often involve research, decision-making skills, and communication.

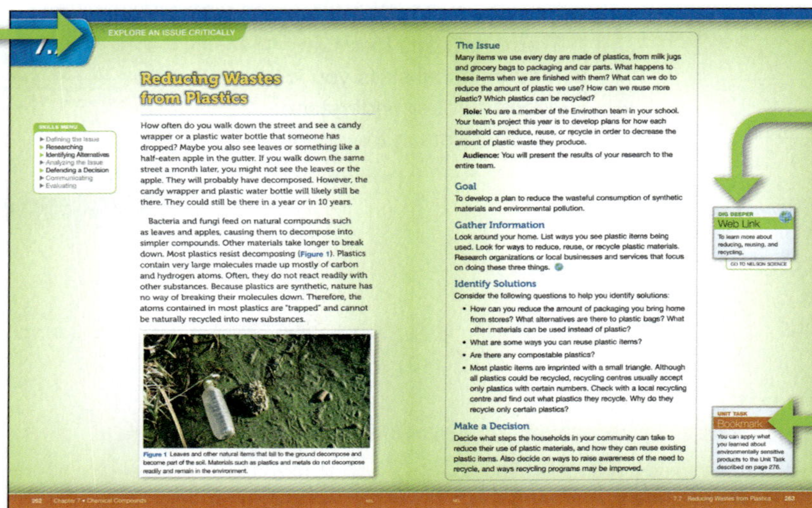

Web Link

When you see this icon, you can visit the Nelson Science website to learn more about the topic, watch a video, do an online activity, or take a quiz.

Unit Task Bookmark

This icon lets you know that the concepts you learned in the section will help you to complete the Unit Task.

Conduct an Investigation

These experimental investigations help you to develop science process skills.

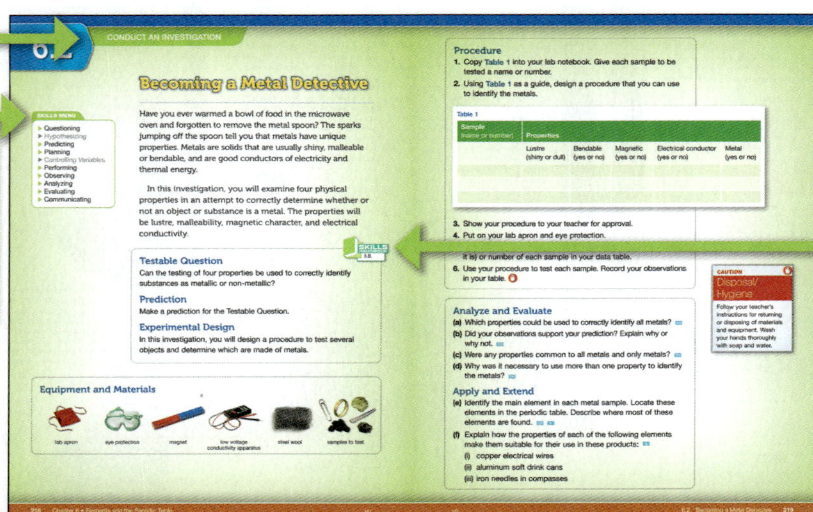

Skills Menu

The Skills Menu in each activity lists the skills that you will use to solve the problem or reach a conclusion.

Skills Handbook Icon

This icon shows you which part of the Skills Handbook has useful information for the activity.

Wrap It Up

What Do You Think Now?
Think about what you learned in the chapter and consider whether you have changed your opinion by agreeing or disagreeing with the statements.

Key Concepts Summary
The Key Concepts Summary feature outlines the main ideas you learned in the chapter.

Big Ideas
Here you can review the Big Ideas of the unit.

Vocabulary
Here you will find a list of all the new terms that were introduced in the chapter and the page number where the term is defined.

Chapter Self-Quiz
The Chapter Self-Quiz is a helpful tool for you to make sure you understand all the concepts you learned in the chapter.

Achievement Chart Icons
All questions are tagged with these icons. They identify the types of knowledge and skills you must use to answer the question.

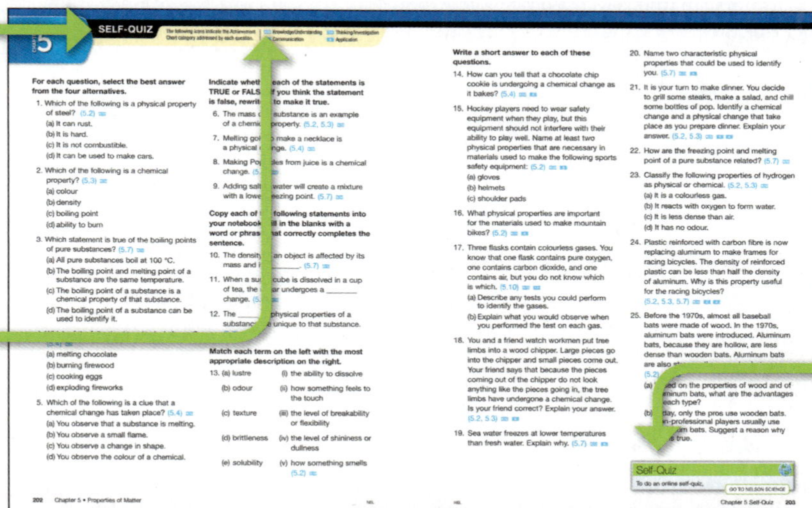

Online Quiz Icon
There is an interactive online quiz for each chapter on the Nelson Science website.

Chapter Review
Complete these questions to check your learning and apply your new knowledge from the chapter.

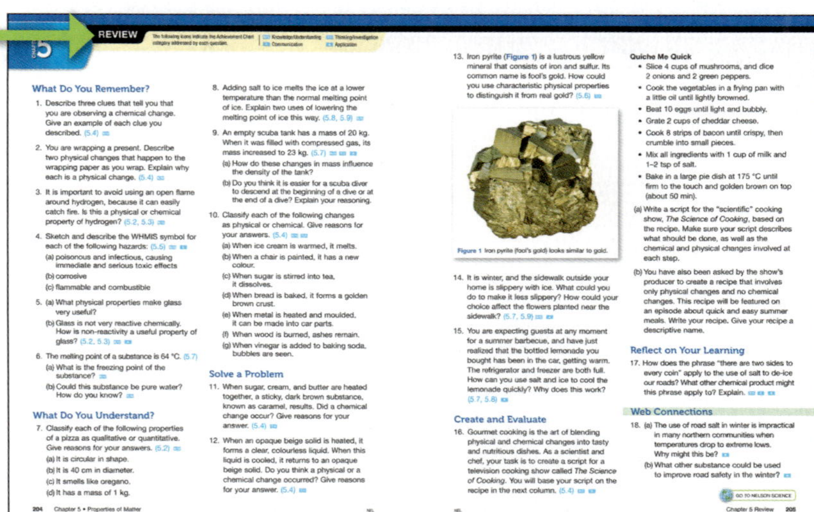

Master Concept Map

This feature brings together the Key Concepts from each chapter and summarizes all the main ideas in the unit.

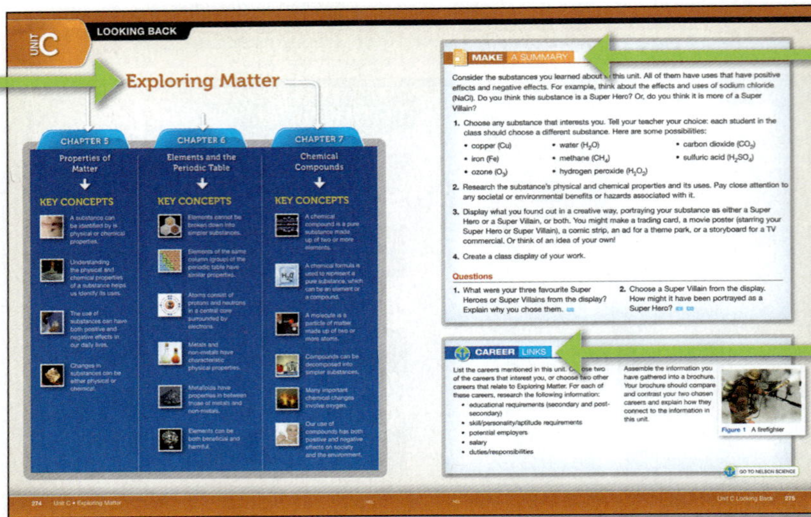

Make a Summary

Summarize what you have learned in the unit by completing the Make a Summary activity.

Career Links

Make connections between what you learned in the unit and future careers by completing the Careers Links activity.

Unit Task

Demonstrate the skills and knowledge you developed in the unit by completing the challenge described in the Unit Task.

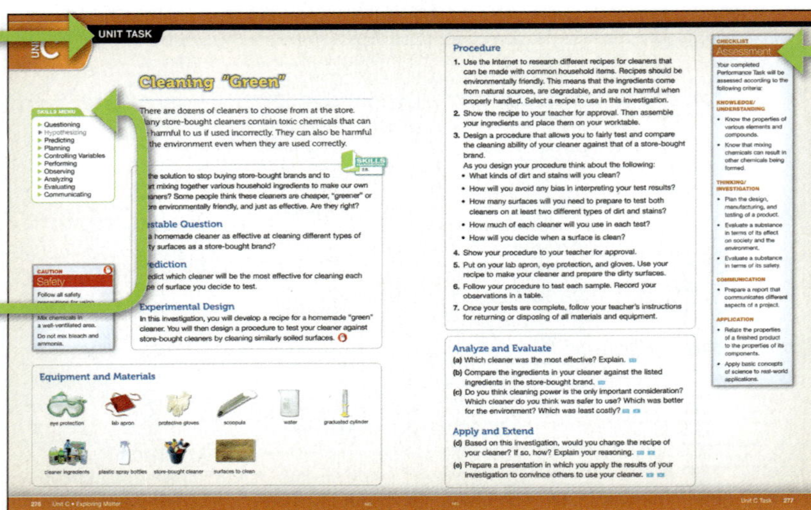

Skills Menu

The Skills Menu identifies the skills you will use to complete the Unit Task.

Assessment Checklist

This checklist lists the criteria that your teacher will use to evaluate your work on the Unit Task. Read this list carefully before completing the task.

Unit Self-Quiz

The Unit Self-Quiz is an opportunity for you to make sure that you understand all the main ideas from the unit.

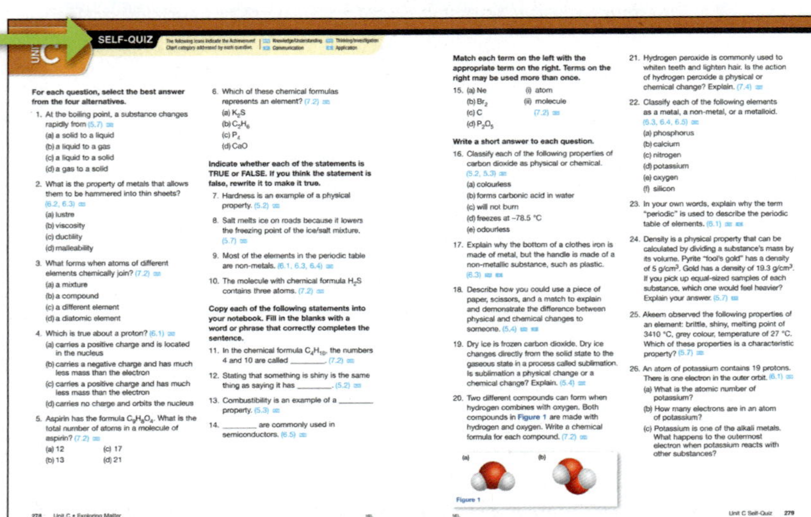

Unit Review

Complete the Unit Review questions to check your learning of all the concepts and skills in the unit.

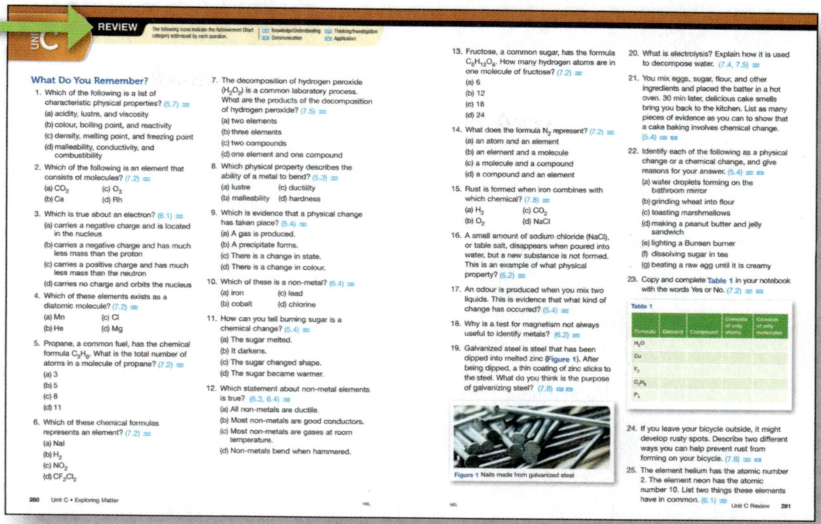

Skills Handbook

The Skills Handbook is your resource for useful science skills and information. It is divided into numbered sections. Whenever you see a Skills Handbook icon, it will direct you to the relevant section of the Skills Handbook.

Glossary

This is a list of all the key terms in the textbook in alphabetical order, along with their definitions. Use the Glossary to check your understanding of any key terms you may need to review.

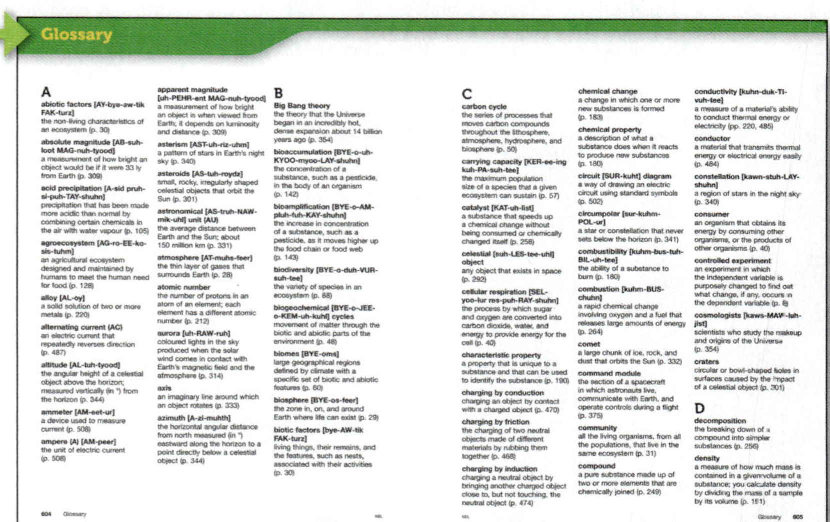

UNIT A Introduction to Scientific Investigation Skills and Career Exploration

GOALS of the Science Program

- To relate science to technology, society, and the environment
- To develop the skills, strategies, and habits of mind required for scientific investigation
- To understand the basic concepts of science

Focus on STSE

Science and Your Life

The goals of science education include much more than just learning scientific facts or knowledge. Your science program and this textbook will help you understand how science affects your everyday life, society, and the environment.

This introductory unit introduces the important scientific investigation skills that you will develop as you study biology, chemistry, space science, and physics in the units that follow.

In these four units, you will have many opportunities to learn through scientific inquiry. Through these activities you will develop, practise, and refine scientific investigation skills. These skills are useful in high school science, and they will continue to be useful in your post-secondary education and in your everyday life. In these units, you will also have opportunities to explore careers that are related to the science topics.

The main purpose of learning science at this level is to make connections. As you progress through this course, you will develop an understanding of how science, technology, society, and the environment (STSE) are related. You will connect the STSE relationships to your everyday life, and you will develop scientific literacy.

Think/Pair/Share

1. List five ways that science and technology directly or indirectly affect your daily life. A
2. Pair up with a partner and share your ideas. Brainstorm to add further examples. C
3. Join another pair and share your lists. Eliminate examples that are the same. Share your list with the whole class. C A

CHAPTER 1
Living and Working with Science

KEY QUESTION

What skills are required to carry out scientific investigations?

KEY CONCEPTS

- Scientific inquiry can be conducted in different ways, depending on the question to be answered.
- All scientific inquiries rely on scientific investigation skills.
- Science and technology are an important part of our everyday lives.
- Scientific literacy is necessary for wise personal decisions, responsible citizenship, and careers.

The Winds of Change!

Wind turbines are becoming a common sight in Ontario, across the country, and around the world. Governments and utility companies are investing in environmentally friendly energy sources and technologies that use the power of the wind.

The Ontario government has made plans to eliminate coal-fired generating stations (**Figure 1**) by 2014. The plan calls for the use of new renewable sources of energy to replace coal. In 2009, approximately 1100 megawatts of electricity was generated from wind energy in Ontario. This is a significant increase in the amount of energy generated from renewable sources but still represents only a little more than 4 % of the greatest demand.

A major offshore wind farm project (**Figure 2**) is planned for Lake Ontario, approximately 17 to 28 kilometres from the shoreline of Prince Edward County. The farm will contain 142 turbines and produce 710 megawatts of electricity—enough to power about 300 000 homes.

Despite the promise of cost-effective energy with no greenhouse gas emissions and no toxic waste, there is a growing opposition to the use of wind farms to generate electrical energy. The local community is concerned about possible negative effects such as noise, physical safety, and danger to migrating birds. They also worry that this wind farm will discourage tourists and decrease real estate values.

Who is right in this issue? Is either side right? How does wind compare to fossil fuels or nuclear power as a source of energy? How does it compare to solar energy or hydro-electricity? Should we be more concerned about saving energy by changing our lifestyle? Where do you stand?

Figure 1 The Nanticoke coal-fired generating station is Canada's largest single source of greenhouse gases.

Figure 2 Offshore wind farms, like this one in Denmark, are being planned for Ontario.

FOCUS ON
READING

How to Read Non-Fiction Text

Non-fiction texts convey ideas and information using words and graphics. Using the following strategies will help you to better understand what you read:

DIG DEEPER

Reading Tip

As you work through the chapter, look for tips like this. They will help you develop literacy strategies.

Before Reading

- Think about the topic. Do you have a personal experience that relates to it? Can you use any knowledge you have acquired to understand the topic?
- Preview the text: Scan titles, headings, bolded words, graphics, and captions.
- Identify the organizational pattern of the text and think about how you have approached this type of text pattern before.
- Ask a purpose-for-reading question: Turn a title or heading into a question that you will answer as you read the text.

During Reading

- Find information that answers your purpose-for-reading question.
- Make connections to your knowledge or personal experiences that are related to the topic.
- Look for relationships among the ideas and information given.
- Ask questions about parts of the text that puzzle you.
- Break the text down into chunks and pause after each one to determine whether or not you understand what you are reading.
- Find the main idea. Sometimes you may have to infer a main idea that is stated indirectly.
- Make jot notes or use sticky notes to keep track of what you understand and what confuses you.

After Reading

- Summarize the main ideas and details that you found important.
- Assess whether you answered your purpose-for-reading question.
- Reflect on what you have learned by forming an opinion, drawing a conclusion, or identifying what you still do not understand.
- Ask yourself how you can apply what you learned from the selection.

1.1

Skills of Scientific Investigation

Scientists assume that the natural world can be explained and understood. Hundreds of years ago, people believed that all of the objects in the sky revolved around Earth. Careful scientific observations and technological advances, such as the invention and use of the telescope, have provided evidence that the Sun is the centre of our Solar System. Scientists now know that the Universe is very different from how they had first thought of it (**Figure 1**).

Figure 1 Observations of the night sky have helped us to better understand the Universe.

DIG DEEPER

Reading Tip

Scan for Specialized Vocabulary

Before reading, scan the text to find the highlighted vocabulary. Check the margins for definitions of these words. Knowing the meaning of these words ahead of time can help you understand the ideas and information they refer to in the text.

Types of Scientific Inquiry

Scientific inquiry is a set of methods that scientists use to gather information and explain the natural world. All scientific inquiry uses similar processes to find answers to questions or to find relationships between variables. A **variable** is any condition that could change in an inquiry. A variable that is deliberately changed by the investigator is called the **independent variable**. A variable that changes in response to the independent variable, but is not directly controlled by the investigator, is called the **dependent variable**.

Two common types of scientific inquiry are the controlled experiment and the observational study.

- **variable:** any condition that changes or varies the outcome of a scientific inquiry
- **independent variable:** a variable that is changed by the investigator
- **dependent variable:** a variable that changes in response to the change in the independent variable

Controlled Experiment

You can perform a **controlled experiment** to determine whether one variable causes an effect on another variable. This means that you can control (change) the independent variable to find out if it causes a change in the dependent variable. For example, suppose you want to determine how the amount of salt in a given volume of water affects the boiling point of the water. A controlled experiment would be the best approach. In this case, the investigator would use different amounts of salt (independent variable) and observe the effect, if any, on the boiling point of the water (dependent variable).

▶ **controlled experiment:** an experiment in which the independent variable is purposely changed to find out what change, if any, occurs in the dependent variable

Observational Study

Often the purpose of a scientific inquiry is to answer a question by gathering scientific information. An **observational study** involves observing a subject or event without interfering with or influencing it. Observational studies start with observations that lead to a question. Sometimes the researchers start with a specific prediction about the answer to the question. Sometimes they also have an explanation for their prediction—but not always. An explanation might be possible only after evidence is collected.

▶ **observational study:** the careful watching and recording of a subject or event to gather scientific information to answer a question

Sciences such as astronomy, geology, and ecology rely heavily on observational studies (Figure 2). Observational studies are done when an investigator cannot control the variables because of safety, time, distance, cost, or ethical considerations.

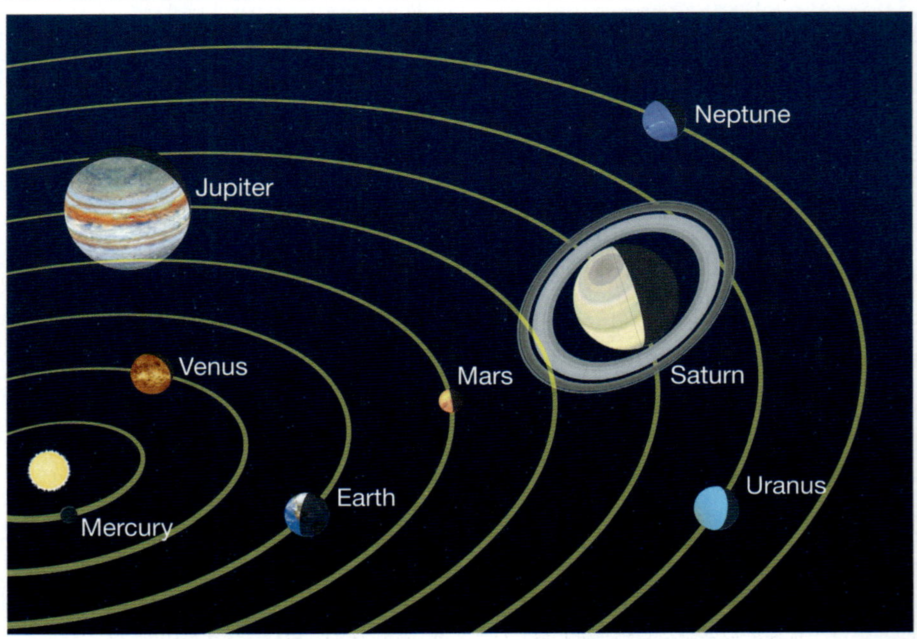

Figure 2 To find out how long it takes for each of the planets to orbit the Sun, scientists had to carefully observe the planets from Earth over long periods of time.

DIG DEEPER

Reading Tip

Scan for Text Features

Text features include headings, sidebars, graphics, and margin definitions. Scan a text before reading to identify the features it uses. This process can help you organize how you read a text and alert you to important information.

TRY THIS QUESTIONS LEADING TO SCIENTIFIC INQUIRY

SKILLS HANDBOOK 3.B.4., 3.B.9.

SKILLS: Planning, Communicating

The type of question asked in a scientific inquiry determines, to a large degree, the most appropriate method to answer the question. In this activity, you will identify the best type of scientific inquiry—controlled experiment or observational study—for answering a series of questions.

Equipment and Materials: notebook or paper; pen or pencil

1. Read the following questions:

- What factors determine the strength of an electromagnet?
- What are the most common types of waste materials dumped at the landfill?
- Will an iron object rust most quickly in air, water, or oil?
- How soon after sunset does the sky become dark enough for us to see stars?
- What kinds of plants and animals are common in the forests in my area?

A. Carefully study each statement and indicate what type of scientific inquiry would be most appropriate.

B. For each question, briefly describe how you would carry out the inquiry.

Scientific Investigation Skills

Certain skills are important for conducting any investigation. These skills can be organized into four categories:

1. initiating and planning
2. performing and recording
3. analyzing and evaluating
4. communicating

Initiating and Planning

All scientific investigations begin with a question. The question may arise from observations of a natural event, from an individual's curiosity (**Figure 3**), or from previous experiments or studies.

Some questions cannot be answered through scientific investigation, so it is important to ask the right questions. A scientific question must be testable. Testable questions have certain characteristics:

- They must be about living things, non-living things, or events in the natural world.
- They must be answerable through scientific inquiries such as controlled experiments or observational studies.
- They may be answered by collecting and analyzing data.

Figure 3 When scientists first observed the northern lights (aurora borealis), they probably wondered what caused them.

hypothesis: a possible answer or untested explanation that relates to the initial question in an experiment

prediction: a statement that predicts the outcome of a controlled experiment

experimental design: a brief description of the procedure by which a hypothesis or prediction is tested

If a controlled experiment should be performed to answer the testable question, you may propose a possible answer. The possible answer is based on existing scientific knowledge. This is called a **hypothesis**. The hypothesis is directly related to the question.

A hypothesis suggests a relationship between an independent variable and a dependent variable. A hypothesis serves two functions: (1) it proposes a possible answer to the question, along with an explanation, and (2) it suggests a method of obtaining evidence that will either support or reject the proposed explanation.

If you do not have a scientific explanation and cannot make a hypothesis, then you can make a **prediction**. A prediction is a statement that predicts the outcome of a controlled experiment, without an explanation. A prediction is not a guess: it is based on prior knowledge and logical reasoning.

A hypothesis usually includes a prediction. It is often written in the form "If ..., then ..., because ...." The "if ... then" part is the prediction; the "because ..." part is the explanation.

A hypothesis or a prediction identifies the independent variable and the dependent variable. It also suggests an experimental design for conducting a fair test. The **experimental design** briefly describes the procedure.

Planning the investigation involves the following:

- identifying the independent and dependent variables
- determining how the changes in the variables will be measured
- specifying how to control the variables not being tested
- selecting the appropriate equipment and materials (**Figure 4**)
- identifying and addressing safety concerns
- deciding on a format for recording observations

Figure 4 The equipment used in space exploration is very specialized.

Performing and Recording

After planning an investigation, it is important to follow the procedures carefully (**Figure 5**). That does not mean that the procedures cannot change. If the procedures present problems, they should be changed, and any changes should be recorded. If the problems cannot be overcome, you may have to go back to the planning stage and start again.

It is important to be aware of possible safety concerns with the procedure. Carefully read the safety guidelines in the Skills Handbook before beginning an investigation, and refer to them frequently.

While performing an investigation, you will need to make accurate observations and record them carefully. Observations are any information that is obtained through the senses. Observations can be quantitative (numerical) or qualitative (non-numerical).

A **quantitative observation** is based on measurements or counting (**Figure 6**). Examples of quantitative measurements include length, mass, temperature, and population counts. Measuring is an important skill in making observations. Not only must the right measuring tool be selected, but it must also be used in such a way as to provide a precise and accurate measurement.

Figure 5 A scientist follows the procedure he has outlined.

▶ **quantitative observation:** a numerical observation based on measurements or counting

Figure 6 Two scientists make quantitative observations on a polar bear.

▶ **qualitative observation:** a non-numerical observation that describes the qualities of objects or events

DIG DEEPER

Learning Tip

Word Origins

The word "quantitative" comes from the word "quantity," which refers to an amount or number. The word "qualitative" comes from the word "quality," which refers to the characteristics of something.

A **qualitative observation** is a description of the qualities of objects and events, without any reference to a measurement or a number. Common qualitative observations include the state of matter (solid, liquid, or gas), texture, and odour. These qualities cannot be measured directly or easily.

The method of recording your observations depends on the type of observation. Quantitative observations are often recorded in a data table. Qualitative observations can be written in words or recorded in pictures or sketches (**Figure 7**). Remember to record your observations clearly and accurately so that you do not have to rely on memory when you report your findings.

Figure 7 These science students record qualitative observations as drawings.

Analyzing and Evaluating

Tables, lists, and drawings of observations are not the final products for the data collected during an investigation. Analyzing, or carefully studying, the observations can provide more information than the raw data itself. You can also plot graphs from the quantitative data to show patterns and trends more clearly. Analyzing observations also helps to identify any errors in measurement.

A very important skill in a scientific investigation is evaluating the evidence that is obtained through observations. The quality of the evidence depends on the quality of the plan, the procedures, the equipment and materials, and the skills of the investigator. In order to evaluate the evidence, you need to evaluate all aspects of the investigation.

The purpose of analyzing and evaluating observations is to answer the question posed at the beginning of an investigation. You may have evidence that you can use to confidently answer the question. Or you may conclude that you do not have enough evidence to answer the question. If your evidence confirms the prediction, then the hypothesis is supported. Note that this does not *prove* the hypothesis to be true. If your evidence does not confirm the prediction, then the hypothesis may not be an acceptable explanation. A scientific investigation is not a failure simply because the hypothesis was rejected. The investigator just has to look at another possible explanation to test.

Scientific research usually requires more than a single experiment. Sometimes other investigators repeat the investigation to see if their results are the same. A scientific investigation often leads to new questions, which then lead to other investigations and other questions.

DIG DEEPER

Reading Tip

Search for the Main Idea

During reading, search for the main idea of the text. Check the first, second, and last sentences of a paragraph. Sometimes, you may have to infer a main idea that is not stated directly.

TRY THIS ANALYZING DATA

SKILLS HANDBOOK 3.B., 6.A., 6.C.

SKILLS: Analyzing, Communicating

Data collection involves the recording of information in an organized way. Analyzing involves studying the data to uncover patterns and trends. In this activity, you will analyze data from a sample investigation. The data in **Table 1** show the distance that Chad can ride his bicycle in a given amount of time.

Equipment and Materials: ruler; graph paper; notebook; pen or pencil

1. The data in Table 1 show the distance covered by Chad as he rides his bicycle along a measured track.
2. Use the data in the table to plot a line graph.
3. Starting at the (0, 0) point on your graph, use a ruler to draw a straight line through your data so that it is touching (or near to) as many points as possible. The data points should be evenly scattered around either side of the line you draw.

A. Make a general statement about the relationship between the time and the distance travelled.

B. Is this relationship easier to see from the data or from the graph? Explain.

C. Use the graph to estimate how far Chad had travelled after 45 s and after 2 min (120 s).

D. Propose an explanation for why all of the points do not fall exactly on the straight line.

E. Look at the angle of the line you drew in Step 3. What do you think this line represents?

Table 1 Distance Travelled by Bicycle

Time (seconds)	Distance (metres)
0	0
10	50
20	120
30	190
40	270
50	350
60	430
70	510
80	580
90	670
100	750

Communicating

One of the important characteristics of scientific investigation is that scientists share their information (**Figure 8**). Before scientific data can be published in a scientific journal, it must be examined by other scientists in a process called peer review. Other experts check that the data are valid and the science is correct.

Clear and accurate communication is essential for sharing information. It is important to share not only the evidence, but also the process by which the evidence was obtained. If the investigation is to be repeated by others, it is just as important to share the design and procedures.

Figure 8 Scientists sharing their information with each other is a key characteristic of scientific investigation.

By sharing their data and the techniques they used to obtain, analyze, and evaluate their data, scientists give others the opportunity to both review the data and use it in future research. The most common method for communicating with others about an investigation is by writing a report or giving a presentation after the investigation is complete.

TRY THIS PASS IT ON!

SKILLS HANDBOOK 3.B.8., 3.B.9.

SKILLS: Evaluating, Communicating

Accurate communication is as important in science as it is in everyday life. In this activity, you will work as a class to demonstrate the importance of accurate communication and the need to develop and refine your skills in this area.

Equipment and Materials: paper; pen or pencil

1. Your teacher will create a simple statement, write it on a piece of paper, and then whisper it to a student in class.
2. That student will write the message on a piece of paper and then whisper the message to another student.
3. The message will be passed on until every person in the class has received the message. Your teacher will record the sequence of students that the message followed.
4. The rules are simple:
   - You must write exactly what you hear.
   - You must state the message only once.
   - You must whisper the message.
   - You cannot show other students what you have written.
5. The last person to receive the message will repeat it out loud to the class.
6. The teacher will then write the original message on the board. All students will then write their version of the message under the previous message.

A. How closely did the final message resemble the original message? Explain.

B. How does this activity demonstrate the need for clear, accurate communication?

C. Give an example from real life of how the message "received" differs from the message "sent."

D. Based on this activity, suggest a set of guidelines to ensure clear, accurate communication.

1.1 Wrap Up

- Curiosity about what we see around us often leads to questions that trigger scientific investigations.
- One type of scientific inquiry is the controlled experiment, in which the researcher keeps all but two variables constant, changes one (the independent variable), and observes the other (the dependent variable).
- A second type of inquiry is the observational study, in which the researcher collects data by observing a situation without affecting it.
- Scientific investigation skills include
  - initiating and planning (asking a question and deciding on the best way to find an answer)
  - performing and recording (carrying out the procedure and making observations)
  - analyzing and evaluating (searching for patterns in the observations)
  - communicating (sharing findings with others)

1.2

Scientific Literacy for Living and Working in Canada

DIG DEEPER

Reading Tip

Make Connections

As you are reading, make connections between information in the text and your own experiences, what you have already read, and facts you know about the topic. For example, think about the technology you use and whether you understand how it works. Do you take this technology for granted?

DIG DEEPER

Web Link

To read famous scientists' statements about the value of scientific literacy,

GO TO NELSON SCIENCE

Science and technology play an increasingly important role in everyday life. Most new technologies are designed so that the average person can use them without understanding how they work. Science, too, is often considered to be too complex to be understood by the average person.

What Is Scientific Literacy?

To make wise personal decisions and to act as a responsible citizen, it is necessary to be scientifically and technologically literate. The Science Teachers' Association of Ontario (STAO) defines a scientifically and technologically literate person as "one who can read and understand common media reports about science and technology, … evaluate the information presented, and … engage in discussions and decision-making activities regarding issues that involve science and technology."

Scientific Literacy for Careers in Science

Look around at your fellow science students. Many of you will find careers in areas that are related to science (**Figure 1**). Generally, employers hire individuals with strong critical-thinking and problem-solving skills and the ability to work as part of a team. These skills are emphasized throughout the entire science program.

Figure 1 Scientific literacy is important in many different careers.

Throughout this book, you will have many opportunities to explore careers that are related to the area of science under study (**Figure 2**). These careers could be in medicine, geology, environmental science, plumbing, culinary arts, cosmetology, or art. Wherever you see a Career icon with the accompanying note in the margin of this book, you will be directed to visit the Nelson website. There you will have an opportunity to research the education and training requirements for these careers and to find out about the roles and responsibilities of these careers.

DIG DEEPER
Career Link
To learn more about careers in science,
GO TO NELSON SCIENCE

Figure 2 (a) Some scientists conduct their research in laboratories. (b) Others conduct their research in the natural world and with (c) living organisms.

Scientific Literacy for Life and Citizenship

Only some people work in jobs that are related to science. However, every one of us is a citizen with certain rights and responsibilities. One of our basic rights is access to a full education. Along with that right comes the responsibility to use that education for the benefit of oneself and society. Science and technology influence our lives in many ways. Therefore, it is important that we understand this influence so that we can make good decisions about issues that affect us as individuals and as a society.

At some point in your life, you will have to make decisions about your own lifestyle. You will need to be aware of the impact your actions have on the world around you. You will likely have to make decisions related to important issues such as climate change, environmental pollution, protection of species, new medical technologies, space exploration, and world hunger (**Figure 3**).

Figure 3 Scientific literacy will help you think critically and form opinions. It will also help you understand the need to take action.

DIG DEEPER

Reading Tip

Summarize the Main Idea

After reading a selection, reinforce your understanding by summarizing the main idea in your own words and drawing a conclusion about it.

As a society, we need to consider both the positive and the negative impacts of developing and using scientific knowledge and new technology. We must be aware of the positive and negative implications of new technologies (**Figure 4**). If we fail to do so, we may experience very unpleasant surprises after the technologies have been adopted.

Figure 4 New computer technology has many positive impacts, but old computers and hard drives often end up in landfills.

It is important to understand the basic concepts of science. However, it is impossible to understand every new scientific discovery and technological advance. It is just as important to learn *about* science as it is to learn science. It is important to know what science can do, to know that knowledge produced by science is reliable, and to know that science is the best way to learn about the natural world. It is equally important to be able to find and evaluate information and to use that information to make decisions.

Achieving scientific literacy is not the same as preparing to be a scientist. It is important whether you are a construction worker, a small-business person, a lawyer, a mechanic, a doctor, an engineer, or a research scientist. Regardless of your plans and ambitions, achieving scientific literacy is an important goal.

1.2 Wrap Up

- Scientific literacy enables people to understand and evaluate information relating to science and technology in the world around us.
- Being scientifically literate allows us to make informed decisions.
- Specific scientific knowledge and skills are necessary for a wide range of careers.
- A general understanding of science is necessary to be an informed citizen.

Chapter 1 KEY CONCEPTS SUMMARY

Scientific inquiry can be conducted in different ways, depending on the question to be answered.

- Controlled experiments determine how an independent variable affects a dependent variable.
- Observations often lead to questions that initiate scientific inquiry.

All scientific inquiries rely on scientific investigation skills.

- All scientific investigations begin with a question.
- Careful analysis of observations helps to make them meaningful.
- Clear communication is important for sharing scientific discoveries and ideas with others.

Science and technology are an important part of our everyday lives.

- All individuals are affected by science and technology every day.
- Most people do not understand the science they encounter.

Scientific literacy is necessary for wise personal decisions, responsible citizenship, and careers.

- Careers in science require scientific knowledge and skills.
- Scientifically literate citizens can benefit society.

UNIT
B Sustainable Ecosystems and Human Activity

UNIT Preview

In this unit, you will

- explore the characteristics of terrestrial and aquatic ecosystems
- investigate factors that affect terrestrial (land) or aquatic (water) ecosystems
- learn how human activity threatens ecosystems and how humans are trying to protect ecosystems

BIG Ideas

- Ecosystems consist of a variety of components, including, in many cases, humans.
- The sustainability of ecosystems depends on balanced interactions between their components.
- Human activity can affect the sustainability of aquatic and terrestrial ecosystems.

Focus on STSE

Trouble for the Cricket Frog

The northern cricket frog is in trouble. This tiny member of the tree frog family spends much of its life in or near water in areas such as marshes, ponds, and streams. The northern cricket frog is currently considered an endangered species in Canada. Cricket frog numbers have declined for three main reasons: loss of habitat, pollution from pesticide, and the introduction of carp that prey on the frogs.

Cricket frogs are amphibians. They have lungs for breathing but also absorb dissolved oxygen directly through their thin skins. When their skin is wet, oxygen from water or air passes through their skin into an extensive network of blood vessels. Because this dissolved oxygen enters their blood directly from water and air, amphibians are especially sensitive to pollution. The health problems and decline in population of amphibians like the cricket frog may be signs that other living things, including humans, are also in trouble from excessive environmental pollution.

Think/Pair/Share

Discuss the following questions with a partner:

1. Why should humans be concerned about a decline in the number of cricket frogs? K/U
2. How might a decline in the number of frogs affect other species in an ecosystem? Use the term "food chain" in your explanation. K/U C
3. What are the three main reasons cricket frogs are endangered? K/U
4. What are some things you could do or avoid doing that might help cricket frogs survive? A

Sustainable Ecosystems and Human Activity

CHAPTER 2
Understanding Ecosystems

Living things are affected by their environments.

CHAPTER 3
Natural Ecosystems and Stewardship

Natural ecosystems are valuable to humans.

CHAPTER 4
Ecosystems by Design

Urban ecosystems depend on natural ecosystems for food and materials.

Unit Task Preview

What Are They Talking About?

The environment is a hot topic in the media. In this Unit Task, you will monitor a variety of media sources for information about the environment or ecology in your community, the country, and the world. You will use these media sources to help you get a better idea of current environmental concerns, documenting your findings on index cards. Then you will present in a creative format what you have learned from these sources.

UNIT TASK
Assessment

You will be assessed on:

- conducting thorough research
- thoughtfully analyzing the stories you find
- communicating the results of your research

UNIT TASK
Bookmark

The Unit Task is described in detail on page 160. As you work through this unit, look for this bookmark and see how the section relates to the Unit Task.

Getting Started: Useful Concepts and Skills

Concepts Review

1. Look at **Figure 1**. K/U
   (a) Which organisms are producers?
   (b) Which organisms are consumers?

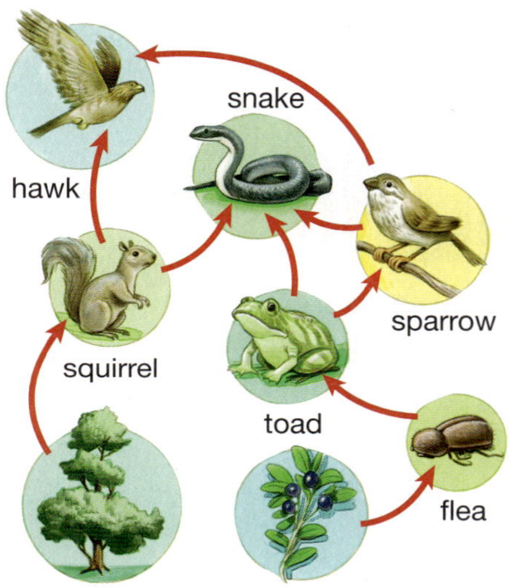

Figure 1

2. **Figure 2** shows a woodland ecosystem. K/U
   (a) List two biotic features of this woodland ecosystem.
   (b) List two abiotic features of this woodland ecosystem.

Figure 2

Skills Review

3. (a) Make a bar graph using the data in **Table 1**. Give your graph a title and label the axes.

Table 1

Year	1996	1998	2000	2002	2004	2006	2008
Number of deer	21	32	36	44	35	18	12

   (b) What does the graph tell you about the deer population?
   (c) What are some factors that may have caused a change in the deer population? T/I C

4. The circle graph in **Figure 3** shows data on species of plants and animals in Ontario that are at risk. T/I
   (a) About what percentage of species is considered endangered?
   (b) About what percentage of species is considered threatened?

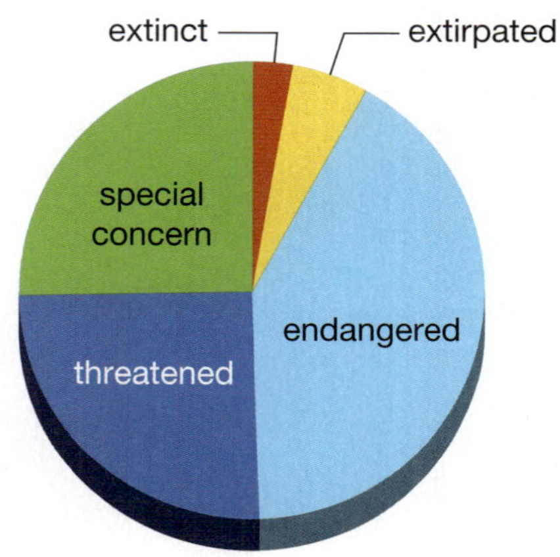

Figure 3

CHAPTER 2

Understanding Ecosystems

KEY QUESTION

How are organisms affected by living and non-living things in their environment?

KEY CONCEPTS

- Energy passes through ecosystems, but matter cycles within ecosystems.
- Photosynthesis and respiration are processes that work together.
- Aquatic and terrestrial ecosystems are strongly influenced by their abiotic environment.
- Human activities disturb the natural cycles of nature.

Humans must be careful not to disturb natural ecosystems.

LITTLE FISH, BIG PROBLEMS

The round goby is a fish about the size of your hand that lives at the bottom of rivers. It also happens to be a big threat to other fish. These grey, brown, and black fish are moving into Great Lakes tributaries such as the Thames and Grand rivers. These waters are home to a variety of aquatic life, including several endangered species.

Until the 1990s, round gobies were not present in the Great Lakes or their tributaries. Scientists believe the round gobies were brought by ships from other parts of the world. Now they are invading the lakes and rivers in the Great Lakes region.

The introduction of a new, aggressive species into an ecosystem can cause problems. Round gobies lay large numbers of eggs several times during a season and have very few predators, which has allowed their population to grow. They aggressively defend their spawning areas, preventing other species from laying eggs in prime areas. Round gobies are greedy bottom feeders that eat small fish, fish eggs, mussels, and insect larvae. It is uncertain how much damage the round gobies will cause.

Round gobies are disturbing some freshwater ecosystems.

WHAT DO YOU THINK?

Many of the ideas you will explore in this chapter are ideas that you have already encountered. You may have encountered these ideas in school, at home, or in the world around you. Some of the statements are not true. Consider each statement and decide whether you agree or disagree with it.

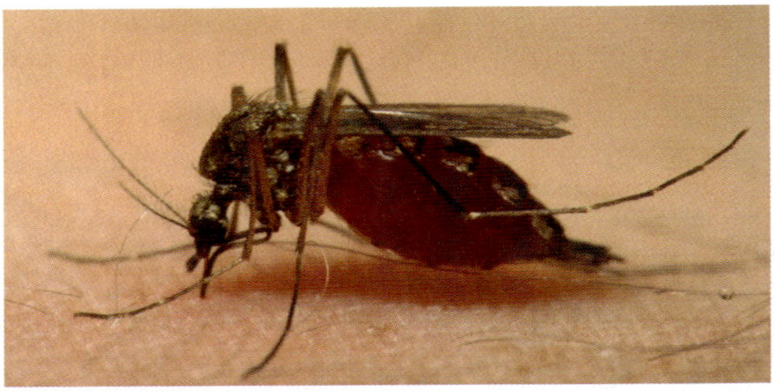

1 All organisms are helpful in the environment. **Agree/disagree?**

2 Human activities such as reducing waste, planting trees, and establishing bird sanctuaries help to improve natural environments. **Agree/disagree?**

3 The health of an ecosystem depends only upon its biotic (living) features. **Agree/disagree?**

4 Humans are able to change the natural environment to suit our needs. **Agree/disagree?**

5 Humans do not compete with other species. **Agree/disagree?**

6 Terrestrial (land) ecosystems are more important to humans than aquatic (water) ecosystems. **Agree/disagree?**

FOCUS ON
Reading

Making Connections

To make connections with a text, join something in a text with something you already know.

Type of connection	How to make this connection
Text-to-Self	Recall an experience you or someone you know has had that is similar to something in the text you are reading.
Text-to-Text	Recall something in a text you have read, viewed, or listened to that is similar to something in the text you are reading.
Text-to-World	Recall a fact you know about science, technology, or the environment that is similar to something in the text you are reading.

Factors That Affect Ecosystems

Human actions, such as cutting down trees or introducing a different plant species can change an ecosystem. Oil spilled into an ocean can harm marine ecosystems. Natural events, such as forest fires or floods, can also drastically change an ecosystem. Ecosystems may be able to recover from changes. Depending on the kind of change, recovery could take from just a few years to thousands of years.

Making Connections *in Action*

Here is how one student made connections to understand information about factors that affect ecosystems in Section 2.3.

Connections I have made to key ideas or information	How that connection helps me understand the text
Text-to-Self: between the photo and my uncle's café that stopped serving Chilean sea bass	helps me form an opinion about the importance of protecting fish stocks
Text-to-Text: between oil spills and a photo of a dying duck covered in oil	helps me visualize how oil spills hurt birds, fish, and plants in the ocean
Text-to-World: between last sentence and the fact that caterpillars are slowing down plant recovery around Mount St. Helens	helps me evaluate the timeline and realize an ecosystem does not recover right away

DIG DEEPER

Reading Tip

As you work through the chapter, look for tips like this. They will help you develop literacy strategies.

Life on Earth

Of all the planets in the Solar System, Earth is the only one teeming with life. Millions of organisms exist on land and in water on our unique planet.

▶ **habitat:** the environment where an organism lives

DIG DEEPER

Reading Tip

Text-to-Text Connections

Brainstorm text-to-text connections by writing down everything that comes to mind when you think about the selection and other texts you have read. You can use a Venn diagram to list similarities and differences between the selection and other texts.

Diversity of Life on Earth

A **habitat** is the place where an organism lives. Terrestrial habitats are found on land. Cacti and rattlesnakes live on land in arid (dry) deserts (**Figure 1**). Enormous trees with swarming insects live in tropical rainforests. Low-lying shrubs and herds of caribou are found in the frozen Arctic. Aquatic habitats are found in water. Colourful fish swim through the waters of the ocean's coral reefs (**Figure 2**). Frogs and crayfish live in small freshwater ponds.

Figure 1 A desert is a terrestrial habitat.

Figure 2 A coral reef is an aquatic habitat.

Earth's Spheres

If you observed Earth from space, you would see that Earth is surrounded by a thin gaseous layer swirling with clouds. If you were close enough, you would see the oceans, land, and ice on Earth's surface.

▶ **atmosphere:** the thin layer of gases that surrounds Earth

The Atmosphere

Earth's **atmosphere** is the thin layer of gases that surrounds the planet. It contains gases such as water vapour, oxygen, and carbon dioxide that are necessary for living things. The atmosphere acts like a blanket wrapped around Earth and prevents Earth from getting too warm or too cool. The atmosphere also blocks some of the Sun's radiation, which can harm organisms, including humans. Without the atmosphere, species would not be able to survive on Earth.

Lithosphere and Hydrosphere

The lithosphere is the rocky outer shell of Earth. Mountains, ocean floors, and the rest of Earth's solid landscape are part of the lithosphere. The hydrosphere is the water on, above, and below Earth's surface. It includes oceans, lakes, ice, clouds, and groundwater.

The Biosphere

Earth's **biosphere** is all the places on Earth where living things can exist. These places are in the atmosphere, lithosphere, and hydrosphere (**Figure 3**).

▶ **biosphere:** the zone in, on, and around Earth where life can exist

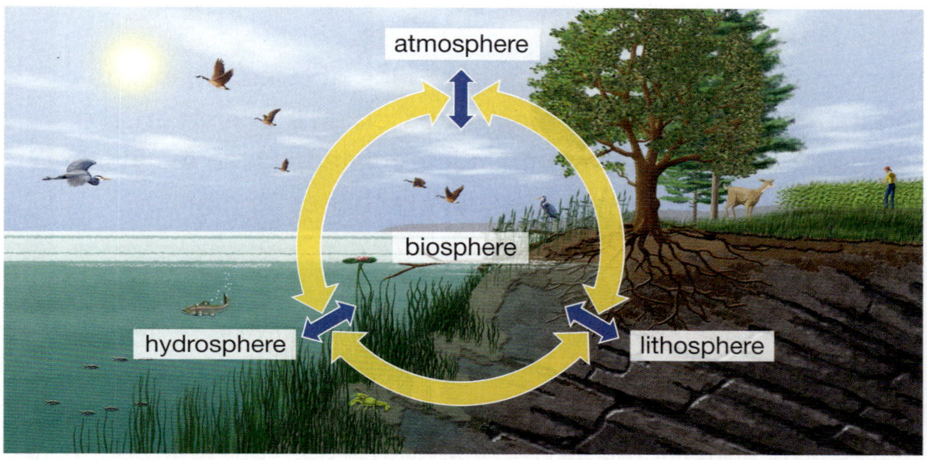

Figure 3 The biosphere is a thin layer that includes land, water, and air.

DIG DEEPER

Learning Tip

Greek Roots

Many scientific terms have roots in the Greek language. *Lithos* means stone, *atmos* means vapour, *hydro* means water, *bio* means life, and *sphere* means ball.

2.1 Wrap Up

- Earth has a variety of organisms living in habitats on land and in water.
- Earth's spheres include the atmosphere, lithosphere, and hydrosphere.
- The atmosphere moderates the temperature and amount of radiation that reaches Earth.
- The biosphere is where life can exist in the atmosphere, lithosphere, and hydrosphere.

CHECK YOUR LEARNING

1. How does the atmosphere help provide the conditions necessary for life to exist on Earth? K/U
2. (a) Describe in your own words the following spheres: lithosphere, hydrosphere, atmosphere, biosphere.
   (b) How are the spheres related? K/U T/I C
3. What did you find interesting about Earth's spheres? Why? T/I
4. Describe an extreme terrestrial habitat in Canada. Explain how organisms are adapted to living there. K/U C

2.2 Introducing Ecosystems

Picture a moose in the woods, moving through the spaces between trees. The moose feeds on shrubs, trees, and aquatic plants around a small pond. You see a beaver dam that has formed the pond. The animals, plants, and pond water are all part of a complex system of living and non-living things.

A Complex System

This complex system, made up of groups of living organisms and their non-living environment, is called an **ecosystem**. The components of an ecosystem are interrelated, which means that changes in one component can affect other components.

- **ecosystem:** all the living organisms in an area and their non-living environment
- **biotic factors:** living things, their remains, and the features, such as nests, associated with their activities
- **abiotic factors:** the non-living characteristics of an ecosystem

Living and Non-Living

Living organisms in an ecosystem are **biotic factors**. Biotic factors include organisms, their waste, their homes, and their remains. Insects, fungi, beavers, moose droppings, and shrubs are all biotic factors. The non-living features of an ecosystem are **abiotic factors**. Weather, water, and rocks are abiotic factors. Table 1 shows some biotic and abiotic factors found in a woodland ecosystem.

Table 1 Components of a Woodland Ecosystem

Biotic	Abiotic

Everything Is Related

Biotic and abiotic factors affect one another, so any change in biotic or abiotic factors can change an ecosystem. For example, when a forest is cut down to make room for houses, the trees that were once home to insects, birds, and squirrels are no longer there. The animals will need to find a new place to live.

Populations and Communities

In most ecosystems, there are many different species. Members of the same species that live in the same ecosystem are a **population**. For example, all the snapping turtles in a pond are a population. All the coho salmon in a stream are a population.

All the populations of different organisms in an area form a **community**. In a pond, the populations of turtles, frogs, fishes, mosquitoes, aquatic plants, algae, and other living things make up the community.

Figure 1 shows how ecosystems are composed of individual organisms, populations, communities, and the physical surroundings in which communities of organisms live.

- **population:** all the members of the same species that live in the same ecosystem
- **community:** all the living organisms, from all the populations, that live in the same ecosystem

Figure 1 Deer are a population in this ecosystem. So are populations of different grass, bird, and tree species. Populations of grasses, deer, birds, and wolves in the same area make up a community.

DIG DEEPER

Did You Know?

In Hot Water!

Thermophiles are organisms that are adapted to live in extremely hot environments in thermal vents or hot spots in the ocean floor. Water shoots up from these vents at high pressure. Water temperatures in the ocean can be as low as 2 °C, but in some thermal vents, temperatures reach 400 °C! The vents are surrounded by ecosystems that include bacteria, clams, mussels, and tube worms.

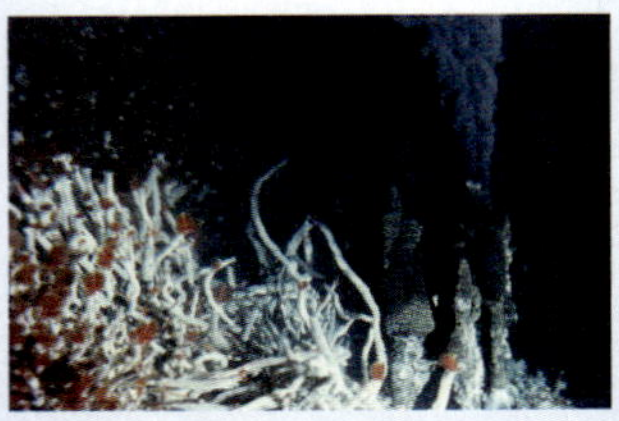

Differences in Ecosystems

Ecosystems can be large or small and can include many or few organisms. An ecosystem can be as large as an entire forest or as small as a rotting log. A large forest ecosystem can contain other smaller ecosystems. A community of bacteria and fungi living in a rotting log is part of a smaller ecosystem.

Each ecosystem has distinct features. You could describe a pond ecosystem by its organisms or by its physical characteristics, such as depth and temperature. **Table 2** shows some examples of large and small ecosystems.

Table 2 Examples of Large and Small Ecosystems

Ecosystem	Characteristic abiotic features	Characteristic biotic features
 coniferous forest	• long, cold winters and warm summers • moderate rainfall • a lot of snow	• large forests and bogs • black bears, red squirrels, moose, and biting flies • short growing season allows fewer organisms to grow
 coral reef	• warm and shallow waters • usually located in tropical regions	• aquatic organisms, including corals, sponges, and many different fish species • easily damaged by changes in water temperature and quality
 beaver pond	• shallow water that is warm in summer; icy cover in winter • may last only a few years • dams could break and cause flooding	• many aquatic species, including insects, fishes, frogs, turtles, muskrats, and beavers • nearby forest is a food source for beavers
 rotting log	• moist environment • very little light • may last only a few years or decades	• decomposing bacteria and fungi • beetles, ants, and other insects • log provides shelter for small animals

TRY THIS ECOSYSTEM ABCs

SKILLS: Analyzing, Communicating

SKILLS HANDBOOK 3.B.7., 3.B.9.

Everything we use comes from the biotic and abiotic parts of the environment.

1. List 10 items you have used or handled in the past two days.
2. List the materials that make up each item. For example, a pencil is made up of wood, graphite, rubber, and metal.
3. Classify each material as abiotic or biotic.
4. Is the item made up of only biotic materials, only abiotic materials, or both?
5. Draw and label a Venn diagram, like the one shown in **Figure 2**, in your notebook. Place the names of the 10 items in your diagram. For example, you would write "pencil" in the overlapping part of the Venn diagram because wood is biotic and the metal is abiotic.

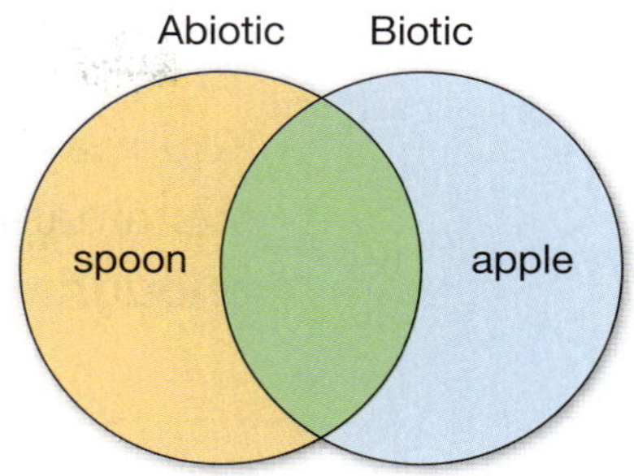

Figure 2 Sample Venn diagram

A. Compare the items in your diagram. What type of items did you use the most of: abiotic or biotic? T/I

2.2 Wrap Up

- Ecosystems are made up of biotic and abiotic factors.
- Biotic factors are the living things, their remains, and features associated with their activities in an ecosystem.
- Abiotic factors are the non-living physical features of an ecosystem.
- A population is made up of all the individuals of the same species in the same ecosystem.
- A community is all the populations in the same ecosystem.

UNIT TASK Bookmark

You can apply what you learned about ecosystems to the Unit Task described on page 160.

CHECK YOUR LEARNING

1. Classify each of the following as biotic or abiotic: temperature, wind, sunlight, dead leaves, spiders, sand, milk, hair, ice, plastic. K/U
2. Does a population include abiotic features, biotic features, or both? Explain. K/U T/I
3. Élise thinks that human activities can change only the biotic features of an ecosystem. Is she correct? Explain your thinking. T/I C
4. People living in northern environments are more likely to suffer from *seasonal affective disorder* (SAD), which is a form of depression associated with low winter light levels. How is SAD an example of the relationship between the biotic and abiotic parts of an ecosystem? C A
5. Reread the Did You Know? on page 32. Identify the biotic and abiotic features of a thermal vent ecosystem. K/U

2.3 Sustainability

You may have heard people talking about achieving a sustainable society. This means living in a way that ensures enough resources will be available for future generations.

▶ **sustainable ecosystem:** an ecosystem that maintains relatively constant biotic and abiotic features forever

Natural and Human-Made Ecosystems

Most natural ecosystems are sustainable. A **sustainable ecosystem** is one that survives and works well over time. Natural ecosystems occur in nature where there is little human influence. A remote forest and a coral reef are both natural ecosystems that may have almost no human influence. A national park, on the other hand, is a natural ecosystem that is protected by humans.

Other ecosystems are strongly influenced by human actions. A playground and a vegetable garden are human-made ecosystems. You will read more about human-made ecosystems in Chapter 4.

DIG DEEPER

Reading Tip

Text-to-Self Connections

Brainstorm text-to-self connections by writing down everything that comes to mind when you think about the selection and your own experiences or those of relatives or friends. You can create a mind map to show the connections.

Figure 1 In 1992, the cod fishery off the Grand Banks, Newfoundland, had to be closed because overfishing caused a huge decline in the cod populations.

Factors That Affect Ecosystems

Human actions, such as cutting down trees or introducing a different plant species, can change an ecosystem. Oil spilled into an ocean can harm marine ecosystems. Natural events, such as forest fires or floods, can also drastically change an ecosystem. Ecosystems may be able to recover from changes. Depending on the kind of change, recovery could take from just a few years to thousands of years.

Managing Our Resources Properly

Humans depend on many resources from natural ecosystems. We depend on forests for resources such as wood and medicines, and we depend on oceans for foods such as fish and shellfish. These resources are renewable and can be harvested for generations to come, but only if the ecosystems are monitored and managed carefully.

Setting Limits

Keeping ecosystems sustainable means that if we cut down trees, we must replace them by planting new seedlings. We must limit the number of fish and other species we take from the oceans so that their populations can be maintained at an appropriate level (**Figure 1**).

Responsible Recreation

Natural ecosystems also provide us with areas for recreation. People must respect the biotic and abiotic features of ecosystems to minimize our impact on the environment. When camping or doing other recreational activities, we should follow this simple advice: take only pictures and leave only footprints.

DIG DEEPER

Web Link

For ideas on how to reduce your impact on the environment when performing outdoor activities,

GO TO NELSON SCIENCE

The Importance of Sustaining Ecosystems

Humans are affected by and influence the ecosystems around them. We use the resources around us to maintain an economy. In northern Ontario, the economy is based on forestry and mining, while the economy of southern Ontario is based on manufacturing and agriculture. Tourism is important to much of the province (**Figure 2**). What would happen if forests were damaged or destroyed? Why is it important to protect natural resources?

Figure 2 Niagara Falls is a popular tourist destination and helps sustain Ontario's economy.

2.3 Wrap Up

- Natural ecosystems are usually sustainable.
- Natural ecosystems must be used by humans in a sustainable way.
- Human-made ecosystems are created by humans and need to be managed.
- Our daily lives influence and are affected by our surrounding ecosystems.

CHECK YOUR LEARNING

1. Name two things that can be done to ensure that we use ecosystem resources in a sustainable way. K/U
2. You and your family are planning a canoeing and camping trip to a nearby provincial park. List five things you should or should not do on your trip so that you "leave no trace" of your visit. A
3. Which of the following ecosystems need human intervention to sustain them? K/U
   (a) farm
   (b) coral reef
   (c) forest
   (d) golf course
4. Minerals, including fossil fuels, are not renewable resources. Suggest two ways that we can use these resources in a sustainable way. T/I A

Factors Affecting an Aquatic Ecosystem

SKILLS MENU

- Questioning
- Hypothesizing
- Predicting
- Planning
- Controlling Variables
- Performing
- Observing
- Analyzing
- Evaluating
- Communicating

Ponds, lakes, and rivers are valuable aquatic ecosystems. Unfortunately, these ecosystems can be harmed by human activities. Salt on roads or fertilizer on lawns can flow into nearby waterways. Acid rain, or rain that contains harmful pollutants, can also enter aquatic ecosystems, affecting plants and animals that rely on these ecosystems. In this investigation, you will alter an abiotic factor in pond water and observe the effect on the ecosystem.

SKILLS HANDBOOK 3.B.2.–3.B.9.

Testable Question

How might the addition of road salt, fertilizer, or acid rain affect the community of organisms in an aquatic ecosystem?

Prediction

Write a prediction (If . . .Then statement) of the expected outcome of the experiment. Base your prediction on what you know about the factors affecting aquatic ecosystems.

Equipment and Materials

eye protection

lab apron

stirring rod

1 L beaker

funnel

hand lens

graduated cylinder

electronic balance

two 2 L plastic pop bottles

marker

2 L pond water with assorted aquatic organisms

liquid plant fertilizer

dropper bottle of dilute sulfuric acid

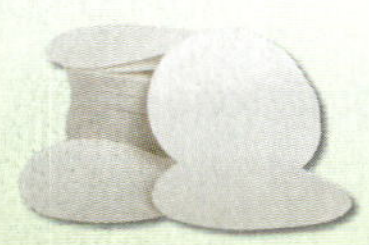
weighing papers

table salt

Experimental Design

Each group will examine the effects of altering an abiotic factor in a model pond ecosystem. The community of organisms in the normal pond water (the control) will be compared with those in which acid, salt, or nutrient content of the pond water has been altered (the treatments). Each group will share its results with the class.

Procedure

1. In your notebook, draw a table similar to **Table 1**. Decide how much of the substance you will add to your model ecosystem. Record your decision in your table. Show the table to your teacher for approval.
2. Label one of your 2 L bottles "control." Label the other bottle with the abiotic factor you are investigating: fertilizer treatment, acid rain treatment, or salt pollution treatment.
3. Put on your eye protection and lab apron.
4. Stir the pond water with the stirring rod.
5. Using the beaker and funnel, measure and pour 1 L of pond water into each 2 L bottle.
6. Use a hand lens to examine each model pond ecosystem. In your notebook, describe or draw the organisms you see. Estimate the abundance of each type.
7. Measure the substance you are testing. Add it to the pond ecosystem labelled with the factor being tested.
8. Design a table to record your observations. Observe your ecosystem every other day for at least three weeks. Record the types of organisms present and estimate their abundance each time.

Table 1 Factors Affecting Model Ecosystem

Treatment	Starting substance	Added substance
control	1 L pond water	none
fertilizer	1 L pond water	___ mL liquid plant fertilizer
acid rain	1 L pond water	___ mL dilute sulfuric acid
salt pollution	1 L pond water	___ g salt

Analyze and Evaluate

(a) Describe the changes observed in each ecosystem. Compare the effects of the test treatment with the control. T/I

(b) How did your results compare with your prediction? T/I

(c) Compare your results with those of other groups who tested the same factor. What was similar or different? Why? T/I

Apply and Extend

(d) Based on the class experiment, how might fertilizer, acid rain, or road salt affect the sustainability of an aquatic ecosystem? A

UNIT TASK

Bookmark

You can apply what you learned about factors affecting an aquatic ecosystem to the Unit Task described on page 160.

2.5

Energy Flow in Ecosystems

The Sun plays an important part in sustaining life on Earth. It is the source of almost all the energy living things need to survive.

▶ **radiant energy:** energy that travels through empty space

The Source of Energy

Most of the Sun's energy is in the form of **radiant energy**, which is energy that travels through empty space. Some of the radiant energy from the Sun is invisible, such as ultraviolet energy. Visible radiant energy is what we see in the form of light.

Where Does the Energy Go?

About 70 % of the Sun's radiant energy is absorbed by Earth's land, atmosphere, and water, and is converted into thermal energy (**Figure 1**). Thermal energy warms the atmosphere and evaporates water. About 30 % of the radiant energy from the Sun is reflected back into space. A very small amount of the radiant energy that reaches Earth's surface is absorbed by living organisms.

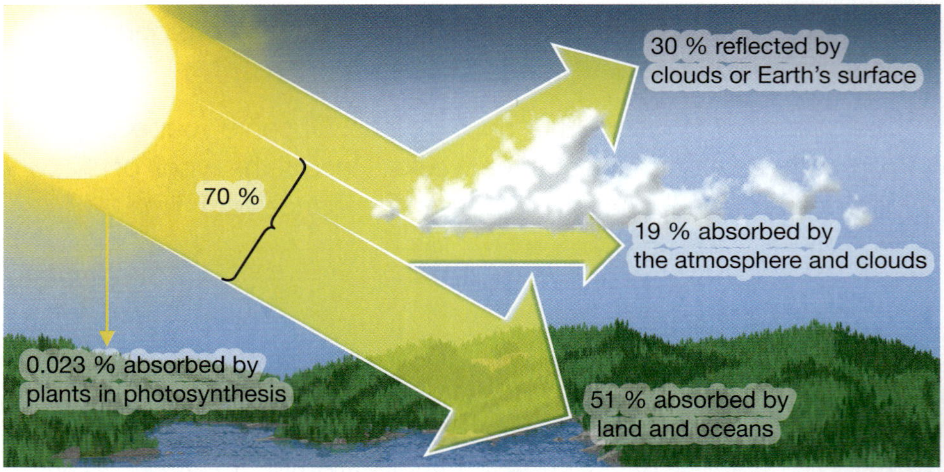

Figure 1 How the Sun's radiant energy is distributed

DIG DEEPER

Did You Know?

Plants in the Dark

Scientists know of only a few plants that do not use photosynthesis—the plants do not need sunlight to grow. Indian pipe is a common plant in Ontario that gets its food from fungi in the soil. Fungi get their food from the roots of nearby trees. The leaves of Indian pipe do not have green chlorophyll, so they are white.

If you do not eat breakfast one morning, you might feel tired or weak because you do not have much energy. All organisms need energy to live and grow, but not all energy works in the same way. Thermal and radiant energy do not directly give organisms all they need to live and grow. Energy from the Sun must first be converted into a form that can be used by living things—chemical energy (**Figure 2**). Chemical energy can be stored in living things and used as needed. Some organisms absorb light energy to make their own chemical energy. Other organisms absorb chemical energy from the foods they eat.

Figure 2 Humans use chemical energy to live, move, and grow. We get the chemical energy we need by eating and digesting food.

Photosynthesis

When you eat an apple, your body gets energy from the apple. But where did the apple get the energy to grow? Some organisms use light to make their own food. An organism that makes its own food is called a **producer**. Plants are producers. They convert light energy to chemical energy in a process called **photosynthesis** (**Figure 3**). The green colour of plants comes from the chemical chlorophyll. Chlorophyll absorbs light energy. The light energy is used to convert two chemicals (carbon dioxide and water) to a different chemical (sugar), which is stored in the plant (**Figure 4**). These sugars contain stored chemical energy. During this process, the plant also releases oxygen into the air. The word equation for photosynthesis describes what happens:

$$\text{carbon dioxide} + \text{water} \xrightarrow{\text{light energy}} \text{sugar} + \text{oxygen}$$

- **producer:** an organism that makes its own energy-rich food using the Sun's energy
- **photosynthesis:** the process in which the Sun's energy is converted into chemical energy

Figure 3 Chlorophyll in leaves traps the light energy used in photosynthesis.

Figure 4 Chemical energy is stored in plant parts.

DIG DEEPER

Learning Tip

How to Read a Word Equation

The + in a word equation can be read as "and." The arrow can be read as "react to form." So, the word equation for photosynthesis could read "carbon dioxide and water, with the addition of light energy, react to form sugar and oxygen."

- **consumers:** organisms that obtain their energy by consuming other organisms, or the products of other organisms
- **cellular respiration:** the process by which sugar and oxygen are converted into carbon dioxide, water, and energy to provide energy for the cell

Cellular Respiration

Unlike producers, which can make their own energy through photosynthesis, humans are **consumers** and need to obtain energy from other sources. Consumers obtain energy by eating other organisms, producers and consumers, or their products.

The food you eat contains stored energy. When you digest, the energy in the foods is released through a process called **cellular respiration**. In this process, the sugar and oxygen are changed into carbon dioxide and water. As this change occurs, energy is released. The word equation for cellular respiration is

$$\text{sugar} + \text{oxygen} \longrightarrow \text{carbon dioxide} + \text{water} + \text{energy}$$

Cellular respiration is different from breathing. Cellular respiration is a chemical process that occurs in the cells of living things. Breathing, or physical respiration, is the process of taking air into lungs to get oxygen into cells. Both animals and plants need oxygen to survive.

Complementary Processes

Photosynthesis and cellular respiration involve the same chemicals—water, carbon dioxide, oxygen, and sugar. Both processes also involve energy. But the reactions are the opposite of each other and produce different products. While photosynthesis produces sugars that contain stored energy, cellular respiration uses the sugars and releases the stored energy so that it can be used for life functions. Both processes need to happen in order to sustain life. **Figure 5** shows how these processes work together.

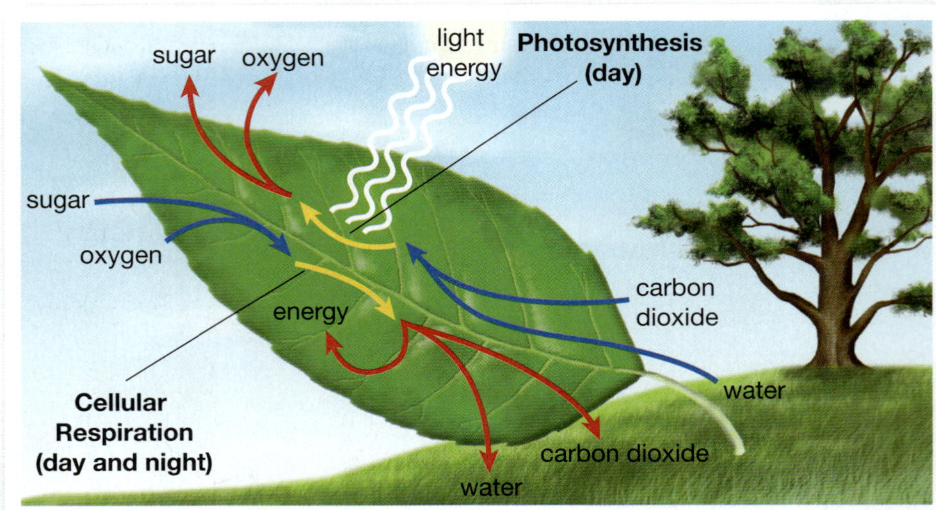

Figure 5 Photosynthesis and cellular respiration are processes that depend on each other.

TRY THIS OBSERVE CARBON DIOXIDE IN CELLULAR RESPIRATION

SKILLS HANDBOOK 3.B.

SKILLS: Questioning, Predicting, Performing, Observing, Analyzing

All cells need a continuous supply of energy. When your lungs take in oxygen, your cells use the process of cellular respiration to convert chemical energy into a usable form. Then you breathe out carbon dioxide.

Equipment and Materials: transparent cup; limewater; drinking straw; scissors; 5 cm by 5 cm cardboard

1. Fill a large cup $\frac{1}{3}$ full of limewater. Limewater turns cloudy when combined with carbon dioxide. Observe and record the appearance of the limewater.
2. Cut an opening in the centre of the piece of cardboard, just large enough to slip a straw through. The cardboard will act as a splash guard. Slide the cardboard halfway along the straw.
3. Place one end of the drinking straw in the limewater. The cardboard should be just above the top of the cup.
4. Slowly breathe out a steady stream of air into the straw. Repeat a few times. Observe and record any changes in the appearance of the limewater.

CAUTION: LIMEWATER
Do not suck on the straw.

A. Explain what you observed about the appearance of the limewater when you breathed into it. T/I C

B. Did this experiment give you any evidence that cellular respiration is going on in your body? Explain. T/I A

2.5 Wrap Up

- Green plants are producers. They use photosynthesis to produce food, which stores the Sun's energy as chemical energy.
- Consumers obtain energy when they eat other organisms or the products of other organisms.
- Plants and animals perform cellular respiration to release energy from food.
- Photosynthesis and cellular respiration are complementary processes.

CHECK YOUR LEARNING

1. How much of the Sun's energy that comes to Earth is used in photosynthesis? What happens to the rest of the energy? K/U
2. Why are photosynthesis and cellular respiration considered complementary processes? K/U C
3. What important substances do plants produce during photosynthesis? How do humans use these substances? K/U C
4. (a) How do producers get the chemical energy they need to live and grow?
   (b) How do consumers get the chemical energy they need to live and grow? K/U
5. Plants can make their own food by the process of photosynthesis. Animals need to find and eat food to obtain energy. What are the advantages and disadvantages of each way of obtaining energy? T/I
6. (a) Why are humans considered to be consumers?
   (b) Name five other organisms that are consumers. K/U
7. Animals are unable to make their own energy, yet you obtain energy when you animal food products. How does this illustrate the flow of energy through an ecosystem? T/I

2.6 Interactions in the Ecosystem

If you ran blindfolded through a forest in Ontario, you would not likely run into a moose or a raccoon. Instead, you would probably hit a tree or trip over plants around your feet. Why are there so many large trees and other plants in the forest but so few large animals? Why do you see hundreds of mosquitoes but not as many frogs to eat them? Each species is limited by the resources it needs.

Ecological Niches

Every species interacts with its environment and other species. These interactions are referred to as the species' **ecological niche**. The ecological niche of a species includes what it eats, what eats it, where it lives, and how it behaves. No two species have the same niche. What factors make up your niche?

▶ **ecological niche:** the function a species serves in its ecosystem, including where it lives, what it eats, what eats it, and how it behaves

Think about the niche of a black bear. Black bears eat plant parts such as nuts and berries (**Figure 1**). They also eat insects and, occasionally, other small animals. They have few predators other than human hunters, but other organisms such as biting insects and parasites feed on them.

Figure 1 Black bears eat a variety of foods.

Types of Consumers

What each species eats has an important impact on an ecosystem. What a species eats helps to define its niche. Consumers fit into the following types based on what they eat and how they behave:

- herbivore – consumer that eats only plants or other producers
- carnivore – consumer that eats only other consumers
- omnivore – consumer that eats both producers and consumers
- scavenger and decomposer – consumers that feed on the remains of other, dead or decaying, organisms

A Community of Eaters

In Section 2.2, you read that there are many populations in an ecosystem and all those populations make up a community. For example, a pond community may include turtles, frogs, minnows, and algae. The organisms that make up this community depend on one another for food. Plants make their own food, and animals eat plants or other animals.

Food Chains

Food chains show the feeding relationships among species in an ecosystem. **Figure 2** shows a typical forest food chain. Food chains illustrate what each species uses for food. The seeds of a pine tree (a producer) are eaten by a red squirrel, which is a herbivore. The squirrel is eaten by the weasel, which is a carnivore. The weasel, in turn, is eaten by another carnivore, a goshawk. Each of these relationships is a link in the food chain.

Carnivores linked in a food chain have a predator–prey relationship with the animals they eat (**Figure 3**). If a hawk eats a weasel, the hawk is known as a *predator*. The weasel is the *prey*.

Figure 3 A hawk is a predator. Its prey are small animals such as this mouse.

Energy in the Food Chain

The forest food chain in **Figure 2** shows that some of the chemical energy stored in the pine seeds is passed to the red squirrel. Some energy passes to the weasel, then it gets passed to the goshawk. In this way, food chains show how energy passes through an ecosystem. Organisms continuously use and release energy.

Figure 2 A food chain shows the feeding relationships in a forest ecosystem.

DIG DEEPER

Learning Tip

Reading Diagrams

In a food chain diagram, the arrows show the direction of energy flow. If the arrow is pointing from the red squirrel to the weasel, it means the energy is passing from the red squirrel to the weasel. The arrow is read "is eaten by." Read the food chain from left to right. For example, the red squirrel is eaten by the weasel.

▶ **trophic level:** the level of an organism in an ecosystem, depending on its feeding position along a food chain

Feeding Levels

The **trophic level** describes the level or position of an organism in a food chain (**Figure 4**).

Figure 4 The trophic levels of a forest ecosystem. Each level describes where the species gets its energy. Organisms are not drawn to scale.

TRY THIS YOU AND YOUR FOOD CHAIN

SKILLS HANDBOOK 3.B.7., 3.B.9.

SKILLS: Analyzing, Communicating

In this activity, you will discover your place in the food chain.

1. Make a list of all the food you eat in one day.
2. Classify each food as an animal or plant product.
3. Use your list to draw three food chains. Include yourself in the food chains.
4. Use your food chains to tell whether you are a herbivore, carnivore, or omnivore.

A. How would you classify your place in each food chain you drew? Explain your thinking. T/I

B. Describe the path of energy in each food chain you drew. Where does your energy come from? T/I

Beyond Food Chains

Food chains show simple feeding relationships, but food chains are part of a larger, more complex system of relationships among species. Think about the forest food chain shown in **Figure 2**. Many herbivores eat pine seeds. Red squirrels eat other foods as well. Red squirrels are food for goshawks, but other consumers also eat red squirrels.

Food Webs

A more complete model of feeding relationships can be seen in a food web. A **food web** shows a series of interconnecting food chains (**Figure 5**). Food webs show that consumers feed on many species. If a consumer feeds on a large number of different species, it will be less affected if one of those populations begins to decrease, or becomes scarce. For example, a snowshoe hare feeds on the leaves and twigs of blueberries, wild grasses, and a variety of trees. If the blueberry bushes become scarce, the snowshoe hare can eat other plants available in its ecosystem. A food web can also include scavengers and decomposers. Scavengers and decomposers are organisms that eat dead or decaying organisms.

▶ **food web:** a representation of the feeding relationships within a community

Figure 5 A food web is still not a complete model of every interaction in an ecosystem. A complete model would show thousands of species.

Effects of Changes to a Food Web

Any change to an ecosystem can affect a food web. For example, in the food web in **Figure 6**, the rusty crayfish disrupts an aquatic ecosystem by competing for the foods eaten by the native crayfish. The rusty crayfish also feeds on eggs and young of bass and pickerel. Fewer worms and snails may cause the number of minnows to decrease. Fewer minnows means less food for the larger fish. Since bass and pickerel do not eat the rusty crayfish invaders, the bigger fish may decrease in numbers.

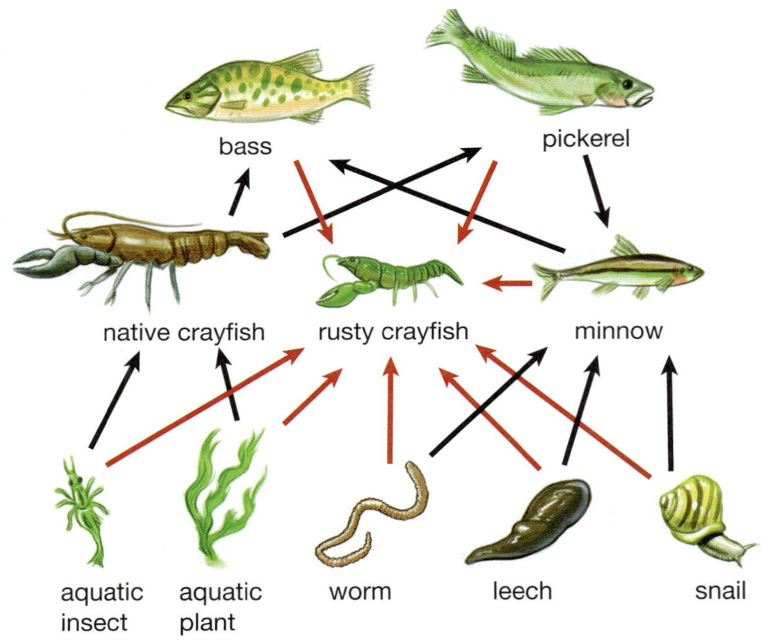

Figure 6 A rusty crayfish invades this ecosystem and competes with native crayfish and minnows for many of the same foods.

Human actions can also disrupt a food web. For example, harmful substances we use may enter an aquatic ecosystem and disrupt or stop the growth of a plant species. Consumers of that species will also suffer because their food source will be reduced or removed.

DIG DEEPER

Reading Tip

Text-to-World Connections

To make text-to-world connections, you sometimes have to conduct research. Gather information about the topic or key ideas. List the three most important points about the topic. Beside each point, make a note of the connection you can make.

Energy Pathways

Have you ever noticed that a crowded room can get warm very quickly? That is because people release thermal energy. Humans, like all organisms, carry out cellular respiration and release energy. When one organism consumes another, it takes the energy it needs to live and releases thermal energy. When this organism is eaten by another organism, the chemical energy stored in its body is passed on to the other organism.

Typically, only about 10 % of the energy taken in by organisms at one trophic level is passed on to organisms at the next trophic level. At each next higher level, the amount of available energy decreases. The bass in **Figure 6** will have less energy available to it than the minnow.

TRY THIS WHAT EATS WHAT?

SKILLS: Analyzing, Evaluating, Communicating

SKILLS HANDBOOK 3.B.7.–3.B.9.

In this activity, you will work in a small group to sort organisms into food chains and food webs.

Equipment and Materials: markers; cards with labelled images of Ontario organisms

1. Arrange the organism cards into three groups: producers, consumers, and decomposers.
2. Use your cards to arrange the organisms into food chains. Create four food chains of three links each. Draw the food chains.
3. Use the cards to arrange three food chains of four links each. Draw the food chains.
4. Use the cards to arrange the organisms into a food web. Draw the food web.

A. Which organisms did you place at the start of each food chain? Why? T/I C

B. Remove one organism from your food web. How does this affect the food web? How would removing this species affect the ecosystem? T/I C

2.6 Wrap Up

- Each species occupies a unique ecological niche.
- Feeding relationships among organisms can be described by food chains and food webs.
- Ecosystems can be disrupted by natural and human actions.
- Only about 10 % of the energy available at any trophic level is transferred to the next higher level.

UNIT TASK

Bookmark

You can apply what you learned about how human activities change ecosystems to the Unit Task described on page 160.

CHECK YOUR LEARNING

1. What is the ecological niche of the black bear? What information helps you describe its ecological niche? K/U
2. How are food chains and food webs similar? How are they different? K/U
3. As you go from the level of a primary consumer such as a hare to the trophic level of a goshawk, the amount of available energy decreases. Explain where the energy has gone. K/U
4. Suppose a family of coyotes builds a den on the edge of a forest. The coyotes begin preying on the hares and mice that both the owl and lynx populations depend on for food. How might that disrupt the balance of the existing forest ecosystem? T/I
5. How would the food web in **Figure 7** change if the numbers of red foxes were greatly reduced by hunters? K/U T/I A

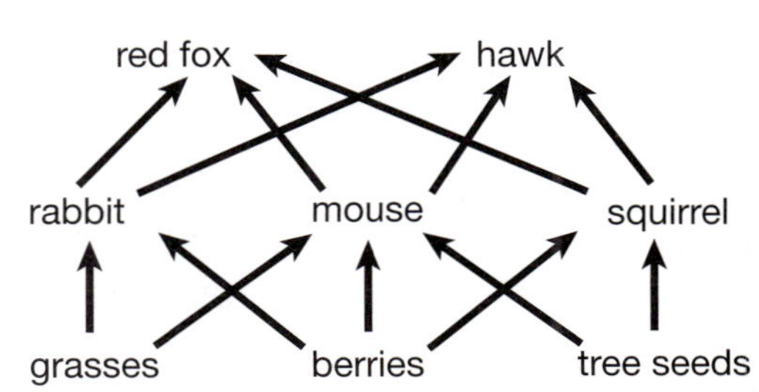

Figure 7

2.7

Cycling of Matter in Ecosystems

Energy flows through ecosystems when organisms consume other organisms. Unlike energy, which is converted into different forms but never returns to its source, matter is recycled through ecosystems. The same matter is transformed many times and used over and over again.

How Old Is Your Body?

All organisms on Earth require water and nutrients. Water and nutrients are made up of matter. Your body gets matter from food, water, and air. However, the particles of matter you get do not stay in your body forever. Your cells are made of matter, and every part of every cell in your body is replaced over time (**Figure 1**). Scientists estimate that every particle in a human body is replaced at least once every seven years!

Figure 1 This woman appears older than the child, but almost every particle in the woman's body has been replaced within the last seven years.

DIG DEEPER

Learning Tip

Word Roots

Because these cycles involve "chemicals" that cycle through living things or organisms (bio) and Earth (geo) processes, they are called biogeochemical cycles.

▶ **biogeochemical cycles:** movement of matter through the biotic and abiotic parts of the environment

Cycles That Maintain Balance

Like your body, an ecosystem undergoes continuous renewal to maintain its delicate balance. But renewal does not mean that new matter is made. All the water and nutrients of organisms are from particles that already exist in the environment. In processes called **biogeochemical cycles**, particles are continuously consumed, rearranged, stored, and released. The four main biogeochemical cycles are carbon, water, nitrogen, and phosphorus. The water and carbon cycles are explained in this section.

The Water Cycle

The **water cycle** is the movement of water as a liquid, solid, or gas around Earth. In the water cycle, shown in **Figure 2**, liquid water evaporates and changes to water vapour. Next, the vapour condenses and forms ice crystals or water droplets. Then, the droplets collect to form clouds and fall to Earth as rain, hail, or snow. Water enters lakes, oceans, the soil, or groundwater, and the process continues. Water that is taken in by plant roots may be released from leaves in a process called transpiration.

▶ **water cycle:** the series of processes that moves water through the environment

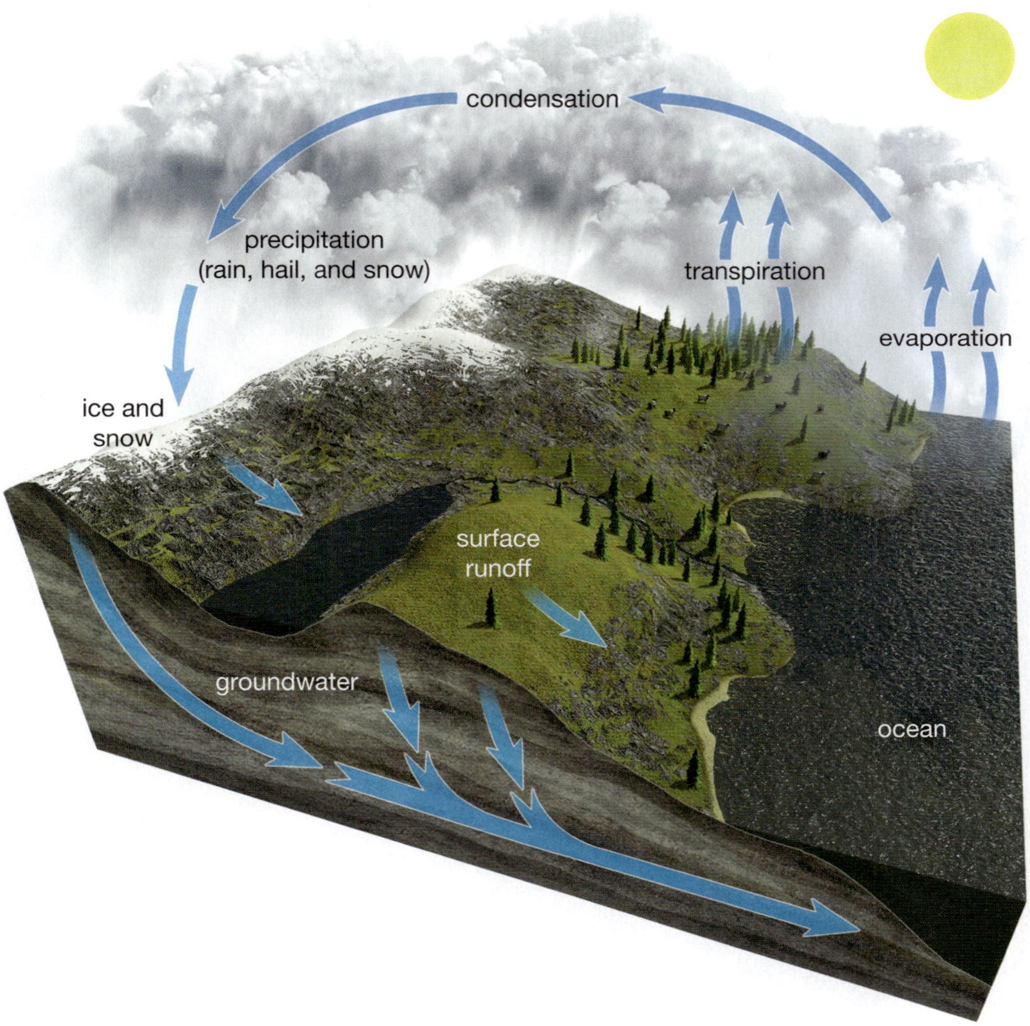

Figure 2 The water cycle is one of the major biogeochemical cycles. Most water is present in abiotic features such as lakes or oceans.

The Carbon Cycle

Carbon is found in both biotic and abiotic features on Earth. Some carbon, in the form of compounds such as carbon dioxide and sugars, is cycled during photosynthesis and cellular respiration. Carbon that moves between the abiotic and biotic parts of an ecosystem is part of the **carbon cycle** (**Figure 3**).

▶ **carbon cycle:** the series of processes that moves carbon compounds throughout the lithosphere, atmosphere, hydrosphere, and biosphere

Carbon Deposits

While a lot of carbon is cycled through the processes of photosynthesis and cellular respiration, most of Earth's carbon is not cycled. It is stored in deposits such as coal, limestone, ocean sediments, the shells of marine animals, oil, and natural gas. These deposits form when decomposed organisms are compressed over millions of years.

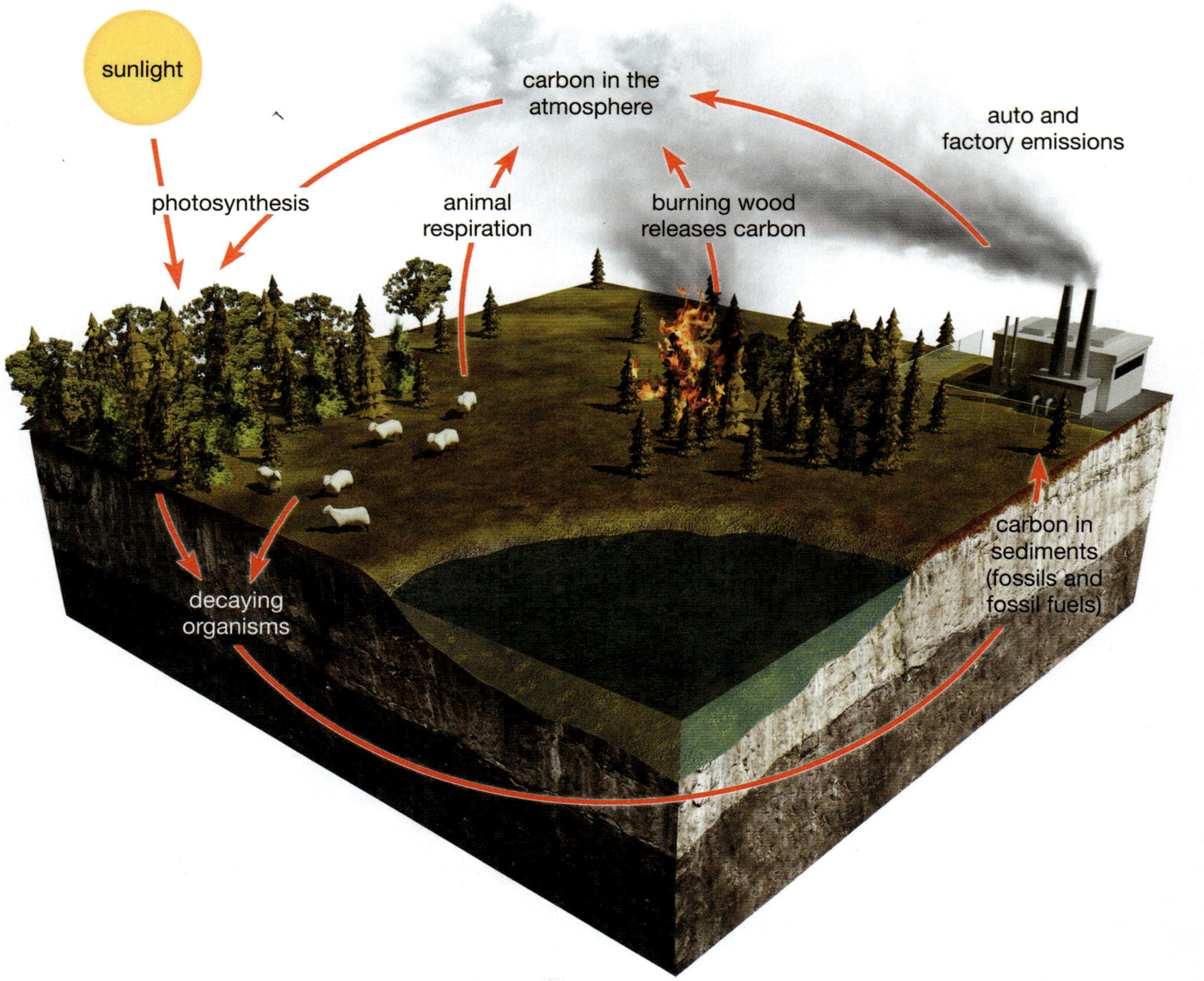

Figure 3 Most carbon moves through the cycle during photosynthesis and cellular respiration.

How Human Activities Change the Carbon Cycle

The actions of humans can greatly affect the carbon cycle.

- **Auto Emissions:** Most of the fuel we burn in automobiles or other vehicles is made from carbon-rich fossil fuels such as oil and natural gas. During burning, the stored carbon is released into the atmosphere as gases such as carbon dioxide and carbon monoxide.
- **Deforestation:** Clearing forests is another human activity that can affect the carbon cycle. Carbon trapped in the bodies of trees is released when they are burned or decompose.

Both burning fossil fuels and cutting down trees remove carbon from long-term storage. Extra carbon dioxide and carbon monoxide in the atmosphere affect the environment. These gases trap energy that would otherwise be radiated into space. The trapped energy causes global warming, which contributes to climate change. Climate change upsets the delicate balance of temperature, rainfall, and resources in habitats, and endangers living things.

2.7 Wrap Up

- Matter moves and cycles through ecosystems in biogeochemical cycles (water, carbon, nitrogen, phosphorus).
- Water moves through the water cycle as a liquid, gas, or solid.
- Most carbon cycling happens during photosynthesis and cellular respiration.
- Human activities can disturb the balance of any of the biogeochemical cycles.

UNIT TASK

Bookmark

You can apply what you learned about how human actions change the carbon cycle to the Unit Task described on page 160.

CHECK YOUR LEARNING

1. Draw a diagram or flow chart showing how water moves through the water cycle. K/U C
2. List the main sources of carbon entering the atmosphere. K/U
3. (a) Give an example of a human activity that affects the carbon cycle.

   (b) How does this activity affect the carbon cycle? K/U C
4. Changes in the carbon cycle cause global temperatures to rise. How might this affect the water cycle? A
5. What steps can people take to slow global climate change? A
6. In this section, you read that all body matter is eventually replaced. How does this influence your understanding of how your own body is involved in biogeochemical cycles? A

FOCUS ON
MATH

Graphing Two Sets of Data

Scientists graph the data they collect to make it easier to analyze. Sometimes, scientists need to graph two sets of data together to interpret their effects on each other.

Population Graphs in Action

The Canadian lynx and the snowshoe hare have a predator–prey relationship that scientists follow closely. You will use two sets of population data and present them in a single graph.

Table 1 Predator–Prey Populations

Year of study	Snowshoe hare population	Lynx population
1	20 000	1 000
2	55 000	1 200
3	65 000	5 500
4	95 000	6 000
5	55 000	2 000
6	5 000	1 500
7	15 000	1 000
8	50 000	6 000
9	75 000	6 000
10	20 000	1 000
11	25 000	500
12	50 000	2 500
13	70 000	4 000
14	30 000	2 500
15	15 000	500

Present the Data on Two Graphs

Step 1: Label the *x*-axis "Year of study." The *x*-axis will be the same for both graphs. Label the *x*-axis with dates from **Table 1**.

Step 2: Determine the range of numbers to be plotted. The snowshoe hare population ranges in size from 5000 to 95 000. Label the *y*-axis "Hare population (in thousands)." Number your *y*-axis from 0 to 100, in intervals of 10.

Step 3: Plot the data from **Table 1**. Join the points with a line.

Step 4: Give your graph a title. A completed graph for the hare population is shown in **Figure 1**.

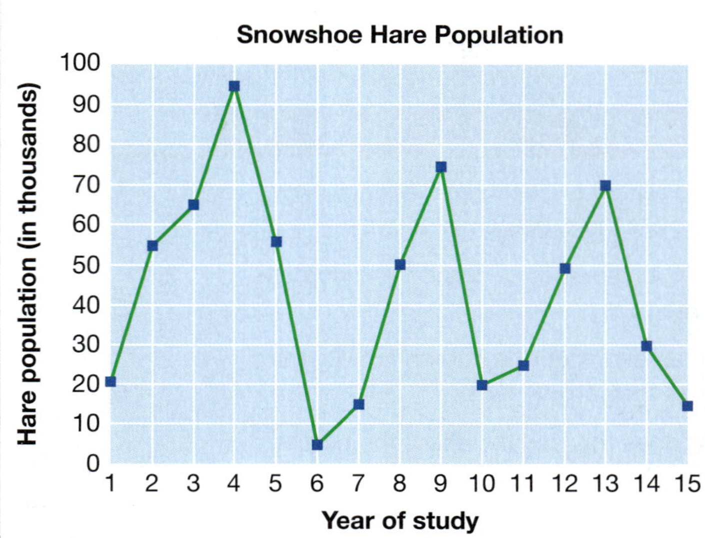

Figure 1

Step 5: Repeat Steps 1 to 4 for the lynx population. Use a different scale for the lynx population. A completed graph for the lynx population is shown in **Figure 2**.

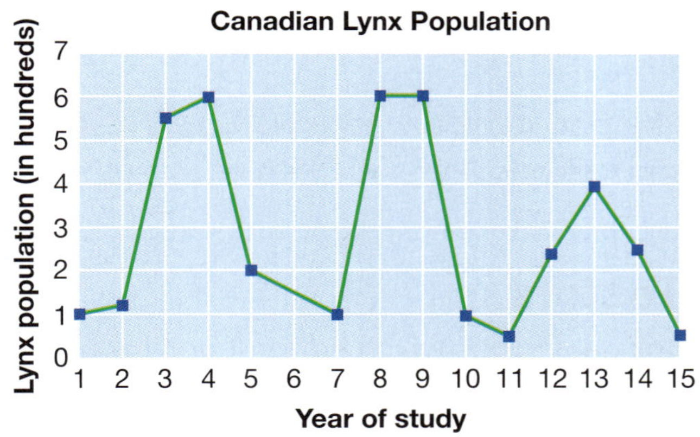

Figure 2

Combine the Two Graphs

Step 1: Create the *x*-axis. This will be the same as in both **Figure 1** and **Figure 2**.

Step 2: Draw and label a *y*-axis on the left side of the *x*-axis that is the same as in **Figure 1**. Plot the hare population data in **Table 1** as a line graph.

Step 3: Draw and label a *y*-axis on the right side of the *x*-axis that is the same as in **Figure 2**. Plot the lynx population data in **Table 1** as a line graph, using a different colour.

Step 4: Give your combined graph a title.

A. Examine the combined population graph shown in **Figure 3**. What population patterns do you notice? T/I

B. How does graphing two sets of data on one graph help you see patterns in the populations? T/I

C. Why do you think the changes in the lynx population so closely follow changes in the hare population? T/I

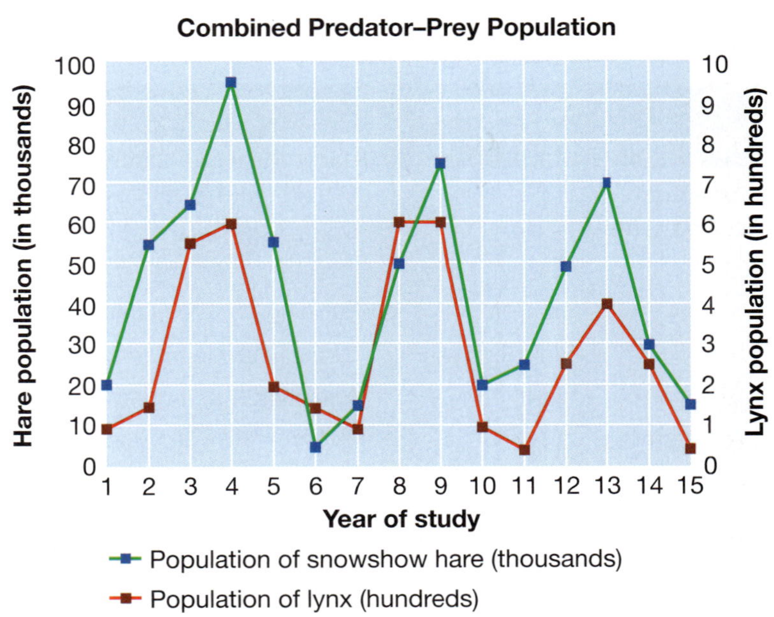

Figure 3

2.8

Biotic and Abiotic Influences on Ecosystems

What factors help us decide where we live? The answers lie in the biotic and abiotic features of an ecosystem.

- **limiting factor:** any factor that restricts the size of a population or where it can live
- **tolerance range:** the range of abiotic conditions within which a species can survive

Limiting Factors

A **limiting factor** determines which types of organisms and how many of each species are able to survive in an ecosystem. These factors can be abiotic, such as the available water sources or the air temperature. Limiting factors can also be biotic, such as plant or animal food sources. Human actions can also act as limiting factors.

Abiotic Limiting Factors

Temperature and precipitation are both abiotic factors that can influence where a species lives (**Table 1**). For example, cacti thrive in dry conditions. Too much rainfall could destroy their shallow roots by flooding them or causing them to rot in unusually wet soil. Every species has a **tolerance range**, which is the range of an abiotic factor that the species can endure. Some species have a wide tolerance range. For example, raccoons can live in tropical forests that are continuously hot as well as in northern forests with very cold winters.

Table 1 Key Abiotic Factors in Ecosystems and Effects of Human Actions

Ecosystem	Key abiotic factor	Human action and result
terrestrial ecosystems	light	Fire and tree cutting can remove shade and expose organisms to more light.
	water	Damming rivers and draining wetlands reduces water availability. Irrigation increases water availability.
	nutrients	Farming practices can change nutrient levels in soil.
	temperature	Global warming can reduce available habitat for cool-adapted species.
aquatic ecosystems	light	Activities that increase erosion or stir up bottom sediments can reduce light at lower depths.
	nutrients	Fertilizer runoff can increase nutrients in water and cause algal blooms.
	acidity	Pollution from burning fossil fuels increases acid rain in aquatic ecosystems.
	temperature	Industries release heated waste water into lakes and rivers, killing fish and other organisms.
	salinity	Salting highways and irrigation can increase the concentration of salt.

Biotic Limiting Factors

While abiotic factors determine where a particular species is able to live, biotic factors often determine how easily a species can survive in an ecosystem. For example, having a lot of a specific food can influence how well a species may survive. A deer can live in dense forests but have more success in open woodlands where there are more of the food plants that they prefer to eat. Biotic factors also include relationships among organisms in an ecosystem. There are several different types of interactions between organisms in ecosystems.

Competition

Many organisms compete with one another for limited resources such as space, food, water, or light. Competition occurs when two or more individuals try to get enough of the same resource. Competition often results in the decline of one species. Plants also compete. For example, they compete for new space and light made available after a forest fire (**Figure 1**).

Predation

Predation is another biotic factor that can affect the success of a species. Predation occurs when a predator kills and eats its prey. Wolves prey on moose (**Figure 2**). Moose populations can grow or shrink depending on the number of wolves present. When more wolves live in an area, fewer moose survive. A larger moose population can support a larger wolf population.

Figure 1 A field of fireweed grows among the remains of trees charred in a forest fire.

Figure 2 Wolves hunt in packs.

Other Interactions

Many key biotic factors describe interactions between individual organisms. Mutualism occurs when two organisms benefit from interacting with each other. Commensalism occurs when one organism benefits and the other is neither harmed nor helped. Parasitism occurs when one organism (the parasite) lives and feeds on or inside another organism (the host). **Table 2** lists and summarizes some of these common interactions.

Table 2 Key Types and Examples of Species Interactions

Relationships	Definitions	Examples
competition	two individuals seek the same resource	• Foxes and coyotes both feed on common prey such as mice and rabbits. • Humans and insects compete for the same crop plants.
predation	one individual feeds on another	• Lynxes prey on snowshoe hares. • Leeches and blackflies are "micro-predators" that feed on humans.
mutualism	two individuals benefit each other	• Nitrogen-fixing bacteria live in the roots of certain plants. The plants provide sugars to the bacteria. The bacteria provide nitrogen to the plants (**Figure 3**).
commensalism	one individual benefits and the other neither benefits nor is harmed	• Many birds nest in particular kinds of trees or in abandoned burrows. • Barnacles live on humpback whales but feed on other organisms in the water (**Figure 4**).
parasitism	one individual lives on or in a host organism and feeds on it	• Tapeworms are parasites of lynxes and wolves (**Figure 5**). • Microbes that cause malaria live within human blood cells.

Figure 3 Bacteria living in nodules on the roots of a soybean plant is an example of mutualism.

Figure 4 The barnacles do not pose any threat to the humpback whale; they feed on organisms in the water.

Figure 5 The tapeworm attaches to the intestinal wall of its host with hooks or suckers. Tapeworms range in length from 1 mm to more than 15 m.

Carrying Capacity

As a population grows, it requires more food, water, and space. At some point, the population reaches the maximum number of organisms that can be continuously supported by the ecosystem. This limit is called the **carrying capacity** of the ecosystem.

When abiotic and biotic factors change, the carrying capacity of the ecosystem also changes. For example, a forest fire reduces the number of trees in that ecosystem. The carrying capacity for blueberry plants then increases since more light and space are available for blueberry plants to grow.

▶ **carrying capacity:** the maximum population size of a species that a given ecosystem can sustain

How Do Human Actions Affect Carrying Capacity?

Human activity can change the carrying capacity of an ecosystem. For example, the decrease in the wolf population due to hunting by humans could increase the carrying capacity for moose in a forest ecosystem. Human actions that increase the productivity and fertility of land, such as irrigation, can increase its carrying capacity for different crops (**Figure 6**). However, actions such as overgrazing can decrease carrying capacity (**Figure 7**).

Figure 6 These green circles are formed from an irrigation project in Malaysia. Irrigation increases the carrying capacity of this desert ecosystem.

Figure 7 As a result of overgrazing, about half of the land in this grassland community is bare soil.

TRY THIS MOOSE AND WOLF POPULATION

SKILLS HANDBOOK 3.B.7., 3.B.9.

SKILLS: Analyzing, Communicating

In this activity, you will examine how predator–prey relationships affect the carrying capacity of an ecosystem.

Table 3 shows the numbers of wolves and moose in an area over a period of 45 years (Figure 8).

Equipment and Materials: graph paper

1. Use a grid similar to the one shown in Figure 9. Label the *x*-axis "Time." Label the *x*-axis with the dates listed in Table 3.
2. Based on the sizes of the numbers for each population, create a scale on the two separate *y*-axes similar to what you observed in the lynx–snowshoe hare graph on page 53.
3. Construct a line graph using the data from Table 3. Use a different coloured line for each population.
4. Give your graph a title. Create a key or legend for your graph so it is easy to read.

Figure 8 What is the relationship between the number of moose and the number of wolves in an ecosystem?

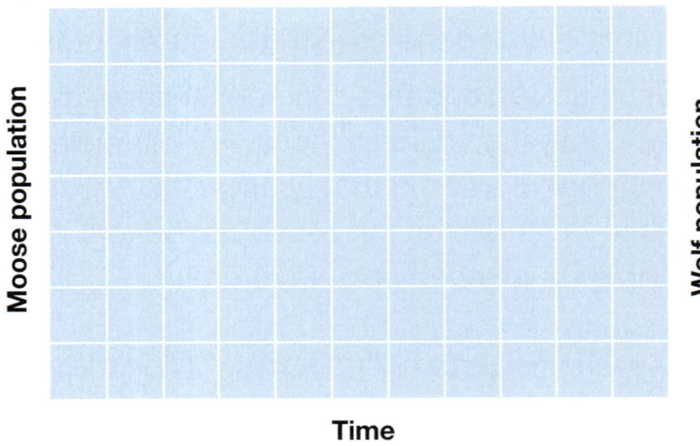

Figure 9

A. Describe the relationship between the number of wolves and the number of moose. T/I C
B. What biotic and abiotic factors might influence the wolf and moose numbers? T/I
C. How would the carrying capacity for wolves be influenced if a large number of moose were killed by hunters? T/I

Table 3 Numbers of Wolves and Moose on Isle Royale

Year	1960	1965	1970	1975	1980	1985	1990	1995	2000	2005
Wolves	22	28	18	41	50	22	15	16	29	30
Moose	610	733	1295	1355	910	1115	1216	2422	850	540

2.8 Wrap Up

- Limiting factors determine the types and numbers of organisms in an ecosystem.
- Limiting factors can be abiotic or biotic.
- Tolerance ranges describe the ranges of physical conditions under which a species can survive.
- Species interactions include competition, predation, mutualism, parasitism, and commensalism.
- Carrying capacity is the maximum population size that an ecosystem can support.
- Human activities can affect the limiting factors for a species and the carrying capacity of an ecosystem.

UNIT TASK

Bookmark

The Unit Task is described in detail on page 160. You can use what you learned about how human actions change the carrying capacity of ecosystems to complete the Unit Task.

CHECK YOUR LEARNING

1. Two species of bird compete for food in a forest. Both eat worms, but one has a narrow beak that allows it to pull worms out of cracks more easily.
   (a) If a drought caused the number of worms to decrease, which species would be more successful?
   (b) Identify the limiting factors. Are they biotic or abiotic? Explain. T/I K/U
2. Fertilizer used by homeowners often ends up in nearby natural ecosystems. What effect do you think this has on the carrying capacity for plant life in these ecosystems? A
3. How are predators and parasites alike? How are they different? T/I
4. A predator is a factor in determining the population of the prey it eats, but a parasite may not be. Why do you think this is so? T/I A
5. Chickadees are birds that have adapted to cold winter weather (**Figure 10**). Birdwatchers provide seeds for them at bird feeders during winter months. K/U T/I
   (a) How do bird feeders influence the carrying capacity of the ecosystem? Explain.
   (b) Suppose birdwatchers observed an increase in the falcon population in recent years. Suggest a reason for this increase.
6. Classify each of these species interactions based on its description. A
   (a) Flatworms live attached to the gills of horseshoe crabs obtaining bits of food from the crabs' meals. The crabs are unaffected.
   (b) Lichens found on tree bark consist of algae and a fungus. The fungus obtains food from photosynthesis of the algae and the algae has a place to live.
   (c) A flea feeds on the blood of a ground squirrel.
   (d) Both bluebirds and starlings nest in holes in trees, poles, and fence posts. Neither species is able to dig holes. Both must rely on holes that already exist.
   (e) A wood turtle lays its eggs on sand or gravel-sand beaches and banks along streams. A raccoon eats the turtle's eggs.

Figure 10

2.9

On Land or in Water?

From colourful coral reefs to frigid mountaintops, Earth has many different environments. Average temperatures in each environment vary, as do precipitation and land or water characteristics. Each environment has distinct life forms living in it, forming unique and complex communities.

Biomes

Areas of the world that have a similar climate and similar organisms are known as **biomes**. There are many biomes on Earth that have easily recognizable ecosystems. Desert, coral reef, tundra, and tropical rainforest biomes are common and have distinct features.

▶ **biomes:** large geographical regions defined by climate with a specific set of biotic and abiotic features

Similar ecosystems appear where there are similar abiotic conditions such as temperature, moisture, and light. These similar ecosystems fit into the categories known as biomes.

Terrestrial Biomes

Climate is the most important factor that determines the location and makeup of a terrestrial (land) biome. Temperature and precipitation greatly influence the formation of terrestrial biomes.

Canada's Biomes

Five major terrestrial biomes are found in Canada: tundra, boreal forest, mountain forest, grassland, and temperate deciduous forest.

Tundra Biome

The tundra is a cold, barren biome in the northernmost part of Canada. Low temperatures and a short growing season limit the kinds of plants that grow here. Many areas in this biome have permanently frozen ground called permafrost. In summer, surface soil thaws so that plant roots can grow. Polar bears and caribou live in the tundra (**Figure 1**).

Figure 1 Caribou's thick fur and fat insulate them against the cold.

Boreal Forest Biome

The boreal forest is the largest biome in Canada. Warm temperatures and rainfall in summer encourage tree growth in northern Ontario's boreal forest. The forest floor is covered with mosses and ferns. Conifer trees are abundant (**Figure 2**). Confers can survive harsh winters while keeping their needle-shaped leaves. The shaded forest floor is covered with mosses and ferns. The boreal forest is home to snowshoe hares, black bears, and grey wolves.

Figure 2 The shape of conifers allow them to discard heavy snow.

Mountain Forest Biome

Temperatures vary in Canada's mountain forest biome. Temperatures decrease with increasing elevation. Winters are cold and summers are moist and cool. The coniferous forest (pine, fir, and spruce trees) provides suitable habitat for many large mammals including elk, grizzly bear, and cougars (**Figure 3**).

Figure 3 The average summer temperature is only about 20 °C in this mountain biome in Alberta.

Grassland Biome

Animals found in southern Canada's natural grassland biome include coyotes, foxes, prairie dogs, and burrowing owls (**Figure 4**). Grasses here grow in areas that get moderate rainfall. Most tree species do not grow in the grasslands. Dry, hot summers provide ideal conditions for fires, which help maintain the grasslands by slowing tree growth. You will learn more about the role of fire in ecosystems in Chapter 3.

Figure 4 Much of the grassland of Canada has been replaced with crops.

Temperate Deciduous Forest Biome

This biome is home to a great variety of deciduous trees such as maples, oaks, and ashes. Agriculture, timber, and forest products industries flourish in southern Ontario because of the climate. The growing season is long and there are many species of small plants (**Figure 5**). The varied plant life supports a rich variety of animals.

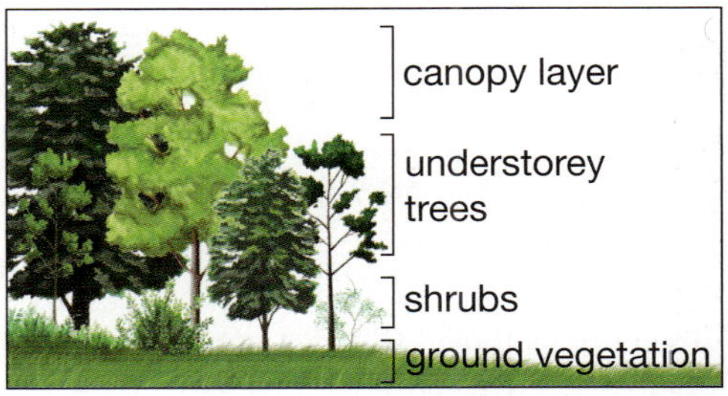

Figure 5 The plants in a temperate deciduous forest biome include canopy trees, understorey trees, shrubs, and ground vegetation.

Figure 6 shows a summary of the characteristics of Canada's five major biomes.

Mountain Forest

Abiotic
- temperatures vary with elevation
- cool summers
- windy conditions
- heavy precipitation on windward side of mountains
- fast-flowing rivers

Biotic
- large coniferous trees
- ferns
- marmots
- squirrels
- elk
- black bears and grizzly bears
- cougar

Tundra

Abiotic
- low temperatures for most of the year
- short growing season
- permafrost layer beneath soil
- precipitation from 0 to 25 cm/year
- poor soil quality

Biotic
- low diversity
- rapid-flowering plants
- mosses and lichens
- caribou
- ptarmigan
- lemmings
- Arctic foxes

Boreal Forest

Abiotic
- warmer than tundra
- no permafrost
- changeable weather
- soil contains some water and is acidic
- precipitation 40 cm/year or more

Biotic
- coniferous trees
- seed-eating birds
- squirrels
- voles
- snowshoe hares
- black bears
- pine martens
- grey wolves

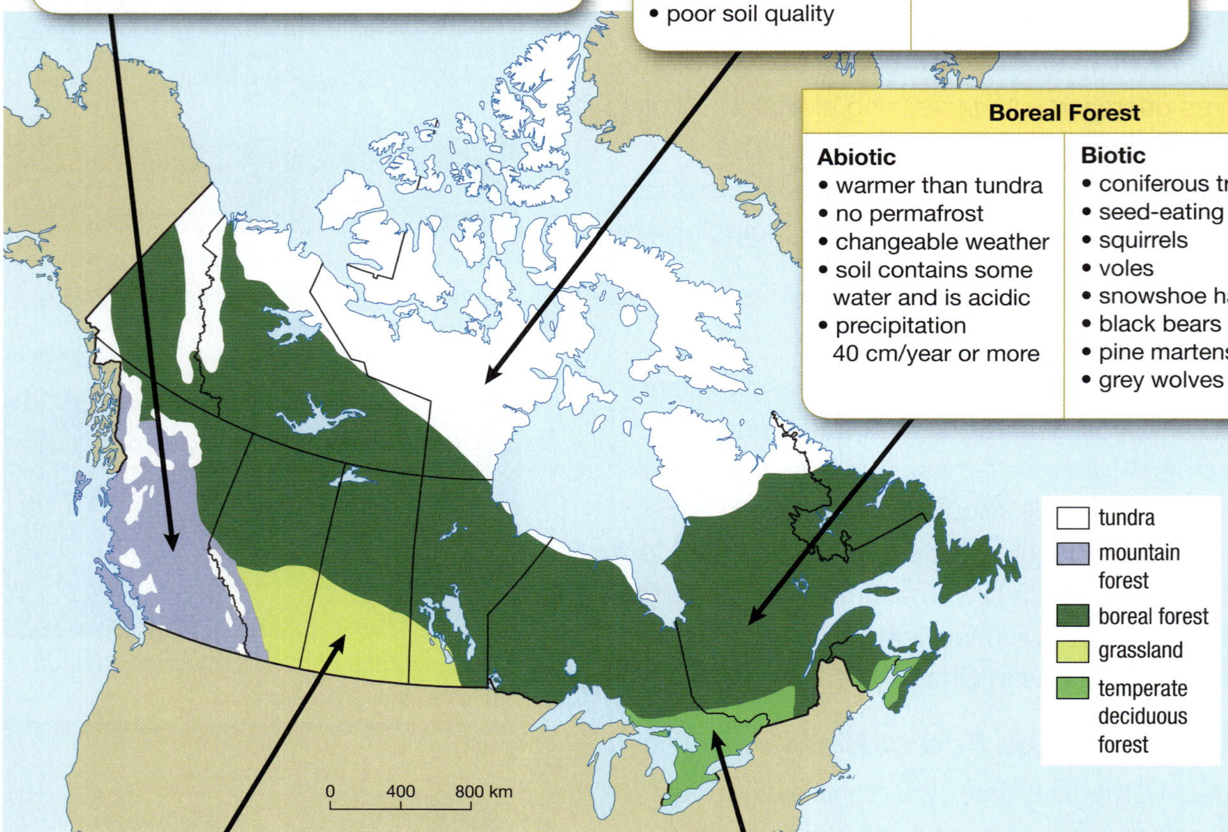

Grassland

Abiotic
- longer growing season than boreal forest
- higher temperatures than tundra or boreal forest
- rich, fertile soil
- precipitation from 25 to 75 cm/year

Biotic
- fescue grasses
- grasshoppers
- bison
- voles and mice
- snakes
- hawks
- coyotes

Temperate Deciduous Forest

Abiotic
- longer growing season than boreal forest
- higher temperatures than tundra or boreal forest
- fertile soil
- precipitation up to 100 cm/year

Biotic
- deciduous trees and other flowering plants
- tree and ground squirrels
- many insects
- shrews and mice
- deer
- black bears
- hummingbirds
- weasels

Figure 6 Canada's biomes

Aquatic Ecosystems

Over two-thirds of our planet is covered with water. 97 % of it is salt water. Aquatic ecosystems are divided into two broad categories: freshwater ecosystems and saltwater or marine ecosystems.

Freshwater Ecosystems

Freshwater ecosystems include rivers, streams, lakes, and ponds. Rivers and streams are unique among ecosystems because they are continuously moving. Organisms living in flowing waters must either swim continuously or attach themselves to the bottom or to an object such as a rock or aquatic plant. Lakes and ponds such as Lake Ontario are more stationary ecosystems. Lakes are often clear and deep. Ponds are often shallow and murky.

Wetlands such as bogs and marshes are large areas of shallow water or saturated soil. Wetlands, such as Oshawa Second Marsh, play an important role in the water cycle by filtering water. They are rich in nutrients and support many species, including birds, fish, amphibians, and insects (**Figure 7**).

Figure 7 Oshawa's Second Marsh drains an area over 100 square kilometres.

DIG DEEPER

Career Link

Conservation authorities protect Ontario's watersheds. To learn more about working for a conservation authority,

GO TO NELSON SCIENCE

Watersheds

Watersheds are an important part of a freshwater ecosystem. A **watershed**, such as the St. Lawrence River basin, is the land area that water travels through as it drains into a river or lake (**Figure 8**). Since water flows downhill, if pollution enters a watershed, it could pollute areas downstream as well.

► **watershed:** the land area drained by a particular river or lake; also called a drainage basin

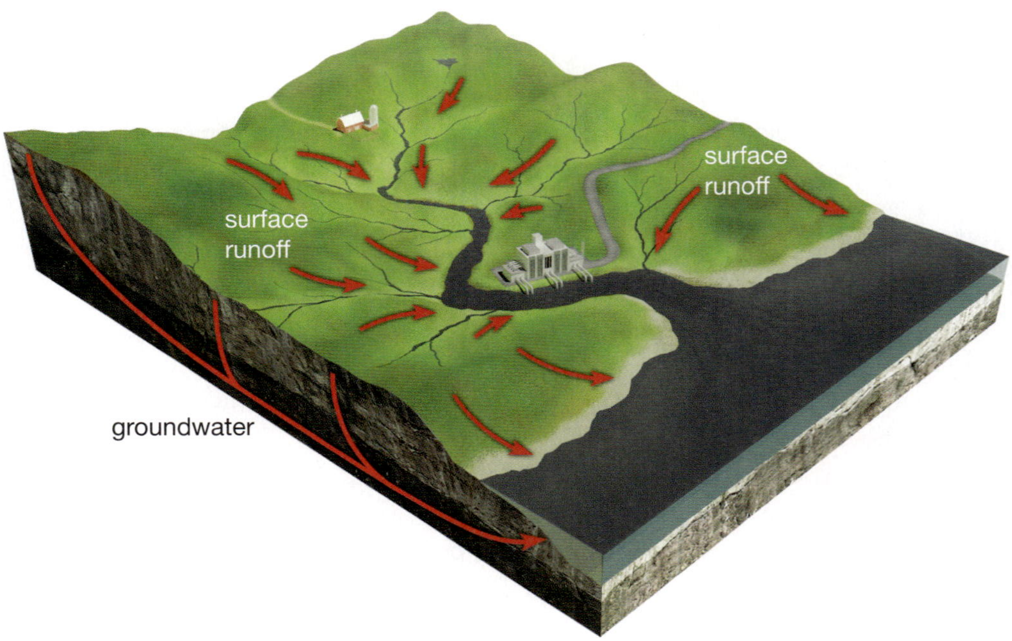

Figure 8 Watersheds must be managed to keep them healthy and pollution-free.

Figure 9 Coral reefs are found in tropical ocean waters near the equator, such as Great Barrier Reef in Australia.

Figure 10 The nutrient-rich waters of the Gulf of St. Lawrence support many species of fish and birds.

Figure 11 The roots of trees in mangrove forests grow out into the water. This helps prevent erosion.

Marine Ecosystems

More than 70 % of Earth's surface is covered with oceans. Most of the water that evaporates into the air in the water cycle comes from oceans. Half of the oxygen produced during photosynthesis comes from organisms living in Earth's oceans.

Coral Reefs

Coral reefs form in warm shallow oceans and they are home to a large number of organisms (**Figure 9**). Coral reefs are extremely sensitive to changes in water temperature, acidity, and pollution because the corals that make up the reefs require specific conditions in order to survive.

Estuaries

Estuaries are partially enclosed bodies of water where fresh water and salt water meet. Estuaries are high in nutrients and often support valuable shellfish such as clams and scallops. The world's largest estuary is the Gulf of St. Lawrence (**Figure 10**).

Mangrove Forests

Mangroves are found along tropical and semitropical shorelines such as those in Honduras or Belize in Central America. Trees that grow there are adapted to live at and beyond the water's edge (**Figure 11**). Mangroves serve as nurseries for young fish. Shoreline developments have destroyed large areas of the mangrove habitat. As a result, storm damage to shorelines has increased.

UNIT TASK

Bookmark

You can apply what you learned about biomes to the Unit Task described on page 160.

2.9 Wrap Up

- A large region with a particular climate and a characteristic community of plants and animals is called a biome.
- Terrestrial ecosystems are on land and have distinctive biotic and abiotic features.
- Aquatic ecosystems can be freshwater or marine.

CHECK YOUR LEARNING

1. What are the two main categories of aquatic ecosystems? K/U
2. List the terrestrial biomes found in Canada. List one biotic and one abiotic feature of each of these biomes. K/U
3. Explain why it is important to protect watershed areas from pollutants. K/U C
4. What terrestrial biome do you live in? List the biotic and abiotic features that you have observed in your biome. K/U C

TECH CONNECT OSSLT

Whales, Darts, and DNA

Crime shows on television often show criminal cases being solved using evidence such as DNA fingerprinting. This same technology is now being used to save endangered species of whales.

For years, scientist Scott Baker, from the University of Auckland, New Zealand, studied endangered humpback whales in the Pacific Ocean (**Figure 1**). Baker was puzzled that humpback whales were not increasing in number after a hunting ban was imposed in 1986. Baker had a hypothesis. He knew that Japan allowed the sale of whale meat from species that were not protected. He suspected that some whale meat being sold was also from whales that were protected.

Gathering the Evidence

Baker needed to gather evidence to prove his suspicions. He designed special darts. When fired from a crossbow, the darts hit the whale and bounced off, removing a small sample of the whale's flesh (**Figure 2**). He used the samples to produce a DNA fingerprint for each whale. Baker gathered DNA samples from different whale species.

Fighting Illegal Whaling

In 1993, Scott Baker made a trip to Tokyo. He bought samples of whale meat being sold at fish markets (**Figure 3**). Baker used his hotel room as a lab and tested the samples to find out which species each sample came from. His results confirmed that many of the samples were from endangered whale species.

Because of Baker's efforts, a new program to protect endangered whale species was started with closer monitoring to prevent illegal whaling.

Figure 3 Whale meat is still sold in Japanese supermarkets.

Figure 1 At maturity, humpback whales are about 15 m long and have a mass of about 36 000 kg. They eat krill, a shrimplike marine invertebrate, and small fish.

Figure 2 Scott Baker used a crossbow to gather DNA evidence from various protected whale species.

2.10 PERFORM AN ACTIVITY

Ecosystem Field Study

SKILLS MENU

- Questioning
- Hypothesizing
- Predicting
- Planning
- Controlling Variables
- Performing
- Observing
- Analyzing
- Evaluating
- Communicating

In this field study, you will visit a local terrestrial ecosystem and observe its abiotic and biotic features.

SKILLS HANDBOOK 3.B.5., 3.B.6., 3.B.9.

Purpose

To visit a natural or artificial ecosystem and collect data on its biotic and abiotic features

Procedure

Part A: Fieldwork—Abiotic Features

1. Spend 10–15 min walking through your ecosystem. Draw a map of the area, and sketch major features that you see.
2. Choose a location in the ecosystem. Mark it on your map. Write a brief description in your notebook of the ecosystem. Is it a wooded or a wetland community? Is there a nearby water source?
3. Use appropriate tools to measure and record each of the following abiotic features: wind speed, temperature, and relative humidity (use a sling psychrometer or hygrometer).
4. Estimate how much canopy (overhead leaf) coverage there is (0 %, 25 %, 50 %, 75 %, or 100 %). This is a way to describe the available light.

Equipment and Materials

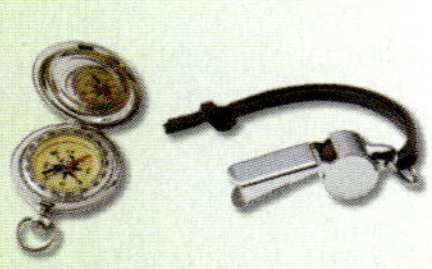
compass and whistle

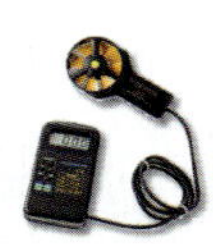
anemometer (for measuring wind speed)

thermometer

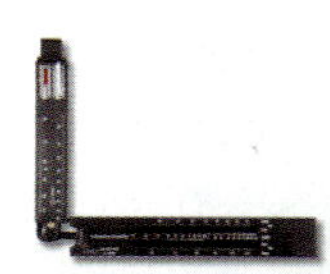
sling psychrometer (or hygrometer)

field guides to local nature

digital camera (optional)

binoculars

magnifying glass

gloves

Part B: Fieldwork—Biotic Features

5. Spend 20–30 min observing organisms in the ecosystem. Record your observations. Use field guides to identify the species. Record unidentified species as Unknown #1, Unknown #2, and so on.
6. In your notebook, write or draw a brief description of each species you observe. Use these techniques to find and observe organisms:
   - Use binoculars to observe as many bird and mammal species as you can.
   - Find and identify as many plant species as you can. Be careful not to overlook small plants. Take note of any fungi you see.
   - Look closely at surfaces of leaves, bark, and the ground for insects.
   - Carefully look under logs or other objects on the ground for amphibians and other organisms. Return any objects that you moved back to where you found them.
   - If a small stream or pond is present, look along the water's edge for plants and animals.
   - Look for any signs of human activity in the surroundings.
7. Make a tally chart in your notebook: How many plant species, bird species, and insects or other kinds of invertebrates were present? How many amphibians or mammals did you see? The health of an ecosystem is reflected, in part, in the size of populations it can support.

CAUTION

Allergies/ Reactions

Take precautions if you have any particular allergies, such as to bee stings. Avoid poison ivy (**Figure 1**). Take a whistle with you in case you become disoriented. Wear gloves.

Figure 1 Poison ivy. Beware!

CAUTION

Safety

Stay at least 1 m from the edge of any pond or stream you encounter.

Analyze and Evaluate

(a) Classify each organism you saw as a producer or a consumer. If it is a consumer, what does it eat? How does it survive in its ecosystem? A

(b) Review your tally chart and your field notes about the organisms you identified. Draw a possible food web based on these organisms. T/I C

(c) Was there any evidence of human influence in your ecosystem? If so, explain what you think that influence was. Was it positive or negative? T/I

(d) Does the ecosystem appear healthy? Use evidence to explain your answer. T/I C

Apply and Extend

(e) Think about the areas surrounding the ecosystem. What do you think are the greatest threats to this ecosystem? Explain. T/I

Chapter 2 KEY CONCEPTS SUMMARY

Energy passes through ecosystems, but matter cycles within ecosystems.

- Energy is passed through ecosystems when organisms eat other organisms.
- Energy is lost to the environment as thermal energy.
- Matter moves through ecosystems in cycles known as biogeochemical cycles.

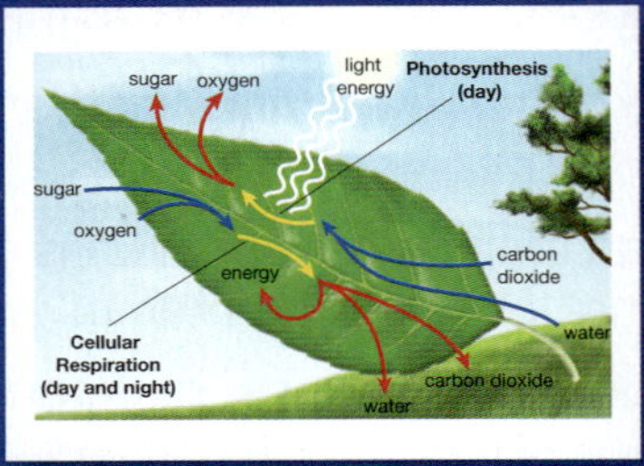

Photosynthesis and respiration are processes that work together.

- In photosynthesis, plants make food in which energy is stored.
- Cellular respiration releases stored energy for use.

Aquatic and terrestrial ecosystems are strongly influenced by their abiotic environment.

- Temperature and water content are important factors in terrestrial ecosystems.
- Temperature and salt concentration are important factors in aquatic ecosystems.

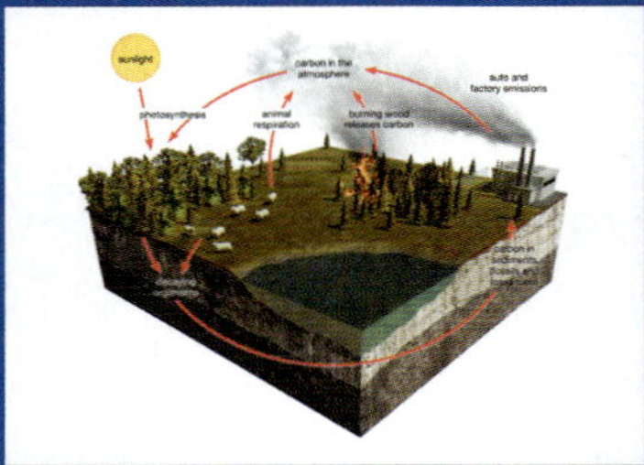

Human activities disturb the natural cycles of nature.

- Human activities change abiotic factors that influence organisms.
- Human activities may remove species from ecosystems or introduce new species to ecosystems.

WHAT DO YOU THINK NOW?

You thought about the following statements at the beginning of the chapter. You may have encountered these ideas in school, at home, or in the world around you. Consider them again and decide whether you agree or disagree with each statement.

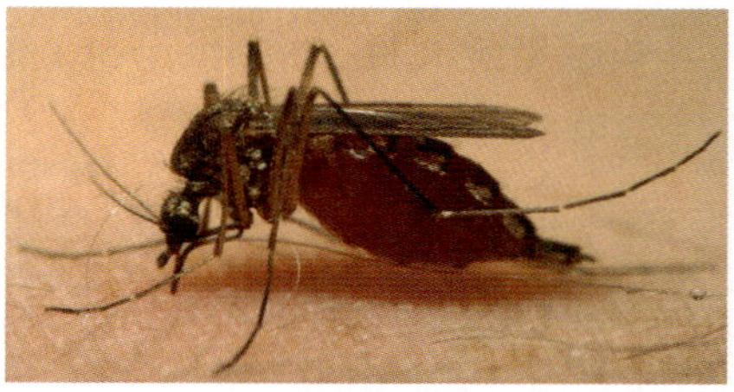

1 All organisms are helpful in the environment.
Agree/disagree?

2 Human activities such as reducing waste, planting trees, and establishing bird sanctuaries help to improve natural environments.
Agree/disagree?

3 The health of an ecosystem depends only upon its biotic (living) features.
Agree/disagree?

4 Humans are able to change the natural environment to suit our needs.
Agree/disagree?

5 Humans do not compete with other species.
Agree/disagree?

6 Terrestrial (land) ecosystems are more important to humans than aquatic (water) ecosystems.
Agree/disagree?

How have your answers changed since then? What new understanding do you have?

BIG Ideas

- Ecosystems consist of a variety of components, including, in many cases, humans.
- The sustainability of ecosystems depends on balanced interactions between their components.
- Human activity can affect the sustainability of aquatic and terrestrial ecosystems.

DIG DEEPER

Vocabulary

CHAPTER 2

SELF-QUIZ

The following icons indicate the Achievement Chart category addressed by each question.

K/U Knowledge/Understanding
C Communication
T/I Thinking/Investigation
A Application

For each question, select the best answer from the four alternatives.

1. Which of these spheres refers to the area where organisms can live? (2.1) K/U
   (a) atmosphere
   (b) hydrosphere
   (c) biosphere
   (d) lithosphere
2. Which of Canada's major biomes has the most rainfall? (2.9) K/U
   (a) tundra
   (b) boreal forest
   (c) grassland
   (d) temperate deciduous forest
3. Which trophic level is made up of producers? (2.6) K/U
   (a) first
   (b) second
   (c) third
   (d) fourth
4. Which of the following is a key abiotic factor for some terrestrial ecosystems but not for aquatic ecosystems? (2.9) K/U
   (a) rainfall
   (b) nutrients
   (c) elevation
   (d) temperature
5. Which term applies to the relationship between two organisms of different species that live together, when one lives and feeds on the other? (2.8) K/U
   (a) commensalism
   (b) competition
   (c) mutualism
   (d) parasitism
6. Which of these factors in a pond ecosystem is an abiotic factor? (2.2) K/U
   (a) bacteria
   (b) algae
   (c) temperature
   (d) fish

Indicate whether each of the statements is TRUE or FALSE. If you think the statement is false, rewrite it to make it true.

7. Natural ecosystems, such as coral reefs and forests, are often not sustainable. (2.3) K/U
8. A carnivore is a consumer that eats only plants. (2.6) K/U
9. An omnivore is a consumer that eats both plants and animals. (2.6) K/U
10. Most of the Sun's energy is absorbed by producers. (2.5) K/U
11. The atmosphere is the sphere made up of all the water on and above Earth. (2.1) K/U
12. Humans obtain chemical energy from the food they eat. (2.5) K/U
13. The main producers in terrestrial ecosystems are microscopic organisms. (2.5) K/U
14. During photosynthesis, plants release carbon dioxide gas into the atmosphere. (2.5) K/U
15. An ecological niche is the role of a species within its ecosystem. (2.6) K/U

Copy each of the following statements into your notebook. Fill in the blanks with a word or phrase that correctly completes the sentence.

16. Plants make food using energy from the Sun in the process called ________ . (2.5) K/U
17. All organisms release energy from sugars in the process called ________ . (2.5) K/U

18. All species living in an ecosystem make up a ________. (2.2) K/U

19. A ________ is all the organisms of one species in an ecosystem. (2.2) K/U

20. ________ occurs when organisms use the same resources in an ecosystem. (2.8) K/U

21. A ________ factor restricts the maximum size of a population. (2.8) K/U

Match each term on the left with the most appropriate description on the right.

22. (a) trophic level — (i) a model showing feeding relationships
    (b) radiant energy — (ii) the place of an organism in a food chain
    (c) mutualism — (iii) energy from the Sun
    (d) food web — (iv) a relationship in which both species benefit (2.5, 2.6, 2.8) K/U

Write a short answer to each of these questions.

23. Are humans producers or consumers? Are they herbivores, carnivores, or omnivores? (2.6) K/U

24. You are planning an overnight camping trip in the temperate deciduous forest. (2.9) K/U A
    (a) What are two biotic factors that you may encounter on your trip?
    (b) What are two abiotic factors that you may encounter on your trip?
    (c) What can you bring on your camping trip that will help you adapt to these factors?

25. Explain how burning fossil fuels affects the carbon cycle. (2.7) K/U

26. You are writing a report comparing the boreal forest and the temperate deciduous forest. You begin by listing the major similarities and differences between the two biomes. (2.9) K/U
    (a) What is one abiotic similarity?
    (b) What is one abiotic difference?
    (c) What is one biotic similarity?
    (d) What is one biotic difference?

27. Examine the partial food web shown in **Figure 1**. (2.6, 2.8)
    (a) Which of the organism(s) in the food web are producers? Which are consumers? K/U
    (b) How would a decrease in the hawk population affect the snake population? T/I C
    (c) If the mouse population decreased due to a disease, how would the other populations be affected? T/I C

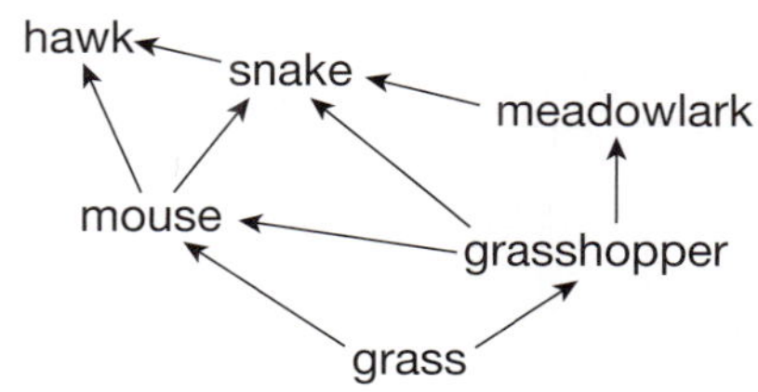

Figure 1

28. Give an example of a human-made ecosystem. What types of human actions are required to manage this ecosystem? (2.3) K/U A

29. A scientist studying the wolf population in an area notices that the number of wolves has increased from 25 to 45 in 10 years. (2.6, 2.8) K/U T/I
    (a) What could be a reason for this increase?
    (b) What other characteristics would you expect to be present in this ecosystem?

30. Abiotic factors are very important in determining the biotic factors in an ecosystem. (2.8, 2.9) K/U C
    (a) What abiotic factors are most significant in determining what organisms can live in an ecosystem? Explain.
    (b) In which Canadian biomes are humans most likely to live? Why?
    (c) In which Canadian biomes are people least likely to live? Why?

Self-Quiz

To do an online self-quiz,

GO TO NELSON SCIENCE

The following icons indicate the Achievement Chart category addressed by each question.

K/U Knowledge/Understanding T/I Thinking/Investigation
C Communication A Application

What Do You Remember?

1. What is the solid part of Earth's surface called? (2.1) K/U
   (a) hydrosphere (c) atmosphere
   (b) lithosphere (d) biosphere

2. During which process is carbon dioxide released? (2.5) K/U
   (a) photosynthesis
   (b) predation
   (c) cellular respiration
   (d) competition

3. Which term refers to all the water on Earth in all its states? (2.1) K/U
   (a) terrestrial (c) hydrosphere
   (b) biosphere (d) ecosystem

4. Which of the following is a sustainable natural ecosystem that can generally survive without regular human maintenance? (2.3) K/U
   (a) a national park
   (b) a coral reef
   (c) a community garden
   (d) a campground

5. What is a population? (2.2) K/U
   (a) all the organisms that live in an ecosystem
   (b) all the organisms of the same species in an ecosystem
   (c) all the biotic and abiotic factors in an ecosystem
   (d) all of the above

6. What is the term used to describe an organism that eats both plants and animals? (2.6) K/U
   (a) carnivore
   (b) herbivore
   (c) omnivore
   (d) producer

7. What is the maximum population that can survive in an ecosystem called? (2.8) K/U
   (a) carrying capacity
   (b) limiting factor
   (c) mutualism
   (d) trophic level

8. Chemical energy is produced during photosynthesis. Where is this chemical energy stored? (2.5) K/U
   (a) vitamins (c) protein
   (b) chlorophyll (d) sugars

What Do You Understand?

9. What human activity is the main source of carbon dioxide in the atmosphere? (2.7) K/U

10. Why do we say that photosynthesis and cellular respiration are complementary processes? (2.5) K/U

11. In what ways is light from the Sun important for all living things on Earth? (2.5) K/U

12. A food chain shows that a grasshopper eats wheat. A frog eats the grasshopper and the frog is eaten by a snake. Sketch the food chain. (2.6) K/U C

13. (a) Which type of relationship is illustrated by each of the following? (2.6) K/U
       (i) a robin eats earthworms
       (ii) winter ticks live and feed on moose
    (b) How are these relationships similar and different? Use the examples to explain. (2.6) K/U C

14. Many animal foods like meat and fat are high in energy content. Where did this energy come from? Use the words "consumer," "producer," and "photosynthesis" in your explanation. (2.5, 2.6) K/U C

Use Figure 1 to answer questions 15 to 17.

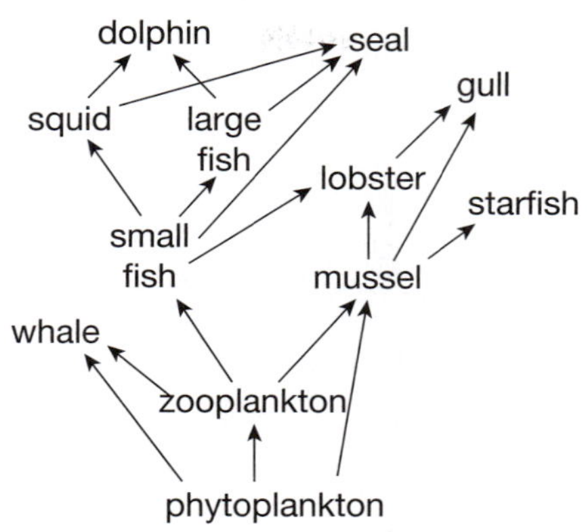

Figure 1 Food web for an estuary ecosystem

15. Name two organisms that would be affected if the population of lobster decreased. Explain. (2.6, 2.8) K/U

16. If whale numbers were reduced by hunting, how would the populations of zooplankton and starfish be affected? (2.6) K/U

17. Suppose mussels began dying off from disease. (2.6) K/U
    (a) How would the gull population be affected?
    (b) How would the population of small fish be affected?

18. Do herbivores and carnivores compete with each other for food? Explain. (2.6, 2.8) K/U

19. What abiotic factors account for the great diversity of organisms in the temperate deciduous forest? (2.9) K/U

20. In what biome is each of the following areas likely to be found? (2.9) K/U
    (a) An area gets about 400 mm of precipitation each year. It has many conifers and wolves.
    (b) An area has low temperatures and a short growing season. Caribou and lichen are among its species.

Solve a Problem

21. A scientist studying predator–prey relationships in a national park notices that the black bear population has increased dramatically in the past 10 years. Since the park was established, no hunting of bears has been permitted. What other conditions would you expect to find in that ecosystem? (2.6, 2.8) T/I A

Create and Evaluate

22. A city is bordered by two lakes, and the city's rainfall drains into the lakes. As the city has grown, there has been more and more algae in the lakes. Suggest reasons for this environmental change. (2.7) T/I

23. Consider the human activities that disrupt natural cycles and ecosystems. Use what you learned in this chapter to make a list of things you or your family can do to minimize the harm you do to these natural systems. (2.6, 2.7, 2.8) C A

Reflect on Your Learning

24. What are the biotic and abiotic factors in your ecosystem? How are you affected by these factors? C A

Web Connections

25. Use a map to find freshwater ecosystems in your area. Use the Internet to find out more about them. What kinds of ecosystems did you research? Are they healthy? Are they experiencing any problems? If so, what are the problems? What is being done to address the problems? T/I A

GO TO NELSON SCIENCE

3 Natural Ecosystems and Stewardship

KEY QUESTION

Why is it necessary to keep ecosystems in balance?

KEY CONCEPTS

- Natural ecosystems are valuable to humans.
- Ecosystems are balanced but can change over time.
- Biodiversity describes the variety and amount of life in an ecosystem.
- Human activities can affect and threaten the sustainability of natural ecosystems.
- Pollution of water, land, and air causes health and economic problems.
- Plant and animal resources should be used in a sustainable way.

The population of large sockeye salmon on the Fraser River dramatically decreased in 2009.

ENGAGE IN

SCIENCE

FISHERIES IN JEOPARDY

A local restaurant menu boasts fresh fish and seafood daily from all around the globe—Canadian sockeye salmon, Arctic char, shrimp from the Gulf of Mexico, sea bass from Antarctica, Hawaiian octopus. What effect does our growing demand for fish and seafood have on these species and the ecosystems in which they thrive?

As the demand for seafood has increased, the number of fish in the oceans and major lakes has decreased. Overfishing and pollution are two factors that have severely reduced populations in many fisheries. Decreased populations affect not only the species being fished, but the entire ecosystem. When one species is lost, an ecosystem is thrown out of balance.

The sockeye salmon fishery in British Columbia is one example of declining fish populations. Each year, sockeye salmon leave the rivers where they were hatched and swim to the Pacific Ocean to mature. Then, they return to the rivers where they were hatched to reproduce.

On average, about 10 million sockeye salmon return to the Fraser River each year. In 2009, most of the salmon did not show up. Only 1 million salmon returned. This was a record low number. The sockeye salmon fishery collapsed.

The collapse of the sockeye salmon population affects more than just the fishing industry in British Columbia. Recreational fishing was stopped. Fewer fishing permits, meaning less food, were available for the First Nations people in the area. Bears and other wilderness predators also had less food to eat.

Scientists do not know what caused this collapse. Why did so many salmon fail to return? Did water temperatures rise to a dangerous level? Was their food supply reduced or destroyed? Did new predators eat the salmon? Were the wild salmon infected with parasites from farm-raised fish or an unknown disease? Was it a combination of factors? These and other questions will need to be answered before the salmon fishery on the Fraser River is restored.

WHAT DO YOU THINK?

Many of the ideas you will explore in this chapter are ideas that you have already encountered. You may have encountered these ideas in school, at home, or in the world around you. Some of the statements are not true. Consider each statement and decide whether you agree or disagree with it.

1 Humans enjoy Ontario's natural ecosystems, but we do not rely on them in our daily lives.
Agree/disagree?

2 Forest fires benefit certain ecosystems.
Agree/disagree?

3 Ecosystems with high biodiversity are more stable.
Agree/disagree?

4 Ontario has a large wildlife population with very few "at-risk" species.
Agree/disagree?

5 An invasive species can have negative effects on an ecosystem into which it has been introduced.
Agree/disagree?

6 Human demand for safe, clean water is a major concern on a global scale, but not in Canada.
Agree/disagree?

FOCUS ON
WRITING

Writing to Describe and Explain Observations

When you write to describe and explain observations, you record two types of properties:

- qualitative properties: properties you use your senses to observe, such as colour, shape, and odour
- quantitative properties: properties you use instruments to measure, such as length, area, volume, and temperature

Write your observations clearly and accurately. Here is one student's observations of a dandelion plant. Beside it are the strategies the student used to write effectively.

> **DIG DEEPER**
> **Writing Tip**
> As you work through this chapter, look for tips like this. They will help you develop literacy strategies.

Observations of a Dandelion Plant

(Use scientific terminology correctly.)

Plant Name: dandelion (*Taraxacum officinale*)

(Use short sentences.)

Plant Location: The habitat of the observed plant was a residential lawn. The plant was located in a sunny, open, and flat area.

(Record measurements accurately.)

Plant Leaves: The observed plant had 14 leaves. The leaves were deep green in colour. The leaf edges were shaped like the teeth of a saw. The longest leaves measured 15.2 cm in length. The shortest leaves measured 6.3 cm in length.

(Describe sensory observations clearly by using concrete nouns, adjectives, verbs, and adverbs.)

Plant Stems: The 5 stems of the observed plant were a mixture of light green, red, and yellow. The stems were shaped like tubes. The longest stem measured 15.1 cm in length. The shortest stem measured 14.5 cm in length.

(Use appropriate units and symbols.)

First Bloom of Flowers: The first bloom of flowers was observed on April 25. According to Environment Canada, the high temperature on April 25 was 14.1 °C. One flower appeared at the end of each of the 5 plant stems. The flowers were bright yellow. The biggest flower...

3.1 Benefits of Natural Ecosystems

Figure 1 Natural ecosystems give people opportunities for recreation and relaxation.

Have you ever canoed down a river or camped in the woods (**Figure 1**)? Maybe you have walked through the woods, listening to birds sing and squirrels rustle through the leaves. Many people like to go on holidays and relax in places where they can experience Earth's natural beauty.

Ecosystem Benefits

Recreational activities and an opportunity to appreciate the beauty of nature are just two of the benefits ecosystems provide to us. Ecosystems offer a wide variety of benefits to humans and for the biosphere in general. **Table 1** lists some of the many types of benefits provided by natural ecosystems.

Table 1 Benefits of Ecosystems

Benefit	Examples
cultural	recreation, education, spirituality
products	lumber, food, medicine
environmental	climate moderation, water and air purification, oxygen supply
economic	job opportunities, tourism products and services
other	erosion reduction, flood prevention, balance of wildlife maintained

Cultural Benefits

The cultural benefits provided by ecosystems include recreation, education, and spirituality. Some people may go to a lake to enjoy canoeing, fishing, or swimming. Some people might go on educational walks led by naturalists to learn more about the birds or plants in an ecosystem. Others use the calm of nature for inspiration and to reduce stress.

Products

Ecosystems provide more than just natural beauty and a place to relax. Each day we use products from ecosystems that help to make our lives easier (**Figure 2**). Terrestrial ecosystems provide us with important products (**Table 2**). Plants provide us with wood, rubber, fabrics, dyes, food, and medicines. Animals provide us with leathers, oils, bones, and beeswax. Some people hunt animals and collect plants for personal and commercial use.

Many Aboriginal communities rely heavily on products and resources from the environment. When the Mohawks of Akwesasne, Ontario, needed more clean drinking water, they built a new water treatment plant. They were careful not to damage any wildlife, fish, or plants. By building a plant that does not have a negative effect on the environment, the people of the Mohawk Nation have helped to ensure their way of life.

Figure 2 We obtain many products from the environment: (a) Trees can be made into paper products; (b) The foxglove plant is a key ingredient in heart medicine; (c) Rubber is made into useful items like protective gloves.

Table 2 Important Products from Terrestrial Ecosystems

Product	Original source	Use
lumber, paper, and cardboard	trees	building construction, furniture, books, paper towels, packaging
maple syrup	maple trees	food, sweetener, flavouring
blueberries	bushes	food
honey	flower nectar (processed by honeybees)	food, sweetener, flavouring
digitalis	foxglove (flowering plant)	treatment of heart disorders
latex (rubber)	various tropical trees	tires, hot water bottles, golf balls, latex gloves

Figure 3 The carbon dioxide absorbed by forests moderates Earth's climate.

Figure 4 Park naturalists help people learn about forests.

Environmental Benefits

Ecosystems provide several environmental benefits:

- **Producing Oxygen:** Microscopic organisms in the ocean produce about 50 % of the oxygen in the atmosphere. These organisms use photosynthesis to make food. Recall that oxygen is a product of photosynthesis.
- **Moderating Climates:** Plants in many different ecosystems, including tropical rainforests, take in carbon dioxide gas for photosynthesis (**Figure 3**). Carbon dioxide helps trap the thermal energy that warms Earth's surface. Changes in the amount of carbon dioxide in the atmosphere can influence climates on Earth.
- **Filtering Pollutants:** Certain plants and organisms can remove and destroy some of the harmful chemicals in the air, water, and soil. This reduces the amount of pollution in the environment.

Economic Benefits

Natural ecosystems can support industries as well as provide valuable products.

- **Forestry:** More than 65 % of Ontario's land surface is covered with forests. Ontario's forests provide $18 billion worth of products each year and employ more than 80 000 people. Another 300 000 people in Ontario have jobs that are related to forest ecosystems, including fishing and trapping, tourism, and parks management.
- **Ecotourism:** Ecotourism, which is a growing industry that relates tourism to the ecology of an area, also creates jobs (**Figure 4**). Hotels, restaurants, transportation services, arts organizations, and companies that organize guided tours all provide jobs in a local area. Goods are needed for tourists, too. Boats, fishing equipment, and specialized clothing are just a few examples of goods that people use to enjoy the outdoors.

Other Benefits

Ecosystems also help to protect us from physical threats, such as erosion and flooding. The roots of plants, which are anchored in soil, help to reduce soil erosion by wind and water. Ecosystems such as wetlands absorb surface water. Then they slowly release the water both underground and at the surface, which prevents flooding after a heavy rain.

DIG DEEPER

Career Link

Ecotourism guides help people to understand and appreciate the beauty and value of forests and other natural landscapes. To learn more about a career in ecotourism,

GO TO NELSON SCIENCE

Valuing Ecosystems

Humans can have positive effects on ecosystems when they make use of ecosystem benefits, or products. For example, popular ecotourism in an area may result in the government recognizing the area as valuable and worth preserving. Public awareness can make it easier to maintain an area or protect species that live there.

Humans can also damage ecosystems.

- Some areas are closed off to help protect nesting birds or other reproducing animals. If hikers, campers, and hunters go into these protected areas, they may damage the ecosystem and the organisms that live there.
- Humans can create litter, which can encourage the growth of bacteria or harm wildlife (**Figure 5**).
- Carelessness with fire is a major concern. Fire can affect air and water quality as well as injure plants and wildlife.
- Construction of roads and buildings disturbs natural habitats. Increased vehicle traffic leads to air pollution and puts animals at risk of being hit by vehicles.

Ecosystems are extremely valuable for many reasons. They are free and renewable systems, but they must be cared for and respected by humans to remain healthy and valuable.

Figure 5 Humans can have a positive or a negative effect on ecosystems.

3.1 Wrap Up

- Ecosystems provide us with many benefits including recreation, food, medicine, and economic opportunities.
- Ecosystems help to protect us from some environmental threats.
- The benefits of an ecosystem are free and renewable.

CHECK YOUR LEARNING

1. Identify one cultural, one environmental, and one economic benefit of an ecosystem. K/U
2. In what ways can humans damage an ecosystem? K/U
3. What are two ways that ecotourism can help an ecosystem? A
4. Describe a benefit that each of the following provides: A
   (a) a river in a wilderness park
   (b) maple trees
   (c) dense forest
5. What kinds of jobs can be created when tourists visit a natural ecosystem? A

3.2

Balance and Change in Ecosystems

What are the characteristics of the tundra biome? The tundra has long, cold winters and cool, short summers. There is a layer of permafrost that limits the types of plants growing there. Lichens, grasses, and small shrubs survive in the tundra. Caribou and Canada geese spend part of each year in the tundra.

A State of Equilibrium

One of the reasons we are able to describe the characteristics of the tundra and other biomes is that these large systems are in a state of **equilibrium**. This means that energy flows through the ecosystem, and nutrients are recycled through food webs in a constant manner. Photosynthesis and cellular respiration are balanced. When an ecosystem is in equilibrium, populations are healthy and stable. The organisms in these biomes have adapted to life in those climates.

▶ **equilibrium:** the state of an ecosystem in which the biotic and abiotic features remain relatively constant over a long period of time

When a tundra ecosystem is in equilibrium, it is consistent and predictable. This predictability allows Aboriginal peoples, such as Inuit, to plan when and where to go hunting and fishing (**Figure 1**). They also know what climate conditions they must be prepared for as the seasons change.

Figure 1 When the tundra is in equilibrium, Inuit can predict when caribou will migrate.

DIG DEEPER

Did You Know?

Caribou Migration

Caribou are large deer-like animals that migrate as far as 1300 km each year. In spring, large caribou herds travel north to give birth and raise their young. They return to the same areas year after year. In early fall, they travel south to spend the winter in warmer climates where more food is available.

Short-Term Changes

Unlike the large system of a biome, smaller ecosystems are in a constant state of change. Both abiotic and biotic factors can cause short-term changes to ecosystems on a local level. For example, volcanic eruptions, forest fires, and disease outbreaks can lead to changes in the number and type of species in a small ecosystem. The population size of each species can also be affected by changes within the ecosystem.

The Process of Change

A volcanic eruption can drastically affect the state of equilibrium of an area. Consider the case of Mount St. Helens, in the state of Washington in the United States (**Figure 2(a)**), which erupted in 1980. Although the damage was extreme, the drastic changes following the eruption were not permanent (**Figure 2(b)**). Within a few years, the ecosystem showed signs of recovery (**Figure 2(c)**).

Figure 2 (a) Before the volcano erupted in 1980, a dense forest ecosystem surrounded the mountain. (b) Immediately after the volcano erupted, the surrounding forest was barren, charred, and lifeless. (c) The recovery has been gradual. Thirty years later, plants, birds, small animals, and larger mammals returned to resettle in the Mount St. Helens region.

Succession

The process of change that has been occurring on Mount St. Helens since it erupted is called ecological succession. **Succession** is a series of changes in the biotic and abiotic features of an ecosystem that occur over time after a disturbance. The disturbance could be natural, such as a forest fire, an earthquake, or a hurricane, or it could be caused by human actions. Succession leads to an increase in the number and variety of species living in an ecosystem.

▶ **succession:** the observed changes over time in an ecosystem, following a disturbance

Ecological succession is a progression from one type of ecosystem to another (**Figure 3**). Succession also occurs in areas that have been drastically altered by human activities, such as deforestation, agriculture, or development. A vacant city lot covered with weeds and small shrubs is a small-scale example of ecological succession. Left undisturbed, it will eventually support a stable ecosystem.

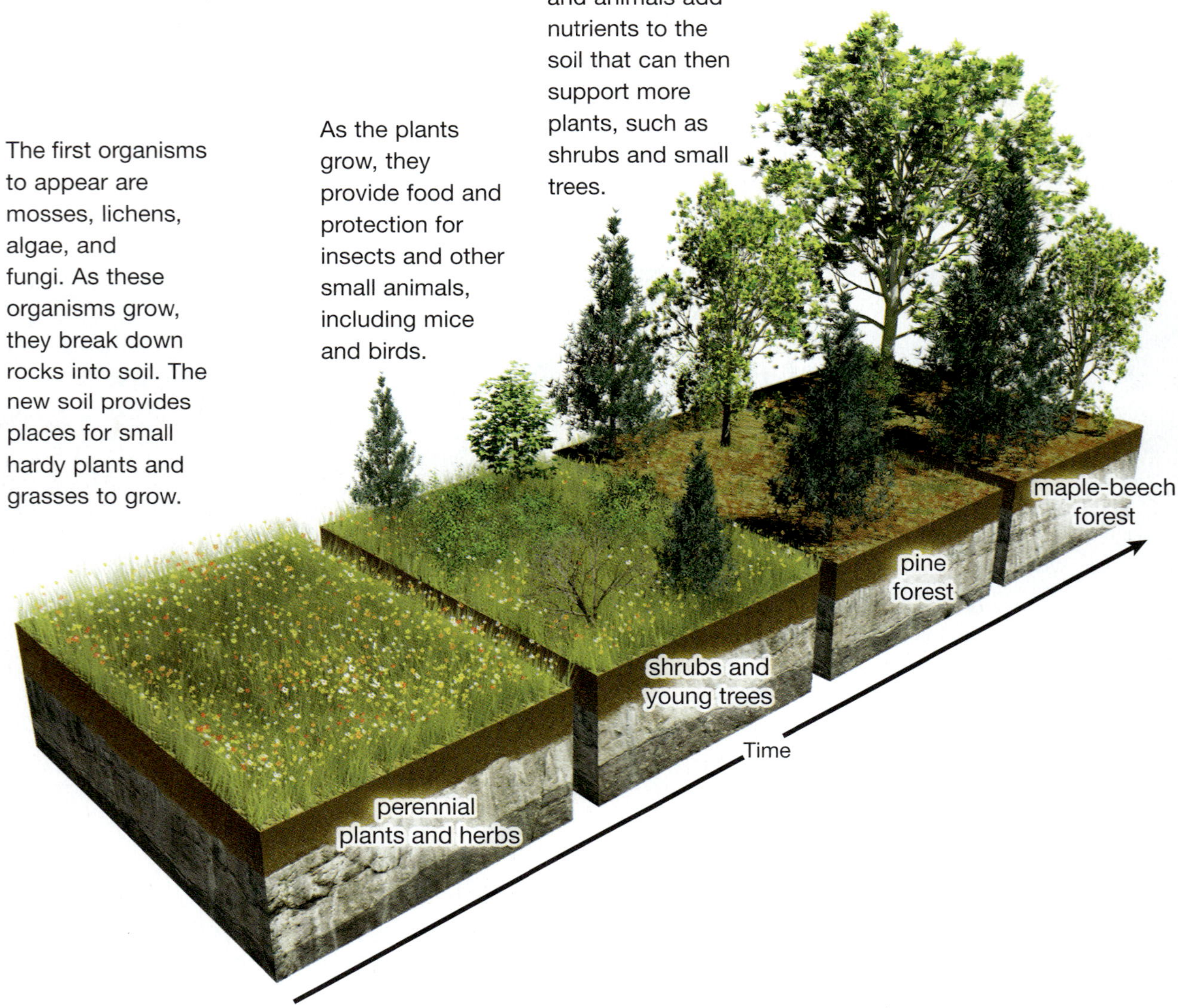

Figure 3 Each square of land in the diagram shows a new time period of the progression from grassland to forest ecosystem.

Fire and Ecosystems

Fire plays an important role in some ecosystems, such as forests and grasslands. Some plants depend on fire for seed release or germination. Other organisms use fire-damaged trees as a habitat. Fire also releases nutrients into the soil. A fire, such as one started by a lightning strike, may clear a forest of large trees. The cleared area gives new plants more light and space to grow. Many animals find renewed sources of food as grasses and other tender plants grow in the clearings caused by fire. Without fires, the diversity of species living in an ecosystem might decline.

Setting Fire Management Policies

At one time, rules were made that all wildfires should be fought and extinguished as quickly as possible. These policies, however, did not consider natural cycles. As the relationships among ecosystems, fire, and succession have become better understood, policies about fire have changed.

When lightning or other natural events cause fires, they may be allowed to burn in a controlled way. This is called the "natural burn" policy. Sometimes fires are started on purpose. In heavily forested areas, forest and wildlife managers may start fires called prescribed burns (**Figure 4**). These "controlled" burns mimic natural fires in areas that have not been burned for a long time.

Figure 4 Prescribed burns create room for new growth and prevent natural fires from burning out of control.

Wildfires and Ecosystem Health

Many Aboriginal peoples and early settlers of prairie ecosystems started grassland fires to remove dead grasses and other plants from the prairie. They understood the relationship between fires and ecosystem health. The removal of dead plants allowed new plants to grow earlier in the spring and in greater numbers. The ashes from fires also contain nutrients that help fertilize the soil. As a result, grassland fires also increased the amount of nutrients in many prairie plants, making them a better source of food for both animals and humans.

DIG DEEPER

Did You Know?

A Succession Example

A volcano called Tseax Cone (SEE-aks Cone) erupted in 1775, killing 2000 Nisga'a First Nations people. Tseax Cone is located in northern British Columbia. Today the lava beds from that eruption are covered with moss and lichens. What does this tell you about how long it can take for succession to occur?

Humans and Succession

Succession allows ecosystems to recover from disturbances and keeps them in balance over the long term. However, not all disturbances caused by humans can be repaired through natural ecological succession. Sometimes humans need to take action to help restore an area that was used for mining or agriculture to its natural state. For example, humans can speed up natural ecological succession by planting an area with small plants or trees and managing the affected area responsibly.

Case Study: The Milton Quarry

Responsible management brought the Milton Quarry, in Ontario, back to life more quickly than if people had not helped. After stone had been mined there for years, engineers reshaped the cliffs in the quarry. The area was restored to encourage the establishment of many different types of habitats, including wetlands, wooded areas, and a lake. Scouts Canada helped plant more than 41 000 trees as part of the restoration effort. Since then, many species of insects, birds, and mammals have returned to the area, and natural ecological succession is occurring (**Figure 5**).

Figure 5 Plant life would not have been able to return to the Milton Quarry if humans had not stopped mining, replanted trees, and restored the landscape.

CITIZEN ACTION

Land Reclamation and Rehabilitation

Caring about the environment and taking action are the responsibility of the government, businesses, and individuals. All citizens should be involved in reclaiming and rehabilitating areas in their communities (**Figure 6**).

Consider how you and your class might get involved.

- Research areas in your community that have been or are being reclaimed.
- Write a short description of what has been done and what is currently being done.
- Explore what you or your classmates can do to help.

Figure 6 Volunteers plant tree saplings at a restoration site.

GO TO NELSON SCIENCE

3.2 Wrap Up

- When the biotic and abiotic parts of an ecosystem remain relatively constant over time, they are in equilibrium.
- Ecological succession is the natural, gradual changes in the biotic and abiotic features of an ecosystem following a disturbance.
- Human activity can upset the state of equilibrium of an ecosystem, and it can interrupt the natural process of succession.
- Fire management policy must consider the relationship between wildfires and ecosystem health.
- Ecological succession allows an ecosystem to recover following a human or natural disturbance.

UNIT TASK Bookmark

You can apply what you learned about equilibrium and succession to the Unit Task described on page 160.

CHECK YOUR LEARNING

1. What does it mean when we describe an ecosystem as being in equilibrium? K/U
2. What factors can upset the equilibrium of an ecosystem? K/U
3. How can an ecosystem be affected by the following changes? K/U
   (a) a volcanic eruption
   (b) building a new housing project
   (c) cutting down all the large trees in a forest
   (d) controlled burns in a prairie
4. How have Aboriginal peoples benefited from their understanding of the following? A
   (a) ecosystem equilibrium
   (b) effects of fire on an ecosystem
5. Construct a flow chart that shows the typical order of ecological succession in an ecosystem that has been destroyed by a volcanic eruption. K/U C
6. How is ecological succession influenced by forest fires? K/U

3.3 Biodiversity

How many species do you think live in the park nearest your home?

▶ **biodiversity:** the variety of species in an ecosystem

Comparing Biodiversity

Biodiversity is the number of different species living in a given area or ecosystem. About 1.5 million species have been identified and studied. Biologists estimate the total number of living species to be more than 5 million and perhaps up to 50 million. Most of these species have not been identified. For example, scientists estimate that Canada has from 140 000 to 200 000 plant and animal species. Only 71 000 of these species have been identified!

Abiotic Factors and Biodiversity

Greater biodiversity exists in ecosystems near the equator than in those near the poles. Abiotic factors such as higher temperatures and a greater amount of rainfall contribute to high biodiversity in ecosystems such as coral reefs and tropical rainforests. In one hectare of land in a Peruvian rainforest, scientists have identified 283 species of trees. In Ontario, one hectare in a deciduous forest would have fewer than 15 species of trees. Ecosystems closer to the equator also have greater diversity of animal species (**Figure 1**).

Figure 1 (a) The ruby-throated hummingbird is the only hummingbird species in Ontario. (b) The booted racket-tail hummingbird is only one of more than 200 species of hummingbird in the Amazon rainforest.

Benefits of Biodiversity

Studies show that ecosystems with high rates of biodiversity, such as a coral reef, are better able to remain healthy and stable when disturbed. The more diverse an ecosystem is, the better able it is to withstand disturbances, such as disease, droughts, or the arrival of a new species. As an ecosystem gets less diverse, it becomes more fragile. For example, if beavers were to disappear from a forest ecosystem, beaver dams would break up, beaver ponds would dry up, and aquatic habitats for a variety of species would be lost.

Protecting Biodiversity

Protecting biodiversity helps to keep ecosystems healthy. The best way for humans to protect biodiversity is to protect ecosystems from disturbances, such as pollution, habitat destruction, and the introduction of new species.

Look at the coral reef ecosystem as an example (**Figure 2**). Reefs attract ecotourists. Corals are very fragile animals. They are easily broken and damaged if boats, anchors, or air tanks worn by divers bump into them. Reefs can become clogged with litter left behind by divers. Gasoline and oil spills from boats can damage the coral reef colony. Spills can also harm numerous fish. People can protect the biodiversity of reefs by controlling and monitoring the number of boats and divers. They can introduce policies to help reduce the amount of trash and pollution near the reefs.

Figure 2 Healthy coral reefs have high biodiversity. However, the biodiversity of a reef can be affected by a change in abiotic factors, such as increased water temperature, or a change in biotic factors, such as a new disease.

3.3 Wrap Up

- Most species on Earth have not been identified or studied.
- Biodiversity refers to the variety of species living in an ecosystem.
- Biodiversity is needed to maintain ecosystem stability.
- Loss of biodiversity makes an ecosystem more fragile.

CHECK YOUR LEARNING

1. What is biodiversity? K/U
2. Where are ecosystems with the highest biodiversity found? K/U
3. Explain the connection between biodiversity and the health of an ecosystem. K/U C
4. What might happen if Ontario's only hummingbird species were to become extinct? A

3.4 PERFORM AN ACTIVITY

Factors That Affect Biodiversity

SKILLS MENU
- Questioning
- Hypothesizing
- Predicting
- Planning
- Controlling Variables
- Performing
- Observing
- Analyzing
- Evaluating
- Communicating

You have already read one example of how biodiversity can be affected by an abiotic factor—the farther you go from the equator, the less biodiversity there is. Other factors, including soil composition, the number and variety of food sources, and the variety of predators, can also affect the biodiversity of an ecosystem.

SKILLS HANDBOOK 3.B.7., 3.B.8.

Purpose

To analyze how biotic factors can affect biodiversity

Procedure

1. **Table 1** contains data linking the number of lizard species in a desert ecosystem to plant variability in the same ecosystem. Plant variability includes differences among plants in terms of height and age. Plant variability was measured on a scale of 0 to 1. As numbers get closer to 1, plant variability in the ecosystem increases.
2. Plot a graph of the data in **Table 1**.

Table 1 Plant Variability and Lizard Species

Plant variability	Average number of lizard species
0.10	3.0
0.20	4.0
0.30	5.0
0.40	6.0
0.50	7.0
0.60	8.0
0.70	9.0
0.80	10.0

Equipment and Materials

graph paper

Analyze and Evaluate

(a) Describe the pattern on the graph. How does plant variability affect diversity in lizard species in this ecosystem? T/I

(b) How do you think plant variability might affect the diversity of other species in the ecosystem? T/I

Apply and Extend

(c) Suppose you want to conduct a similar study within a local ecosystem. A T/I

(i) How would you measure plant variability in your ecosystem?

(ii) For which species would you collect data?

FOCUS ON
MATH

Interpreting Circle Graphs

Graphs help us to visualize data that can sometimes be complicated or difficult to understand in words or numbers alone. A circle graph is a circle divided into sections. Each section represents one category of data. All circle graphs compare parts of a whole.

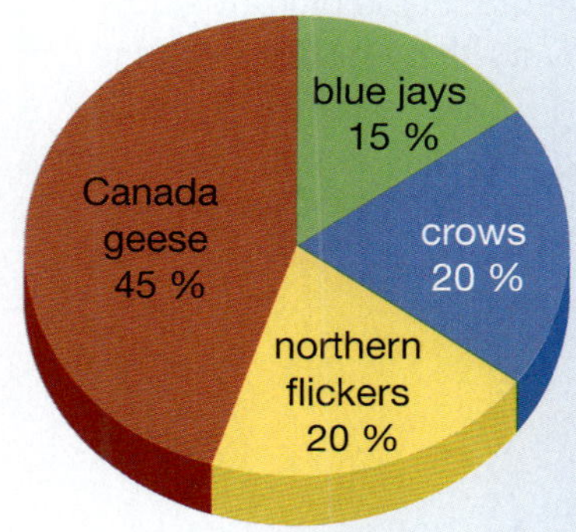

Figure 1 Birds counted in a forest

Use Figure 1 to answer questions A to D.

A. Which section of the graph represents the largest part of the whole?

B. Which section of the graph represents the smallest part of the whole?

C. Which two sections of the graph could be added together to make up more than half?

D. Can you tell from this graph how many birds were counted altogether? Explain your thinking.

Use Figure 2 to answer questions E to G.

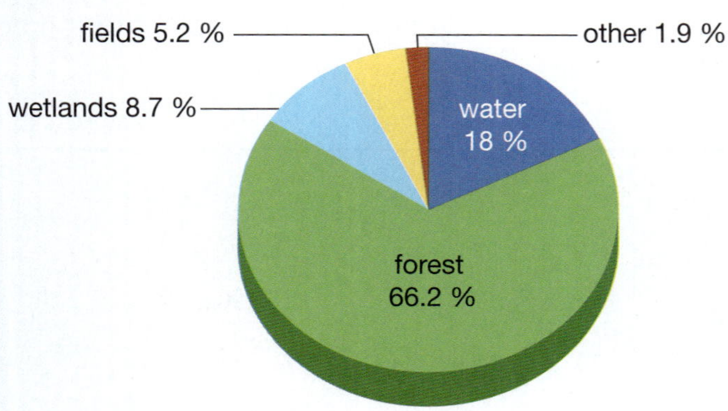

Figure 2 Land and water areas in Ontario

E. What type of area makes up the largest part of Ontario?

F. How much of Ontario is wetlands?

G. How much of Ontario is water?

3.5 Biodiversity under Attack!

Figure 1 Chameleons are a type of lizard known for their ability to change colour to adapt to their environment.

Out of the 180 species of chameleons on Earth, 59 of them can be found only in Madagascar, an island in the Indian Ocean off the eastern coast of Africa (**Figure 1**). The chameleons' habitats are under threat from forestry practices and other human activity. What do you think that means for the chameleons?

Biodiversity Hotspots

Areas of very high biodiversity, called hotspots, are found in natural ecosystems that are mostly undisturbed. These ecosystems are often home to species that are rarely seen outside those areas. Researchers have identified 25 small areas on the planet as biodiversity hotspots, including Madagascar (**Figure 2**). Hotspots are a precious part of our planet's biodiversity.

Many of these hotspots are threatened by human activity. Removing all trees from part of a forest destroys animal habitats. Pollution from chemicals such as industrial waste or fertilizers can reduce water quality, threatening the survival of aquatic organisms. Strip mining destroys habitats and removes other resources that are important to terrestrial organisms. These biodiversity hotspots need extra care to protect them from such threats and to keep them healthy and sustainable.

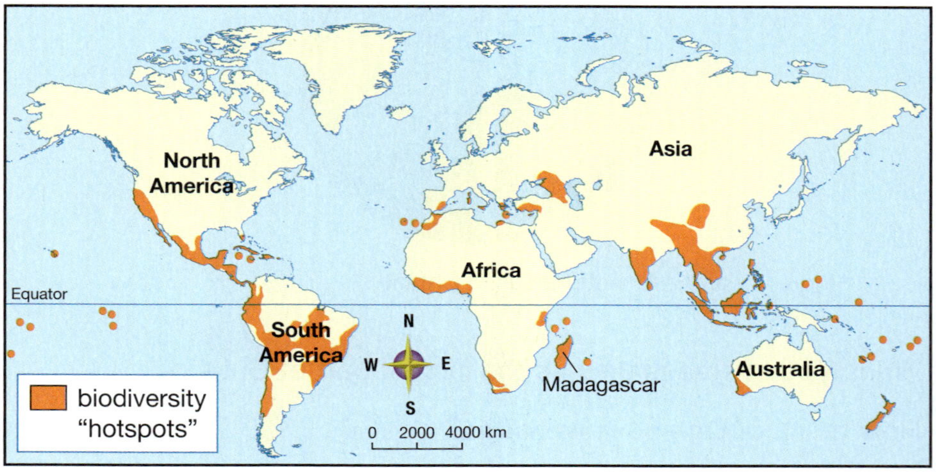

Figure 2 Terrestrial biodiversity hotspots cover more than 1.4 % of land on Earth. Yet, more than half of all the known plant species and terrestrial vertebrates live in these hotspots.

The Range of Biodiversity in Ontario

Recall from Section 3.3 that ecosystems closer to the equator tend to have higher biodiversity than those that are closer to the poles. In Ontario, there are more species in the south and fewer species in the north. Humans tend to populate regions of Earth in the same way. There are more people in southern Ontario than in northern Ontario. More people means that

- more natural ecosystems are cleared for farming, housing, or road construction
- more pollution is created, which can damage ecosystems and the organisms that live in them

For these reasons, there is a greater threat to biodiversity in southern Ontario, where more people live.

Species at Risk

The greatest threat to biodiversity is that a species can become extinct. An **extinct** species is one that has died out and no longer exists anywhere on Earth. The Committee on the Status of Endangered Wildlife in Canada (COSEWIC) monitors the status of species. Its members include governments, universities, other agencies, and Aboriginal groups. COSEWIC uses data on existing species to separate organisms at risk into four categories, shown in **Table 1**. The table also includes examples of extinct organisms.

▶ **extinct:** no longer existing on Earth

Table 1 Status of Wildlife in Canada

Classification	Definition	Number of species (2009)	Examples
special concern	may become threatened or endangered because of various factors	160	polar bear, great blue heron, Atlantic cod
threatened	likely to become endangered if no action is taken	145	humpback whale, wood bison, Kentucky coffee-tree
endangered	in immediate danger of becoming extirpated or extinct	244	barn owl, swift fox, northern cricket frog
extirpated	no longer exists in the wild in a specific area	23	paddlefish, grey whale
extinct	no longer exists on earth	13	great auk, passenger pigeon, sea mink

Species Extinction

Extinction is a process that may result from natural catastrophes or gradual changes to a species' surroundings. Asteroid impacts or volcanic eruptions are rare, but they can wipe out entire species. The climate of a region may change over thousands of years, killing off species that are unable to adapt.

Human Activity and Extinction Rates

The greatest threat to today's species is human activity. The rate at which species become extinct has been steadily increasing because of human disruptions to ecosystems. Humans have been known to exploit new species when they have come across them. About 80 species of birds and mammals became extinct shortly after humans reached North America over 10 000 years ago. These include the sabre-toothed cat, the mammoth, and the Stellar's sea cow (**Figure 3**).

Figure 3 In 1741, the Stellar's sea cow was discovered. Humans settled areas nearby and the Stellar's sea cow became extinct only 27 years later.

Species at Risk in Canada

Species living today may become extinct in your lifetime. For example, Ontario's Blanding's turtle is currently a threatened species because humans have destroyed their habitats, collected them to sell as pets, and run over them on highways. Polar bears are listed in the "special concern" category. The melting of sea ice due to an increase in Earth's average surface temperature has made it more difficult for the bears to find food. Some scientists predict that as much as 66 % of the world polar bear population could disappear by 2050.

More than 700 species of vertebrates are known to have become extinct in the past 400 years. Twelve of the extinct species had lived in Canada. Organisms are still becoming extinct. Within the deciduous forests of southern Ontario, as many as 167 species are at risk of becoming extinct (**Figure 4**).

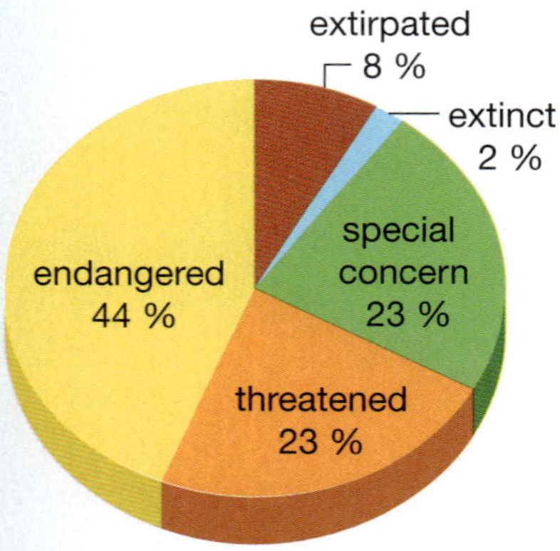

Figure 4 Of the 167 species at risk in Ontario's deciduous forest, 4 species are extinct, 13 are extirpated, 72 are endangered, 39 are threatened, and 39 are of special concern.

DIG DEEPER
Web Link
To learn more about Canada's and Ontario's species at risk,
GO TO NELSON SCIENCE

Reasons to Protect Biodiversity

Recall that ecosystems with high biodiversity are healthier and more stable. Keeping ecosystems healthy and sustainable is just one reason to protect biodiversity on Earth. In addition, biodiversity is needed to

- ensure ecosystem stability and sustain all life forms
- provide natural processes to produce oxygen, carbon dioxide, and nitrogen that protect our planet
- help species adapt to changes in their environment
- provide raw materials for food, clothing, medicine
- enhance our quality of life by adding variety to our surroundings
- help shape our culture and inspire poets, painters, and writers

3.5 Wrap Up

- Biodiversity hotspots are under attack by human activity.
- COSEWIC categorizes at-risk species as extirpated, endangered, threatened, or of special concern.
- Extinction rates have increased due to human disruption of ecosystems.
- Protecting biodiversity ensures ecosystem stability and sustains all life forms.

UNIT TASK

Bookmark

You can apply what you learned about threats to biodiversity to the Unit Task described on p. 160.

CHECK YOUR LEARNING

1. What is the biggest threat to biodiversity? K/U
2. What are two causes of extinction? Give an example of each. K/U
3. Why is biodiversity in Ontario at greater risk in the south than in the north? K/U
4. Choose an ecosystem that you spend time in. C A
   (a) Describe one action that you can take to help to protect biodiversity in this ecosystem.
   (b) How will your action help to protect biodiversity?
5. Give an example of a species that may become extinct in your lifetime. Explain the factors that could lead to this extinction. K/U C
6. What are biodiversity hotspots? Where are they usually found? K/U
7. Why is Madagascar considered a biodiversity hotspot? K/U
8. List three examples of how human activity threatens hotspots. K/U
9. (a) What does COSEWIC stand for?
   (b) What is the purpose of the committee?
   (c) How does it categorize organisms? K/U
10. What is the difference between a species that is extirpated and a species that is extinct? Give an example of each from Canada. K/U C

3.6

Loss of Ecosystems

What do you think the advantages and disadvantages of city growth might be? An advantage may be that your school and the shopping centre are closer to your home than they might have been decades ago. One of the disadvantages is habitat loss. Look out your classroom window to see an example of habitat loss. The school's parking lot and sports fields have replaced what were once natural habitats. If you look at Toronto from the top of the CN Tower, you will see a human-dominated landscape (**Figure 1**).

Figure 1 With the growth of cities, habitat loss becomes a more serious problem.

Loss of Terrestrial Ecosystems

When large areas of natural ecosystems are converted into farmland or urban areas, habitat loss can become extreme. For example, the lynx needs a large area of natural habitat to hunt for its food (**Figure 2**). Since the arrival of European settlers, areas that were once natural and untouched by humans have been altered for agriculture and urban development. The areas where the lynx roams have become smaller and smaller as a result.

Figure 2 The natural ecosystems of animals such as the lynx are affected when humans build roads, buildings, and highways.

Fragmentation

This process of breaking once large ecosystems into smaller ecosystems is called fragmentation. Areas that have been fragmented are more likely to be affected by flooding, pollution, or wind damage. A fragmented ecosystem is not as sustainable as an undisturbed one.

Have you ever seen or heard about a deer, black bear, or coyote wandering through a neighbourhood? A backyard is not a bear's natural habitat. The bear's ecosystem has been altered, and the bear is making an effort to adapt to its surroundings. When a large ecosystem becomes fragmented, animals may be forced to cross highways to get access to food or water. This can result in the death of animals hit by vehicles.

Building with an Ecosystem in Mind

Imagine that a planner is considering how to design a large housing development. The new development will require clearing a large piece of land within a natural ecosystem. What should the planner and builder think about before they begin clearing any land? **Table 1** shows the key factors that can help preserve and sustain a fragmented ecosystem and its biodiversity. The green areas represent natural ecosystems.

Habitat Loss Around the World

Southern Ontario has had severe habitat loss. However, there is less habitat loss occurring now than there was when cities were first being built. Threatened areas still exist, but now they are being monitored and managed more carefully.

DIG DEEPER

Did You Know?

Wildlife Crossings

A long section of the Trans-Canada Highway cuts through Alberta's Banff National Park. To reduce the fragmentation effects of the four-lane road, Parks Canada has built 24 wildlife crossings to join ecosystems that were fragmented by the highway. These overpasses and underpasses are used by more and more animals, resulting in fewer being hit by vehicles and increased sustainability.

Table 1 Factors That Help Protect Wildlife Habitat Fragments

Factor	Poorer option	Better option	Explanation
size		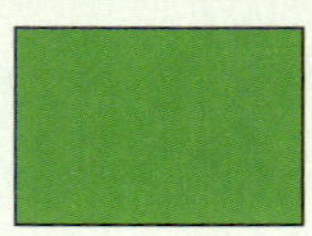	Large blocks support larger and more stable populations and communities.
number	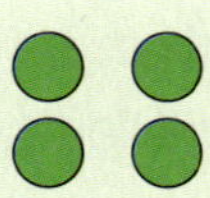	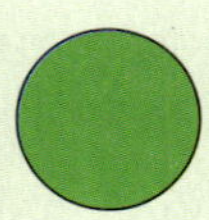	One large area is better than any equal area composed of many smaller areas because there is less outside influence.
proximity	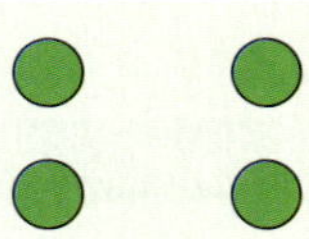	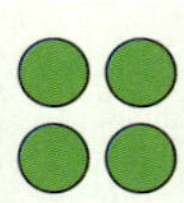	The closer ecosystem fragments are to each other, the greater the chance populations will be able to use the entire area.
connectedness	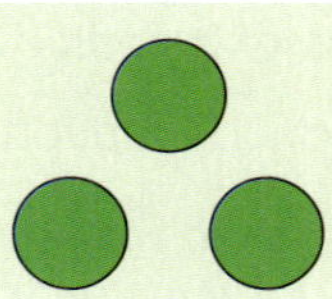		Interconnected areas provide wildlife corridors and permit migration between larger blocks.
integrity		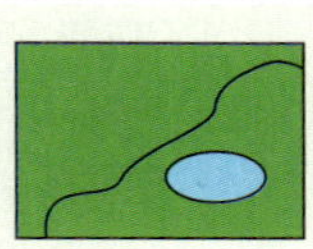	Access by roads and trails can increase pollution, hunting, and fishing.

Habitat destruction is a worldwide issue. The most serious rates of habitat loss today are in Africa, Latin America, and the Caribbean where the human population and industry is growing rapidly (**Figure 3**). You read in Section 3.5 that habitat loss in Ontario is greater in the south because that is where most of the people live. The same is true of habitat loss in other parts of the world. Because of this growth, areas of natural ecosystems, including tropical rainforests and grasslands, are being cleared for resources, agriculture, and development.

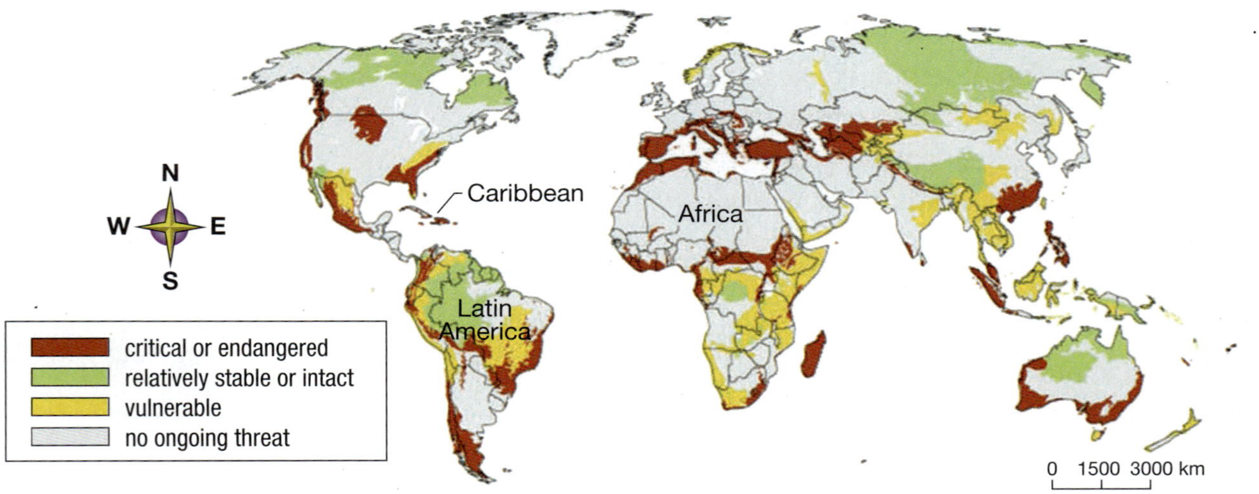

Figure 3 Many of Earth's ecosystems are vulnerable or critically endangered.

Loss of Wetlands and Aquatic Ecosystems

Human activity threatens wetlands and aquatic ecosystems just as it affects terrestrial ecosystems. Damage to ecosystems occurs along lake and ocean shorelines. Freshwater ecosystems around the world are in even greater danger than terrestrial ecosystems. **Table 2** lists some of the human activities that cause damage to coastlines and aquatic habitats around the world.

Table 2 Impact of Human Activities on Wetlands and Aquatic Ecosystems

Human activity	Impacts on ecosystem
replacing or removing natural vegetation along coastlines and waterfronts	habitat destruction, increased shoreline erosion, species loss, loss of breeding areas
removing sand and sediment to create deeper water for boats	habitat destruction, disruption of organisms living at the bottom
increased sediment runoff from land-clearing, agricultural, or mining operations	sediment may smother organisms and habitats; increased sediment reduces water quality
commercial fishing	bottom trawlers and drag lines injure or kill organisms that live on the bottom
draining wetlands for urban development and agriculture	loss of wetland habitats and species

One of the most damaging practices is draining wetlands for the building of cities or farmland. As you read in Chapter 2, wetlands are unique freshwater ecosystems that support many plant and animal species that cannot survive in any other conditions. Natural wetlands are flat, and often have deep, nutrient-rich soil and a lot of water. This makes wetland environments ideal for agriculture.

Commercial Fishing

As shown in **Table 2**, commercial fishing can have a negative impact on an ecosystem. Aside from species loss, habitat destruction is also a concern. Some fishing practices involve using a trawl line, or net, that drags along the bottom of the aquatic habitat to catch everything in its path (**Figure 4**). This dragging causes physical damage to the environment and disrupts organisms that live and lay eggs on the ocean floor. The lines can injure or kill aquatic organisms, whether they were the type of organism being fished for or not. A turtle excluder device or TED is a specialized device that allows a captured sea turtle to escape when caught in a fisherman's net. The use of the device allows all animals larger than 10 cm to escape the nets unharmed.

Figure 4 Dragging nets damage the environment and catch much more than just the fish for which they were used. Air-breathing animals such as turtles drown when caught in nets.

Restoring Wetlands

In recent years, there has been a push to stop draining wetlands in order to keep these fragile ecosystems healthy. People are working to re-establish the natural environment and to protect the many species that live there. Restoring wetlands creates valuable habitats for wildlife.

One success story in restoring wetlands is the 725 hectare Hilliardton Marsh in northeastern Ontario. The marsh had once been turned into farmland, but it has now returned to a wetland breeding location for many species, including two species of at-risk birds—the coot and the black tern (**Figure 5**).

Figure 5 The wetland restoration of Hilliardton Marsh in northeastern Ontario has provided habitat to threatened birds such as (a) the coot and (b) the black tern.

RESEARCH THIS SWEET GRASS GARDENS

SKILLS HANDBOOK 4.C.2., 4.C.4., 4.C.6.

SKILLS: Researching, Analyzing the Issue, Communicating

The Six Nations live in the Grand River region of Ontario. They have a close connection with the native species of the area and are committed stewards of their local natural ecosystems. Their area includes the largest remaining deciduous forest in Canada. In 1996, two members of the Six Nations started the first Native-owned and -operated plant nursery in North America (**Figure 6**).

1. Use the Internet to research Sweet Grass Gardens and the efforts of the Six Nations.

A. Sweet Grass Gardens specializes in growing native plant species. What makes these plants special to the Six Nations? T/I

B. Write a summary describing why you think Sweet Grass Gardens is so successful. T/I C

C. What are at least two ways the Six Nations and their eco-centre are helping the community and the environment? T/I

Figure 6 Sweet Grass Gardens is owned and operated by the Six Nations people.

GO TO NELSON SCIENCE

UNIT TASK Bookmark

You can apply what you learned about habitat loss and restoration to the Unit Task described on page 160.

3.6 Wrap Up

- Habitat loss is a serious threat to terrestrial and aquatic ecosystems.
- Fragmentation reduces the sustainability of ecosystems.
- Wetlands have been drained for farming because of the rich soil.
- Restoring wetland habitats to their natural conditions is an important way to protect the biodiversity of the area.

CHECK YOUR LEARNING

1. Why is fragmentation of ecosystems such a serious problem? K/U
2. Describe three human activities that threaten terrestrial habitats. K/U
3. Describe three human activities that threaten aquatic habitats. K/U
4. Developers are planning to build a shopping complex on land containing a small pond. They would like to preserve the natural ecosystem somehow. What advice would you give them? A
5. Do any of your own behaviours or habits contribute to habitat loss or fragmentation? Explain. A
6. In some places, wildlife crossings have been built under busy highways. What problem is this practice designed to help? T/I A
7. How can digging out and removing material from the bottom of a harbor to make water deeper impact the ecosystem? K/U

3.7

Invasive or Exotic Species

Have you ever seen European starlings or carp in Ontario (**Figure 1**)? If so, you have seen species that are not native to our natural environment. The introduction of non-native species to ecosystems by humans is a major cause of species loss.

Figure 1 (a) The European starling and (b) six species of carp are non-native species in Ontario. They compete with native species and damage valuable natural resources.

Non-native Species

Non-native species arrive in ecosystems in a variety of ways. Boats or travellers from other countries may carry some organisms unknowingly. Some people may bring seeds or cuttings of exotic plants to grow in their gardens. Sometimes non-native species were intentionally introduced to help solve a problem.

The introductions of many non-native, or exotic, species fail because few species can tolerate an entirely new environment. Once in a while, an introduced species is successful in its new environment. There may be few predators or diseases to limit the reproduction of the new species.

Native species might not be able to compete with the introduced species. In cases like this, a new species can increase quickly and become invasive. **Invasive species** are introduced species with growing populations that spread and have a negative impact on their new environment.

DIG DEEPER

Did You Know?

Chestnut Blight

Chestnut blight, a fungus that infects trees, was accidentally introduced to North America during the 1900s when infected trees from Asia were imported. Within a few decades of its arrival, the fungus had killed billions of American chestnut trees. Researchers still do not know how to control the spread of the disease, but they have successfully bred trees that are resistant to the fungus that causes the disease.

▶ **invasive species:** a non-native species that has a negative impact on the natural environment

Figure 2 Invasive beavers have caused much damage to trees in Argentina.

Figure 3 The emerald ash borer has killed millions of ash trees in southwestern Ontario, Michigan, and surrounding areas since it was first noticed in 2002.

A Plan Gone Wrong

The Canadian beaver was introduced to South America during the 1940s for the value of its fur. However, the warmer climate prevented their fur from growing thick enough to be valuable in trading, so the beavers were not trapped or hunted. As a result, the population grew out of control. Beavers caused damage to native ecosystems in Argentina by damming rivers. When a non-native species of fox was introduced to prey on the beavers, the results were even worse. The foxes fed on native rabbits instead of beavers. The lack of rabbit droppings, in turn, affected the soil and left the whole ecosystem out of balance. Descendants of the original beaver population have continued to spread, invade other habitats, and damage forests (**Figure 2**).

Invasive Species in Canada

There are more than 3000 invasive species in Canada. More than 185 are in the Great Lakes region. Invasive species such as the emerald ash borer beetle, native to eastern Asia, have negative impacts on the ecology, economy, tourism, and health of a region (**Figure 3**). **Table 1** lists some of these impacts.

Canada spends million of dollars each year to control invasive species. The damage caused by invasive species around the world is estimated to be $1.4 trillion annually.

Table 1 Effects of Invasive Species

Type of effect	Consequences	Examples
ecological	• population decline of native species	• sea lamprey • Japanese knotweed
economic	• damage to forestry, agriculture, and fisheries	• zebra mussel • emerald ash borer beetle
tourism	• invasive aquatic plants make waterways impassable by boats; negative impacts on fishing and recreation	• European green crab • jointed oat grass • kudzu vine
health	• invasive species bring diseases • pesticides used on invasive species cause pollution and health risks	• mosquito (West Nile virus) • Chinese mitten crab • carpet burweed

Controlling Introduced Species

There are three approaches to controlling an invasive species:

1. **Chemical control** is the use of pesticides to protect trees, crops, fisheries, and human health. Pesticides are effective but may pollute the environment and kill native species as well.
2. **Mechanical control** includes human-made barriers to prevent invasive species from moving into or out of an area (**Figure 4**). Pulling or digging up invasive plants or trapping insects are also examples of mechanical control.
3. **Biological control** means that a predator or infectious agent is used to kill or harm the invasive species.

Figure 4 These barriers prevent invasive fish species from entering Cootes Paradise wetland in Hamilton Harbour.

RESEARCH THIS WE DO NOT BELONG!

SKILLS HANDBOOK 4.C.2., 4.C.6.

SKILLS: Researching, Communicating

1. Use the Internet and other electronic and print resources to research an invasive species in Ontario. Then answer these questions about the species.

A. Where is the species from originally? T/I

B. How did it arrive? Was it introduced accidentally or intentionally? T/I

C. What is the population of this species in Ontario now? Where in Ontario can it be found? T/I

D. What is the main concern with this species? T/I

E. What, if anything, is being done to control the invasive species? T/I

GO TO NELSON SCIENCE

3.7 Wrap Up

- Invasive species are non-native species whose introduction is harmful to ecosystems.
- Invasive species may negatively affect our health and the economy.
- Invasive species can sometimes be controlled using chemical, mechanical, or biological methods.

UNIT TASK Bookmark

You can apply what you learned about invasive species to the Unit Task described on p. 160.

CHECK YOUR LEARNING

1. Are all introduced species invasive? Explain. K/U C
2. Why might a new species be intentionally introduced into an ecosystem? K/U
3. What are some consequences of introducing an invasive species? K/U
4. How do you think the European green crab might affect fishing in Canada? Use the Internet to find additional information about the European green crab and its impact on Canadian fisheries. T/I
5. Female emerald ash borers produce about 50 to 100 eggs, which are laid individually on the bark surface or within bark cracks and crevices. As larvae hatch, they tunnel into the tree, where they feed and spend the winter. Why do you think areas near where the insect has been found have restricted the sale and movement of firewood? T/I

Attack of the Beetles

Canada has become the new home to thousands of non-native species. Each one was brought from somewhere else by humans accidentally or intentionally. Either way, these unwelcome inhabitants were not meant to live in Canada.

The Asian longhorned beetle, shown in **Figure 1**, is a species that accidentally entered North America in 1996. It arrived in packing materials shipped from Asia. The beetle is an insect that bores into hardwood trees, such as maples, elms, poplars, and willows, and damages tissues in the trees (**Figure 2**). The trees eventually die.

It is hard to tell when the beetle larvae are being transported from one place to another. The larvae live deep inside the tree wood for most of the year and are not easily seen. They are often moved around in untreated lumber, firewood, and living trees. In 2003, the beetle was confirmed to be in Ontario.

The Asian longhorned beetles can be controlled mechanically and chemically. Aside from cutting and destroying infested trees and their wood, scientists are now using a pesticide to help fight infestations. The pesticide is injected into the trunk of the tree or into the ground, where it is absorbed by the roots of the tree. When the adult beetles and the larvae feed on these tissues, they die.

The beetle's first appearance in North America was in Chicago, Illinois, in 1998. Over the next decade, the battle against the beetle was waged. Finally, in April 2008, the U.S. Department of Agriculture announced that the Asian longhorned beetle had been eliminated from the state of Illinois. Efforts are still underway in other parts of North America to eliminate this pest.

Figure 1 The Asian longhorned beetle is 2.5 cm to 4.0 cm long. Its body is shiny black with white spots. Its antennae are banded with black and white stripes.

Figure 2 As larvae, the beetles burrow into a tree, feeding on and damaging tree tissues. Adult beetles burrow out of a tree, creating round holes, one sign that a tree has been infested.

3.8

Pollution

Canadian Aboriginal peoples lived in a sustainable way for thousands of years. They used of all of their resources carefully (**Figure 1**).

Figure 1 Cedar trees provided material for baskets, tools, shelter, clothing, and canoes.

Modern Waste Producers

Since those days, two factors have changed humans' relationship with their environment. First, population has increased dramatically. Second, we now use many synthetic and toxic substances. Each person now has a greater impact on the environment than each person did in the past.

The release of toxic substances into the environment has increased, affecting land, water, and air (**Figure 2**). Some pollutants, such as pesticides and fertilizers, were invented with a specific purpose. Other pollutants, such as automobile exhaust and packaging from products, are results, or by-products, of our consumer society. Even biodegradable natural products can cause pollution when they are placed into the environment in high concentrations.

Figure 2 Humans pollute (a) land, (b) water, and (c) air.

Acid Precipitation

When fossil fuels are burned, sulfur dioxide and nitrogen oxides are released into the air. These pollutants combine with water vapour in the atmosphere and form acids. The acids move through the water cycle, sometimes over very long distances. Finally, they return to Earth in the form of **acid precipitation**.

▸ **acid precipitation:** precipitation that has been made more acidic than normal by combining certain chemicals in the air with water vapour

Acid Precipitation and Aquatic Ecosystems

Acid precipitation, often called acid rain, affects both aquatic and terrestrial ecosystems. The effects of acid rain on terrestrial and aquatic ecosystems are shown in **Figure 3**. As rivers and lakes become more acidic, species' numbers may decline disrupting food webs and even causing extinction.

Not all rivers and lakes are equally affected by acid precipitation. Limestone mineral deposits near lakes tend to lessen the effects of acid rain. Southwestern Ontario is made up of mostly limestone, but northern and southeastern Ontario are made up of mostly granite, which does not lessen the effects of acid precipitation. For this reason, acid rain causes more damage in northern and southeastern Ontario.

Acid Precipitation and Terrestrial Ecosystems

Acid rain can have serious effects on terrestrial ecosystems. As acids build up in the soil, the chemical structure of the soil is altered. Bacteria and other micro-organisms living in the soil are sometimes not able to adjust to the higher acid levels and die. The altered soil provides fewer or different nutrients for plants. Some plants are not able to grow and stay healthy in the altered soil. Acid rain has less impact on food crops in agroecosystems than on natural terrestrial ecosystems. This is because lime and fertilizers are added to lower the acid levels in the soil in which crops are grown.

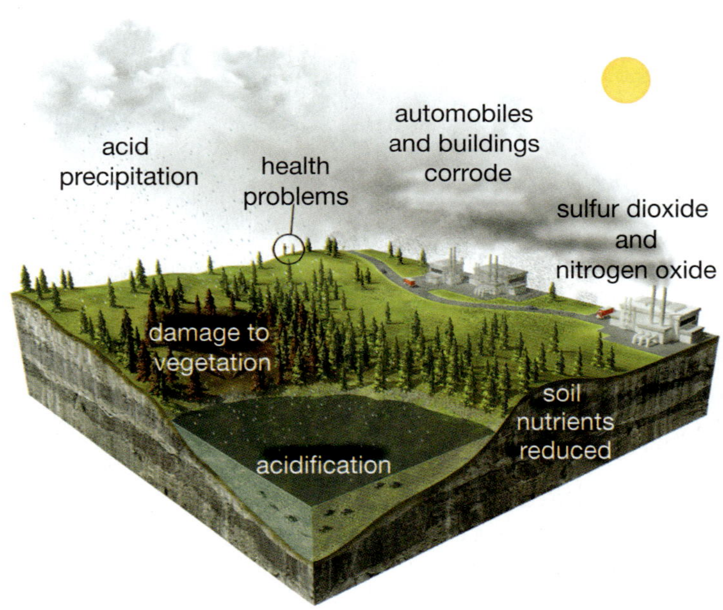

Figure 3 The diagram shows how acid rain forms and damages both aquatic and terrestrial ecosystems.

Acid Precipitation and Humans

Sulfur dioxide is toxic and causes respiratory problems such as asthma and bronchitis in humans. Acid rain speeds up the corrosion of metal, which makes cars and iron structures rust faster than they would with non-acidic precipitation. Acid rain also dissolves concrete and marble, which damages statues and buildings (**Figure 4**). Acid rain and its effects cause billions of dollars in medical and economic costs each year.

Figure 4 Acid rain damage on buildings

Reducing Acid Precipitation

In eastern Canada, sulfur dioxide emissions declined over 60 % between 1980 and 2001. However, even today, the rain in the Muskoka-Haliburton area of Ontario is still 40 times more acidic than non-acidic precipitation. Canada has adopted new technologies and laws regarding pollution emissions. However, much of our acid rain is caused by pollution that comes from the United States, where high-sulfur coal is still widely used for generating electricity.

Even if acid emissions were eliminated, it could still take hundreds of years for ecosystems to recover. Until then, forests in affected areas must rely on poorer soils. This situation threatens the sustainability of our forests.

3.8 Wrap Up

- Acid rain is precipitation that has been made more acidic than usual by the combination of sulfur dioxide and nitrogen oxides with water vapour in the atmosphere.
- Acid rain can travel long distances throughout the atmosphere.
- Acid rain pollutes aquatic and terrestrial ecosystems, damages human health, and corrodes buildings and statues.

UNIT TASK

Bookmark

You can apply what you learned about the effects of acid precipitation to the Unit Task described on p. 160.

CHECK YOUR LEARNING

1. Why do people in today's society create more pollution than people who lived long ago? K/U
2. (a) Name two chemical pollutants that are responsible for acid precipitation.
   (b) How do these pollutants end up in Earth's atmosphere? K/U
3. (a) What are some of the negative effects of acid precipitation on aquatic ecosystems? K/U
   (b) What are some of the negative effects of acid precipitation on terrestrial ecosystems? K/U
4. (a) How does acid precipitation affect humans? K/U
   (b) What can humans do to reduce acid precipitation? K/U

How Do Oil Spills Affect Marine Life?

SKILLS MENU
- Questioning
- Hypothesizing
- Predicting
- Planning
- Controlling Variables
- Performing
- Observing
- Analyzing
- Evaluating
- Communicating

In this activity, you will make a model of an ocean and shoreline and observe what happens in an oil spill. Use cooking oil to represent an oil spill, a cork to represent an animal, and shaking motions to represent a storm or waves.

SKILLS HANDBOOK 3.B.4.– 3.B.9.

Purpose

To demonstrate what happens in an oil spill

Procedure

Part A: Model Oil Spill

1. Fill a glass jar $\frac{2}{3}$ full of water. Add blue food colouring to make the "ocean."

2. Pour 2 tbsp of cooking oil into the jar. Drop a cork stopper into the jar.

3. Put on the jar lid securely. Shake the jar vigorously. Observe what happens to the oil and the cork stopper. Record your observations.

Equipment and Materials

lab apron

gloves

glass jar with lid

tablespoon

toothbrush

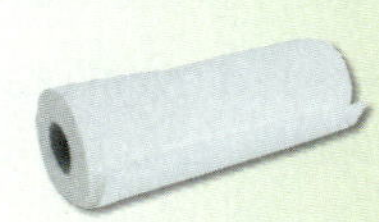
paper towels

hair dryer

2 dishpans of water

sponges

blue food colouring

cooking oil

cork stopper

sand

bird cage wood chips

bird feather

detergent

Part B: Cleaning Up Oil Spills

4. Fill one pan with water. Firmly pack a mound of sand at one end to model a sandy shore. Add 2 tbsps cooking oil to the water in the pan. Gently slosh the mix of water and oil to simulate waves.
5. Choose two or three different clean-up materials to test. Make a plan for how you will use each material to clean up the oil spill.
6. Test your plan. Which materials worked best? Record your observations.

Part C: Birds and Oil

7. Fill the second pan with clean water. Drop a bird feather into the water. Shake it off. Allow the feather to dry naturally, or dry it using a hair dryer on the lowest heat setting. Does the feather still fluff up? Record your observations.
8. Add 1 tbsp of oil to the water. Drop the same bird feather into the water and oil. What happens to the feather?
9. Try to clean the feather.
10. Drop the cleaned, dried feather into a pan of water. How does the feather react? Record your observations.

Analyze and Evaluate

Part A

(a) Where does the oil in the jar of water collect? T/I

(b) What happened to the cork dropped into the jar? T/I

(c) What happened when you shook the jar? T/I

(d) In a real ocean ecosystem during an oil spill, what will happen to surface organisms such as birds? What will happen to animals that come to the surface to breathe such as whales or turtles? A

(e) Over time, spilled oil becomes denser than ocean water. It sinks to the bottom. What will happen to sea urchins, crabs, and other bottom-dwellers? A

Part B

(f) How well might your methods work on an actual spill? C A

(g) What animals are most affected by an oil spill? A

Part C

(h) What happened to the feather in the pan of water and oil? T/I

(i) How might oily feathers affect a bird? A

How Do We Clean Up This Mess?

How would you feel if you learned that getting your hair cut could help the environment?

A non-profit organization in San Francisco, California, called *Matter of Trust* is working on a new way to clean up oil spills in the future. The process involves a few odd ingredients, including hair compost, and worms!

First, the organization collects cut fibres from animals, such as human hair, animal fur, and wool. Why hair? Hair clippings normally go to waste; think about how good hair is at soaking up oil from our scalp! The organization uses clippings collected from the floors of salons, pet groomers, and wool farmers and forms the clippings into mats (**Figure 1**). The toxic oil clings to the mats, just like it does to hair on your head.

This is where the worms come into the picture. The organization puts the oil-soaked hair mats through a composting process in which bacteria help to break down the mats into harmless substances. In the final step, worms eat the composted hair mats (**Figure 2**). The worms' waste makes excellent natural fertilizers. These fertilizers are added to the soil to provide nutrients to growing plants. This is a complete recycling project with a valuable clean-up job in the middle!

Not only are the hair mats useful in cleaning up oil, but creating these mats is a new source of green jobs. Imagine using the scraps of your hair after a haircut to help clean the world of toxic oil spills!

Figure 1 Volunteers use the hair mats to soak up an oil spill on a beach.

Figure 2 Hair mats soak up more oil than other materials, and they are then used as nutrients for worms, leaving rich compost behind.

3.10

Managing Our Forests

Most of the wood you see and use every day comes from natural forests. Natural forest ecosystems provide us with valuable resources of wood and wood fibre. Canada is a large source of wood. Approximately 60 % of our original forests have been cut at least once (**Figure 1**).

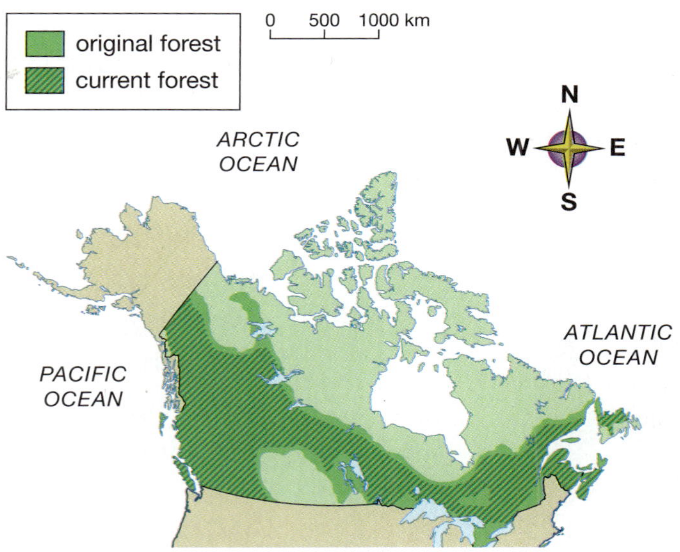

Figure 1 The map shows the location of original forests and the current forests in Canada. The current forests have been cut or cleared at least once.

Forestry Practices

There are three basic methods of forest harvesting: clear-cutting, shelterwood cutting, and selective cutting (**Table 1**). Clear-cutting, or the removal of all or most trees in an area, is the most common method, accounting for 90 % of the trees harvested in Canada. Although this method is the most profitable, it also has drawbacks. Nutrients are more easily lost from the soil after an area is clear-cut. The lack of roots causes soil to erode by being blown away by the wind or washed into streams and lakes. Soil buildup can harm fish by clogging their gills so they cannot breathe. Nutrients in the soil may affect the water by causing the growth of algae. Trees are often replanted in place of the harvested ones or allowed to regrow naturally. When planted, only a few types of trees grow in an area that once was a diverse forest ecosystem.

Table 1 Three Main Forestry Practices

Forestry practice	What is it?	Advantages	Disadvantages
clear-cutting	the removal of all or most of the trees in an area	• least expensive and quickest method • allows for new growth • mimics fire patterns	• has the greatest impact on the ecosystem • soil is washed or blown away • soil health suffers
shelterwood cutting	mature trees are left standing to provide shelter for new growth	• not all trees in an area are cut down	• the area is not cleared for new growth
selective cutting	selected trees are harvested	• has the least impact on the ecosystem	• the most expensive method

DIG DEEPER

Career Link

Forestry technicians help manage forest resources, including replanting trees in logged areas and working to help keep forest fires under control. To learn more about being a forestry technician,

GO TO NELSON SCIENCE

Figure 2 This logo means that wood fibre used to make a product came from forests managed in a responsible way.

The Importance of Forest Management

Because forestry is such an important industry in Canada, it is important to have practices that keep the forests as sustainable as possible. In 1993, the International Forest Stewardship Council was founded. It set guidelines for forestry practices to ensure that the forests remain sustainable. This was a recent positive trend in the industry. If the forests are lost, the resources as well as the industry are lost.

By 2009, more than 100 million hectares, or 7 %, of the world's forests were certified. This means that they are managed in a sustainable way. The number of new trees planted and the number of trees cut down is in balance. Wildlife is protected in the areas. Waterways are safe and protected as well, and the biodiversity is kept in balance. When buying lumber or forest products, you may see a logo to show that the lumber was taken from a certified forest (**Figure 2**).

Supply and Demand of Forest Goods

In our economy, the idea of supply and demand influences the price of goods. When there is not enough of one product to go around, the product is in high demand. This makes the price of the item rise, because people are willing to pay more for something they really want. But this pattern is dangerous when it comes to producing and selling goods from natural resources. If the demand for a certain type of rare wood goes up, people will be willing to pay high prices for it. Logging companies will feel pressure to cut down more trees than they should or to cut down rare tree species (**Figure 3**). Soon the species may become over-harvested or threatened. Even though there may be a high demand for lumber and people may be willing to pay high prices for it, the supply must be maintained so that the resources can continue to grow in a healthy ecosystem.

Figure 3 Trees such as mahogany are valuable and in high demand. This increases logging pressure and threatens the biodiversity of the forest.

Wildlife Stewardship

Stewardship means taking responsibility for managing and protecting the environment. Maintaining forests that continue to produce wood products and provide habitats for wildlife requires action on the part of humans. If we take a resource from the ecosystem, we must accept the responsibility for keeping the ecosystem healthy.

Stewardship refers to a wide range of actions that Canadians can take to care for our environment. Forest stewardship can protect the ecosystem, and it can make the forest more productive for the economy. While we care for and protect the environment, we also use it to meet our own needs and make the planet a good place to live. With stewardship of Earth's resources, we can live in harmony with the ecosystems that we rely on.

Today we know what behaviours are responsible and which ones will cause permanent damage to an ecosystem. As we act as stewards for our planet, we can teach others to conserve our forest resources, including timber, wildlife, soil, and water.

▶ **stewardship:** taking responsibility for managing and protecting the environment

3.10 Wrap Up

- The three main forest harvesting methods are clear-cutting, shelterwood cutting, and selective cutting.
- Forests that are harvested must be managed to reduce ecological impacts such as soil erosion and decreased biodiversity.
- Sustainable forestry practices protect ecosystems and maintain biodiversity.

CHECK YOUR LEARNING

1. Compare the advantages and disadvantages of clear-cutting and selective cutting. K/U
2. In what ways do humans manage forest ecosystems? K/U
3. How does the idea of supply and demand influence forestry practices? K/U
4. How would you explain to an executive of a logging company the importance of forest sustainability? C A
5. What are the advantages and disadvantages of shelterwood cutting? K/U
6. What does it mean if a forest is "certified"? K/U

3.11 PERFORM AN ACTIVITY

Comparing Forest Management Practices

Modern forestry practices often use management practices to increase the amount of wood and wood fibre that can be obtained from natural forests. The managed forests can cause ecological changes that may have long-term consequences on the ecosystem.

SKILLS MENU

- Questioning
- Hypothesizing
- Predicting
- Planning
- Controlling Variables
- Performing
- Observing
- Analyzing
- Evaluating
- Communicating

SKILLS HANDBOOK 3.B.5.–3.B.9.

Purpose

To assess some impacts of forestry management practices

Procedure

1. Examine **Table 1**. These values represent the average numbers of individual animals caught per 100 days of trapping in the three different forest-types in Nova Scotia.
2. Make a bar graph that shows the number of individual animals caught per 100 days of trapping versus the type of forest for each species listed in **Table 1**.

Table 1 Small Mammal Counts in a Maple-Birch Forest in Nova Scotia

Species	Average number caught per 100 days of trapping		
	Uncut	Shelterwood	Clear-cut
short-tailed shrew	3.9	3.4	3.4
masked shrew	2.2	2.3	4.9
red-backed vole	7.0	4.9	6.0
deer mouse	1.0	3.0	0.7
meadow vole	0.2	0.6	2.8
eastern chipmunk	0.2	0.4	0.2
jumping mouse	2.9	4.7	0.3

DIG DEEPER

Learning Tip

Data Tables

Data tables, such as **Table 1**, help organize information. A data table helps you sort information into groups. It can also help you see patterns in the data. Every data table should have a title that tells what it is about. The rows and columns should be labelled so that someone who is not familiar with the data will be able to read the table.

3. Examine **Table 2**. These values represent the number of pairs of each bird species per square kilometre in an uncut (mature) forest and clear-cut forest. The clear-cut forests are 3 to 5 years old.

4. Make a Venn diagram to compare the species found in each type of forest. Include species in the following categories: C
   (i) found only in the uncut forest
   (ii) found only in the clear-cut forest
   (iii) found in both the uncut and clear-cut forests

Table 2 Average Number of Breeding Pairs of Birds in Mature and Clear-Cut Forests

Bird species	Number per square kilometre	
	Mature maple-birch forest	Clear-cut (3–5 years old)
ruby-throated hummingbird	0	23
least flycatcher	137	0
hermit thrush	43	0
red-eyed vireo	53	0
black-throated green warbler	37	0
common yellowthroat	0	152
American redstart	65	0
white-throated sparrow	7	127

Analyze and Evaluate

(a) Which forest had the biggest difference in trappings compared to an uncut forest? T/I

(b) How many more woodland jumping mice were trapped in the shelterwood forest than in the clear-cut forest? T/I

(c) What was the difference between the total number of animals trapped in the uncut forest and in the clear-cut forest? T/I

(d) Compare shrew populations in the three forest types. Were the differences significant? T/I

(e) Voles, mice, and chipmunks are all rodents that feed mostly on seeds and other plant matter. Were they all influenced in the same way by the different cutting methods? T/I

(f) Which mammal species benefited most from the forestry practices? T/I

(g) How did clear-cutting affect bird species in the study represented in **Table 2**? Compare the effects on overall numbers with the effects on individual species. T/I

Apply and Extend

(h) Select two bird species that were found only in the uncut forest and two that were found only in the clear-cut forest. Use library or Internet resources to find out what they eat and where they like to nest. T/I

(i) Why is it important to know what an organism needs to survive in order to predict how it will respond to a change in its environment? Give examples to support your answer. A C

DIG DEEPER
Web Link

For more information about bird species, what they eat, and where they nest,

GO TO NELSON SCIENCE

Chapter 3 KEY CONCEPTS SUMMARY

Natural ecosystems are valuable to humans.

- Ecosystems provide us with products and services.
- Ecosystems are free and renewable if we can keep them sustainable for the future.
- Natural ecosystems provide jobs, attract tourists, and improve the economy.

Ecosystems are balanced but can change over time.

- Large ecosystems are usually in equilibrium. The biotic and abiotic factors in the ecosystem are relatively constant over time.
- Succession is the series of changes that happens over time after an ecosystem is disturbed.

Biodiversity describes the variety and amount of life in an ecosystem.

- Biodiversity of many ecosystems is threatened by human activity.
- In Canada, at-risk species are categorized as extirpated, endangered, threatened, or of special concern.

Human activities can affect and threaten the sustainability of natural ecosystems.

- Most biodiversity hotspots are near the equator.
- The major causes of species loss are habitat destruction and fragmentation.
- Invasive species can damage natural ecosystems.

Pollution of water, land, and air causes health and economic problems.

- Acid rain affects terrestrial and aquatic ecosystems.
- Acid rain is harmful to human health and infrastructure such as buildings.

Plant and animal resources should be used in a sustainable way.

- Forestry practices can damage ecosystems by causing soil erosion and polluting waterways.
- Forest management is necessary to keep forests sustainable.

WHAT DO YOU

THINK NOW?

You thought about the following statements at the beginning of the chapter. You may have encountered these ideas in school, at home, or in the world around you. Consider them again and decide whether you agree or disagree with each statement.

1 Humans enjoy Ontario's natural ecosystems, but we do not rely on them in our daily lives.
Agree/disagree?

2 Forest fires benefit certain ecosystems.
Agree/disagree?

3 Ecosystems with high biodiversity are more stable.
Agree/disagree?

4 Ontario has a large wildlife population with very few "at-risk" species.
Agree/disagree?

5 An invasive species can have negative effects on an ecosystem into which it has been introduced.
Agree/disagree?

6 Human demand for safe, clean water is a major concern on a global scale but not in Canada.
Agree/disagree?

How have your answers changed since then? What new understanding do you have?

BIG Ideas

- Ecosystems consist of a variety of components, including, in many cases, humans.
- The sustainability of ecosystems depends on balanced interactions between their components.
- Human activity can affect the sustainability of aquatic and terrestrial ecosystems.

DIG DEEPER

Vocabulary

equilibrium p. 82
succession p. 83
biodiversity p. 88
extinct p. 93
invasive species p. 101
acid precipitation p. 105
stewardship p. 113

CHAPTER 3

SELF-QUIZ

The following icons indicate the Achievement Chart category addressed by each question.

K/U Knowledge/Understanding | T/I Thinking/Investigation | C Communication | A Application

For each question, select the best answer from the four alternatives.

1. Which statement describes an ecosystem in equilibrium? (3.2) K/U
   (a) an island that has just had a large volcanic eruption
   (b) an established forest that supports a variety of plants and animals
   (c) a region in the Arctic where the average temperature is steadily increasing
   (d) an area of grassland just after a damaging fire
2. Which of the following actions is an example of chemical control of invasive species? (3.7) K/U
   (a) removing the species by hand
   (b) applying pesticides to the area where the species is
   (c) learning about the species and where it comes from
   (d) introducing another species to control the invasive one
3. Which of the following is an economic benefit of ecosystems? (3.1) K/U
   (a) climate moderation
   (b) education
   (c) ecotourism
   (d) oxygen supply

Indicate whether each of the statements is TRUE or FALSE. If you think the statement is false, rewrite it to make it true.

4. Most non-native species that are introduced to a new environment become invasive in that new environment. (3.7) K/U
5. Human actions have caused all of the species extinctions on Earth. (3.5) K/U
6. All fires are harmful to a forest ecosystem. (3.2) K/U

Copy each of the following statements into your notebook. Fill in the blanks with a word or phrase that correctly completes the sentence.

7. A hurricane hits an island and washes away the surface soil and plant and animal life. A year later, small plants and insects begin to reappear on the island. This is an example of ecological _______. (3.2) K/U
8. Ecosystems in a state of _______ will show little change over long periods of time. (3.2) K/U
9. When fossil fuels are burned, sulfur dioxide and nitrogen oxides can combine with water vapour in the air and eventually return to Earth as _______. (3.8) K/U
10. An area with very high biodiversity is referred to as a(n) _______. (3.5) K/U

Match each lettered term in the first list with the most appropriate numbered description from the second list.

11. **Ecosystem Disruption**
    (a) acid precipitation (3.8)
    (b) fragmentation (3.6)
    (c) clear-cutting (3.10)
    (d) invasive species (3.7) K/U

    Effect on Ecosystem
    (i) increases rate of soil erosion
    (ii) damages statues and buildings
    (iii) habitat loss
    (iv) interferes with natural food web of ecosystem

Write a short answer to each of these questions.

12. Write a newspaper ad that promotes sustainable practices that could be accomplished in an ecosystem in Ontario. (3.1) C A

13. Describe why it is important for young people to practise stewardship when interacting with their ecosystem. (3.10) C A

14. Stewardship is something that we should all be concerned with. (3.10) K/U C A
    (a) Explain what it means to you to be a good steward of your natural environment.
    (b) What example of stewardship have you demonstrated?

15. A forest near your home will be used to provide lumber to the community. Which method would you prefer that forest management teams use to take the lumber: clear-cutting, shelterwood cutting, or selective cutting? Explain your choice. (3.10) C A

16. Some schools in Ontario are fumigated with pesticides to kill pests such as earwigs—an invasive insect species. (3.7) C A
    (a) Do you agree with the chemical spraying of schools for this purpose?
    (b) Do you think people should fumigate their homes to control earwigs?

17. Some parks that have fragile natural features, such as wetlands, limit the number of people who can visit them each day. Some people think this is a good way to preserve natural features. Others think these parks should be open to whoever wants to visit them. What do you think? Explain why you think so. (3.2) C

18. Which do you expect to have a higher biodiversity: an uncut, old-growth forest or a managed forest owned by a lumber company that plants new a new tree for each tree it cuts down? Explain. (3.10) K/U C

19. A nature conservancy buys an old farm to serve as research area for studying succession. Describe the steps the area will go through as it changes from a farm to a forest. (3.2) K/U C

20. Explain why a forest manager may order certain parts of the forest she manages to be deliberately burned. (3.2) K/U C

21. Continued exploration has shown that there is more biodiversity in deserts than was originally thought. List some steps people can take to protect biodiversity in deserts (3.3) K/U C

22. What makes Madagascar a biodiversity hotspot? (3.5) K/U

23. Give two examples of natural events that have caused extinctions. Give two examples of ways people have caused animals to become extinct. (3.6) K/U

24. A new town is to be established in a heavily forested area. The builders want to preserve as much of the biodiversity of the forest as possible. Draw a sketch of what the roads and housing areas of the town might look like. (3.6) C A

25. Why is it difficult for Ontario to stop acid rain pollution from continuing to damage the environment? (3.8) K/U

Self-Quiz

To do an online self-quiz,

GO TO NELSON SCIENCE

CHAPTER 3

REVIEW

The following icons indicate the Achievement Chart category addressed by each question.

K/U Knowledge/Understanding
C Communication
T/I Thinking/Investigation
A Application

What Do You Remember?

1. In your notebook, write the word(s) needed to complete each of the following sentences.
   (a) ________ is the variety of species in an ecosystem. (3.3)
   (b) The changes following events such as a forest fire are called ecological ________. (3.2)
   (c) ________ forms when chemicals from industry mix with water vapour in the air. (3.8)
   (d) Loss of biodiversity in an ecosystem makes that ecosystem become less ________. (3.5)
   (e) ________ is taking responsibility for managing and protecting the environment. (3.10) K/U

2. Match the term on the left with the correct definition on the right. (3.5) K/U

(a) extinct	(i) a species that may become endangered or threatened
(b) extirpated	(ii) a species facing immediate danger of extinction or extirpation
(c) endangered	(iii) a species that no longer exists on Earth
(d) special concern	(iv) a species that no longer lives in a particular region
(e) threatened	(v) a species likely to become endangered if factors do not change

3. (a) Compare the biodiversity of 10 hectares of deciduous forest in Ontario with that of 10 hectares of rainforest.
   (b) Explain what might be responsible for differences in the level of biodiversity in the two ecosystems. (3.5) K/U C

4. What are some effects of acid precipitation? (3.8) K/U

What Do You Understand?

5. Describe the three main ways of controlling invasive species. (3.7) K/U

6. Habitat loss is not as rapid in Ontario as it once was. Why is this the case? (3.10) K/U

7. Why do modern societies have more problems managing waste and maintaining ecosystems than Aboriginal peoples of long ago? (3.8) K/U

8. Classify each of the following as a cultural benefit, an economic benefit, or other benefit of an ecosystem. (3.1) K/U
   (a) hiking trail
   (b) lumber for building
   (c) clean air
   (d) medicines made from plants
   (e) wilderness campsite
   (f) lake for canoeing
   (g) lobsters for eating

9. Match the human activity on the left with the possible ecological impact on the right. (3.6, 3.8, 3.10) K/U

(a) wetland drainage	(i) increased acid rain
(b) release of sulfur oxides into the air	(ii) loss of biodiversity of aquatic ecosystems
(c) clearing of land next to rivers	(iii) increased erosion

10. (a) What are three benefits of ecotourism? (3.1) K/U
    (b) Are there any possible disadvantages of ecotourism? Explain your reasoning. (3.1) T/I

11. A grassy field borders an oak-maple forest. Which one of these ecosystems is more likely to be undergoing succession? Why? (3.2) T/I

Solve a Problem

12. Your school wants to create a new sports field on the property next door. The property consists of a wetland that is home to several species of special concern. You and a group of students have been asked for your thoughts on whether or not the sports field should be created. (3.5, 3.6) T/I C A
    (a) Why is it important to preserve the wetlands and the species that live there?
    (b) What arguments would you make to convince the school authorities that they should preserve the wetland ecosystem?
    (c) What alternative uses of the land would you suggest to the school?

Create and Evaluate

13. A lightning-caused forest fire is burning out of control near a very sensitive breeding habitat for endangered barn owls. The financial cost of fighting the fire is great, and there is a personal risk to firefighters. (3.2) C A
    (a) Give two possible advantages and disadvantages of letting the fire burn.
    (b) If you were a decision maker, would you allow the fire to burn naturally? Or, would you ask firefighters to put out the fire? Justify your answer.
14. Research a restoration effort such as the one undertaken to preserve the Milton Quarry described in Section 3.2. (3.2) T/I C A
    (a) What caused the original damage to the ecosystem you selected?
    (b) What efforts have been underway to restore the ecosystem to its natural status?
    (c) What restoration efforts used for this ecosystem could be applied to the restoration of a similar ecosystem?
15. Some people think hunting and fishing should be banned as recreation. Should there be different rules for those who hunt or fish for food compared with those who do so only for sport? Explain your answers. (3.10) T/I C

Reflect on Your Learning

16. How might what you learned in this chapter affect your behaviour in terms of stewardship of natural ecosystems? A
17. In this chapter, you learned that Ontario has species that are identified as at-risk, either of becoming threatened or endangered, or of extinction. K/U T/I
    (a) How many at-risk species could you name before reading this chapter?
    (b) What resources could you use to find out more about at-risk species in your region of the province?

Web Connections

18. The Galapagos Islands are home to some of the most interesting species of plants and animals on Earth. Like other locations, the islands have suffered from the introduction of invasive species, including goats and black rats. Research more information about the islands and the impacts of invasive species there. Choose one invasive species on the islands and describe its history. T/I C
    (a) How did the species arrive on the islands?
    (b) What ecological niche does the species occupy?
    (c) How has the species affected native species and ecosystems on the islands?
19. Choose one of the issues in this chapter that is of special interest to Canadians. Research Canadian private or government organizations that are working on the issue. If possible, find an organization that is run by or uses the help of high school students. T/I A
    (a) What actions is each organization taking to solve the problem or improve the situation?
    (b) Make a list of questions you would ask if you had an opportunity to speak to a person from each of the organizations.

GO TO NELSON SCIENCE

CHAPTER 4 Ecosystems by Design

KEY QUESTION

Are human-designed environments sustainable?

KEY CONCEPTS

- Some ecosystems are designed, created, and maintained by humans.
- Agricultural practices disrupt natural biogeochemical cycles.
- It is important to keep Earth's soil healthy.
- Pesticides are used to reduce crop losses due to insects, weeds, and other pests.
- Pesticides have ecological costs.
- Urban ecosystems depend on natural ecosystems for food and materials.

This illustration shows an artist's concept of the walled city of Masdar.

The city of Masdar is in the United Arab Emirates.

MASDAR, THE ECO-FRIENDLY CITY

For the first time in history, half of the world's population lives in cities. By 2025, nearly 60 % of people will live in cities. To provide a good quality of life for residents, city planners around the world are aiming to create more "eco-friendly" cities. Eco-friendly cities would use sustainable methods that cause little harm to the environment. They would rely less on fossil fuels and more on alternative sources of energy.

That is the plan for the city of Masdar, which is being built in the United Arab Emirates. The goals for the city are to have zero carbon emissions and fully recyclable wastes. Its planners hope that when Masdar is completed in 2016, it is a shining example of a city of the future.

Masdar's location on the coast of the Persian Gulf will allow it to take advantage of the cool sea breezes to produce wind energy. The city will use solar panels in front of buildings to capture solar energy while providing shade from the Sun. Solar energy and wind farms will provide most of Masdar's electrical energy.

The city will also be car-free. People will use a railway system on magnetic tracks for transportation. The absence of cars will allow for narrow streets. These narrow streets will help funnel breezes from one side of the city to the other.

The city will also minimize waste. The wastes from plants and animals, including humans, will be used to create nutrient-rich soil and fertilizer. Industrial wastes, such as plastics and metals, will be recycled or reused. Waste water will be used to grow crops that will be used to produce renewable fuels made from plant matter (biofuel).

Would you like to live in this city? What would be the benefits? What potential problems might the city have? What would need to be changed to make your city or town more eco-friendly?

WHAT DO YOU

THINK?

Many of the ideas you will explore in this chapter are ideas that you have already encountered. You may have encountered these ideas in school, at home, or in the world around you. Some of the statements are not true. Consider each statement and decide whether you agree or disagree with it.

1 Farming and other agricultural activities are similar to processes that take place naturally in ecosystems.
Agree/disagree?

2 Organic farming can benefit the environment.
Agree/disagree?

3 Pesticides are needed to control pests.
Agree/disagree?

4 Many of the plants and animals farmed in Ontario are native to this region.
Agree/disagree?

5 The actions of people who live in cities do not affect the surrounding natural environments.
Agree/disagree?

6 We can use as many resources as we need because we can always recycle or reuse them.
Agree/disagree?

FOCUS ON
READING

Asking Questions

Before reading, ask yourself questions that help you remember facts you already know about the topic. Ask questions as you read to help you understand the ideas and information in the text.

Type of question	How to get an answer
Literal ("On the Lines")	Look for statements that answer your question.
Inferential ("Between the Lines")	Add a clue in the text to a fact you know to make your own idea.
Evaluative ("Beyond the Lines")	Add a statement in the text to something you know to make a judgment or form an opinion.

Figure 3 Fertilizer runoff can cause algal blooms in water.

Environmental Impact of Fertilizers

Both synthetic and natural fertilizers can have a negative effect on the environment. In general, more problems are associated with the use of synthetic fertilizers.

Fertilizers can enter the soil too rapidly and change the number and types of organisms found in the soil. As a result, the soil has less of the organic matter that provides a natural supply of nutrients. The soil becomes dependent on the application of fertilizers to replace nutrients. This cycle is not sustainable.

During a heavy rain or spring thaw, fertilizers can flow from the land in runoff into aquatic ecosystems. The excess nutrients can stimulate the growth of algae, causing an uncontrolled algal bloom (Figure 3). Algal blooms reduce the amount of oxygen in the water. The lack of oxygen kills fish and other aquatic organisms.

DIG DEEPER

Reading Tip

As you work through the chapter, look for tips like this. They will help you develop literacy strategies.

Asking Questions *in Action*

Here is how one student asked questions to understand information about the environmental impact of fertilizers.

Questions I asked about the text	How that question helps me understand the text
Literal: Why do algal blooms kill fish?	helps me understand that algae uses up the oxygen that fish need
Inferential: Why is fertilizer lost during heavy rains?	helps me understand that fertilizer sits on top of the soil, not inside it
Evaluative: How useful is this information to me?	helps me form an opinion that I should keep my aquarium at home clean so my fish will not die

4.1

Engineered Ecosystems and Modern Agriculture

Unlike other species, most humans spend little time in natural ecosystems. We have the ability to change our surroundings to suit our needs. Lawns, golf courses, and gardens are places that humans have designed and constructed. They are artificial, or engineered, ecosystems.

Comparing Natural and Engineered Ecosystems

Engineered ecosystems are different from natural ecosystems in several ways. In a natural ecosystem, the human influence is no greater than the influence of any other species. Natural ecosystems tend to have a great variety of organisms. Natural processes and relationships among organisms keep the ecosystem in balance.

In contrast, the features of engineered ecosystems do not vary as much as those of natural systems. This is because features have been added or removed on purpose. Engineered ecosystems tend to have lower biodiversity than natural ecosystems, meaning that engineered ecosystems have fewer types of organisms (**Figure 1**). Engineered ecosystems need humans to maintain them.

Figure 1 The number of species in this small ecosystem is very limited.

Modern Agriculture

What did you have for breakfast today? Your food probably came from a store. It may have been brought to that store from a number of farms, perhaps thousands of kilometres from your home.

Agriculture in Canada

Agriculture is a major industry in Canada. From the cranberry bogs of British Columbia to the potato fields of PEI, this country produces a great variety of foods (**Figure 2**). Most people do not think about where their food comes from or how its production affects the environment.

Figure 2 This dairy farm in Ontario is one of the many types of Canadian agriculture.

Foods from Afar

Consider the toppings you may enjoy on pizza. For example, the toppings for Hawaiian style pizza include tomato sauce, cheese, ham, and pineapple. Most of these foods are grown in Canada, but none are from species native to Canada. **Figure 3** shows where these foods were first raised or grown.

Figure 3 Most of the ingredients in this pizza are native to other parts of the world.

Nearly all of the foods produced in Canada were introduced from other parts of the world. Using non-native plants and animals allows us to enjoy a variety of nutritious foods. **Table 1** shows some other native and non-native foods that are grown and raised in Canada.

Table 1 Agricultural Products of Canada

	Non-native species	Native species
plant crops	corn, wheat, potatoes, beans, barley, rye, oats, soybeans, canola, tomatoes, apples, squash, beans, peanuts, carrots	strawberries, maple, wild rice, blueberries
livestock species	cattle, pigs, chickens, sheep, goats	turkeys

DIG DEEPER

Did You Know?

Grass Seeds Again?

About 90 % of the world's food comes from just 15 plant and 8 animal species! The top three food sources are the seeds of rice, wheat, and corn—all of which are grasses.

- **agroecosystem:** an agricultural ecosystem designed and maintained by humans to meet the human need for food
- **pest:** any plant, animal, or other living organism that causes illness or harm and is an annoyance to humans

Farmlands as Engineered Ecosystems

An engineered ecosystem is designed to fulfill a specific function. An **agroecosystem**, or farm, is the most common example of an engineered ecosystem. Ranches and tree plantations are other examples. Agroecosystems are designed to provide us with foods and other products we need.

Comparing Natural Ecosystems and Agroecosystems

In a natural ecosystem, many different species interact, maintaining a natural balance in the ecosystem. For example, when bees eat the nectar of flowers, they pollinate the flowers. Then, pollen makes flowers grow. Rabbits eat grass, and their waste fertilizes the soil, making the grass healthier (**Figure 4(a)**).

In an agroecosystem, a very limited number of species interact. A greenhouse might grow only tomatoes. Corn might be the only crop grown in a large field (**Figure 4(b)**). Because the number of species has been limited, the natural cycles no longer work. Without natural cycles to keep it in balance, the ecosystem becomes unsustainable. Humans must maintain a balance of abiotic and biotic factors. For instance, farmers must manage the water and nutrient supply, and work to eliminate pests and diseases. A **pest** is a living thing that causes harm to another living thing and is a nuisance to humans.

DIG DEEPER

Reading Tip

Ask Evaluative Questions

Sometimes ideas or information in texts makes you ask an evaluative question. For example, you might wonder about engineered ecosystems and ask, *Is it worth it to convert natural ecosystems into monocultures? Is there a more sustainable solution?* You may need to conduct further research in order to make an informed judgment.

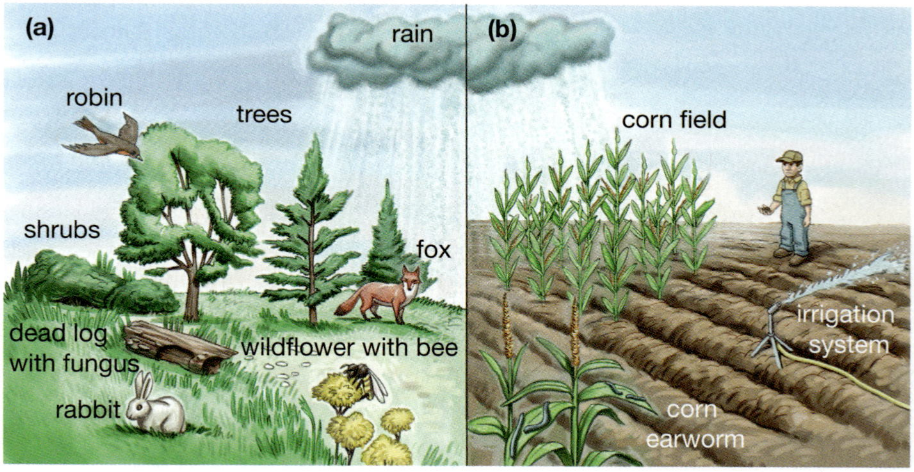

Figure 4 (a) A natural ecosystem maintains its own balance of biotic and abiotic factors. (b) In an agroecosystem, humans work to maintain the balance.

Monocultures

On many farms, people grow crops in **monocultures**, where only one crop is grown in large fields. Sometimes one crop covers thousands of acres. There is very little biodiversity in monocultures.

- **monocultures:** agroecosystems (farms) consisting of single crops in an area

Benefits of Monocultures

Growing only one type of plant in a field helps farmers get the highest crop yields per hectare of land. Without competition from other species, the plants are able to grow larger. Crops can be harvested quickly and easily because the same kinds of equipment can be used for the entire crop.

Drawbacks of Monocultures

Whole fields of monoculture crops can be wiped out by a single disease. For example, the Great Irish Famine of the 1800s happened partly because of potato blight, a water mould that spreads rapidly in warm, wet conditions (**Figure 5**). It can cause whole plants to collapse in a short period of time. In 2008, insect pests known as plant hoppers destroyed thousands of tonnes of the rice crop in Vietnam and China, causing a shortage of food.

In addition, monocultures place an enormous strain on soil nutrients. When a field is full of one type of plant, those plants compete with each other for the same nutrients. Therefore, monoculture crops require widespread use of fertilizers.

Figure 5 Potatoes infected with blight have grey or dark patches that quickly decay. The potatoes may also rot later when they are stored.

4.1 Wrap Up

- Humans change their surroundings to suit their needs.
- Engineered ecosystems have low biodiversity.
- Many engineered ecosystems are unsustainable because they lack the natural cycles necessary to keep the ecosystem in balance.
- The most common type of engineered ecosystem is an agroecosystem.
- Modern farming often involves planting fields of only one type of crop, known as monocultures.
- Most of the crop and livestock species in Canada are non-native species.

UNIT TASK

Bookmark

You can apply what you learned about agroecosystems and other engineered ecosystems to the Unit Task described on page 160.

CHECK YOUR LEARNING

1. What is an agroecosystem? K/U
2. Why are agroecosystems that contain only a few types of organisms unsustainable? K/U
3. Explain what is meant by the following statement: "Most of Canada's crops consist of non-native species." K/U C
4. Make a T-chart to compare natural ecosystems with engineered ecosystems. K/U C
5. What are the advantages of a monoculture? What are the disadvantages? K/U
6. List three engineered ecosystems that you are familiar with. How do you know these systems are engineered? K/U
7. (a) Describe how you use engineered ecosystems.
   (b) Describe how you use natural ecosystems. C A

4.2 Why Soil Health Matters

We depend on farmers to provide us with food. Farmers, in turn, depend on healthy soil. Land plants depend on soil not only for physical support, but also for nutrients and water. Soil also provides oxygen to the roots of a plant.

DIG DEEPER

Reading Tip

Ask Questions in Action

If you do not understand something while reading a text, stop reading and ask a question about what you don't understand. Try rereading the previous sentence to see if you missed information that answers your question. Or, you can continue reading to see if information in the next sentences or paragraphs answers your question.

Components of Soil

The four main components of soil are minerals, organic matter, air, and water. The minerals form when wind, water, chemicals, and living things gradually cause rock to break up into tiny particles. The small bits of organic matter in soil are called humus (HYOO muhs). These bits come from the decay and waste products of living things. The humus provides nutrients that plants need in order to grow. Between the minerals and humus in soil are spaces called pores. Pores contain air and water.

Besides plants, soil supports other organisms. Look closely at a handful of soil and you will likely find fungi, insects, worms, and bacteria (**Figure 1**). These organisms live in the soil, and help decompose, or rot, the organic matter. Through their movements, the larger organisms help keep the soil loose and well mixed.

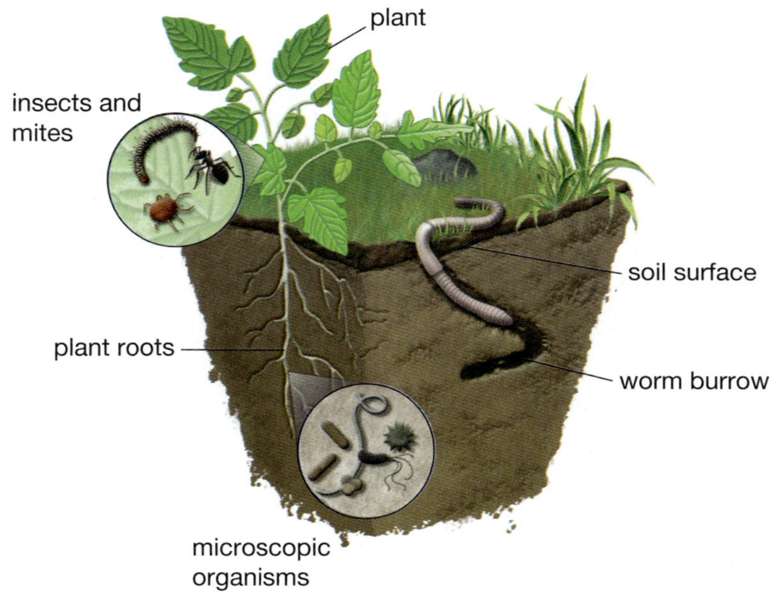

Figure 1 The bacteria, fungi, worms, and insects in soil make nutrients available to plants.

Managing Soil

Anyone who engineers an ecosystem, including gardeners and landscapers, needs to consider soil health.

DIG DEEPER

Career Link

To learn more about a career in landscaping,

GO TO NELSON SCIENCE

Soil Nutrient Cycles

The amount and availability of nutrients is the number one factor in keeping soil healthy. The most important nutrients for plants are nitrogen, phosphorus, and potassium. Nutrients are cycled through natural ecosystems with the help of producers (plants), consumers (animals), and decomposers.

However, when crops grow on farms, some of the nutrients stored in the plant tissues are removed from the ecosystem (**Figure 2**). If nutrients are not returned to the original farmland, the soil will gradually run out of nutrients.

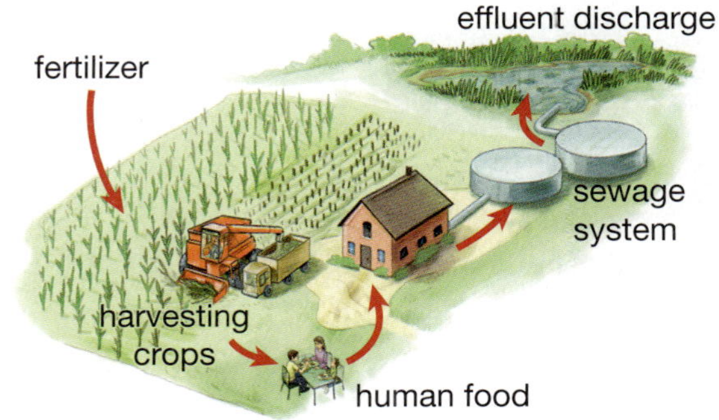

Figure 2 Nutrients move through farmlands much differently than through natural ecosystems. In this illustration, the nutrients are shown as arrows.

Fertilizers

The most common method that farmers use to replace soil nutrients is to apply fertilizers. A **fertilizer** is any substance that helps plants grow.

Until the early 1900s, farmers could use only **natural fertilizers**, such as manure and seaweed. Manure and seaweed both contain nutrients that plants need to grow such as nitrogen, phosphorus, and potassium. Farmers would mix these into the soil. Natural processes would break down the fertilizer so that the nutrients could be used by the crops.

Synthetic fertilizers go through chemical processes, but they contain the same nutrients as natural fertilizers. They are even made from natural sources. Synthetic fertilizers are easy to use because they can be sprayed or spread on fields and often do not need to be worked into soil. In addition, the nutrients are immediately available to plants, so the fertilizers can quickly improve crop growth. The use of synthetic fertilizers became widespread in the 1960s. As a result, both food production and farm profits increased.

- **fertilizer:** a substance applied to an area to provide the nutrients plants need for growth
- **natural fertilizers:** plant nutrients that have been obtained from natural sources and have not been chemically changed by humans
- **synthetic fertilizers:** plant nutrients that are produced by subjecting natural materials to chemical processes

Figure 3 Fertilizer runoff can cause algal blooms in water.

Figure 4 Drainage pipes remove excess water from fields.

Environmental Impact of Fertilizers

Both synthetic and natural fertilizers can have a negative effect on the environment. In general, more problems are associated with the use of synthetic fertilizers.

Fertilizers can enter the soil too rapidly and change the number and types of organisms found in the soil. As a result, the soil has less of the organic matter that provides a natural supply of nutrients. The soil becomes dependent on the application of fertilizers to replace nutrients. This cycle is not sustainable.

During a heavy rain or spring thaw, fertilizers can flow from the land in runoff into aquatic ecosystems. The excess nutrients can stimulate the growth of algae, causing an uncontrolled algal bloom (**Figure 3**). Algal blooms reduce the amount of oxygen in the water. The lack of oxygen kills fish and other aquatic organisms.

Controlling the Flow of Water in Soil

All farms need a reliable supply of water. Many farmers add water to their fields using irrigation. They get the water from surface sources, such as lakes or rivers, or from groundwater. To remove excess water from their fields, many farmers use drainage pipes (**Figure 4**).

Although irrigation and drainage practices are helpful to farmers, they can cause problems. Removing too much water from a lake or reservoir can result in a water shortage. In addition, removing water from wetlands can damage habitats and wildlife. More than half of the wetlands in southern Ontario have been harmed by human activities. Wetlands have been drained to irrigate crops on farms, and filled for building projects. They are often polluted by runoff from nearby farms and businesses. Farms, land developers, and businesses near wetlands must balance their needs for water with the need to maintain the wetlands. They also need to protect wetlands from harmful runoff.

Soil Needs Space

The air spaces, or pores, in soil allow water and dissolved nutrients to pass through the soil. They also provide oxygen to plant roots and soil organisms. However, soil particles can be compacted, or squeezed together, by heavy equipment or by people walking or driving over the soil. Compacted soil has less air space. As a result, plant roots may not get the oxygen they need. Water will not drain properly through compacted soil. Over time, compacted soils become unable to support life.

TRY THIS WEIGHING IN ON SOIL COMPACTION

SKILLS HANDBOOK 3.B.2., 3.B.5.–3.B.8.

SKILLS: Controlling Variables, Performing, Observing, Analyzing, Evaluating

In this activity, you will observe how soil compaction affects soil health.

Equipment and Materials: 2 identical flowerpots with drainage holes; tray; stopwatch; measuring cup or graduated cylinder; potting soil; water

1. Add loose potting soil to both flowerpots until they are $\frac{3}{4}$ full.
2. In the first pot, *gently* push down on the soil surface to remove any large air spaces and even out the soil surface.
3. In the second pot, *firmly* push down to squeeze the soil into the bottom of the pot. Make sure to compact the entire soil surface.
4. Place the pots on the tray. Using a stopwatch, begin timing as you slowly pour 150 mL of water onto the soil surface in the first pot. Record the time it takes for *all* of the water to soak below the surface of the soil.
5. Repeat Step 4 for the second pot, containing the compacted soil.

A. How long did it take for the loose soil to soak up the water? How long did it take for the compacted soil to soak up the water? T/I

B. Based on your observations, do you think soil compaction affects the amount of water that gets to plant roots? Why or why not? T/I

C. How might soil compaction affect the growth of plant roots? Explain your reasoning. T/I

D. Gardeners usually use a small spade to break up the soil before adding a new plant to a garden. Why do you think they do this? T/I

Alternative Farming Practices

There are ways to reduce the negative environmental effects of farming. These include no-tillage farming, crop rotation, and crop selection.

No-tillage Farming

In no-tillage farming, farmers leave the ground undisturbed after the crop is harvested. They do not plough, or till, the remaining vegetation into the soil (**Figure 5**). This helps retain soil nutrients and improves soil quality. The remaining plant stalks protect the soil from erosion.

Figure 5 In no-tillage farming, the plant stalks are left on the field after the crop is harvested.

Crop Rotation

In crop rotation, farmers rotate or change the crops they plant on a certain area of land (**Figure 6**). Different crops use different amounts of nutrients. By rotating crops, farmers can reduce their use of fertilizers.

Figure 6 Soybeans, corn, and wheat are often grown in rotation in southern Ontario.

Crop Selection

Farmers can also choose to grow crops that are suited to local growing conditions. For example, farmers may choose to grow drought-resistant and heat-tolerant plants in an area with hot, dry summers. This would reduce or eliminate the need for irrigation and lessen the impact on water sources.

UNIT TASK

Bookmark

You can apply what you learned about soil health to the Unit Task described on page 160.

4.2 Wrap Up

- Healthy soil is vital to plant growth.
- The four main components of soil are minerals, organic matter, air, and water.
- Soil nutrient cycles are different in engineered ecosystems than they are in natural ecosystems.
- Lost nutrients can be replaced by applying synthetic or natural fertilizers.
- Fertilizers can cause problems for the environment.
- Farmers change the natural locations of water supplies when they add or remove water from their fields.
- Soil compaction can reduce the ability of soil to support life.
- Practices such as no-tillage farming, crop rotation, and crop selection reduce the impacts of agriculture on the soil health.

CHECK YOUR LEARNING

1. Identify the four main components of soil. Where does each of these components come from? K/U
2. List the kinds of organisms that live in soil. Explain how these organisms affect the soil. K/U
3. Many gardeners mix compost into their garden soil. Compost is decaying organic matter, such as the items shown in **Figure 7**. Why is compost a valuable addition to soil? A

Figure 7 Kitchen waste, such as vegetable and fruit food scraps, coffee grounds, and egg shells make good compost.

4. Plants use their leaves rather than their roots to take in carbon dioxide from the air. Yet, air is an important component of soil. Explain why. K/U
5. Draw a diagram that shows the path of nutrients in a farmland. Include fertilizers in your diagram. K/U C
6. Make a T-chart that shows the advantages and disadvantages of using fertilizers. K/U C
7. What is no-tillage farming and why would farmers use this practice? K/U
8. Do you consider soil to be a renewable or a non-renewable resource? Explain your answer. A
9. When a new home is built in a wooded area, heavy machinery must move between trees to bring building materials to the site. Within a few years, homeowners often find that some of the trees surrounding their new home are dead. What do you think caused the trees to die? A
10. How do soil nutrient cycles in engineered ecosystems and in natural ecosystems differ? K/U

AWESOME SCIENCE OSSLT

Black Gold

Scientists are excited about an ancient technique for enriching soil.

More than 1500 years ago, Indigenous peoples living in the rainforest developed a sustainable, low-tech agricultural practice. They added charcoal to their soil. This charcoal-rich soil is known today as "biochar."

These Indigenous peoples cleared forests by cutting and then burning the trees. They used the harvested wood as cooking fuel and to make charcoal fertilizer. The result was an extremely nutrient-rich, human-engineered black soil.

Adding charcoal to soil has a number of benefits. It allows the soil to hold more water, it helps hold minerals, and it enhances plant roots' ability to absorb nutrients. The benefits of charcoal on soil health are so dramatic that some soil scientists refer to the biochar as "black gold." Scientists describe biochar as having the ability to improve soil fertility. It may be better than compost, animal manure, and other fertilizers (**Figure 1**).

While useful as a fertilizer, biochar's greatest potential may be to store huge amounts of carbon. Charcoal is composed of almost pure carbon that was removed from the atmosphere in the form of carbon dioxide gas. Carbon dioxide in the atmosphere is currently believed to be largely responsible for recent climate change. Referring to the technique as "black gold agriculture," scientists say this revolutionary technique could provide a cheap method of reducing atmospheric carbon dioxide by simply trapping the carbon in soil (**Figure 2**). Unlike burning, the process of producing biochar does not produce carbon dioxide, so the carbon is not released into the atmosphere. Rather it is trapped in the charcoal and eventually in the soil where it will stay for hundreds or even thousands of years.

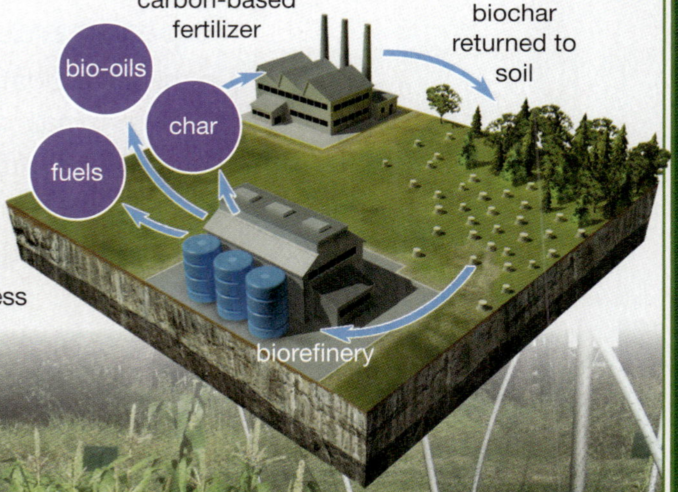

Figure 2 The biochar process

Figure 1 The corn plants on the left were grown in soil containing charcoal. The corn plants on the right were grown in soil without charcoal.

4.3 CONDUCT AN INVESTIGATION

Understanding Soil

SKILLS MENU

- Questioning
- Hypothesizing
- Predicting
- Planning
- Controlling Variables
- Performing
- Observing
- Analyzing
- Evaluating
- Communicating

Soils have different textures, based on their mineral components. Smaller mineral particles, such as clay, do not allow as much water to pass through. So soils with smaller particles have smaller spaces for air and water. Larger particles, such as sand, allow water to drain but do not hold as many nutrients. Soil scientists use simple tests to determine the amount of each type of particle in soil.

SKILLS HANDBOOK 3.B.2., 3.B.5.–3.B.9.

Testable Question

How does the composition of soil affect the percentage of seeds that will germinate?

Hypothesis/Prediction

Write a hypothesis for the Testable Question. Your hypothesis should include a prediction as well as a reason for your prediction, based on your knowledge of different types of soil and the needs of plants.

Experimental Design

In this investigation, you will determine what type of soil is best for seed germination. The independent variable in this investigation is the composition of the soil. The dependent variable is the percentage of seeds that germinate. Variables that should be controlled include water, temperature, light, size of pots, and spacing of seeds.

Equipment and Materials

eye protection

lab apron

gloves

clear plastic cups

permanent marker

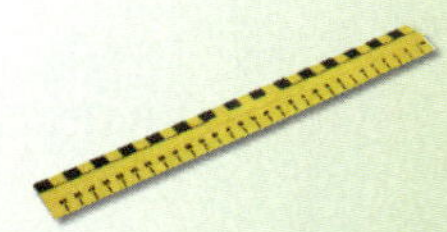
ruler

potting soil

water

sand

fine gravel

seeds of various types

Procedure

1. Prepare 9 cups: 3 with potting soil, 3 with a mixture of potting soil and sand, and 3 with a mixture of potting soil and gravel. Label the cups with a permanent marker.
2. Choose 3 types of seeds, and obtain 30 of each type.
3. Plant 10 of each type of seed in each type of soil, according to the directions on the seed package. Make sure you plant seeds of the same type with the same spacing and at the same depth in each cup. Label each cup with the type of seed you planted.
4. Water the newly planted seeds. Give each cup the same amount of water.
5. Make a table in which to record the number of seeds that germinate in each cup each day for 10 days.
6. Place the cups where each cup is the same temperature and gets the same amount of light.
7. Each day, check the seeds to see if any have germinated. Record the number of seeds that germinate in each cup. Remember to give each cup a little water each day, and to give each cup the same amount of water.

CAUTION

Hygiene

Wash your hands after handling soil.

Analyze and Evaluate

(a) In which cup(s) did the first seed(s) germinate? What type of soil was in the cups? T/I

(b) Which soil had the highest percentage of overall germination? T/I

(c) Did you notice any patterns in your data? Did seeds germinate better in one type of soil than in the others? Explain. T/I

Apply and Extend

(d) A family wants to have a garden, but is not sure their soil is good for growing vegetables. Based on your investigation, describe the "ideal" soil. C A

(e) What other factors might affect seed germination? How could you design an investigation to test these variables? T/I

DIG DEEPER

Learning Tip

Calculating Percentage

To determine the percentage of seeds that germinated in each type of soil, divide the number of seeds that germinated by the number of seeds planted, then multiply by 100. For example, if you planted 20 seeds in each type of soil, and 15 of them germinated, then you would figure out the percentage this way:

$15 \div 20 = 0.75$

$0.75 \times 100 = 75\ \%$

4.4 Pests and Poisons

Humans are in competition with many other organisms. Some of them eat the same foods we do. For instance, have you ever chased a rabbit out of your garden for eating the vegetables you've grown? Other plants compete with the plants we grow. For example, gardeners are constantly pulling weeds out of their flowerbeds. Still other organisms, such as mosquitoes and lice, actually feed on us! How do we control these organisms?

Changing Food Webs

Recall from Section 4.1 that a pest is a living thing that causes illness or harm to another living thing, or is simply a nuisance to humans. However, the term "pest" refers only to how humans see the relationship. It is important to remember that there are no pests in nature. There are simply producers, consumers, and decomposers.

In natural food webs, each organism has a specific role, and the relationships are complex. Producers support entire communities that may include hundreds or thousands of consumer species. In engineered ecosystems, such as agroecosystems, these relationships are absent or drastically changed. **Table 1** compares the food web in a grassland with the food web in a potato field.

Table 1 Comparing Food Webs

Type of organism	Natural grassland	Potato field
producers	50 to 100 species of wildflowers and other plants coexist	monoculture of potato plants; the few other plant species that occur are considered weeds
herbivores	numerous species of herbivorous insects, rodents, birds, and so on	a few species such as the Colorado potato beetle and the flea beetle; both are considered pests
carnivores	numerous species of carnivorous insects, spiders, snakes, foxes, hawks, and so on	very few carnivores, if any
food web	complex web with hundreds of species	food web effectively eliminated, leaving a single producer with some pest species

When farmers plant a monoculture, they often create the ideal environment for pests. A known pest in Canada is the European corn borer, shown in **Figure 1**. When a population of these insects finds a field of corn, they begin to feed and reproduce rapidly. In a natural food web, predators help to control the insect population. But on a farm, the natural food web that includes corn borers has been changed. There may be fewer of the corn borers' natural predators. As a result, the population of corn borers can grow, take over, and easily destroy the entire crop.

Figure 1 Young European corn borers (larvae) feed on leaves, creating small pinholes. They may even eat the corn kernels or carry diseases into the plant.

Controlling Pests

As long as there has been farming, people have used some form of pest control to keep their crops free of damage. After all, most people do not like to buy cabbage with holes in the leaves or apples with brown spots.

Pesticides

One of the most common ways to control or eliminate pests is to use a substance called a **pesticide** (**Figure 2**). There are different names for pesticides, depending on their target. Herbicides kill plants. Insecticides kill insects. Fungicides kill fungi.

▶ **pesticide:** a substance used to kill a pest

Compared to killing pests by hand, using pesticides requires less labour. Therefore, pesticides are considered to be an effective way to maintain a crop or protect a forest.

How Pesticides Work

There are many types of pesticide, and each one kills pests in a different way. For instance, diatomaceous (dy uh tahm AY shuhs) earth is a natural pesticide. It is made up of the fossil remains of tiny algae called diatoms. The substance is ground into a fine, abrasive powder. The powder scratches the waxy outer coating of small organisms, such as insects. As a result, the organisms dry out and soon die.

Other pesticides interfere with biological processes, such as photosynthesis. Still others kill pests by causing damage to vital organs of the body.

Figure 2 Many people use pesticides to kill pests. A number of pesticides are toxic to humans, requiring warnings such as this one.

How Pesticides Are Applied

Some pesticides are applied by contact, which means the pesticide must touch the pest. Other pesticides are indirectly applied, usually by spraying (**Figure 3**). For instance, herbicides are sprayed on the soil and are taken up through the roots of the weed. An insecticide can be sprayed on a plant and later be consumed by an insect.

Figure 3 Because most pesticides are toxic, the people applying them must wear protective gear.

Once they are applied, pesticides vary greatly in how long they remain active, or persist, in the environment. Some are persistent and keep working for many years. Others last only a few days before their chemical structure breaks down and they lose their effect.

Natural or Synthetic?

Just as fertilizers can be natural or synthetic, so can pesticides. Natural pesticides are obtained from sources found in nature. One example is the insecticide called pyrethrum (py REE thruhm). Pyrethrum is obtained from the flowers of certain chrysanthemum plants (**Figure 4**). However, most pesticides are chemical, or synthetic, which means they are manufactured by humans through chemical processes.

Figure 4 This chrysanthemum is known as the Dalmatian flower. Its flowers are the source of a natural pesticide called pyrethrum.

The pesticide obtained from chrysanthemum flowers is just as poisonous as a pesticide that has been manufactured. Pyrethrum will kill all insects it comes into contact with—pests or not. Generally, pesticides obtained from natural sources are less persistent than synthetic pesticides. In other words, they do not last as long in the environment, especially when rain or wind carries them away. Pyrethrum breaks down quickly in sunlight. It usually loses its effect within 24 h. On the other hand, the synthetic version of this pesticide can stay where it is applied for two months or longer. Still, today's synthetic pesticides are less persistent than those developed 30 years ago.

Alternatives to Pesticides

Organic farming is an alternative to pesticide use. This kind of farming does not use synthetic pesticides or synthetic fertilizers. Instead, organic farmers sometimes have to accept crop losses due to naturally occurring pests. However, their losses may be offset by the higher prices they get for their organic products. They also save money by not purchasing synthetic chemicals so the cost of production is lower.

DIG DEEPER
Career Link
To learn more about a career in organic farming,
GO TO NELSON SCIENCE

RESEARCH THIS MEET A PESTICIDE

SKILLS: Researching, Communicating, Evaluating

SKILLS HANDBOOK 4.A.4., 4.A.6., 4.A.7.

1. Use the library, the Internet, or other sources to find out which pesticides are used in Canada.
2. Select one of these pesticides, natural or synthetic. Learn more about it.
   - How is it used?
   - What pest(s) does it target?
   - How toxic is it?
   - What warnings are associated with its use?
   - Are there alternatives?

A. Share your findings with the class. You may want to use a poster, brochure, or oral presentation to communicate your findings. T/I C

B. As a class, decide on a way to organize your findings. Will you sort the pesticides into categories? Will you rank them from safest to most dangerous? Present your findings to another class. T/I C

GO TO NELSON SCIENCE

4.4 Wrap Up

- The term "pest" is only used to show a relationship between a type of organism and humans.
- One of the most common ways to control pests is to use pesticides.
- Pesticides cause physical harm or interfere with biological processes.
- Some pesticides are delivered by contact, while others are applied indirectly.
- Persistent synthetic pesticides are more likely to cause harm to the environment.
- Organic farming is an alternative to using pesticides.

UNIT TASK
Bookmark
You can apply what you learned about pesticides to the Unit Task described on page 160.

CHECK YOUR LEARNING

1. (a) Why are "pests" more likely to be found on a farm than in a natural ecosystem?
   (b) Explain why pests do not exist from an environmental point of view. K/U C
2. (a) Name three different types of pesticides.
   (b) What are their target pests? K/U
3. Do you think that natural pesticides are better than synthetic pesticides? Explain your reasoning. C A
4. Do you think that organic farming is a good alternative to using pesticides? Why or why not? C A

4.5 Problems with Pesticides

Although pesticides are helpful to farmers and other people, the use of pesticides has many drawbacks.

Pollution

When pesticides are sprayed, some of the poison never reaches the intended pest. Instead, winds carry it away to other areas. The pesticide may then cause air, soil, or water pollution. If a pesticide contaminates the groundwater, people might drink the contaminated water and get sick. Pesticides that pollute aquatic ecosystems can damage essential food webs.

Figure 1 DDT prevented the eggshells of peregrine falcons from developing properly. The resulting eggshells were too thin and broke. Many baby birds died.

Non-target Organisms

Pesticides can harm organisms that are not the intended target. For example, spraying pesticides at the wrong time of year may kill honeybees, which are vital for pollinating flowers and fruit crops. As a result, less fruit will be produced.

Bioaccumulation

Some pesticides build up in the bodies of individual organisms. The organism may not be able to break down the chemicals, or eliminate them with other wastes. As a result, the pesticide collects, or accumulates, in the tissues or organs. This is known as **bioaccumulation**. Bioaccumulation occurs when a pesticide cannot dissolve in water, but does dissolve in fats and oils. Animals use their fats and oils to store energy, build tissues, and in many other important processes.

▶ **bioaccumulation:** the concentration of a substance, such as a pesticide, in the body of an organism

Many species of birds, including bald eagles and peregrine falcons, declined in numbers because the pesticide DDT bioaccumulated in their bodies (**Figure 1**). DDT interfered with the female birds' ability to produce strong eggshells. The use of DDT was completely banned in Canada in 1985. Since then, populations of these once threatened birds have rebounded significantly. In 2009, Ontario banned the use of pesticides for cosmetic—or visual—reasons.

Bioamplification

When pesticides build up in the tissues of an organism, they can also affect other organisms. Remember, all individuals in an ecosystem are a part of a food chain. At high concentrations, a pesticide can harm or kill the organism. Pesticides become more concentrated as they move up the food chain. This is because the toxins stored in an organism at one trophic level are passed on to the organisms at the next trophic level. This process is called **bioamplification**.

▶ **bioamplification:** the increase in concentration of a substance, such as a pesticide, as it moves higher up the food chain or food web

Resistance to Pesticides

When a pesticide is used for a long period of time, some species may become resistant to it. The pesticide can no longer control the pest. Since the individuals that resist the pesticide survive longer, they can produce offspring. They pass their resistance on to their offspring. Over time, the population may become highly resistant to the pesticide. This means that the farmer needs to apply more pesticide or a stronger pesticide to control the pests.

4.5 Wrap Up

- Although pesticide use has its benefits, it can also cause harm.
- Pesticides can pollute the air, water, and soil.
- Pesticides can harm species that are not the intended target.
- Pesticides can build up in the bodies of individual organisms and be passed along food chains.

UNIT TASK
Bookmark

You can apply what you learned about the problems associated with pesticides to the Unit Task described on page 160.

CHECK YOUR LEARNING

1. Make a graphic organizer showing the benefits and drawbacks of using pesticides. K/U C
2. How can spraying pesticides on crops cause problems? K/U
3. What is meant by the term "non-target organism?" K/U
4. Why do certain pesticides build up in the body of an organism? K/U
5. If a pesticide is effective but is found to accumulate in organisms, what can people do to prevent harm to the environment? A
6. Explain the difference between bioaccumulation and bioamplification. K/U
7. DDT has some potential health and environmental problems but is still a valuable pesticide in eliminating insects that carry diseases such as malaria in some parts of the world. Do you think DDT should be banned everywhere? Explain your position. T/I A

Make Your Own Natural Pesticide

SKILLS MENU

- Questioning
- Hypothesizing
- Predicting
- Planning
- Controlling Variables
- Performing
- Observing
- Analyzing
- Evaluating
- Communicating

More and more people are using natural pesticides to control pests (**Figure 1**). This activity provides a recipe for a natural pesticide—garlic spray. The ingredients are items that many people have in their homes and are inexpensive to buy.

Figure 1 Houseflies are common household pests.

SKILLS HANDBOOK 3.B.

Purpose

To prepare and evaluate a natural insecticide made from common materials

Equipment and Materials

eye protection

lab apron

disposable gloves

mixing bowl

mixing spoon

spray bottle

1 L water

medium onion, chopped

garlic bulb, minced

cayenne pepper

liquid dish soap

Procedure

Part A

1. Develop criteria you can use to judge the effectiveness of your insecticide. T/I
2. Use your criteria to design a procedure to evaluate the insecticide. T/I

Part B

3. Put on your eye protection, lab apron, and gloves.
4. Place the minced garlic and chopped onion in the mixing bowl.
5. Add 1 tbsp cayenne pepper and 1 L water. Use the spoon to mix well.
6. Allow the mixture to sit undisturbed for one hour.
7. Add 1 tbsp liquid dish soap. The mixture is now ready for use (**Figure 1**).
8. Transfer the mixture to a spray bottle. The mixture can be stored in a refrigerator for up to one week. Always use gloves when spraying.
9. Test your insecticide. Be sure to get permission to use the pesticide before applying it. T/I

CAUTION

Protective Clothing

Always wear personal protection (safety goggles, gloves, and apron) when applying any pesticides, even natural ones.

Figure 1 Use a spray bottle to apply this natural pesticide.

Analyze and Evaluate

(a) How effective was your pesticide? Explain your thinking. T/I

(b) Report the results of your test to others. Explain how you made the insecticide, how you tested it, and what the outcome of the test was. T/I C

4.7 EXPLORE AN ISSUE CRITICALLY

The Ontario Pesticide Law

SKILLS MENU

- Defining the Issue
- Researching
- Identifying Alternatives
- Analyzing the Issue
- Defending a Decision
- Communicating
- Evaluating

Everyone enjoys a well-groomed lawn or city park. However, it should not take regular use of pesticides to maintain them. The government of Ontario says a more natural lawn is preferred. According to the Organic Landscape Alliance, chemical-dependent lawns are more likely to be affected by pests and diseases (**Figure 1**). However, a healthy, natural lawn is less likely to be damaged by pests (**Figure 2**). It is also less affected by drought, temperature extremes, and general wear and tear.

Figure 1 A chemical-dependent lawn is unnatural and unsafe.

Figure 2 A natural lawn is healthier and safer.

SKILLS HANDBOOK 4.C.

The Issue

On Earth Day 2009, a cosmetic pesticides ban went into effect for all of Ontario. This ban says that pesticides cannot be used for cosmetic purposes on lawns, gardens, parks, and schoolyards. Some pesticide use is still allowed for public health and safety reasons, such as stopping the spread of the West Nile virus.

Role: You are a community member of a small town. You may want to take on the role of a parent, teacher, business owner, or someone else who might have an interest in this issue. You will need to identify your role, and describe the point of view of that person. You have been invited to share your opinion on the pesticide ban at a town meeting. Government representatives will be at the meeting.

Audience: Your audience is made up of other community members and a group of representatives from the provincial government.

Goal

To persuade the community members and the representatives of your point of view

Gather Information

- Learn more about the cosmetic pesticides ban in Ontario.
- Learn about the effects of pesticides on humans.
- Identify the benefits of the ban as well as the problems it may create.

Identify Solutions

- From the point of view of your chosen role, describe how the ban will affect you personally and the community as a whole.
- Identify what you see as positive and negative aspects of the ban.
- Are there any changes that would make the ban more acceptable to you? Explain.

Make a Decision

Do you agree or disagree with the cosmetic pesticides ban? What changes do you think would make the ban better for the community as a whole?

Communicate

Present the information you gathered to the audience at the town meeting. Tell the group if you agree or disagree with the ban, and why. Propose and explain changes you think would make the ban more acceptable.

DIG DEEPER

Web Link

For more information on the Ontario pesticide law,

GO TO NELSON SCIENCE

4.8

City Life

Do you live in an urban, suburban, or rural area (**Figure 1**)? Urban areas, such as Toronto, are usually too crowded for people to grow their own food. Instead, their food has to be transported from outside of the city. Suburban communities, such as Orleans, Ontario, are located just outside of cities. People who live in suburbs often travel to the city for work and for cultural events. In a rural area, such as the area around Omemee, Ontario, buildings are often spread out. Natural ecosystems, such as wetlands and forests, are common.

Figure 1 More than half of the world's population lives in cities such as Toronto (a). Fewer people live in suburbs (b), and fewer still live in rural areas (c).

The Urban Ecosystem

Urbanization is the buildup of rural or natural lands into urban areas as more and more people move to existing cities or suburbs. Urbanization can create several problems. For one thing, it can produce "heat islands." These areas are warmer than nearby rural areas. The air temperature of a city with one million people can be 1 °C to 3 °C warmer than less-densely populated surrounding areas. Urban heat islands tend to decrease air quality by increasing the production of pollutants such as ozone. Water quality decreases as warmer water flows into nearby streams, stressing the ecosystems.

The differences between urban and natural areas are striking. Urban areas are often classified as "urban ecosystems," but an urban ecosystem is unlike a natural ecosystem. Natural terrestrial ecosystems have "living carpets" of grass, trees, and other plants. Urban ecosystems are built on a layer of concrete and asphalt. They do combine artificial and natural elements, but there are often many more artificial elements than natural ones.

DIG DEEPER

Reading Tip

Read Between the Lines

If you read a word or term that you are unfamiliar with seeing in print, say the word aloud and ask yourself if you have heard the word before or used it when speaking. Also, check the margin and the context of the sentence or paragraph to see if a definition is given.

In a large city, much of the natural vegetation has been replaced by human structures, such as buildings, roads, and sidewalks. There are not as many different types of plant life as in a natural ecosystem. Even if there are grasses and trees in parks, they are often monocultures of non-native plants. The cycles of water and nutrients that take place in natural ecosystems are replaced by underground pipes, sewers, and waste treatment plants.

The information in **Table 1** illustrates that humans cannot survive in an isolated urban environment. Unlike living things in a natural ecosystem, humans in urban areas rely on outside ecosystems.

Table 1 Comparison of Urban and Natural Ecosystems

Feature	Urban ecosystems	Natural ecosystems
surface features	• dominated by buildings, paved roads, and parking areas	• dominated by plants
plant species	• low density of plants • mostly non-native species such as grasses, trees, shrubs, and small flowering plants	• high diversity of native species
common non-human animal species	• low diversity and abundance • mostly non-native species such as dogs, cats, house mice, Norway rats, pigeons, starlings, and house sparrows	• high density and diversity of native species
water cycle	• surface water rapidly enters sewers and drainage systems • most of the surface cannot absorb water • human population consumes large quantity of water obtained from surface or groundwater supplies • waste water returned to surface water	• surface water in ponds, streams, rivers, and lakes • most rainfall absorbed at surface • water moves through soil and enters groundwater
flow of materials	• massive incoming shipments of materials to support human demand for consumer goods • human body wastes collected, treated, and disposed of in surrounding environment • household and commercial wastes disposed of in surrounding environment • usually minimal recycling and composting	• natural nutrient cycles occur throughout ecosystem
food webs	• dependent on importing food from outside agroecosystems	• natural nutrient cycles occur throughout ecosystem
sustainability	• cannot be sustained without large shipments of materials to and from other ecosystems	• independently sustainable

Greener Cities

The word "green" no longer refers to just a colour. This adjective is now used to describe anything that is environmentally sound. Cities have advantages that make "greening" them easier. For one thing, living close together allows people to save energy and space. When people travel shorter distances, and when they have access to large public transportation systems, they should drive their cars less.

A green city has planners and politicians who develop strategies to make it more sustainable. These planners work to reduce the **ecological footprint** of the city—the land and resources needed to sustain the city's human population.

▶ **ecological footprint:** a measure of the land area required to sustain a given population of humans

Case Study: Vancouver, British Columbia

Located between the mountains and the ocean, Vancouver is British Columbia's largest city (**Figure 2**). It is also a very green city. For one thing, Vancouver has about 1.6 million trees and is dedicated to taking good care of this "urban forest." It also has 200 parks and about 30 km of waterfront. The city draws 90 % of its energy from renewable sources. The main source of energy is hydroelectricity. Vancouver also plans to increase its use of renewable energy sources such as wind and solar energy.

Figure 2 Vancouver is a shining example of a "green city."

Vancouver also has something that few cities have—a 100-year plan for sustainability. The plan calls for the city to reduce its emissions of greenhouse gases by 80 % by 2050. There is also a program called One Day to encourage residents and businesses to reduce their use of energy.

Case Study: Iqaluit, Nunavut

Iqaluit, on Baffin Island, is the capital of Nunavut (**Figure 3**). It is Canada's newest and most northerly capital. It is also a fast-growing community.

Figure 3 Iqaluit is developing a long-term plan to become a world-class "green city."

Since Arctic areas are significantly affected by global warming, Iqaluit is developing a long-term plan for sustainability. A program called Save 10 educates individuals in ways they can reduce their personal energy use by 10 %. The planned Plateau Subdivision is a sustainable housing project where planners have aligned many of the roads with the most common winds so that winter snow is less likely to accumulate on them. This reduces the need to use fossil fuels and snow removal equipment, which can be very costly for a small city in a cold climate such as Iqaluit.

TRY THIS DO YOU SUFFER FROM NDD?

SKILLS: Planning, Analyzing, Evaluating

SKILLS HANDBOOK 3.B.4., 3.B.7., 3.B.8.

Have you heard of nature deficit disorder, or NDD? It is not a recognized medical condition. However, people use this term to describe a negative trend in the well-being of children. NDD results from not spending enough time in nature. Children with NDD are more likely to have behaviour and social problems.

1. Create a survey to find out how familiar your classmates are with the natural world. Your survey should include at least five questions. They can be yes-or-no, multiple-choice, or short-answer questions.

   A few sample questions are listed below:

   - When was the last time you spent time in a natural ecosystem, such as hiking in the forest, swimming at the beach, or canoeing on a river?
   - What is your favourite natural setting?
   - Name as many different kinds of amphibians, reptiles, fish, and birds as you can.

2. Give the survey to at least five students. Compile their answers and compare them with your own.

A. How knowledgeable about natural ecosystems are the people you surveyed—very knowledgeable, somewhat knowledgeable, or not very knowledgeable? T/I A

B. Make a list of activities people could do while spending time in a natural environment. Include some activities that are free or inexpensive. T/I A

C. How might spending more time in a natural ecosystem influence how a person values natural ecosystems? A

4.8 Wrap Up

- Urban areas are places in which homes and other buildings are crowded together, with little open space.
- People who live in urban areas rely on outside ecosystems for most of their resources.
- Living in urban ecosystems can benefit the environment because many people living close together can reduce consumption of energy, resources, and space.
- A city becomes more "green" as it reduces its impact on the environment.

UNIT TASK

Bookmark

You can apply what you learned about urban areas to the Unit Task described on page 160.

CHECK YOUR LEARNING

1. Make a graphic organizer to compare urban areas, suburban areas, and rural areas. K/U C
2. (a) What is a "heat island"?
   (b) Why is it a problem for urban areas? K/U
3. Compare the ecosystem you live in with a natural ecosystem. Is the ecosystem you live in sustainable? K/U A
4. (a) Explain the term "green city" in your own words.
   (b) Why is Vancouver called a green city? K/U C
5. List three actions, in addition to those discussed in this section, that cities could take to become more sustainable. A C
6. What is Iqaluit doing to make itself more sustainable? K/U

Chapter 4 KEY CONCEPTS SUMMARY

Some ecosystems are designed, created, and maintained by humans.

- Some humans spend little time in natural ecosystems.
- Humans can change natural ecosystems, but we cannot change our dependence on them.
- A farmland is an engineered ecosystem.

Agricultural practices disrupt natural biogeochemical cycles.

- On a farm, a limited number of species interact.
- Humans must manage the water and nutrient supply for farms.
- Farmers commonly apply fertilizers to replace soil nutrients.

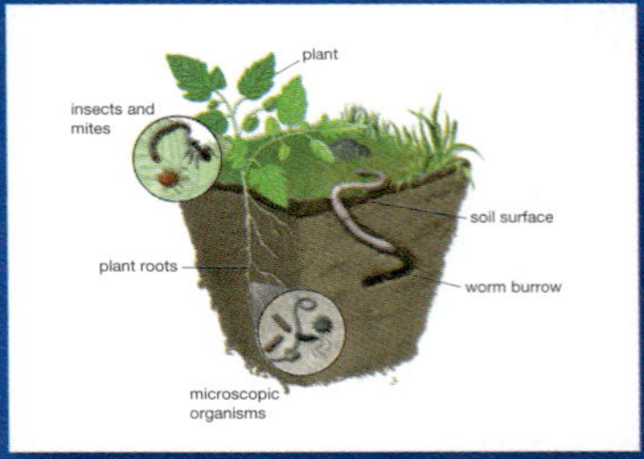

It is important to keep Earth's soil healthy.

- The four main parts of soil are minerals, organic matter, air, and water.
- Soil health depends on the amount and availability of nutrients.
- Fertilizers are beneficial but can create problems for the environment.

Pesticides are used to reduce crop losses due to insects, weeds, and other pests.

- There are no "pests" in nature. There are simply producers, consumers, and decomposers.
- Most pesticides are poisons that control or eliminate pests.

Pesticides have ecological costs.

- Pesticides can pollute the soil, air, groundwater, and aquatic ecosystems.
- Pesticides can harm organisms that are not the intended target.
- Some pesticides build up in the bodies of individual organisms and move up through the food chain.

Urban ecosystems depend on natural ecosystems for food and materials.

- Urban ecosystems combine artificial and natural elements.
- In a large city, most of the natural vegetation is replaced by human structures.

WHAT DO YOU

THINK NOW?

You thought about the following statements at the beginning of the chapter. You may have encountered these ideas in school, at home, or in the world around you. Consider them again and decide whether you agree or disagree with each statement.

1 Farming and other agricultural activities are similar to processes that take place naturally in ecosystems.
Agree/disagree?

2 Organic farming can benefit the environment.
Agree/disagree?

3 Pesticides are needed to control pests.
Agree/disagree?

4 Many of the plants and animals farmed in Ontario are native to this region.
Agree/disagree?

5 The actions of people who live in cities do not affect the surrounding natural environments.
Agree/disagree?

6 We can use as many resources as we need because we can always recycle or reuse them.
Agree/disagree?

How have your answers changed since then?
What new understanding do you have?

BIG Ideas

- Ecosystems consist of a variety of components, including, in many cases, humans.
- The sustainability of ecosystems depends on balanced interactions between their components.
- Human activity can affect the sustainability of aquatic and terrestrial ecosystems.

DIG DEEPER

Vocabulary

agroecosystem p. 128
pest p. 128
monoculture p. 128
fertilizer p. 131
natural fertilizers p. 131
synthetic fertilizers p. 131
pesticide p. 139
bioaccumulation p. 142
bioamplification p. 143
ecological footprint p. 150

CHAPTER 4

SELF-QUIZ

The following icons indicate the Achievement Chart category addressed by each question.

K/U Knowledge/Understanding
C Communication
T/I Thinking/Investigation
A Application

For each question, select the best answer from the four alternatives.

1. Which statement best describes an engineered ecosystem? (4.1) K/U
   (a) It is made entirely of abiotic parts.
   (b) It uses alternative energy sources.
   (c) It uses fertilizer and pesticides.
   (d) It is designed for a specific function.

2. What are the main components of soil? (4.2) K/U
   (a) rock, minerals, and humus
   (b) humus, minerals, water, and air
   (c) minerals, chemicals, and organic matter
   (d) space, air, water, and minerals

3. Which of the following does soil provide for plants? (4.2) K/U
   (a) oxygen, sunlight, and nutrients
   (b) water, physical support, and monocultures
   (c) physical support, nutrients, water, and oxygen
   (d) fertilizer, water, oxygen, and pesticides

4. What do farmers typically do to replace soil nutrients? (4.2) K/U
   (a) add fertilizers
   (b) use pesticides
   (c) plant monocultures
   (d) reroute water

5. The concentration of a pesticide in the body of an eagle is higher than the level of the same pesticide in the body of the salmon that it eats. What concept does this example illustrate? (4.5) K/U
   (a) bioamplification
   (b) bioaccumulation
   (c) biofeedback
   (d) pesticide resistance

6. Which word is most closely associated with the word "city"? (4.8) K/U
   (a) rural
   (b) green
   (c) urban
   (d) suburban

7. What type of ecosystem is a farmland? (4.1) K/U
   (a) an engineered ecosystem
   (b) an urban ecosystem
   (c) a green ecosystem
   (d) a native ecosystem

Indicate whether each of the statements is TRUE or FALSE. If you think the statement is false, rewrite it to make it true.

8. Agriculture is a major industry in Canada. (4.1) K/U

9. An urban ecosystem has many different types of grass, trees, and other plants that make a "living carpet." (4.8) K/U

10. Wetlands are in danger because human activities often add water to them. (4.2) K/U

11. If a pesticide stays in the environment for too long, it can pollute the groundwater. (4.5) K/U

12. Substances such as manure and seaweed are synthetic fertilizers. (4.2) K/U

Copy each of the following statements into your notebook. Fill in the blanks with a word or phrase that correctly completes the sentence.

13. In a(n) _______ ecosystem, the human population is too crowded for people to grow their own food. (4.8) K/U

14. An ecosystem designed for the production of crops is called a(n) _______. (4.1) K/U

15. A single crop growing in a large area is called a(n) _______. (4.1) K/U

16. The amount of _______ in soil determines how much of the soil's nutrients are available to living things. (4.2) K/U

17. Chemicals or other substances that we use to kill unwanted organisms are called _______. (4.4) K/U

Match each pesticide on the left with its target organism on the right.

18. (a) rodenticide (i) beetles
    (b) herbicide (ii) crabgrass
    (c) fungicide (iii) rats
    (d) insecticide (iv) mushrooms (4.4) K/U

Write a short answer to each of these questions.

19. Why are many engineered ecosystems unsustainable? (4.1) K/U

20. How does what you eat for breakfast matter to the health of your ecosystem? (4.1, 4.8) A

21. Why would a farmer choose to grow one crop instead of growing several different crops at once? (4.1) K/U

22. Why is it important for farmers to monitor the health of their soils? (4.2) K/U

23. Create a flow chart to show the flow of nutrients in a farmland ecosystem. (4.2) K/U C

24. What kind of weather should be avoided when fertilizing your lawn? Why? (4.2) A

25. How does soil become compacted? Why is this a problem? (4.2) K/U

26. Suppose your friend is allergic to bee stings. He sees bees as pests that threaten his health. Your friend wants to apply a pesticide to his yard so that the bees will not bother him. What advice would you give to your friend, and why? (4.4, 4.5) A C

27. Give an example that was not used in the text of a food chain in which pesticides could build up in the tissues of an organism and move up the food chain. (4.5) A

28. A lumber company plants a tree for every tree they cut down in a forest. Is that forest a natural or an engineered ecosystem? Explain. (4.1) K/U A

29. A homeowner digs up some of his lawn to plant a vegetable garden. Why does he add some earthworms to the soil before he plants any vegetables? (4.2) K/U A

30. A farmer has been growing the same crop in a field for several years. After the last harvest, she left the bare stalks in the field and did not plant anything for two years. Why did the farmer do this? (4.2) K/U

31. An apple orchard owner decides to switch to organic methods to grow his next crop of apples. Name one risk and one benefit of switching to organic methods. (4.4) K/U A

32. Describe four things people who live in a suburban area can do to make their area "more green." (4.8) C A

33. Explain the ways in which people who live in an urban ecosystem with public transportation are benefiting the environment more than a suburban ecosystem in which everyone must drive everywhere. (4.8) A

34. Use examples to describe how the nutrient cycle in the soil of an engineered ecosystem is different from the nutrient cycle in a natural ecosystem. (4.2) K/U C

Self-Quiz

To do an online self-quiz, GO TO NELSON SCIENCE

CHAPTER 4

REVIEW

The following icons indicate the Achievement Chart category addressed by each question.

K/U Knowledge/Understanding T/I Thinking/Investigation
C Communication A Application

What Do You Remember?

1. Which is a disadvantage of using fertilizers? (4.2) K/U
   (a) They can build up in the tissues of organisms.
   (b) They remove nutrients from an ecosystem.
   (c) They can cause algal blooms.
   (d) They may kill species that are not the target.
2. Do most Canadians rely on native plants and animals for their food? Explain, using examples. (4.1) K/U
3. What is the difference between a synthetic fertilizer and a natural fertilizer? (4.2, 4.4) K/U
4. Why do some pesticides build up in the tissues of organisms while others do not? (4.5) K/U
5. List five actions that a city could take to become "greener," or more sustainable. (4.8) K/U
6. Give two examples of native species grown as agricultural products in Canada. Give two examples of non-native species grown as agricultural products in Canada. (4.1) K/U
7. Why does a field planted with only one type of crop often need more fertilizer than a field planted with a variety of crops? (4.1) K/U
8. What is humus and what does it do for plants growing in healthy soil? (4.2) K/U
9. What trait of natural fertilizers makes it necessary to reapply them more often than synthetic fertilizers? (4.2) K/U
10. What three elements are the most important nutrients found in soil for plants? (4.2) K/U
11. How can too much fertilizer used on a farm lead to a lack of oxygen in the waters of nearby ponds or lakes? (4.2) K/U
12. Why can pests like corn borers do more damage to a field of corn on a farm than they would to corn growing in the wild? (4.4) K/U
13. Explain how the statement "Organic farmers do not use fertilizers or pesticides" is incomplete. How can you make it complete and correct? (4.5) K/U

What Do You Understand?

14. Do you have to be physically located in an ecosystem to make use of its services? Explain. (4.1, 4.8) K/U
15. Use a chart to compare and contrast the natural ecosystem and the agroecosystem shown in **Figure 1**. (4.1) K/U C

Figure 1

16. How might soil compaction influence nutrient and water cycles in a farmland ecosystem? (4.1, 4.2) K/U

17. The term "pest" is never used to refer to natural ecosystems. Why? (4.1, 4.4) K/U

18. Herbicides are commonly used in agriculture and in residential areas. Herbicides kill weeds. What characteristics do you think make weeds good at competing with other organisms? (4.1, 4.5) K/U

Solve a Problem

19. When a pesticide is used for a long period of time, some pests develop resistance to the chemical. As a result, the pesticide is no longer able to control the pest. **Figure 2** shows how many species of insects and mites became resistant to pesticides over a period of 80 years. (4.5) T/I C
    (a) Describe the trend that is shown for insects and mites.
    (b) Explain reasons for this trend.

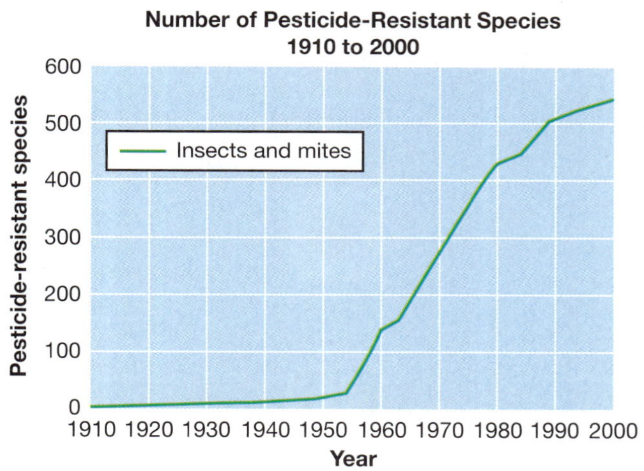

Figure 2

Create and Evaluate

20. You just purchased a farm, with six different fields, and you would like to grow corn, soybeans, and wheat. You are interested in using crop rotation to reduce the amount of pesticides and fertilizers you will need. You find the following short article while researching crop rotation.

The Basics of Crop Rotation

Farmers may want to rotate between soybeans, corn, and wheat. During a soybean year, the nitrogen in a field's soil increases. This reduces the need to add nitrogen fertilizer the following year for corn. In the third year of the rotation, wheat is planted. Because wheat is not planted in rows, it crowds out weeds and requires less herbicide.

    (a) Make a three-year plan for your farm. For each year, make a sketch of your farm showing what will be planted in each of the six fields.
    (b) Write a caption for each sketch that briefly explains the reasoning behind your choices. (4.2) T/I C A

Reflect on Your Learning

21. What aspects of human-engineered ecosystems do you think are the least sustainable? Explain. K/U C

22. What choices do you think are most important in making communities more sustainable and environmentally friendly? Explain your reasoning. A

Web Connections

23. Garbage, or solid waste, disposal is a very visible example of how humans can change the natural environment. Humans produce solid waste when they make and use products. We generate more waste when we no longer value the product and do not reuse or recycle it. T/I A
    (a) What methods of solid waste disposal are used in Ontario?
    (b) What are the advantages and disadvantages of each method?
    (c) How does each method influence the ecosystem?

GO TO NELSON SCIENCE

Sustainable Ecosystems and Human Activity

CHAPTER 2

Understanding Ecosystems

KEY CONCEPTS

Energy passes through ecosystems, but matter cycles within ecosystems.

Photosynthesis and respiration are processes that work together.

Aquatic and terrestrial ecosystems are strongly influenced by their abiotic environment.

Human activities disturb the natural cycles of nature.

CHAPTER 3

Natural Ecosystems and Stewardship

KEY CONCEPTS

Natural ecosystems are valuable to humans.

Ecosystems are balanced but can change over time.

Biodiversity describes the variety and amount of life in an ecosystem.

Human activities can affect and threaten the sustainability of natural ecosystems.

Pollution of water, land, and air causes health and economic problems.

Plant and animal resources should be used in a sustainable way.

CHAPTER 4

Ecosystems by Design

KEY CONCEPTS

Some ecosystems are designed, created, and maintained by humans.

Agricultural practices disrupt natural biogeochemical cycles.

It is important to keep Earth's soil healthy.

Pesticides are used to reduce crop losses due to insects, weeds, and other pests.

Pesticides have ecological costs.

Urban ecosystems depend on natural ecosystems for food and materials.

MAKE A SUMMARY

Forests are an important natural resource in Canada. Suppose you are in charge of managing one of Canada's natural forests. Your job is to make sure that all of the activities that take place in this forest are sustainable. You have a list of requests from various groups wanting to conduct activities in the forest you manage. Here is the list:

A. A local company wants to harvest berries and sell them in local markets.
B. A group of developers wants to build a ski resort on the mountainside.
C. A photography club wants to drain a wetland area so they can build an observation platform.
D. A hiking club wants to cut down trees to clear a path for hiking.
E. A mountain-biking club wants to widen existing bike trails by removing brush and other plants.
F. Some biology students want to install insect traps to identify insects in the forest.
G. An Aboriginal community is asking for exclusive rights to harvest specific plants.
H. A hunting club wants to build cabins for its members in an area of the forest that contains rare plants.
I. A fishing club wants to stock a lake with a new fish species so that its members can catch more fish.

Questions

1. For each request, decide whether to approve or reject it. Provide your reasoning for every decision. Explain why the activity is or is not sustainable. T/I A
2. For each activity that you determined to be unsustainable, brainstorm ways to make the activity sustainable. A
3. For each activity that you determined to be sustainable, provide two ways to make sure the activity remains sustainable. A
4. Prepare a report to the Canadian Forest Services that summarizes your results. C

CAREER LINKS

List the careers mentioned in this unit. Choose two of the careers that interest you, or choose two other careers that relate to Sustainable Ecosystems and Human Activity. For each of these careers, research the following information:

- educational requirements (secondary and post-secondary)
- skill/personality/aptitude requirements
- potential employers
- salary
- duties/responsibilities

Assemble the information you have discovered into a poster. Your poster should compare and contrast your two chosen careers and explain how they connect to sustainable ecosystems and human activity.

Figure 1 What skills are needed to be a fisher?

GO TO NELSON SCIENCE

What Are They Talking About?

SKILLS MENU

- Defining the Issue
- Researching
- Identifying Alternatives
- Analyzing the Issue
- Defending a Decision
- Communicating
- Evaluating

"Blocked Rivers Threaten Livelihood of Brazilian Tribes"
"Bankrupt Textile Mill Leaves Its Poison Behind"
"Norway Whale Catches Fall to Lowest in a Decade"
Believe it or not, these headlines are real. They all came from recent news stories about the environment.

The Issue

Every day, we hear so much about our changing environment. Global warming, the loss of biodiversity, disappearing habitats—the news media seem to bombard us with stories about these and other issues (**Figure 1** and **Figure 2**). How can we make sense of it all? How can we communicate our understanding to others?

Figure 1 Invasive species are a hot news topic. The sea lampreys attached to this trout are a species that has invaded the Great Lakes.

Goal

In this Unit Task, you will monitor a variety of media for stories about environments and ecosystems. The stories can be about local, regional, or global issues. You will use these media sources to better understand current environmental concerns. Then you will present what you learn from these sources to others in a creative format.

Figure 2 Many news stories are about pesticides. Dandelions are agricultural pests, and many farmers use herbicides to control them.

Gather Information

During the time you study this unit, you will collect stories about the environment by

- reading local, regional, and national newspapers or magazines
- searching the Internet for "late-breaking" news
- watching news programs and environmental documentaries
- listening to radio news programs and talk shows

Select 10 stories that you find most interesting or important. As you listen to, watch, or read the stories, take careful notes. The following template will help you collect your information. Copy this template into your notebook. Complete a separate one each time you come across an environmental news story.

Environmental News Story

Today's date: ____________

Media type: ☐ Newspaper ☐ Magazine
☐ Television ☐ Radio
☐ Internet ☐ Other: ____________

Specific source and date (name of newspaper or magazine, Internet site and URL, television show and channel, radio show and station, etc.):

____

Headline or title: ______________________________

Notes:

____

Identify Solutions

Review your notes. Select six to eight stories of the ten you collected that you think are the most interesting or important. Make sure that your selections represent at least three types of media sources.

Make a Decision

For each story you selected, prepare a summary card that contains the following information:

- title of the article
- the main points or features of the article
- the social effects and challenges related to this issue
- any further questions you may have about the issue
- your ideas for solution(s) to any problems presented in the article

Communicate

Choose the article that was most interesting to you and produce your own report about the issue. Using a format of your choice, such as a video, a news report, a graphic novel, or a poster, present and discuss the issue in more detail. Include information about what is currently being done to minimize human impact on the situation. Share your final product with your classmates.

CHECKLIST

Assessment

Your completed Performance Task will be assessed according to the following criteria:

KNOWLEDGE/ UNDERSTANDING

- Gather 10 different news stories from at least three different media types.
- Complete an information chart for each story.

THINKING/ INVESTIGATION

- Select six to eight stories that are most interesting or important to you.
- Answer the summary questions for your selected stories.

COMMUNICATION

- Share a report on the issue, in a format of your choice, with your class.

APPLICATION

- Complete one summary card for each story that you selected.

SELF-QUIZ

The following icons indicate the Achievement Chart category addressed by each question.
K/U Knowledge/Understanding T/I Thinking/Investigation
C Communication A Application

For each question, select the best answer from the four alternatives.

1. Which is a terrestrial habitat? (2.1) K/U
   (a) coral reef
   (b) forest
   (c) pond
   (d) river
2. Which is an example of a population? (2.2) K/U
   (a) all of the organisms in a lake
   (b) all of the plants in a garden
   (c) all of the seals on a beach
   (d) all biting insects
3. Which is an example of ecological succession? (3.2) K/U
   (a) After trees lose their leaves in winter, the leaves grow back in spring.
   (b) After a tsunami seriously damages a village, the people begin to move back in.
   (c) After a sea star loses an arm, another one grows in its place.
   (d) After a volcanic eruption, lichens start to appear on bare rock.
4. Which is the best definition of biodiversity? (3.3) K/U
   (a) the variety of species in an ecosystem
   (b) the total number of animals that live in a biome
   (c) all of the individuals in a population
   (d) all of the populations in a community
5. Which is a chemical that helps plants grow? (4.2) K/U
   (a) biotic factor
   (b) fertilizer
   (c) pesticide
   (d) soil
6. What is the main source of energy for organisms on Earth? (2.5) K/U
   (a) energy from photosynthesis
   (b) radiant energy from the Sun
   (c) electric energy
   (d) ecological energy
7. A farmer plants cabbage in an area of his farm with poor drainage. Cabbage can grow well in wet soil. This practice is an example of (4.2) K/U
   (a) crop rotation
   (b) crop irrigation
   (c) crop selection
   (d) no-tillage farming

Indicate whether each of the statements is TRUE or FALSE. If you think the statement is false, rewrite it to make it true.

8. Predators are never found on the first trophic level of a food chain. (2.6) K/U
9. The water cycle cannot be affected by human activities. (2.7) K/U
10. Invasive species are species that are native to the ecosystem they are invading. (3.7) K/U
11. The persistence of a pesticide refers to how long it stays in the environment. (4.5) K/U
12. Acid rain is caused by the buildup of pollutants such as sulfur dioxide in Earth's atmosphere. (3.8) K/U
13. A monoculture is an example of an engineered ecosystem. (4.1) K/U
14. A wetland is an example of a biome. (2.9) K/U
15. The spheres of Earth include the atmosphere, the ecosphere, the hydrosphere, and the lithosphere. (2.1) K/U
16. The ecosystems with the most biodiversity are found near the equator. (3.3) K/U

Copy each of the following statements into your notebook. Fill in the blanks with a word or phrase that correctly completes the sentence.

17. The non-living components of an ecosystem are called ________. (2.2) K/U

18. The conversion of the Sun's energy into chemical energy that is then stored in food is called ________. (2.5) K/U

19. When a species no longer exists on Earth, it is said to be ________. (3.5) K/U

20. Soil contains small bits of organic matter called ________. (4.2) K/U

21. A factor that determines how many organisms can live in an ecosystem is called a(n) ________ factor. (2.8) K/U

22. In a food chain, the organisms that eat producers are called ________. (2.5) K/U

Match each organism on the left with its most appropriate biome on the right.

23. (a) maple tree (i) tundra
(b) pine tree (ii) boreal forest
(c) polar bear (iii) grassland
(d) bison (iv) temperate deciduous forest (2.9) K/U

Write a short answer to each of these questions.

24. What ecological benefits do wetlands provide? (2.9) K/U

25. Why is it important for soil to have pores? (4.2) K/U

26. How do pesticides work? (4.4) K/U

27. What is an ecotourist? (3.1) K/U

28. Draw and label a simple diagram that shows the flow of water in the water cycle. (2.7) K/U C

29. What does it mean to say that an ecosystem is in a state of equilibrium? (3.2) K/U

30. A local lake has been contaminated with pesticides for years. Which would you expect to have a higher concentration of pesticides in its tissues: a top predator such as a pike or a primary consumer such as a sunfish? Explain your reasoning. (4.5) K/U

31. A researcher is studying frogs in a tropical rainforest. (2.9) T/I
(a) What are two biotic factors the researcher may encounter during the study?
(b) What are two abiotic factors the researcher may encounter during the study?
(c) What can the researcher bring along to help in adapting to these factors?

32. Algal blooms can decrease the amount of oxygen available for other organisms in a lake. Why is this dangerous for the other organisms? (4.2) K/U

33. What impact can changing the number and type of organisms found in soil have on a soil's ability to help plants grow? (4.2) K/U

34. Humans and bears are both omnivores. What does this mean we have in common? (2.6) K/U

35. Explain the advantages of a pesticide that breaks down quickly in sunlight. What, if any, are the disadvantages? (4.4) K/U

36. How can people living in an urban environment reduce the amount of acid rain produced? (3.8) K/U

37. Why would a city prevent a chemical plant from being built in its watershed? (2.9) A

UNIT B

REVIEW

The following icons indicate the Achievement Chart category addressed by each question.

K/U Knowledge/Understanding | T/I Thinking/Investigation | C Communication | A Application

What Do You Remember?

1. What is a threatened species? (3.5) K/U
   (a) a species that is extirpated
   (b) a species that no longer exists
   (c) a species that no longer lives in a particular region
   (d) a species that may become endangered

2. Which ecosystem has the greatest biodiversity? (3.3) K/U
   (a) a desert
   (b) a rainforest
   (c) a pond
   (d) a grassland

3. What is the solid part of Earth's surface called? (2.1) K/U
   (a) hydrosphere
   (b) biosphere
   (c) atmosphere
   (d) lithosphere

4. Oxygen is required by almost all living things for the process of (2.5) K/U
   (a) cellular respiration
   (b) transpiration
   (c) photosynthesis
   (d) condensation

5. What is one disadvantage of pesticides? (4.5) K/U
   (a) They can build up in the tissues of an organism.
   (b) They interfere with nutrient cycles.
   (c) They can cause algal blooms.
   (d) They require too much water.

6. What type of organism is on the first level of every food chain? Why is it at this level? (2.5) K/U

7. Which choice best describes Canada's boreal forest? (2.9) K/U
   (a) permafrost
   (b) deciduous trees
   (c) hot, dry summers
   (d) conifers

8. Name two abiotic factors that are important for both terrestrial ecosystems and aquatic ecosystems. (2.2, 2.8) K/U

9. In mutualism, how do organisms of different species interact? (2.8) K/U

10. A classmate insists that extinctions can only be caused by human activities. Is he correct? Explain. (3.5) K/U C

11. In what type of ecosystem do most humans tend to spend their time? Give an example. (4.1) K/U

12. How can fertilizers harm nearby bodies of water? (4.2) K/U

13. Why do we use the term "pest" to refer to some plants, animals, and insects? (2.6, 4.1, 4.4) K/U

14. What may happen when an invasive species enters an ecosystem? (3.7) K/U

15. What level in most food chains is held by carnivores? Explain. (2.6) K/U C

16. What is continuously rearranged and used in a biogeochemical cycle? Give examples. (2.7) K/U

17. A forest fire destroys the biotic factors in a local ecosystem. Name and describe the process that will occur in the ecosystem during the years following the fire. (3.2) K/U

18. Where would you be likely to find a monoculture? Give a reason why. (4.1) K/U

19. Give one example of a natural fertilizer and one benefit farmers gain by using it. (4.2) K/U

20. What is an ecosystem's carrying capacity? (2.8) K/U

21. Describe a watershed. (2.9) K/U

22. What makes an estuary unique? (2.9) K/U

23. How do producers use light energy? (2.5) K/U

24. What effect does acid rain have on ecosystems? (3.8) K/U

25. In the 1700s farmers used to bury fish on top of seeds that they planted. Explain why they would do this. (4.2) A

26. What can form when chemicals released by automobiles and industry mix with water vapour? (3.8) K/U

27. What term is used to describe all of the water on Earth in all of its forms? (2.1) K/U

28. What gas is released by a cell as a result of respiration? (2.5) K/U

29. Classify each of the following as a cultural benefit, an economic benefit, or other benefit of an ecosystem. (3.1) A
    (a) stocking a local lake with young fish
    (b) clean water
    (c) allowing lumber cutting in wilderness area
    (d) preserving a wetland
    (e) reintroducing wolves into a forest
    (f) removing an old dam from a river
    (g) feeding stranded bison during a severe winter

What Do You Understand?

30. How can a fire be beneficial to a forest ecosystem? (3.2) K/U

31. Compare what happens to a forest after clear-cutting to what occurs after shelterwood cutting. (3.10) C

32. Explain how a tropical rainforest is a sustainable natural ecosystem that can survive without regular human maintenance. (2.3) K/U

33. If a top predator such as a wolf is removed from a food web, what can happen to the population of grasses and rabbits in the area? (2.6) K/U

34. How can an ecosystem provide economic benefits for an area? Give a specific example. (3.1) K/U

35. What terrestrial biomes are found in Canada? List one abiotic and one biotic factor of each of these biomes. (2.9) K/U

36. List the most important abiotic factors in an aquatic ecosystem. (2.2, 2.4, 2.8) K/U

37. Describe two ways of controlling an invasive species. (3.7) K/U

38. Which biome has more biodiversity—a grassland or a deciduous forest? Explain. (2.9, 3.3) K/U

39. Are abiotic factors more important than biotic factors in an ecosystem? Explain your answer. (2.2) C A

40. Compare urban ecosystems and agroecosystems in terms of
    (a) biodiversity
    (b) nutrient cycling (2.7, 3.3, 4.1, 4.8) C A

Solve a Problem

41. **Figure 1** shows a food web in an aquatic ecosystem. List three different food chains in this ecosystem. Tell whether each member of each food chain is a carnivore, omnivore, herbivore, or producer. (2.6) K/U

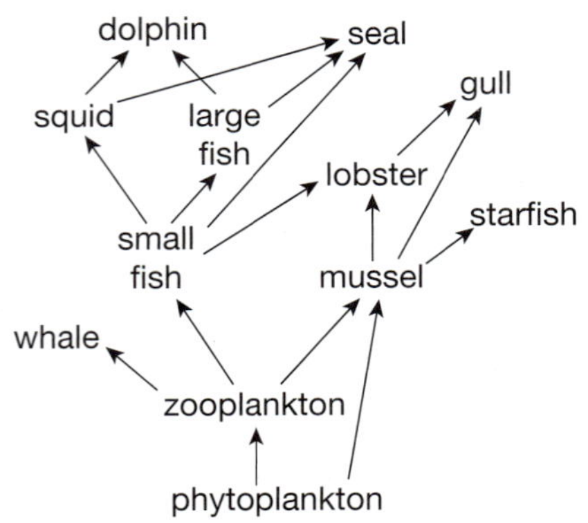

Figure 1

42. Habitat loss is not as rapid in Ontario as it once was. Yet, the human population in Ontario continues to rise. Suggest a possible reason for this observation. (3.6) A

43. A large farming corporation wants to import a type of wasp from overseas. This wasp feeds on caterpillars that eat the leaves of the crops they grow. Why would a government advisory panel suggest that permission should not be granted? (3.7) K/U A

44. Suppose you are member of the neighbourhood stewardship group. Describe at least two types of projects your group can sponsor to spread the idea of stewardship. (3.10) A

45. How would planting trees on city streets affect the carbon cycle in an area? (2.7) C

46. It has been proposed that certain areas be allowed to spray DDT on bed netting to help control mosquitoes that spread malaria. Describe the benefits and risks of this plan. (4.5) A

47. A large number of species are considered to be at risk of extinction in Ontario. How could you find out more about the plants and animals that are at risk in your region of the province? (3.5) T/I

48. Construct a Venn diagram, like the one in **Figure 2**, to compare the characteristics of engineered ecosystems and natural ecosystems. (4.1) K/U C

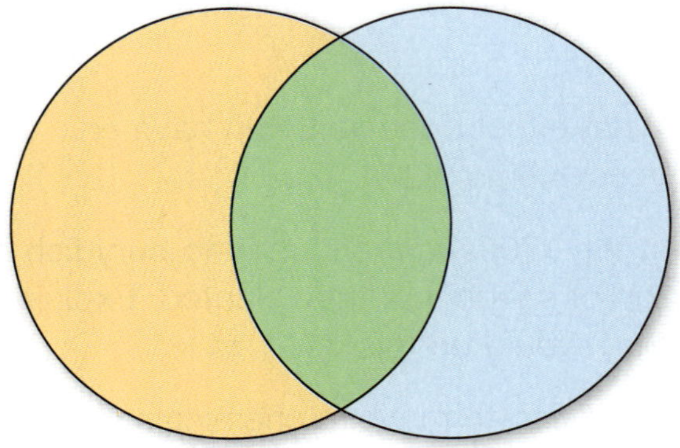

Figure 2

49. Engineered ecosystems, such as golf courses, require regular maintenance by humans. Many golf courses use synthetic fertilizers and pesticides to keep the fairways and greens in good condition. Your local golf course wants to have a "greener" operation. Write a letter to the manager and to the clients with five suggestions for making the golf course more environmentally friendly. (4.1) C A

50. Your neighbour runs an organic farm and uses the manure from her animals as fertilizer. She claims that since manure is a natural fertilizer it has no negative impact on the environment, so she can add as much as she likes to her crops. What would you say to your neighbour to explain that her claim is not necessarily accurate? (4.2) C A

Create and Evaluate

51. Jet skiing in Canada's lakes and coastal waters is an activity that many people enjoy during summer (**Figure 3**). (3.1) C A
    (a) Brainstorm the ecological benefits of jet skiing.
    (b) Brainstorm the ecological costs of jet skiing.
    (c) Should jet skiing be allowed in Canada's waters? State your opinion and justify it using your answers from (a) and (b).

Figure 3

52. Organic farming is becoming more and more popular in Ontario and elsewhere. If organic produce is grown without the addition of expensive pesticides and fertilizers, why do these products cost more than the non-organic variety? (4.2) A

Reflect on Your Learning

53. How might the information that you learned in this unit influence your behaviour regarding a more sustainable lifestyle? C

54. Which concept in this unit did you find difficult to understand? Why? C

55. What are you most curious about as a result of reading this unit? How could you go about learning more about that topic? T/I C

Web Connections

56. At one time, a large population of walruses lived along the coast of Eastern Canada. Today, this species is extirpated. T/I C A
    (a) What happened to this population of walruses?
    (b) Do you think walruses should be reintroduced into Eastern Canada? Why or why not?
    (c) How might the Gulf of St. Lawrence and Maritime provinces be different if the walruses were still there today?

57. Select one ecological issue discussed in this unit. Find out what Canadian organizations might be involved in this issue. Look for one government agency, one non-governmental organization, and one environmental group that includes high school students. T/I C A
    (a) What are the organizations?
    (b) What actions or projects have they been involved with?
    (c) Have their actions benefited the environment? If so, how?

58. Use a map to identify an aquatic ecosystem in your area. Find out more about this ecosystem. T/I C A
    (a) Describe the biotic and abiotic characteristics of this ecosystem.
    (b) Identify any environmental problems associated with this ecosystem. Briefly describe these problems and any attempts at solving them.

59. Research the Norfolk County Alternative Land Use Services Pilot Project. T/I A
    (a) What is the purpose of this project?
    (b) What are farmers doing to help protect ecosystems?
    (c) What are some examples of projects completed through this program?

GO TO NELSON SCIENCE

UNIT C Exploring Matter

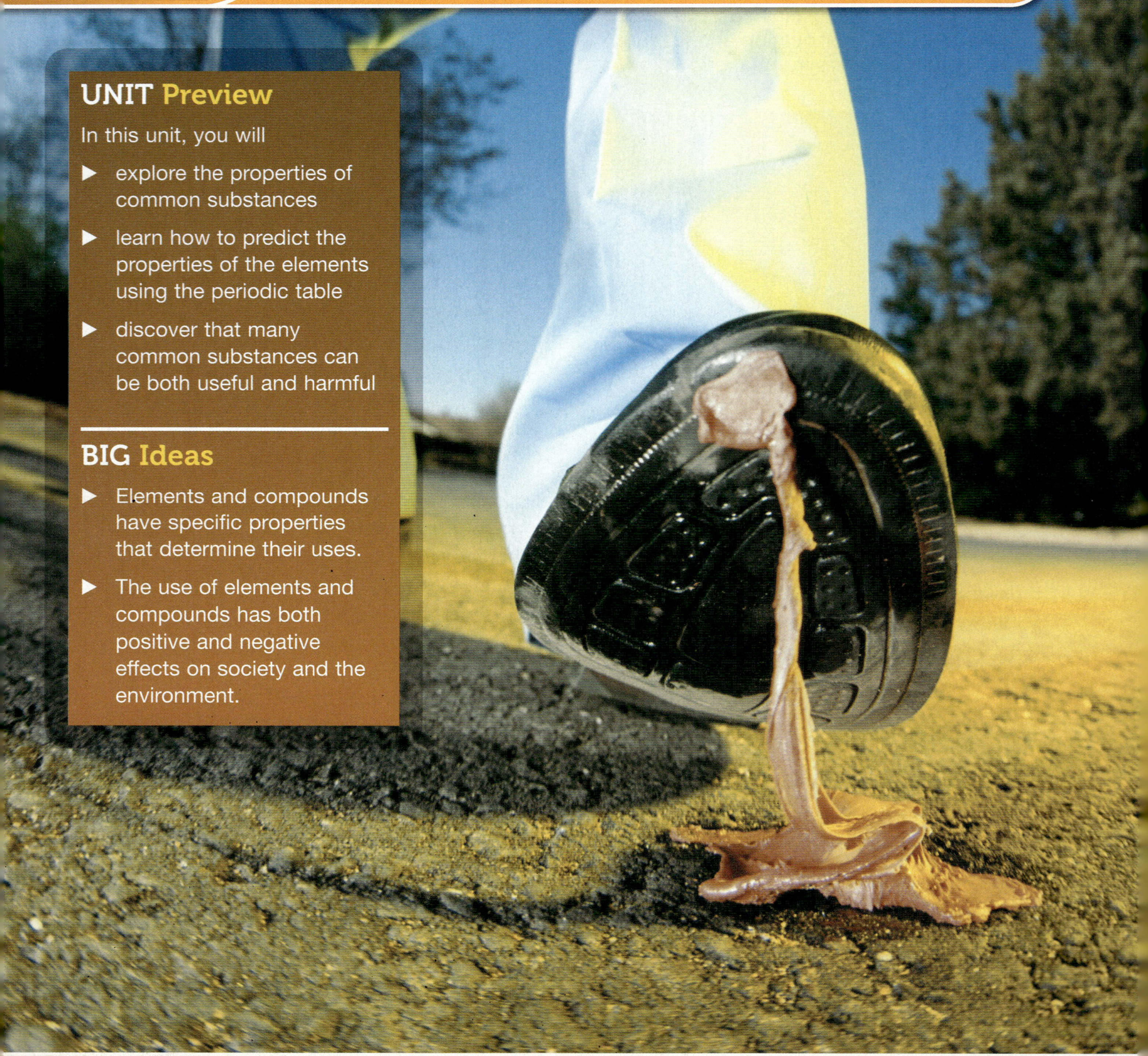

UNIT Preview

In this unit, you will

- explore the properties of common substances
- learn how to predict the properties of the elements using the periodic table
- discover that many common substances can be both useful and harmful

BIG Ideas

- Elements and compounds have specific properties that determine their uses.
- The use of elements and compounds has both positive and negative effects on society and the environment.

Focus on STSE

The Best Bubble Gum

What makes a chewing gum so good? Is it the taste, the texture, or the size of the bubble you can blow? People have been chewing gum for thousands of years. Stone Age people chewed birch bark tar, which forms when the bark is heated. Native peoples of North America chewed the sap of spruce trees.

Today, the chewy, rubbery substance in gum no longer comes from natural sources. Instead, it is a synthetic material made from the chemicals found in oil. The scientists who develop these synthetic "gum bases" design them for the type of gum that they want to make. For example, the gum bases used for regular chewing gum have different properties than those used in bubble gum, which needs to be more elastic.

But, there are more substances in gum than just the rubbery stuff. Look at the ingredients on the package. Carnauba wax, for example, is an ingredient in many types of chewing gum. It comes from tree leaves, and it gives the gum a dry, shiny coating. Carnauba wax is also found in car wax, shoe polish, and dental floss! Shellac, which comes from a type of insect, is used to give wood furniture a polished finish. It is also used to give many types of chewing gum a shiny appearance.

Substances are added for flavouring. For example, sorbic acid and malic acid give gum a sour apple flavour. Chocolate-flavoured chewing gum cannot be made because the cocoa butter in chocolate makes the gum too soft to be chewed.

Think/Pair/Share

1. List the properties of the ideal chewing gum. Then, compare lists with a partner. K/U
2. Think of sweeteners in chewing gum. Are there drawbacks, or problems, associated with these sweeteners? A

Exploring Matter

CHAPTER 5

Properties of Matter

A substance's properties determine its uses.

CHAPTER 6

Elements and the Periodic Table

Many properties of the elements can be predicted using the periodic table.

CHAPTER 7

Chemical Compounds

Chemical compounds play important roles at home and at work.

Unit Task Preview

Cleaning "Green"

In this Unit Task, you will use what you learned about the properties of substances to develop a recipe for a homemade, environmentally friendly cleaner. Then, you will conduct a test to compare the effectiveness of your cleaner with store-bought cleaners. Finally, you will communicate the results of your test with others.

UNIT TASK

Assessment

You will be assessed on:

- using your understanding of the properties of substances to design and produce a "green" cleaner
- designing and carrying out a test to evaluate the effectiveness of the product in comparison to store-bought products
- communicating the effectiveness, safety, and environmental effects of your product

UNIT TASK

Bookmark

The Unit Task is described in detail on page 276. As you work through this unit, look for this bookmark and see how the section relates to the Unit Task.

Getting Started: Useful Concepts and Skills

Concepts Review

1. Which states of water are shown in **Figure 1**? K/U

Figure 1

2. Classify each of the following as either a pure substance or a mixture: K/U
   (a) hot chocolate
   (b) water
   (c) oxygen

3. (a) Describe the differences between a solid and a liquid.
   (b) Describe what the particles in the liquid are doing as the liquid evaporates. K/U

4. Liquid flowing into the stream in **Figure 2** comes from an abandoned farm. What possible environmental effects could this have? Give reasons for your answer. T/I A

Figure 2

Skills Review

5. (a) Name the measuring devices shown in **Figure 3**.
   (b) Which device measures mass? Which device measures volume? K/U

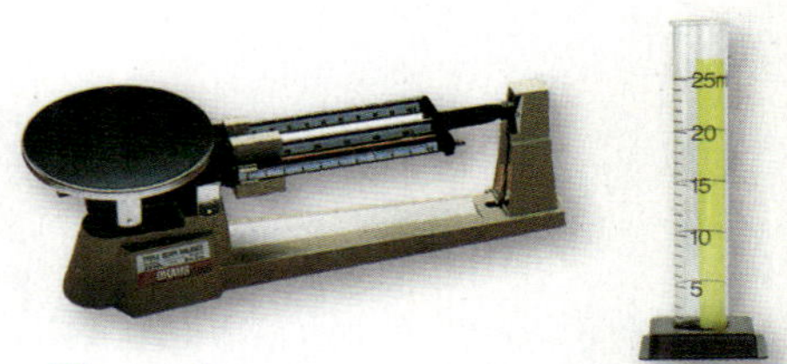

Figure 3

6. What safety precautions is the scientist in **Figure 4** taking? K/U

Figure 4

7. Give reasons for the following safety procedures. K/U
   (a) Wash hands after handling chemicals.
   (b) Close containers of unused chemicals promptly after use.
   (c) Read safety labels of all chemicals you use.
   (d) Check with your teacher before you dispose of any chemicals in the sink.
   (e) Never pour unused chemicals back into the stock bottle.

CHAPTER 5

Properties of Matter

KEY QUESTION

How are our lives affected, both positively and negatively, by the substances found in everyday products?

KEY CONCEPTS

- A substance can be identified by its physical or chemical properties.
- Understanding the physical and chemical properties of a substance helps us identify its uses.
- The use of substances can have both positive and negative effects in our daily lives.
- Changes in substances can be either physical or chemical.

This rotating drill bit is embedded with diamonds and is used to drill into Earth's crust beneath the Pacific Ocean.

This drill tip is embedded with tiny diamond chips.

Back Forward Stop Refresh Address

HOME ABOUT NEWS SCIENCE

THE WORLD OF GEMSTONES

Figure 1 England's crown jewels use a variety of jewels, including sapphires and rubies.

Figure 2 Amethyst is made up of the mineral quartz. Scientists are not sure if the colour is due to iron or manganese.

Figure 3 The Hope Diamond gets its colour from the presence of boron.

Strange as it may seem, ruby and sapphire gemstones are made up of the same mineral. So, why is a ruby bright red, while a sapphire is blue? Rubies and sapphires are both made of the mineral corundum, which is colourless in its pure form. However, these gems are brightly coloured because they contain metal impurities. **Figure 1** shows the many-coloured gemstones in England's crown jewels. The large ruby on the crown is red because it contains small amounts of chromium. Sapphires are blue because they contain iron.

Other gemstones also get their colour from metal impurities. Amethyst, for example, is mined in Ontario's Thunder Bay region (**Figure 2**). Amethyst, which is made up of the mineral quartz, gets its violet colour from the presence of iron. Emeralds, which are made of the mineral beryl, are green because they contain small amounts of chromium. The same metal impurity can sometimes produce different coloured gemstones, depending on the mineral that make up the stones. For example, chromium colours emeralds green, and rubies red.

Surprisingly, not all diamonds are colourless. In fact, they come in various colours including yellow, orange, olive, and even black. Perhaps the most famous coloured diamond is the Hope Diamond. This large gemstone, shown in **Figure 3**, is a brilliant blue. A diamond of this size and colour is extremely rare and valuable. The Hope Diamond is worth well over $100 million! It is currently located at the Smithsonian's Natural History Museum in Washington, D.C.

WHAT DO YOU THINK?

Many of the ideas you will explore in this chapter are ideas that you have already encountered. You may have encountered these ideas in school, at home, or in the world around you. Some of the statements are not true. Consider each statement and decide if you agree or disagree with it.

1 Some chemicals can be both useful and harmful to us.
Agree/disagree?

2 In a physical change, only the appearance of the substance changes.
Agree/disagree?

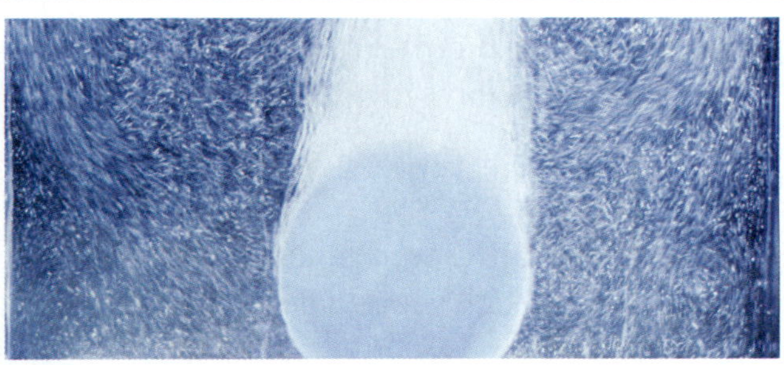

3 Only scientists make use of chemical changes.
Agree/disagree?

4 Gold is valuable because people like the way it looks.
Agree/disagree?

5 An edible substance such as salt does not pose any harm to the environment.
Agree/disagree?

6 The properties of a material depend on what it is made of.
Agree/disagree?

FOCUS ON WRITING

Writing a Summary

Writing a summary can help you to remember what you read. You can also use a summary as a study guide. A summary is a condensed or shorter version of the text. It only contains the main ideas. Personal opinions are not included. Before writing a summary make sure you read and reread the text so you are familiar with the content. Here are some tips on how to write a summary. This example is from Section 5.4.

> **DIG DEEPER**
> **Writing Tip**
> As you work through the chapter, look for tips like this. They will help you develop literacy strategies.

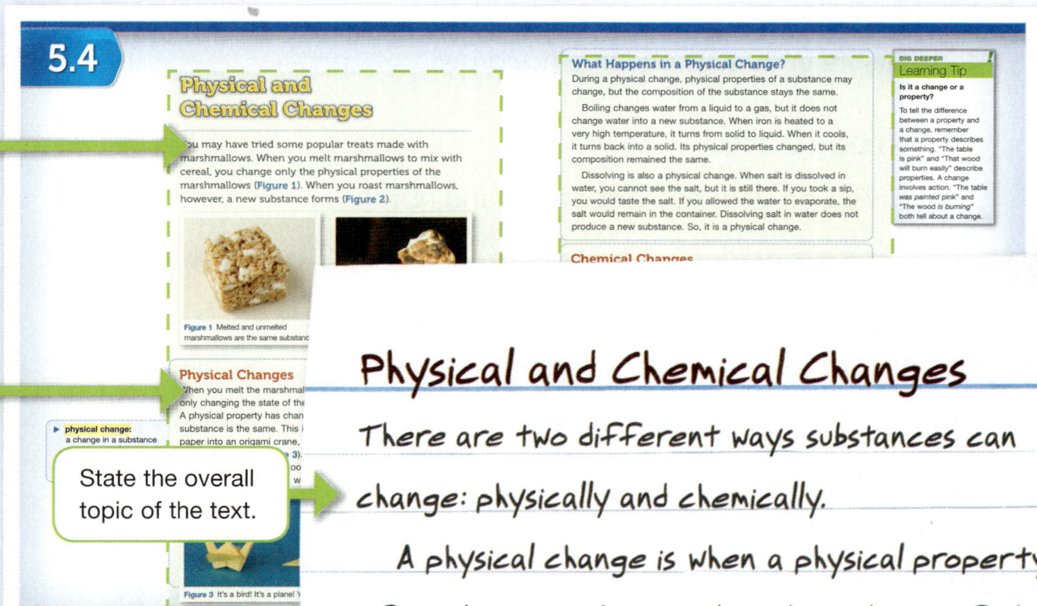
5.4

Physical and Chemical Changes

…ou may have tried some popular treats made with marshmallows. When you melt marshmallows to mix with cereal, you change only the physical properties of the marshmallows (Figure 1). When you roast marshmallows, however, a new substance forms (Figure 2).

Figure 1 Melted and unmelted marshmallows are the same substanc…

Physical Changes

…hen you melt the marshmal… only changing the state of the… A physical property has chan… substance is the same. This i… paper into an origami crane, …

physical change: a change in a substance …

Figure 3 It's a bird! It's a plane! …

What Happens in a Physical Change?

During a physical change, physical properties of a substance may change, but the composition of the substance stays the same.

Boiling changes water from a liquid to a gas, but it does not change water into a new substance. When iron is heated to a very high temperature, it turns from solid to liquid. When it cools, it turns back into a solid. Its physical properties changed, but its composition remained the same.

Dissolving is also a physical change. When salt is dissolved in water, you cannot see the salt, but it is still there. If you took a sip, you would taste the salt. If you allowed the water to evaporate, the salt would remain in the container. Dissolving salt in water does not produce a new substance. So, it is a physical change.

Chemical Changes

DIG DEEPER
Learning Tip
Is it a change or a property?
To tell the difference between a property and a change, remember that a property describes something. "The table is pink" and "That wood will burn easily" describe properties. A change involves action. "The table was *painted* pink" and "The wood *is burning*" both tell about a change.

Physical and Chemical Changes

There are two different ways substances can change: physically and chemically.

A physical change is when a physical property of a substance changes, but the makeup of the substance does not change. The substance may have a different state, shape, or temperature, but the substance is the same.

When a substance changes chemically, a different substance is formed. In a chemical change, the original substance changes into a new substance.

The main difference between the two types of changes is that a physical change does not result in a new substance, but a chemical change does.

> State the overall topic of the text.

> Organize ideas and information in the same order as in the original text.

> Notice that examples, descriptions, and explanations that were in the original text above have been left out.

> Write one clear sentence for each major point.

> Write a closing sentence that connects the ideas together or restates the big idea from the text.

5.1 PERFORM AN ACTIVITY

Grow Your Own Gemstone

SKILLS MENU

- Questioning
- Hypothesizing
- Predicting
- Planning
- Controlling Variables
- Performing
- Observing
- Analyzing
- Evaluating
- Communicating

Gemstones are crystals. The different colours of these crystals are due to substances that are present in the crystal as it grows. For example, adding potassium chromium sulfate will cause an alum crystal to change colour. In this activity, you will grow your own gemstone.

CAUTION
Safety
Report any broken glassware to your teacher.

SKILLS HANDBOOK 1.A., 3.B.5.–3.B.7.

Purpose

To influence and observe the growth of crystals

Procedure

1. Put on your eye protection and apron.
2. Use a graduated cylinder to measure 120 mL of warm water.
3. Pour the water into a clean glass container.

Equipment and Materials

eye protection

apron

warm water

graduated cylinder

2 clean glass containers

2 to 3 tablespoons of alum

spoon or scoopula

paper towel

wooden stick

piece of thread

food colouring or potassium chromium sulfate

4. Add the alum, a little at a time. Stir the mixture to dissolve the alum after each addition. Do this until it stops dissolving. Cover the container with a paper towel. Wait one day.
5. Observe the mixture. Some crystals will have formed. Pour the alum solution from the first container into the second container, leaving any solid crystals behind. Describe the crystal(s).
6. Choose the largest crystal to be your "seed" crystal to grow a bigger crystal. Set aside the first container until Step 8.
7. Add about 2 mL of food colouring or a few crystals of potassium chromium sulfate to the alum solution in the second container.
8. Tie one end of the thread to your seed crystal. Tie the other end to a wooden stick and place it across the top of the second container containing the alum solution. Adjust the length of the thread so the crystal is suspended in the mixture. Cover the container with a paper towel.
9. Observe your crystal each day for up to two weeks. Record your observations in a table such as **Table 1** below.

Table 1 Observations

Day	Size	Shape	Colour
1			
2			

CAUTION

Hygiene

Wash your hands after handling the solution. Clean the containers according to your teacher's directions. Clean up any spills immediately.

Analyze and Evaluate

(a) Describe the shape, size, and colour of your crystal. T/I C

(b) What do you think would happen if you used more than one seed crystal? T/I

(c) Why do you think it was important to cover the container? T/I

Apply and Extend

(d) Do you think any of the crystals that are found in nature formed in solutions? Use the Internet to explore this possibility. T/I A

5.2 Physical Properties: What We Can Observe

Figure 1 Kevlar vests have saved the lives of police officers and their canine helpers.

The police officer in **Figure 1** is wearing a bulletproof Kevlar vest. These bulletproof vests serve much the same purpose as the armour worn by knights in medieval times. Why don't we use medieval armour today? What properties do you think Kevlar has that make it more useful in modern times?

What Are Physical Properties?

All substances have properties that can be used to identify them. A **physical property** is a characteristic of a substance that can be observed or measured without forming a new substance. *Qualitative* physical properties are those you can observe with your senses, such as physical state, odour, colour, texture, and lustre (shiny appearance). *Quantitative* physical properties are those that you measure and give a number value, such as an object's mass, volume, and temperature.

▶ **physical property:** a characteristic of a substance that you can observe or measure without forming a new substance

Brittleness and hardness are other physical properties. Brittleness is the breakability, or flexibility, of a substance. Glass is brittle, while modelling clay is flexible. Hardness is the relative ability to be scratched by another substance. Steel is very hard, while candle wax is soft and easily scratched or dented. All of these physical properties can be determined without forming a new substance.

Useful Physical Properties

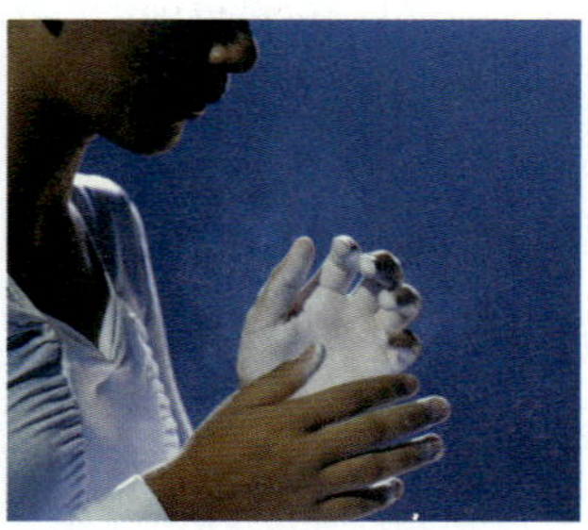

Figure 2 Chalk is useful because of its physical properties.

Like the Kevlar used in bulletproof vests, many substances are useful because they have specific physical properties. You probably have used chalk on cement or on a blackboard. Chalk is brittle, so it breaks easily and can be ground into a powder. It is hard enough to hold but soft enough to leave a mark on a surface.

In addition, chalk is not soluble in water, so it will not dissolve when it gets wet. The **solubility** of a substance describes how well it dissolves in another substance. Chalk is also very good at absorbing moisture. This property is very useful to gymnasts, rock climbers, and people who want to keep their hands dry and free of perspiration while handling equipment (**Figure 2**).

▶ **solubility:** the ability to dissolve in another substance

TRY THIS CLOSE-UP OF A RUNNING SHOE

SKILLS HANDBOOK 3.B.

SKILLS: Observing, Analyzing, Evaluating, Communicating

How do you choose a new running shoe? Are comfort and support more important than breathability and weight? How important are style and colour? In this activity, you will examine the physical properties of the materials that make up a running shoe. These properties determine the function of these materials in the shoe.

Equipment and Materials: running shoe

1. Make a four-column table, like **Table 1** below. Then, complete the first three columns.

Table 1 Parts of a Running Shoe

Shoe part	Material used	Physical properties	Function
laces	polyester	flexible	

2. With a partner, complete the Function column by listing the function of each shoe part you listed. For example, you might complete the "laces" row in the table by writing "keeps shoe on" in the Function column.

A. Which of the physical properties you listed are most useful in a running shoe? T/I A

B. Which materials used in the shoe might present a problem to the environment when the shoes are thrown away? Explain your thinking. A

5.2 Wrap Up

- A physical property is a characteristic of a substance that can be observed without forming a new substance.
- Examples of physical properties include physical state, colour, texture, solubility, brittleness, hardness, shape, temperature, mass, and volume.
- Qualitative physical properties are those you can observe with your senses, while quantitative physical properties are those that you measure.

UNIT TASK

Bookmark

You can apply what you learned about physical properties to the Unit Task described on page 276.

CHECK YOUR LEARNING

1. What are three physical properties of chocolate? K/U
2. Uncooked spaghetti noodles are hard, brittle, and measure 25.5 cm. K/U
   (a) Which of these properties are qualitative?
   (b) Which of these properties are quantitative?
3. (a) What properties of chalk make it useful to a gymnast?
   (b) What properties of chalk make it useful to a teacher? K/U A
4. (a) What are some of the physical properties of sandpaper? T/I
   (b) Which of these physical properties make sandpaper useful? A
5. (a) What are some uses of duct tape?
   (b) What physical properties of duct tape make it useful? T/I A
6. What physical properties of copper make it useful in a variety of applications?

5.3 Chemical Properties: How Substances Change

You may have seen someone mix baking soda and vinegar. What happens? The mixture fizzes and bubbles. Yet, this fizzing has nothing to do with the physical properties of baking soda or vinegar.

What Is a Chemical Property?

A **chemical property** describes a substance's ability to undergo changes and produce one or more new substances. The process that the substance goes through is called a chemical change. When a match burns, a chemical change occurs because the matchstick is coated with a substance that burns easily. This ability to burn, or its **combustibility**, is a chemical property of the match.

- **chemical property:** a description of what a substances does when it reacts to produce new substances
- **combustibility:** the ability of a substance to burn

Useful Chemical Properties

The usefulness of many substances is determined by their chemical properties. Many substances, such as wood and coal, are useful because they burn easily. Other substances are useful because they do *not* burn easily. These substances, such as liquid and powder fire retardants, are used to make fire-resistant building materials or flame-resistant clothing (**Figure 1**). Firefighters have to understand the properties of combustible materials so that they can put out fires quickly and safely.

Figure 1 Protective suits help keep firefighters safe.

DIG DEEPER

Career Link

For more information about a career in fire safety, fire prevention, or firefighting,

GO TO NELSON SCIENCE

DIG DEEPER

Writing Tip

Restating the Main Idea

When you restate the author's main idea in your summary, you say it in words that you understand. For example, if the text says "the usefulness of many substances is determined by their chemical properties," you might restate this idea by saying "chemical properties of a substance determine what you can use it for."

Metals are used to make jewellery, but only some metals are used for body piercing (**Figure 2**). Metals that can be safely used for a piercing have certain chemical properties. Stainless steel, titanium, and some types of gold can be used in body piercing. These metals do not react easily with substances in the body. Silver and copper are not good choices, because they react with substances in the body, on the skin, and in the air.

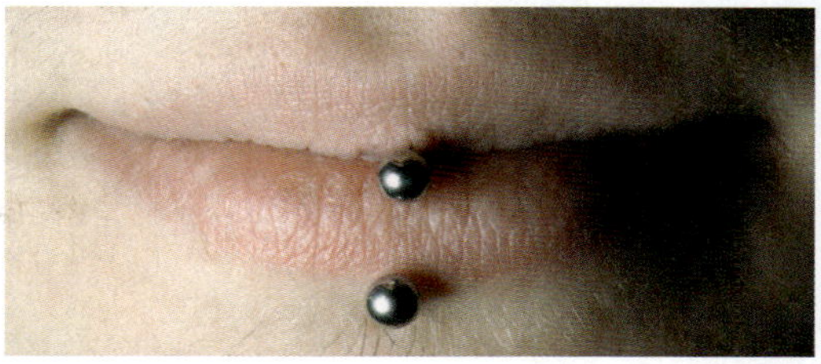

Figure 2 Metals used for body piercing must not react easily with substances in the body.

TRY THIS POLISHING PENNIES

SKILLS HANDBOOK 3.B.

SKILLS: Performing, Observing, Analyzing

Over time, shiny copper pennies become dull and dark. That is because the copper reacts with oxygen in the air to form a new substance—copper oxide. You can remove the copper oxide with a mixture of vinegar and salt.

Equipment and Materials: several dull pennies; $\frac{1}{4}$ cup white vinegar; 1 tsp salt, spoon; plastic or glass bowl

1. Mix the vinegar and salt in the bowl.
2. Dip pennies in the mixture using a spoon.
3. Wait a few minutes, and then remove the pennies.

A. What did you observe when you removed the copper pennies from the salt–vinegar mixture? T/I

5.3 Wrap Up

- A chemical property describes what a substance does when it reacts to produce new substances. Combustibility and reactivity are two examples.
- Many products are useful because of the chemical properties of the substances they contain.

UNIT TASK Bookmark

You can apply what you learned about chemical properties to the Unit Task described on page 276.

CHECK YOUR LEARNING

1. Classify each description as a physical or chemical property: K/U
   (a) Wood burns.
   (b) Clay is brown.
   (c) Steel wool rusts.
   (d) A paper clip is 3 cm long.
2. Use an example to describe how the combustibility of a substance can be both useful and dangerous. K/U A
3. (a) What chemical properties of nail polish remover makes this product useful? K/U
   (b) What chemical property makes nail polish remover dangerous? K/U
4. Surgeons sometimes use metal plates and screws to hold broken bones together. What chemical properties do you think the metal should have? K/U A

5.4

Physical and Chemical Changes

You may have tried some popular treats made with marshmallows. When you melt marshmallows to mix with cereal, you change only the physical properties of the marshmallows (**Figure 1**). When you roast marshmallows, however, a new substance forms (**Figure 2**).

Figure 1 Melted and unmelted marshmallows are the same substance.

Figure 2 Roasting marshmallows causes a new substance to form.

Physical Changes

When you melt the marshmallows for the bars in **Figure 1**, you are only changing the state of the marshmallows from a solid to a liquid. A physical property has changed, but the composition of the substance is the same. This is called a **physical change**. If you fold paper into an origami crane, the paper has changed shape, but it is still paper (**Figure 3**). If you saw wooden planks to build a wooden tree house, the wood is in a new shape. But, it is still the same substance—it is still wood.

▶ **physical change:** a change in a substance that does not involve the formation of a new substance

Figure 3 It's a bird! It's a plane! Yes, but it is still paper.

What Happens in a Physical Change?

During a physical change, physical properties of a substance may change, but the composition of the substance stays the same.

Boiling changes water from a liquid to a gas, but it does not change water into a new substance. When iron is heated to a very high temperature, it turns from solid to liquid. When it cools, it turns back into a solid. Its physical properties changed, but its composition remained the same.

Dissolving is also a physical change. When salt is dissolved in water, you cannot see the salt, but it is still there. If you took a sip, you would taste the salt. If you allowed the water to evaporate, the salt would remain in the container. Dissolving salt in water does not produce a new substance. So, it is a physical change.

DIG DEEPER

Learning Tip

Is it a change or a property?

To tell the difference between a property and a change, remember that a property describes something. "The table is pink" and "That wood will burn easily" describe properties. A change involves action. "The table *was painted* pink" and "The wood *is burning*" both tell about a change.

Chemical Changes

A change that results in the formation of new substances is called a **chemical change**. For example, roasting a marshmallow on a campfire produces a brown, flaky substance on the surface of the marshmallow (**Figure 2**). This substance is clearly different from the gooey centre of the marshmallow.

chemical change: a change in which one or more new substances is formed

You can tell a chemical change has taken place because a new substance formed. For example, some metal jewellery might leave a green stain on your skin. The green stuff is a new substance produced when the metal reacts with sweat.

Permanent hair colouring also involves chemical changes (**Figure 4**). For example, a bleaching chemical such as hydrogen peroxide are first used to change the natural pigment in your hair into new substances that have no colour. Hair colour technicians use their understanding of chemical changes every day to produce the exact colour their client wants.

DIG DEEPER

Career Link

For more information about a career as a hair colour technician,

GO TO NELSON SCIENCE

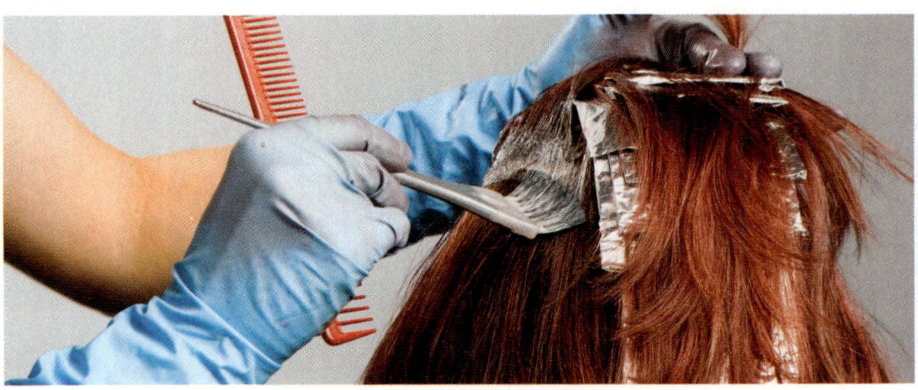

Figure 4 The hair dye is changing chemicals inside each strand of hair.

DIG DEEPER

Writing Tip

Identify the Purpose

When writing the second sentence of a summary, consider three common purposes of science texts: to inform, to describe, to explain. What is the purpose of the section "Chemical Change Clues?"

Chemical Change Clues

The four observations below are hints that a chemical change may have occurred. One way to find out exactly what new substances have formed is to conduct tests in a lab.

1. Energy is produced: Changes in temperature or light are signs that a chemical change has taken place (**Figure 5**).
2. Change of colour: A new substance has formed that has a different colour than the original substance (**Figure 6**).

Figure 5 When something is burning and giving off light and thermal energy, a chemical change is taking place.

Figure 6 Colourless egg whites turning white during cooking is evidence of a chemical change.

3. Change of odour: A new substance has formed that has a detectable odour (**Figure 7**). In scientific language, all smells are called odours, whether they are pleasant or unpleasant.
4. Bubbles of a gas are produced that are not caused by heating (**Figure 8**).

Figure 7 The new products formed from chemically changed garbage are not always fragrant.

Figure 8 As the antacid tablet dissolves, its chemicals react with water to form bubbles of carbon dioxide gas.

5. A new solid called a **precipitate** forms when two liquids are mixed (**Figure 9**). For example, adding vinegar to milk causes milk to form curdles. These curdles are a new, solid substance.

▶ **precipitate:** a solid that forms when two liquids are mixed

Figure 9 Precipitates can be many different colours. Some are white, like the curdles in milk, while others may be blue, orange, yellow, green, or black.

5.4 Wrap Up

- A physical change does not involve forming a new substance. Cutting wood, boiling water, and dissolving salt in water are examples of physical changes.
- A chemical change results in the formation of a new substance. Burning a candle and cooking eggs are examples of chemical changes.
- Evidence of a chemical change includes colour change, odour change, gas production, precipitate production, and production of light or thermal energy.

CHECK YOUR LEARNING

1. Classify each as a physical or chemical change: K/U
   (a) folding a paper airplane
   (b) water freezing on a windshield
   (c) an car rusting
   (d) dissolving flavour crystals in water
   (e) making a campfire
2. Explain your thinking for each of your answers in question 1. K/U C
3. A candle burns for 15 min, and some wax drips and collects at the base of the candle. The candle becomes shorter. K/U
   (a) What changes were physical changes? Explain.
   (b) Was there any evidence of a chemical change? Explain.
4. What evidence suggests that these changes are chemical changes? K/U C
   (a) Bubbles form when baking soda is mixed with lemon juice.
   (b) Cookies baking in the oven give off a delicious aroma.
   (c) A match is struck and ignites.
   (d) Bleach turns a pink towel white.
5. When a frozen pizza is placed in the oven, the cheese begins to darken and the crust becomes brown on the edges. Are these physical or chemical changes? How do you know? K/U A
6. Why is dissolving salt in water a physical change, not a chemical change? K/U

5.5 PERFORM AN ACTIVITY

Practise Safety in Science

SKILLS MENU

- ▶ Questioning
- ▶ Hypothesizing
- ▶ Predicting
- ▶ Planning
- ▶ Controlling Variables
- ▶ Performing
- ▶ Observing
- ▶ Analyzing
- ▶ Evaluating
- ▶ Communicating

Who is responsible for your safety in a science laboratory? You are! Everyone who shares the lab shares this responsibility. This means following safety precautions exactly as you are directed. In this activity, you will learn about the safety rules, symbols and procedures necessary to work safely in the lab.

SKILLS HANDBOOK 1.A.–1.D.

Purpose

To review safety procedures and symbols in the science laboratory

Procedure

Part A: Floor Plan of Your Classroom

1. In your notebook, sketch a floor plan of your classroom.

2. On your floor plan, label the locations of all safety equipment and emergency exits.

3. Below your floor plan, make notes on the correct use of each piece of safety equipment.

Part B: General Safety Rules

4. In your notebook, create a two-column table. Use the Skills Handbook or information provided by your teacher. List laboratory safety rules in the first column, and write the reason for the rule in the second column. Be sure your table includes safety rules about each of the following:

- eye protection and lab aprons
- loose clothing
- food and drink
- long hair and hats
- personal items
- working alone

Part C: Safety Symbols

5. In your notebook, copy each of the WHMIS symbols in Section 1.D.1 of the Skills Handbook. Beside each symbol, write the hazard indicated.

Part D: Safety Techniques

6. Work with a partner. Demonstrate to your partner how you would do each of the following:
   - Smell a chemical holding your breath while wafting the gas towards you.
   - Pour a liquid.
   - Light a Bunsen burner.
   - Heat a test tube over a Bunsen burner.
   - Unplug an electrical cord.

Analyze and Evaluate

(a) What does WHMIS stand for? K/U C

(b) Describe the WHMIS symbol for each of the following: K/U C

(i) flammable and combustible material

(ii) poisonous and infectious material causing immediate and serious toxic effects

(iii) corrosive material

(iv) biohazardous infectious material

(c) What should you do in each of the following situations? Give reasons for your answers. T/I A

(i) You get hungry during class and consider munching on a granola bar over a clean lab bench.

(ii) You spill some water on the floor.

(iii) You break a beaker.

(d) **Figure 1** shows two students performing an experiment in the lab. Find five safety errors that the students made. A

Apply and Extend

(e) Research WHMIS. What is the purpose of WHMIS? T/I A

(f) What is an MSDS and what is its role in WHMIS? A

UNIT TASK

Bookmark

You can apply what you learned about safety in the lab to the Unit Task described on page 276.

Figure 1 What is wrong with this picture?

Identify Physical and Chemical Properties

SKILLS MENU
- Questioning
- Hypothesizing
- Predicting
- Planning
- Controlling Variables
- Performing
- Observing
- Analyzing
- Evaluating
- Communicating

The properties of a substance can be used to identify that substance. In this activity, you will discover some physical and chemical properties of common substances.

SKILLS HANDBOOK 1.A., 1.B., 2.B.

Purpose

To identify physical and chemical properties of some common substances

Procedure

Part A: Colour, Lustre, Hardness, and Brittleness

1. Put on your eye protection and lab apron.
2. Prepare a table to record your observations (**Table 1**).
3. For these tests, gather samples of the following substances: aluminum foil, copper wire, rock salt, antacid tablet, and candle wax.

Equipment and Materials

4. Observe the colour and lustre (shine) of each substance.
5. Compare the hardness and brittleness of each substance as directed by your teacher. Record your observations.

Table 1 Observations

Substance	Observation

Part B: Solubility in Water

6. For this test, gather samples of the following substances: sugar, salt, baking soda, and cornstarch.
7. Half fill four wells in a well plate with distilled water.
8. Use a toothpick to stir a small amount of one substance into one well. Use just enough to cover the end of the toothpick. Repeat with the three other substances in the other wells using a new toothpick each time. Save these mixtures for Part C.
9. Which substances dissolved? Record your observations.

Part C: Electrical Conductivity

10. Test each mixture from Part B for electrical conductivity. Rinse and dry the ends of the conductivity apparatus before each test. Record your observations.

Part D: Reaction with an Acid

11. Add 10 drops of vinegar, a weak acid, to each substance already in the well plate.
12. Look for signs of a chemical change. Record your observations.

Part E: Reaction with Iodine

13. Rinse and dry the well plate. Repeat Steps 6 to 8.
14. Add a small drop of iodine to each substance in the well plate.
15. Which substance(s) changed colour? Record your observations.

CAUTION

Stains

Iodine stains skin. Avoid skin contact.

Analyze and Evaluate

(a) Which substances were soluble? How can you tell? T/I C

(b) Which substances reacted with iodine? How can you tell? T/I C

(c) Which substances reacted with acid? How can you tell? T/I C

(d) Which parts of the activity tested the physical properties of substances? Which parts tested the chemical properties of substances? T/I

Apply and Extend

(e) Why is baking soda the active ingredient in some antacids? A

UNIT TASK

Bookmark

You can apply what you learned about the physical and chemical properties of substances to the Unit Task described on page 276.

5.7

Unique Properties

Your fingerprint is a unique characteristic that distinguishes you from others. Fingerprints are commonly used to identify people. Your fingerprint will remain the same, even if other physical properties—such as your height or weight—change.

Characteristic Physical Properties

Substances also have unique characteristics that can be used to identify them. A **characteristic property** is a physical or chemical property that is unique to a particular substance.

▶ **characteristic property:** a property that is unique to a substance and that can be used to identify the substance

Freezing Point, Melting Point, and Boiling Point

The temperatures at which common physical changes like freezing, melting, and boiling take place are characteristic properties of a substance because they do not change.

- Freezing point is the temperature at which a substance turns from liquid to solid. The freezing point of a substance is also the same temperature as its melting point.
- Melting point is the temperature at which a substance turns from a solid into a liquid. The freezing/melting point of pure water is 0 °C.
- Boiling point is the temperature at which a substance forms bubbles of vapour that escape into the air. Pure water boils at 100 °C.

Pure Substances

Carbon has the highest melting point of all pure substances. It melts at 3550 °C. Helium has the lowest melting point of all pure substances. It melts at –272.2 °C. Tungsten has the highest boiling point. It boils at 5927 °C.

Gallium is a metal that is so soft, it can be cut with a knife. Its melting point of 29.8 °C is unusually low for a metal. The temperature of the human body is 37 °C. That means that if you hold gallium in the palm of your hand, as shown in **Figure 1**, the gallium will melt. Despite its low melting point, gallium has an extremely high boiling point of 2403 °C. The range of temperatures at which gallium is a liquid is the greatest of any pure metallic substance.

Figure 1 The unusual melting and boiling points of gallium make it useful for special high-temperature thermometers.

Mixtures

The freezing/melting point of a mixture of substances is different than the freezing/melting point of a pure substance. For example, dissolving another substance into water forms a mixture with a lower freezing point than pure water. For example, a salt and water mixture can have a freezing point as low as –10 °C. This is why salt is used to make icy roads safer to drive on (**Figure 2**). Spreading salt on the roads allows the saltwater solution to remain liquid below 0 °C. Liquid water is safer to drive on than ice.

Liquid water is also important in a car engine. You could use water to cool your car's engine, as long as its temperatures stayed between the freezing and boiling points of water. However, once the temperature drops below 0 °C, the cooling system freezes solid. Adding antifreeze to the water forms a mixture with a lower freezing point, and a higher boiling point, than pure water. This prevents the cooling system from freezing in the winter and "boiling over" in the summer.

Figure 2 Icy roads are dangerous. A salt spreader adds salt to icy roads. This creates a mixture with a freezing point lower than 0 °C, making roads safer at low temperatures.

Density

Density is also a characteristic physical property. **Density** is the mass in a given volume of a substance. Density is expressed as mass per volume, or mass/volume. Each substance has its own characteristic density. For example, the density of pure water is 1 g/mL. This means that 1 mL of pure water has a mass of 1 g. Densities are quite variable. For example, the density of lead is 11.36 g/cm^3 while the density of aluminum is only 2.7 g/cm^3. Gold has a density of 19.32 g/cm^3.

▶ **density:** a measure of how much mass is contained in a given volume of a substance; you calculate density by dividing the mass of a sample by its volume

Lifejackets

The density of the average human is just about the same as that of water. We can sink or float, depending on how much air we hold in our lungs. A lifejacket is very helpful when we are unable to hold enough air in our lungs.

An inflatable lifejacket like the one in **Figure 3** has a tiny cylinder of compressed carbon dioxide gas attached inside it. When you pull the ripcord, shown on the lower left in the photo, the compressed gas rushes out of the cylinder and inflates the lifejacket. The mass of the jacket does not change, but its volume increases significantly. That means the density of the lifejacket is much lower than the density of the water you are in. This fact helps to keep you safely floating on the water instead of sinking to the bottom.

Figure 3 In Canada, the law requires every boat to carry a personal flotation device (PFD) for each passenger.

TRY THIS SALTY ICE

SKILLS HANDBOOK 3.A., 3.B.

SKILLS: Performing, Observing, Analyzing, Communicating

Try this activity to see how an impurity, such as salt, affects the freezing point of water.

Equipment and Materials: beaker; ice cubes; piece of thread; salt; stirring rod; thread

Part A: Adding Salt to Ice Water

1. Half-fill a beaker with cold water.
2. Fill the beaker of water with ice cubes.
3. Place a thermometer into the beaker. The temperature of the ice water will decrease. Wait until the temperature is near 0 °C.
4. Add about a tablespoon of salt to the ice water, and stir gently.
5. Look at the thermometer and observe what happens to the temperature.

Part B: Adding Salt to Ice

6. Place the end of a piece of thread on top of a new ice cube.
7. Sprinkle a small amount of salt (a few pinches) on top of the thread and ice cube.
8. Wait about 20 s, then gently lift the thread.

A. What happened to the temperature of the ice water when salt was added in Step 4? T/I

B. Use what you know about how salt affects the freezing point of the water–salt mixture to explain what you observed in Step 5. T/I

C. What did you observe in Step 8? Why do you think this happened? T/I

5.7 Wrap Up

- A characteristic property is a property that is unique to a substance.
- Characteristic properties can be used to identify a substance.
- Freezing point, melting point, boiling point, and density are characteristic physical properties.
- Adding another substance to a pure substance produces a mixture with a lower freezing point.

CHECK YOUR LEARNING

1. Explain why the boiling point of water is a characteristic physical property, but the temperature of water is not. K/U T/I
2. Describe what happens to the freezing point of pure water when salt is added to it. K/U
3. Is an inflated life vest more or less dense than an uninflated life vest? Explain. Use the terms *mass* and *volume* in your explanation. K/U C
4. Why is using antifreeze in a car engine better than using water alone? K/U A
5. What characteristic property of steel can be used to separate steel cans from aluminum cans at a recycling depot? K/U T/I A
6. A drinking glass at a crime scene contains a clear, colourless liquid that could be pure water. You are the investigator. Before you send the liquid to a lab for testing, you want to rule out the possibility that the glass contains water. Use what you know about the characteristic properties of water to design a simple test. T/I C
7. Megan tested several properties of isopropyl alcohol, sometimes called rubbing alcohol, in a lab investigation. She found that its volume was 50 mL, its freezing point was –89 °C, and it was colourless. Which of these is a characteristic property? K/U
8. Aluminum has a density of 2.7 g/cm^3. Copper has a density of 8.96 g/cm^3. Which metal would you choose to build a model airplane? Why? C A

AWESOME SCIENCE OSSLT

Antifreeze in Fish

The ocean waters off Newfoundland and Labrador can get as cold as –2 °C. Fish are cold-blooded animals. This means that their bodies are the same temperature as their surroundings. Surprisingly, fish in this region survive these frigid temperatures without freezing. How? Their bodies produce a protein that acts as a natural antifreeze.

When water begins to freeze, it first forms tiny solid crystals. These crystals grow as more water molecules collect on them. Some fish in a cold saltwater environment have a defense against ice crystal growth (**Figure 1**). When winter arrives, these fish produce an antifreeze protein in their cells. The protein attaches to the surfaces of new ice crystals as they form. By coating the ice crystals, the proteins prevent any more water molecules from attaching to them. The crystals cannot grow!

Scientists have discovered how to extract and use the antifreeze protein in other ways. One use is to help preserve tissues used for organ transplants. Organs, such as the heart, kidneys, and lungs, must be kept cold while in transit (**Figure 2**). However, if the organs get too cold, water in the cells may freeze to form ice crystals. This would damage the cells, making the organs useless. Storing the organs in fish antifreeze makes it possible to safely transport them over greater distances.

Without knowing it, you may have encountered fish antifreeze proteins yourself. Cosmetics companies have added these proteins to their creams and lotions to make them smoother.

And, even stranger, fish proteins are being used to make low-fat ice creams. When ice cream contains less fat, it often contains more water than regular ice cream. As a result, ice crystals form in low-fat ice cream, giving the product a grainy, undesirable texture. So, the edible and tasteless fish antifreeze proteins are added to them to keep ice crystals from forming. This keeps the ice cream smooth and creamy.

GO TO NELSON SCIENCE

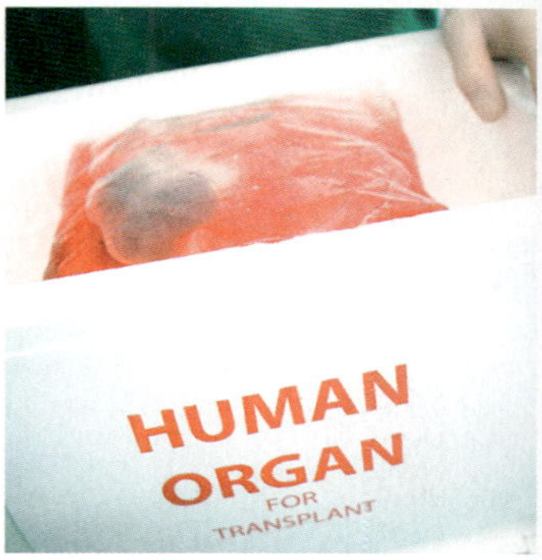

Figure 2 Organs for transplant must be delivered quickly to preserve the delicate tissues.

Figure 1 An ocean eelpout produces its own antifreeze.

5.8 PERFORM AN ACTIVITY

We Scream for Ice Cream

SKILLS MENU

- Questioning
- Hypothesizing
- Predicting
- Planning
- Controlling Variables
- **Performing**
- **Observing**
- **Analyzing**
- Evaluating
- **Communicating**

Ice cream is a mixture of frozen cream, flavourings, and tasty additives like fruit and chocolate. In this activity you will use a mixture of salt and ice to freeze the cream. This is how it works.

- Adding salt to ice produces a mixture with a lower freezing point than the freezing point of water, 0 °C. The result is a cold, slushy mixture.
- Melting requires energy. As the ice in this mixture continues to melt, it absorbs energy from its surroundings, including the cream used in this activity.
- Eventually, the temperature of the ice–salt mixture becomes so low the cream freezes.

SKILLS HANDBOOK 3.B.5.–3.B.7., 3.B.9.

Purpose

To use an ice–salt mixture to freeze cream into ice cream

Equipment and Materials

small sealable plastic bag

large sealable plastic bag

thermometer

measuring spoons

measuring cup

$\frac{1}{2}$ cup cream

2 tbsp sugar

$\frac{1}{4}$ tsp flavouring

2 cups ice

2 tbsp salt

Procedure

1. Place $\frac{1}{2}$ cup cream, 2 tbsp sugar, and $\frac{1}{4}$ tsp flavouring into the small bag and seal it.
2. Place 2 cups of ice inside the large bag. Use the thermometer to measure the temperature of the ice. Record the temperature.
3. Add 1 tbsp of salt to the ice, and mix. Measure and record the temperature of the ice–salt mixture.
4. Add a second tablespoon of salt to the ice–salt mixture, and mix. Measure and record the temperature.
5. Place the sealed small bag inside the large bag. Seal the large bag securely.
6. Shake the bags gently for about 10 min. Hold the large bag only by its top corners when shaking. Avoid touching the rest of the bag.
7. When the cream mixture is frozen, open the bags. Be careful to keep the salt and ice out of the ice cream.

Analyze and Evaluate

(a) What happened to the temperature of the ice after salt was added to it? T/I

(b) What happened to the state of the ice after salt was added to it? T/I

(c) Why do you think the cream mixture needed to be shaken? What would happen if you did not shake the mixture? T/I

(d) In Step 6, why do you think you were instructed to hold only the top corners of the bag? T/I

Apply and Extend

(e) Ice cream manufacturers often blend air into ice cream as it is being made. What effect would this have on the hardness and texture of the ice cream? Is this a physical or a chemical change? A

5.9 EXPLORE AN ISSUE CRITICALLY

Are We Salting or Assaulting Our Roads?

SKILLS MENU

- Defining the Issue
- Researching
- Identifying Alternatives
- Analyzing the Issue
- Defending a Decision
- Communicating
- Evaluating

About 50 % of the salt produced in the world is used for de-icing roads in cold climates. Using salt on roads improves driving conditions, but it can also cause problems. Salt speeds up rusting on cars. Salt also causes ice in cracks in the roads to thaw. This can create a cycle of repeated freezing and thawing throughout the winter, which widens the cracks and creates potholes. Salt that is washed off the road can contaminate nearby soil and damage vegetation (**Figure 1**). Eventually salt may enter the groundwater or surface streams, rivers, and lakes, where it can threaten aquatic ecosystems.

Figure 1 Salt use on roads can create potholes, which make roads treacherous in any season.

SKILLS HANDBOOK 4.C.2.–4.C.7.

The Issue

Some members of the town council want town crews and homeowners to stop using road salt in winter. They are worried about the harmful effect of road salt on the environment. People who live in the area have been invited to express their opinions at the next council meeting.

Role: Work in a small group to explore the issue. Each member of the group will take one of the following roles:

- town homeowner
- town parks commissioner
- town roads commissioner
- environmentalist
- lawn care company owner
- school bus driver
- senior citizens' group president
- Grade 9 student

Audience: The audience will be the mayor and other members of the town council.

Goal

To communicate your viewpoint for or against the use of road salt in winter

Gather Information

Use the library, the media, the Internet, or other sources to learn more about

- your town's policies on this issue
- statistics on road accidents or pedestrian injuries related to road conditions
- the financial cost of the salt
- damage to plants caused by road salt
- the financial cost of repairing damage to roads caused by repeated freezing and thawing of ice
- the effect of road salt on aquatic ecosystems

Identify Solutions

Research and propose alternatives to using road salt. Use the following questions as a guide:

- Can we use other chemicals or substances to melt the ice that are more environmentally friendly?
- Can we use other methods to provide better traction on icy roads?
- Can we increase snow removal?

Make a Decision

In your community role, choose one of the possible solutions you think is best. Gather facts that support your opinion. List reasons for your choice.

Communicate

Role-play a discussion in which group members present their viewpoints and supporting reasons to the mayor and town council members. Listen as others in your group present their viewpoints. Then, come to an agreement on what should be done. Write a press release, blog, or radio announcement, telling the town about the final decision.

DIG DEEPER

Web Link

For information about the positive and negative effects of road salt,

GO TO NELSON SCIENCE

5.10 PERFORM AN ACTIVITY

What Is This Gas?

SKILLS MENU

- Questioning
- Hypothesizing
- Predicting
- Planning
- Controlling Variables
- Performing
- Observing
- Analyzing
- Evaluating
- Communicating

Oxygen, hydrogen, and carbon dioxide are colourless, odourless, and tasteless gases. Yet, each gas has different chemical properties.

- Oxygen is necessary for burning. Objects burn brighter and faster when pure oxygen is present.
- Hydrogen burns so quickly in air that it often produces a loud "pop" when it is ignited.
- Carbon dioxide gas does not burn and puts out a flame. It also reacts with limewater to produce a white solid.

SKILLS HANDBOOK 1.D., 3.B.5.–3.B.9.

Purpose

To produce different gases and identify each gas

Procedure

1. Put on your eye protection and lab apron.

Equipment and Materials

eye protection

lab apron

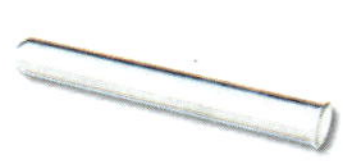
5 test tubes

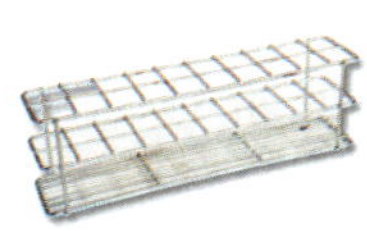
test-tube rack

stopper for test tube

retort stand with clamps

10 mL graduated cylinder

Bunsen burner

spark lighter

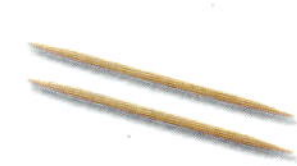
toothpicks

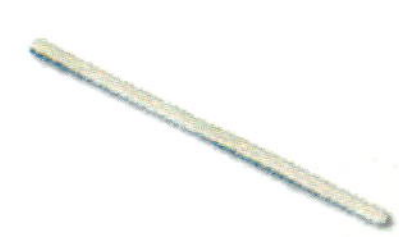
wooden splints

candle

limewater (saturated calcium hydroxide solution)

manganese dioxide

hydrogen peroxide (3 % solution)

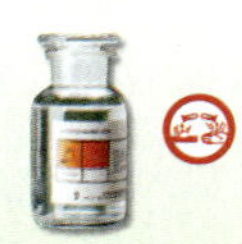
hydrochloric acid (5 % solution)

sodium hydrogen carbonate (baking soda)

magnesium ribbon

2. Copy **Table 1** into your notebook. Use the table to record your observations at each step of the activity.

Table 1

Chemicals	Burning splint test	Glowing splint test	Limewater test	Candle test	Identify the gas

3. Set up five test tubes in the test-tube rack and label them "A," "B," "C," "D," and "limewater." Fill a the test tube labelled "limewater" about $\frac{1}{3}$ full of limewater.

4. Add to test tubes A to D a tiny amount of manganese dioxide powder on the end of a toothpick. Then, add 4 mL of hydrogen peroxide each to test tubes A to D. Allow the reaction to proceed for about 15 s.

5. Light a wooden splint with the Bunsen burner and bring the burning splint to the mouth of test tube A.

6. Light another wooden splint and blow out its flame. Insert the glowing splint halfway into test tube B.

7. Carefully pour the gas produced in test tube C into the test tube of limewater. Be careful not to pour out any of the hydrogen peroxide and manganese dioxide. Place a stopper in the limewater test tube and shake to mix the gas and liquid.

8. Light the candle and carefully pour the gas produced in test tube D onto the flame.

9. Use clean test tubes. Repeat Steps 3 to 8, using about 1 mL of baking soda and 4 mL of hydrochloric acid.

10. Use clean test tubes. Repeat Steps 3 to 8, using 3 cm of magnesium ribbon and 4 mL of hydrochloric acid.

CAUTION

Flames

Tie back long hair and loose clothing. Clamp the Bunsen burner to a utility stand and light it using a spark lighter. Safely discard the used splint used in each step.

CAUTION

Corrosive

Hydrochloric acid is corrosive. Any spills on the body or on clothing should be washed immediately with cold water. Report any spills to your teacher.

Analyze and Evaluate

(a) Identify each gas produced in the reactions. Record this information in the last column of **Table 1**. T/I

(b) For each test, were the changes physical or chemical? How can you tell? T/I

Apply and Extend

(c) From your observations in this activity, suggest reasons why birthday balloons are filled with helium, not hydrogen. A

CHAPTER 5

LOOKING BACK

Chapter 5 KEY CONCEPTS SUMMARY

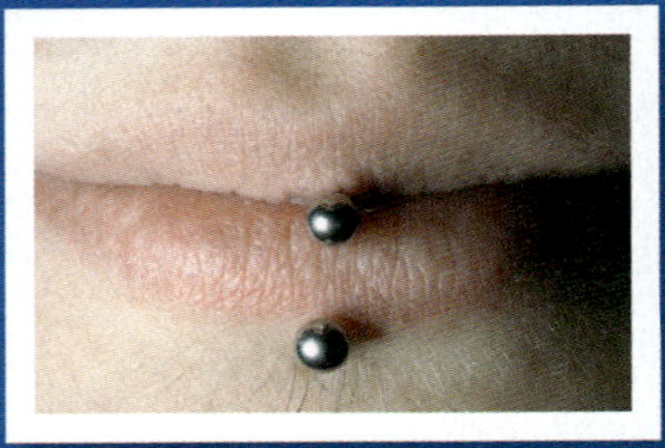

A substance can be identified by its physical or chemical properties.

- Physical properties can be observed without forming a new substance.
- Physical properties may be qualitative or quantitative.
- Chemical properties describe what the substance does when it reacts to form new substances.

Understanding the physical and chemical properties of a substance helps us identify its uses.

- Characteristic physical properties are unique to each substance.
- Characteristic physical properties are density, freezing/melting point, and boiling point.

The use of substances can have both positive and negative effects in our daily lives.

- Substances can have properties that help people and keep them safe.
- Substances may have unintended effects on the environment or on the health of living things.

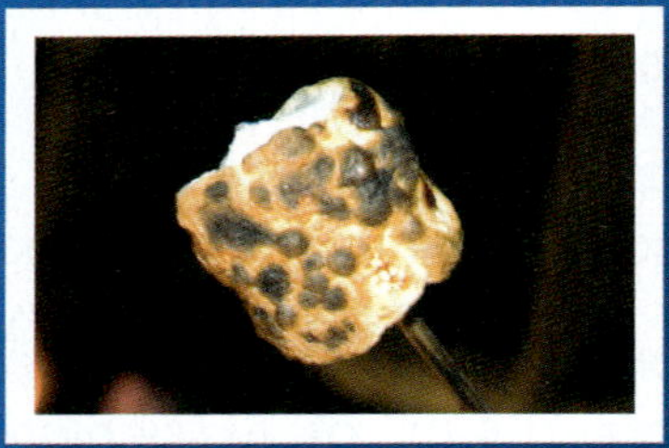

Changes in substances can be either physical or chemical.

- Physical changes are changes in which no new substances are formed.
- Chemical changes are changes in which one or more new substances are formed.

WHAT DO YOU THINK NOW?

You thought about the following statements at the beginning of the chapter. You may have encountered these ideas in school, at home, or in the world around you. Consider them again and decide whether you agree or disagree with each statement.

1 Some chemicals can be both useful and harmful to us. **Agree/disagree?**

2 In a physical change, only the appearance of the substance changes. **Agree/disagree?**

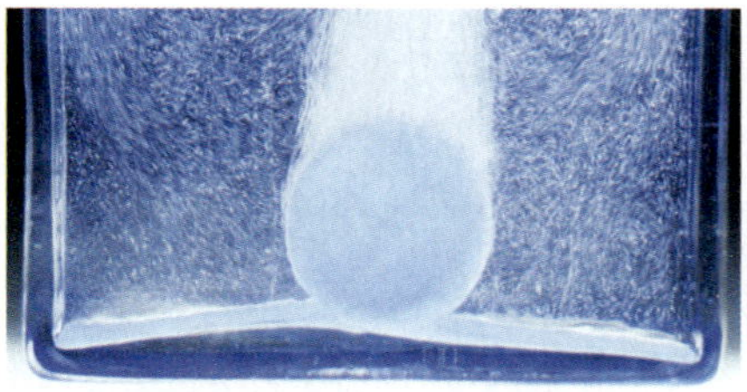

3 Only scientists make use of chemical changes. **Agree/disagree?**

4 Gold is valuable because people like the way it looks. **Agree/disagree?**

5 An edible substance such as salt does not pose any harm to the environment. **Agree/disagree?**

6 The properties of a material depend on what it is made of. **Agree/disagree?**

How have your answers changed since then? What new understanding do you have?

BIG Ideas

- Elements and compounds have specific properties that determine their uses.
- The use of elements and compounds has both positive and negative effects on society and the environment.

DIG DEEPER

Vocabulary

physical property p. 178
solubility p. 178
chemical property p. 180
combustibility p. 180
physical change p. 182
chemical change p. 183
precipitate p. 185
characteristic property p. 190
density p. 191

CHAPTER 5

SELF-QUIZ

The following icons indicate the Achievement Chart category addressed by each question.

K/U Knowledge/Understanding | T/I Thinking/Investigation | C Communication | A Application

For each question, select the best answer from the four alternatives.

1. Which of the following is a physical property of steel? (5.2) K/U
   (a) It can rust.
   (b) It is hard.
   (c) It is not combustible.
   (d) It can be used to make cars.

2. Which of the following is a chemical property? (5.3) K/U
   (a) colour
   (b) density
   (c) boiling point
   (d) ability to burn

3. Which statement is true of the boiling points of pure substances? (5.7) K/U
   (a) All pure substances boil at 100 °C.
   (b) The boiling point and melting point of a substance are the same temperature.
   (c) The boiling point of a substance is a chemical property of that substance.
   (d) The boiling point of a substance can be used to identify it.

4. Which of the following is a physical change? (5.4) K/U
   (a) melting chocolate
   (b) burning firewood
   (c) cooking eggs
   (d) exploding fireworks

5. Which of the following is a clue that a chemical change has taken place? (5.4) K/U
   (a) You observe that a substance is melting.
   (b) You observe a small flame.
   (c) You observe a change in shape.
   (d) You observe the colour of a chemical.

Indicate whether each of the statements is TRUE or FALSE. If you think the statement is false, rewrite it to make it true.

6. The mass of a substance is an example of a chemical property. (5.2, 5.3) K/U

7. Melting gold to make a necklace is a physical change. (5.4) K/U

8. Making Popsicles from juice is a chemical change. (5.4) K/U

9. Adding salt to water will create a mixture with a lower freezing point. (5.7) K/U

Copy each of the following statements into your notebook. Fill in the blanks with a word or phrase that correctly completes the sentence.

10. The density of an object is affected by its mass and its ________. (5.7) K/U

11. When a sugar cube is dissolved in a cup of tea, the sugar undergoes a ________ change. (5.4) K/U

12. The ________ physical properties of a substance are unique to that substance. (5.7) K/U

Match each term on the left with the most appropriate description on the right.

13. (a) lustre — (i) the ability to dissolve
    (b) odour — (ii) how something feels to the touch
    (c) texture — (iii) the level of breakability or flexibility
    (d) brittleness — (iv) the level of shininess or dullness
    (e) solubility — (v) how something smells
    (5.2) K/U

Write a short answer to each of these questions.

14. How can you tell that a chocolate chip cookie is undergoing a chemical change as it bakes? (5.4) K/U A

15. Hockey players need to wear safety equipment when they play, but this equipment should not interfere with their ability to play well. Name at least two physical properties that are necessary in materials used to make the following sports safety equipment: (5.2) K/U A
    (a) gloves
    (b) helmets
    (c) shoulder pads

16. What physical properties are important for the materials used to make mountain bikes? (5.2) K/U A

17. Three flasks contain colourless gases. You know that one flask contains pure oxygen, one contains carbon dioxide, and one contains air, but you do not know which is which. (5.10) K/U T/I
    (a) Describe any tests you could perform to identify the gases.
    (b) Explain what you would observe when you performed the test on each gas.

18. You and a friend watch workmen put tree limbs into a wood chipper. Large pieces go into the chipper and small pieces come out. Your friend says that because the pieces coming out of the chipper do not look anything like the pieces going in, the tree limbs have undergone a chemical change. Is your friend correct? Explain your answer. (5.2, 5.3) K/U C

19. Sea water freezes at lower temperatures than fresh water. Explain why. (5.7) K/U C

20. Name two characteristic physical properties that could be used to identify you. (5.7) K/U A

21. It is your turn to make dinner. You decide to grill some steaks, make a salad, and chill some bottles of pop. Identify a chemical change and a physical change that take place as you prepare dinner. Explain your answer. (5.2, 5.3) K/U C A

22. How are the freezing point and melting point of a pure substance related? (5.7) K/U

23. Classify the following properties of hydrogen as physical or chemical. (5.2, 5.3) K/U
    (a) It is a colourless gas.
    (b) It reacts with oxygen to form water.
    (c) It is less dense than air.
    (d) It has no odour.

24. Plastic reinforced with carbon fibre is now replacing aluminum to make frames for racing bicycles. The density of reinforced plastic can be less than half the density of aluminum. Why is this property useful for the racing bicycles? (5.2, 5.3, 5.7) K/U C A

25. Before the 1970s, almost all baseball bats were made of wood. In the 1970s, aluminum bats were introduced. Aluminum bats, because they are hollow, are less dense than wooden bats. Aluminum bats are also stronger than wooden bats. (5.2) T/I A
    (a) Based on the properties of wood and of aluminum bats, what are the advantages of each type?
    (b) Today, only the pros use wooden bats. Non-professional players usually use aluminum bats. Suggest a reason why this is true.

Self-Quiz

To do an online self-quiz, GO TO NELSON SCIENCE

CHAPTER 5

REVIEW

The following icons indicate the Achievement Chart category addressed by each question.

K/U Knowledge/Understanding | T/I Thinking/Investigation | C Communication | A Application

What Do You Remember?

1. Describe three clues that tell you that you are observing a chemical change. Give an example of each clue you described. (5.4) K/U

2. You are wrapping a present. Describe two physical changes that happen to the wrapping paper as you wrap. Explain why each is a physical change. (5.4) K/U

3. It is important to avoid using an open flame around hydrogen, because it can easily catch fire. Is this a physical or chemical property of hydrogen? (5.2, 5.3) K/U

4. Sketch and describe the WHMIS symbol for each of the following hazards: (5.5) K/U C
   (a) poisonous and infectious, causing immediate and serious toxic effects
   (b) corrosive
   (c) flammable and combustible

5. (a) What physical properties make glass very useful?
   (b) Glass is not very reactive chemically. How is non-reactivity a useful property of glass? (5.2, 5.3) K/U A

6. The melting point of a substance is 64 °C. (5.7)
   (a) What is the freezing point of the substance? K/U
   (b) Could this substance be pure water? How do you know? K/U

What Do You Understand?

7. Classify each of the following properties of a pizza as qualitative or quantitative. Give reasons for your answers. (5.2) K/U
   (a) It is circular in shape.
   (b) It is 40 cm in diameter.
   (c) It smells like oregano.
   (d) It has a mass of 1 kg.

8. Adding salt to ice melts the ice at a lower temperature than the normal melting point of ice. Explain two uses of lowering the melting point of ice this way. (5.8, 5.9) K/U

9. An empty scuba tank has a mass of 20 kg. When it was filled with compressed gas, its mass increased to 23 kg. (5.7) K/U T/I A
   (a) How do these changes in mass influence the density of the tank?
   (b) Do you think it is easier for a scuba diver to descend at the beginning of a dive or at the end of a dive? Explain your reasoning.

10. Classify each of the following changes as physical or chemical. Give reasons for your answers. (5.4) K/U T/I
    (a) When ice cream is warmed, it melts.
    (b) When a chair is painted, it has a new colour.
    (c) When sugar is stirred into tea, it dissolves.
    (d) When bread is baked, it forms a golden brown crust.
    (e) When metal is heated and moulded, it can be made into car parts.
    (f) When wood is burned, ashes remain.
    (g) When vinegar is added to baking soda, bubbles are seen.

Solve a Problem

11. When sugar, cream, and butter are heated together, a sticky, dark brown substance, known as caramel, results. Did a chemical change occur? Give reasons for your answer. (5.4) T/I

12. When an opaque beige solid is heated, it forms a clear, colourless liquid. When this liquid is cooled, it returns to an opaque beige solid. Do you think a physical or a chemical change occurred? Give reasons for your answer. (5.4) T/I

13. Iron pyrite (**Figure 1**) is a lustrous yellow mineral that consists of iron and sulfur. Its common name is fool's gold. How could you use characteristic physical properties to distinguish it from real gold? (5.6) T/I

Figure 1 Iron pyrite (fool's gold) looks similar to gold.

14. It is winter, and the sidewalk outside your home is slippery with ice. What could you do to make it less slippery? How could your choice affect the flowers planted near the sidewalk? (5.7, 5.9) K/U A

15. You are expecting guests at any moment for a summer barbecue, and have just realized that the bottled lemonade you bought has been in the car, getting warm. The refrigerator and freezer are both full. How can you use salt and ice to cool the lemonade quickly? Why does this work? (5.7, 5.8) A

Create and Evaluate

16. Gourmet cooking is the art of blending physical and chemical changes into tasty and nutritious dishes. As a scientist and chef, your task is to create a script for a television cooking show called *The Science of Cooking*. You will base your script on the recipe in the next column. (5.4) T/I C

Quiche Me Quick

- Slice 4 cups of mushrooms, and dice 2 onions and 2 green peppers.
- Cook the vegetables in a frying pan with a little oil until lightly browned.
- Beat 10 eggs until light and bubbly.
- Grate 2 cups of cheddar cheese.
- Cook 8 strips of bacon until crispy, then crumble into small pieces.
- Mix all ingredients with 1 cup of milk and 1–2 tsp of salt.
- Bake in a large pie dish at 175 °C until firm to the touch and golden brown on top (about 50 min).

(a) Write a script for the "scientific" cooking show, *The Science of Cooking*, based on the recipe. Make sure your script describes what should be done, as well as the chemical and physical changes involved at each step.

(b) You have also been asked by the show's producer to create a recipe that involves only physical changes and no chemical changes. This recipe will be featured on an episode about quick and easy summer meals. Write your recipe. Give your recipe a descriptive name.

Reflect on Your Learning

17. How does the phrase "there are two sides to every coin" apply to the use of salt to de-ice our roads? What other chemical product might this phrase apply to? Explain. T/I C A

Web Connections

18. (a) The use of road salt in winter is impractical in many northern communities when temperatures drop to extreme lows. Why might this be? A

(b) What other substance could be used to improve road safety in the winter? A

GO TO NELSON SCIENCE

CHAPTER 6

Elements and the Periodic Table

KEY QUESTION

Are the properties of elements predictable?

KEY CONCEPTS

- Elements cannot be broken down into simpler substances.
- Elements of the same column (group) of the periodic table have similar properties.
- Metals and non-metals have characteristic physical properties.
- Metalloids have properties in between those of metals and non-metals.
- Elements can be both beneficial and harmful.
- Atoms consist of protons and neutrons in a central core surrounded by electrons.

Fireworks—Colourful Explosions

Imagine the surprise of the first person who exploded a firecracker by accident. Historians think that person may have been a Chinese cook who lived about 2000 years ago. The cook accidently mixed some common substances that were known to burn. The ingredients were probably ones used in Chinese cooking and medicines. To the cook's surprise, the mixture burned quickly, producing a bright orange flame. He had invented gunpowder! Then, according to legend, the cook packed this mixture into a hollow tube of bamboo wood. Igniting the mixture caused the bamboo to blow apart in an incredibly loud explosion. This was the first firecracker.

Thus, the science of fireworks, or pyrotechnics, began with a bang in China. Ancient Chinese fireworks did not have the bright colours we enjoy today. Over time, people learned that certain substances give off bright colours when they burn. Compounds with metals such as barium, iron, or strontium make the bright colours of fireworks.

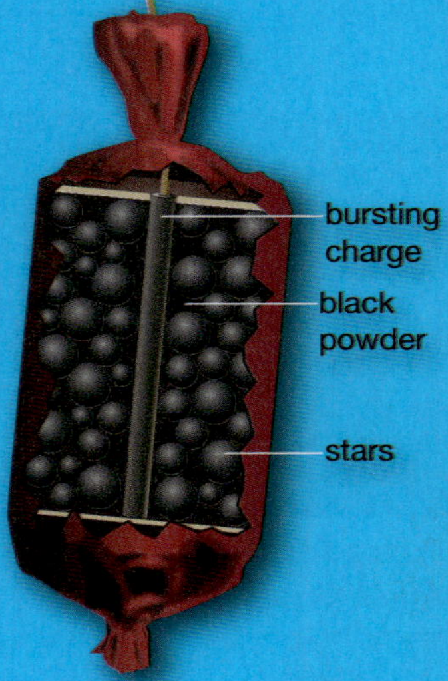

The inside of a fireworks tube contains black powder (gunpowder), a bursting charge, and "stars" containing chemicals that produce different colours.

WHAT DO YOU

THINK?

Many of the ideas you will explore in this chapter are ideas that you have already encountered. You may have encountered these ideas in school, at home, or in the world around you. Some of the statements are not true. Consider each statement and decide whether you agree or disagree with it.

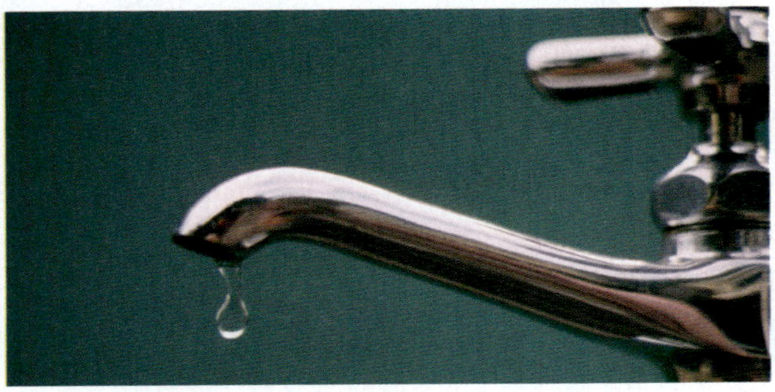

1 Water is a common element.
Agree/disagree?

2 Pure gold makes the best gold jewellery.
Agree/disagree?

3 Most of the substances around us are made from metals.
Agree/disagree?

4 Most metals are magnetic.
Agree/disagree?

5 Using hot tap water to prepare food can be bad for your health.
Agree/disagree?

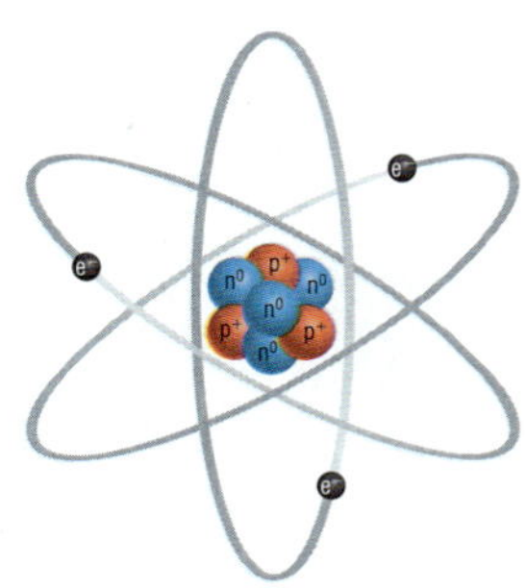

6 An atom is the smallest particle of an element.
Agree/disagree?

FOCUS ON
READING

Making Inferences

When you make inferences, you read "between the lines" to figure out what the author is suggesting. Use the following strategies to make inferences:

- use clues such as key words, examples, or explanations
- think about what you already know about the topic
- combine the clues with what you already know about the topic to form an idea of your own
- revise your idea if you find new information or clues that alter or disprove your idea

> **DIG DEEPER**
> **Reading Tip**
> As you work through this chapter, look for tips like this. They will help you develop literacy strategies.

The Great Canadian Diamond Rush

Diamonds were first discovered in the Northwest Territories in the 1980s. Now, Canada is one of the top three diamond producers in the world. Mines in the Northwest Territories, Nunavut, and northern Ontario produce more than ten million carats of Canadian diamonds each year. That is more than 15 % of the world's supply!

Benefits and Drawbacks

The diamond rush has benefited the Canadian North in many ways. However, any form of mining disturbs the environment. The large open-pit mines in which companies dig for diamonds can scar the landscape and can lead to harmful runoff (**Figure 3**). Aboriginal peoples are also displaced by diamond mining. About 70 % of Canada's Native reserves are on or near areas that mining companies are interested in.

Figure 3 The Diavik Diamond Mine on Lac de Gras, Northwest Territories

NEL 6.6 Charcoal Pendants or Diamond Pencils? 231

Making Inferences *in Action*

Here is how one student made inferences as she read about Canadian diamonds in Section 6.6.

Clues from the text	+	What I already know	=	My idea (inference)
• recent industry in Canada • mines in northern Canada • Canada produces 15 % of the world's diamonds		My aunt's diamond ring has a polar bear stamp to show it's from Canada.		Since Canada is just getting started in diamond mining, we might lead the world in the future.
• open pit diamond mines are near 70 % of Native reserves		I saw a documentary about how open pit mining harms the environment.		Harmful runoff could affect Aboriginal peoples more than anybody else.

6.1 Atoms and Elements

Think about how many words are in an English dictionary. All these words are formed from only the 26 letters in our alphabet. In a similar way, three particles form the many kinds of atoms that make up everything around you.

- **proton:** a positively charged particle in an atom
- **neutron:** a particle in an atom that has no electrical charge
- **electron:** a negatively charged particle found in an orbit outside the nucleus of an atom

Parts of an Atom

Each atom contains three kinds of particles: protons, neutrons, and electrons. A **proton** is positively charged. A **neutron** has no charge and is similar in mass to a proton. Both protons and neutrons are found in the tiny central part of an atom called the nucleus. An **electron** is negatively charged and has much less mass compared to a proton and a neutron. Electrons orbit the nucleus, much like planets orbit the Sun.

The numbers of protons and electrons in an atom are equal. The positive charges of the protons and negative charges of the electrons cancel each other out, making each atom neutral (neither positive nor negative). **Table 1** summarizes some important information about each kind of particle found in an atom. We use atomic models to show the number of each kind of subatomic particle and where it is located (**Figure 1**).

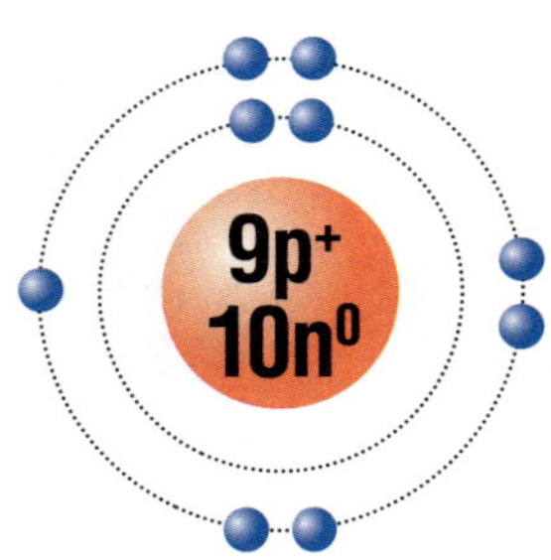

Figure 1 This model of a fluorine atom shows that it has 9 protons ($9p^+$), 9 electrons (blue dots), and 10 neutrons ($10n^0$). There are 2 electrons in the first orbit and 7 electrons in the second orbit.

Electron Orbits

As shown in **Figure 1**, electrons are located in orbits outside the nucleus of the atom. The first orbit can hold 2 electrons. The second orbit can hold up to 8 electrons. The third orbit can also hold up to 8 electrons.

Table 1 Particles in an Atom

Particle name	Charge	Location	Symbol
proton	+	nucleus	p^+
neutron	neutral	nucleus	n^0
electron	–	orbits	e^-

The Importance of Organization

If we put things in a logical order, or organize them, it is easier to find those things later. Imagine looking for items in a grocery store where items are arranged in alphabetical order. Pickles would be next to Popsicles and lettuce would be next to light bulbs! Pickles do not need to be frozen and lettuce wilts if it is not stored in a cool, damp place, so a different organizing system is needed. Putting groceries in order by the type of item and the type of storage they need makes more sense (**Figure 2**).

Organizing the Elements

All matter is made up of elements. An **element** is a substance that cannot be broken down into simpler substances. Gold, copper, and oxygen are common examples of elements. Each element is made up of only one kind of atom. This is why gold has exactly the same properties no matter where in the world it is mined. Gold and copper have different properties because they are made of different kinds of atoms.

You may recall that scientists often record data in tables. A table helps organize data so people can make sense of it. All the elements are arranged in the **periodic table** so that elements with similar properties are grouped together.

- **element:** a pure substance that cannot be broken down into simpler substances, either physically or chemically
- **periodic table:** the table that lists all the known elements in rows and columns based on patterns of similar properties

Figure 2 Items in a grocery store are placed in aisles according to shared properties.

The Modern Periodic Table

Elements in the periodic table are arranged in order of the number of protons their atoms have (**Figure 3**). The number of protons is called the element's **atomic number**. Each element has a different atomic number (**Figure 4**).

- Each element in the periodic table has its own box. In the box is the element's chemical symbol, name, and atomic number.
- The first letter of every chemical symbol is a capital letter (e.g., N for nitrogen). If there is a second letter, it is lowercase (e.g., Na for sodium).
- Each row in the periodic table is called a **period**. The properties of the elements that are in the same period can be very different.
- The elements are also placed in columns according to how their electrons are arranged. Each column in the periodic table is called a **group** or **chemical family**. Elements in the same group have similar properties. Within each group, all atoms have the same number of electrons in their outermost orbits.
- Most elements are either metals (blue boxes) or non-metals (pink boxes). They are separated by a dark line that looks like a staircase. Hydrogen is the only exception. Elements along the staircase have properties in between those of metals and non-metals. These elements are called metalloids (green boxes).

▶ **atomic number:** the number of protons in an atom of an element; each element has a different atomic number

▶ **period:** a horizontal row in the periodic table of elements

▶ **group** or **chemical family:** a vertical column in the periodic table of elements

Figure 4 The atomic number is at the top left-hand corner of each element on the periodic table.

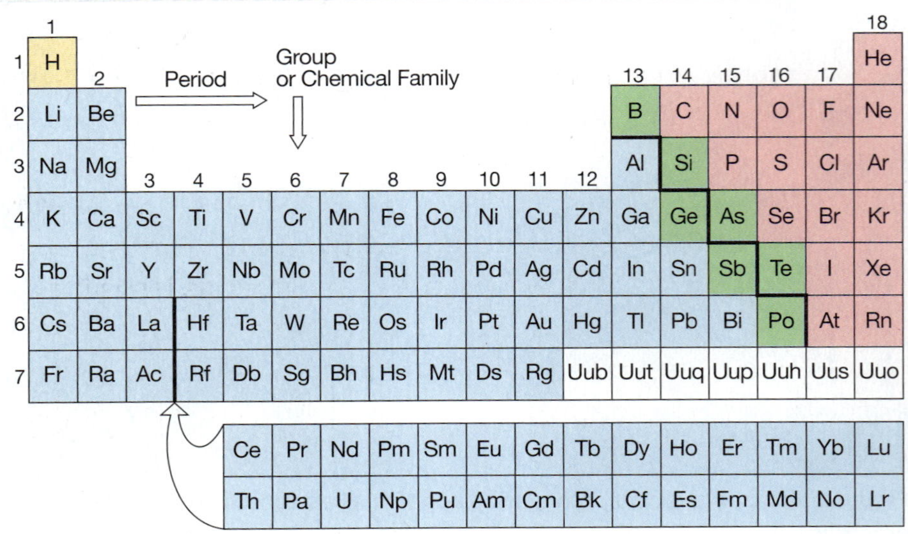

Figure 3 The elements of the periodic table are arranged in rows (also called periods) and columns. Each group of elements has similar properties.

Chemical Families

Group 1

The structure of atoms explains why elements have similarities and differences. The alkali metal group consists of the elements below hydrogen in the first column of the periodic table, starting with lithium. Lithium (Li) has one outer electron, as do sodium (Na), potassium (K), and other elements in the group. When an alkali metal reacts, each of its atoms loses the single electron in its outermost orbit. The alkali metals have similar reactions because they all have one electron in their outermost orbits (**Figure 5**). However, the reactivity of alkali metals increases from the top of Group 1 to the bottom.

Group 2

The elements in Group 2 are the alkaline earth metals. Beryllium (Be), magnesium (Mg), calcium (Ca), strontium (Sr), barium, and radium (Ra) all have two electrons in their outermost orbits. The reactivity of Group 2 metals also increases from the top of the group to the bottom.

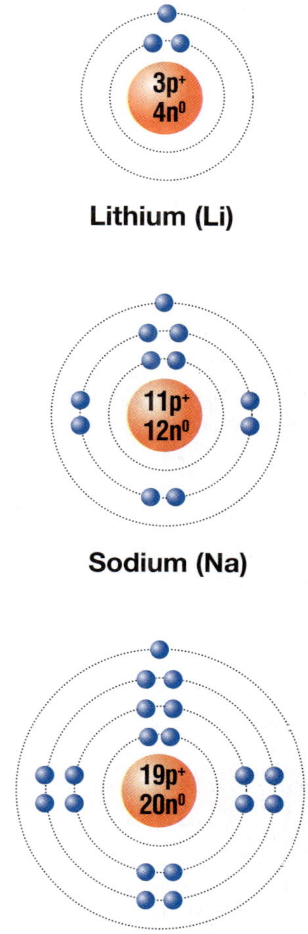

Figure 5 These elements react in a similar way because each atom has one electron in its outermost electron orbit.

Periodic Trends

Elements in the same group or family always share some chemical and physical properties. Elements in the same period also show some trends in their reactivity. For example, Group 1 alkali metals are more reactive than the elements in Group 2. That means that sodium (Na) is more reactive than magnesium (Mg), and potassium (K) is more reactive than calcium (Ca). The elements in Group 17 halogens are more reactive than Group 16 elements in the same row. You will learn more about Group 17 and Group 18 in Section 6.4. The rows on the periodic table are called periods because they show the repeating nature of these trends in reactivity. The word "period" refers to cycles, such as the classes that recur daily in your school schedule. **Figure 6**, on the next page, shows properties and common uses of the elements.

Periodic Table of the Elements

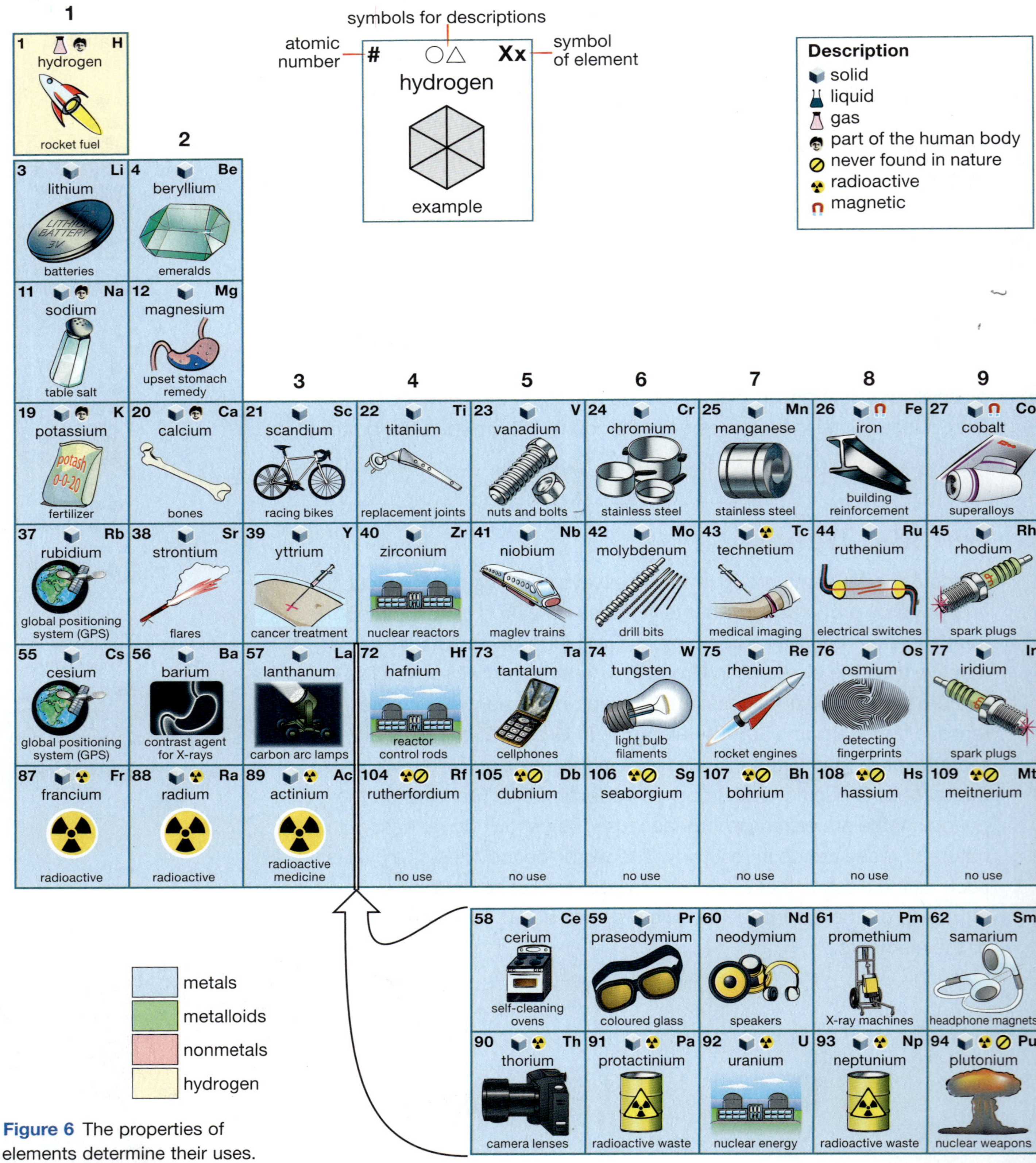

Figure 6 The properties of elements determine their uses.

10	11	12	13	14	15	16	17	18
								2 He helium blimps
			5 B boron sports equipment	6 C carbon diamonds	7 N nitrogen fertilizer	8 O oxygen emergencies on airplanes	9 F fluorine dental health	10 Ne neon neon signs
			13 Al aluminum cans and foil	14 Si silicon computer chips	15 P phosphorus safety matches	16 S sulfur food preservatives	17 Cl chlorine making drinking water safe	18 Ar argon light bulbs
28 Ni nickel stainless steel	29 Cu copper water pipes	30 Zn zinc galvanized steel	31 Ga gallium light-emitting diodes (LEDs)	32 Ge germanium solar panels	33 As arsenic crop insecticide	34 Se selenium dandruff shampoo	35 Br bromine flame retardant materials	36 Kr krypton headlights
46 Pd palladium dental crowns	47 Ag silver jewellery	48 Cd cadmium Ni-Cd batteries	49 In indium liquid crystal displays (LCDs)	50 Sn tin bronze	51 Sb antimony paints	52 Te tellurium tinting glass	53 I iodine disinfectant	54 Xe xenon stadium lamps
78 Pt platinum jewellery	79 Au gold jewellery	80 Hg mercury dental amalgams	81 Tl thallium ant killer and rat poison	82 Pb lead radiation shield for X-rays	83 Bi bismuth pearly effect in cosmetics	84 Po polonium radioactive	85 At astatine cancer medicine	86 Rn radon earthquake prediction
110 Ds darmstadtium no use	111 Rg roentgenium no use	112 Uub copernicium no use	113 Uut ununtrium no use	114 Uuq ununquadium no use	115 Uup ununpentium no use	116 Uuh ununhexium no use	117 Uus ununseptium no use	118 Uuo ununoctium no use

63 Eu europium luminous paint	64 Gd gadolinium magnetic resonance imaging (MRIs)	65 Tb terbium laser material	66 Dy dysprosium compact discs	67 Ho holmium reactor control rods	68 Er erbium colourant in glasses	69 Tm thulium laser eye surgery	70 Yb ytterbium stainless steel	71 Lu lutetium determining the age of meteorites
95 Am americium smoke detectors	96 Cm curium energy source for the Mars Rover	97 Bk berkelium radioactive waste	98 Cf californium aircraft safety check	99 Es einsteinium no use	100 Fm fermium no use	101 Md mendelevium no use	102 No nobelium no use	103 Lr lawrencium no use

TRY THIS FLAMING COLOURS

SKILLS HANDBOOK 3.B.5.–3.B.7., 3.B.9.

SKILLS: Performing, Observing, Analyzing, Communicating

Fireworks and sparklers give off colours when they are heated or burned. You can heat substances that contain different metal elements by holding samples of them in a flame (**Figure 7**). This is called a flame test. The bright colours of light that result are due to the structure of the metal atoms. Light is given off when electrons of the atoms gain and then lose energy. The flame test can be used to identify unknown metals in paint samples.

CAUTION: FLAME
Never heat anything without your teacher's permission. Always wear eye protection when you are working with fire. Keep yourself, and anything else that can burn, away from heat or flames. Tie back long hair and loose clothing. Never reach across a flame. Do not leave the flame unattended.

Figure 7 The colour given off in a flame test can be used to identify a substance.

Equipment and Materials: eye protection; lab apron; plastic vials; samples of metal salts; matches or striker; Bunsen burner; retort stand and clamp; wire loops; distilled water

1. Put on your lab apron and eye protection.
2. Obtain a small sample of each salt provided as well as the wire loop that goes with the sample.
3. Make a chart to record the names of your salt samples and the colour of flame each produces.
4. Clamp the burner to the retort stand. Light your Bunsen burner. Adjust the flame so that it is mostly blue.
5. Dip the end of the wire loop into the water and then into the first sample. Close the vial.
6. Hold the wire loop in the flame of the Bunsen burner until the flame changes colour. Record the name of the sample and the colour of the flame in your chart. Rinse the wire loop well in a stream of water.
7. Repeat Steps 5 and 6 for each sample.
8. Follow your teacher's instructions for the disposal of the materials and equipment.
9. Wash your hands thoroughly with soap and water.

A. Which element is responsible for the flame colours in each substance observed? T/I

B. If your teacher gave you an unlabelled salt, and it turned the flame colour violet, what would this tell you? T/I

C. Why should you clean off the wire loop each time or use a different wire loop for each test? T/I

D. How might you use a flame test to determine which metal is in an unknown sample? T/I

E. Look at **Figure 8**. How might you use a flame test to determine which metal is responsible for this green firework display? T/I

Figure 8 The colour of the explosion tells you what metal is in the fireworks.

6.1 Wrap Up

- Protons and neutrons are found in the nucleus of the atom. Electrons orbit the nucleus.
- Protons are positively charged and have about the same mass as neutrons, which have no charge. Electrons are negatively charged and have very little mass compared to protons and neutrons.
- The numbers of protons and electrons in an atom are equal. The positive and negative charges cancel out each other, giving the atom a neutral charge.
- Elements cannot be broken down into simpler substances.
- The elements in the periodic table are in order by increasing atomic number.
- Elements in the same group have similar properties.

CHECK YOUR LEARNING

1. Name the three subatomic particles. K/U
2. How many electrons would an atom with the atomic number 12 have? T/I
3. Why does an atom have a neutral charge? K/U
4. (a) Why do potassium oxide and potassium chloride give off the same colour of light in a flame test?

   (b) How does a fireworks technician make use of this property? K/U C A
5. Write the chemical symbol for each element listed below. K/U

   (a) aluminum
   (b) fluorine
   (c) phosphorus
   (d) titanium
   (e) carbon
   (f) gold
   (g) potassium
   (h) zinc
6. Write the name for the element indicated by each symbol below. K/U C

   (a) As
   (b) B
   (c) Cu
   (d) Pb
   (e) Ag
   (f) Ca
   (g) Fe
   (h) W
7. What is an atomic number? Why is it important? K/U
8. An investigator suspects that water found at a crime scene was poisoned with barium nitrate. What experiment could be done to test the suspicion? A
9. The model in **Figure 9** shows an atom of the element nitrogen. K/U

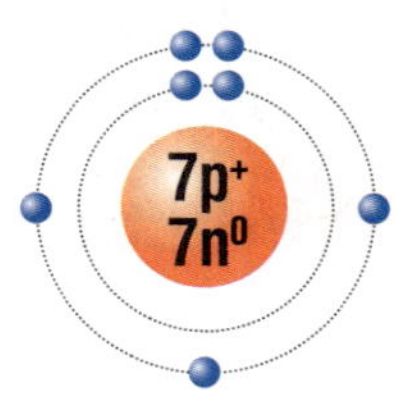

Figure 9

   (a) How many protons are in one atom of nitrogen?
   (b) How many electrons are in one atom of nitrogen?
   (c) How many electrons are in the outer orbit of an atom of nitrogen?

10. Why do all alkali metals have similar reactions? K/U
11. Copy **Table 2** into your notebook. Use the periodic table to complete it. T/I C

Table 2

Element	Symbol	Number of protons	Number of electrons	Atomic number
hydrogen		1		
	Ca			20
phosphorus				

6.2

CONDUCT AN INVESTIGATION

Becoming a Metal Detective

SKILLS MENU

- Questioning
- Hypothesizing
- Predicting
- Planning
- Controlling Variables
- Performing
- Observing
- Analyzing
- Evaluating
- Communicating

Have you ever warmed a bowl of food in the microwave oven and forgotten to remove the metal spoon? The sparks jumping off the spoon tell you that metals have unique properties. Metals are solids that are usually shiny, malleable or bendable, and are good conductors of electricity and thermal energy.

In this investigation, you will examine four physical properties in an attempt to correctly determine whether or not an object or substance is a metal. The properties will be lustre, malleability, magnetic character, and electrical conductivity.

SKILLS HANDBOOK 3.B.

Testable Question

Can the testing of four properties be used to correctly identify substances as metallic or non-metallic?

Prediction

Make a prediction for the Testable Question.

Experimental Design

In this investigation, you will design a procedure to test several objects and determine which are made of metals.

Equipment and Materials

lab apron

eye protection

magnet

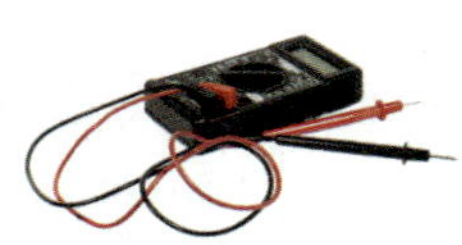
low voltage conductivity apparatus

steel wool

samples to test

Procedure

1. Copy **Table 1** into your lab notebook. Give each sample to be tested a name or number.
2. Using **Table 1** as a guide, design a procedure that you can use to identify the metals.

Table 1

Sample (name or number)	Properties				
	Lustre (shiny or dull)	Bendable (yes or no)	Magnetic (yes or no)	Electrical conductor (yes or no)	Metal (yes or no)

3. Show your procedure to your teacher for approval.
4. Put on your lab apron and eye protection.
5. Obtain samples to be tested. Record the name (if you know what it is) or number of each sample in your data table.
6. Use your procedure to test each sample. Record your observations in your table.

Analyze and Evaluate

(a) Which properties could be used to correctly identify all metals? T/I

(b) Did your observations support your prediction? Explain why or why not. T/I

(c) Were any properties common to all metals and only metals? T/I

(d) Why was it necessary to use more than one property to identify the metals? T/I

Apply and Extend

(e) Identify the main element in each metal sample. Locate these elements in the periodic table. Describe where most of these elements are found. T/I C

(f) Explain how the properties of each of the following elements make them suitable for their use in these products: A

(i) copper electrical wires

(ii) aluminum soft drink cans

(iii) iron needles in compasses

CAUTION

Disposal/ Hygiene

Follow your teacher's instructions for returning or disposing of materials and equipment. Wash your hands thoroughly with soap and water.

6.3

Metals

What do fighter jets, racing bicycles, and artificial hips have in common (Figure 1)? They all have metal parts. Metal is used in many different kinds of products.

Figure 1 Mixtures containing the same light, strong metal element—titanium—that is used in (a) the fan blades of the jet engine are also used in (b) racing bicycles and (c) joint replacements.

Properties of Metal Elements

What is a **metal**?

- They are usually shiny.
- They are good conductors of electricity and thermal energy. This property is called **conductivity**.
- Except for mercury, they are solids at room temperature.
- Most are **malleable** (can be hammered into different shapes) and **ductile** (can be pulled into a long, thin wire).
- Most react with acids.

Examples: gold (Au), silver (Ag), potassium (K), calcium (Ca), copper (Cu), tin (Sn), iron (Fe), lead (Pb)

- **metal:** a shiny substance that usually is malleable (bendable) and is a good conductor of electricity
- **conductivity:** a measure of a material's ability to conduct thermal energy or electricity
- **malleable:** a material's ability to be hammered into different shapes
- **ductile:** a material's ability to be pulled into a wire
- **alloy:** a solid solution of two or more metals

Metal Mixtures

Most metal objects are mixtures of metallic elements from the periodic table. Stainless steel, for example, is a mixture of mostly iron (Fe), nickel (Ni), and chromium (Cr). The addition of nickel and chromium makes stainless steel almost rustproof. A metal produced by blending different metal elements together is called an **alloy**. Brass, for example, is an alloy of the metal elements copper (Cu) and zinc (Zn). Bronze is another alloy that contains copper. In this case, the alloy is copper mixed with tin.

Designer Metals

Jewellery

Metals have many useful properties, which is why they are used so often. Gold and silver are the main metals used to make fine jewellery. Gold's colour and lustre make gold jewellery desirable. However, pure gold is too soft to be used for jewellery. Instead, pure gold is alloyed with other metals like silver and copper to harden it.

Art

Glass blowers and artists who work with glass use different metals to create a rainbow of glass colours (**Figure 2**).

Everyday Uses

Metals such as gold and aluminum are so malleable that they can be hammered and rolled into foil. Copper is prized for its ductility and electrical conductivity. Copper is ideal for electrical wiring because it is an excellent conductor, it does not corrode easily, and it is inexpensive.

Special Uses

Metals have many other uses, some of which might surprise you. For example, the torch for the Sydney Olympics in 2000 was carried underwater for 3 min along Australia's Great Barrier Reef (**Figure 3**). To remain lit underwater, the flame had to be extremely hot. The fuel used to produce this flame was a mixture of oxygen-generating chemicals and finely powdered magnesium metal.

Figure 3 Because the temperature of burning magnesium is well over 2000 °C, the Olympic torch could remain lit underwater.

DIG DEEPER

Did You Know?

Karats

The amount of gold in a piece of jewellery, is defined by the measurement of karats. The purest gold is 24-karat gold. This tends to make a ring so soft that it can be bent. An 18-karat gold ring is much stronger. It has 18 parts gold and 6 parts other metal. A 14-karat gold ring has 14 parts gold and 10 parts other metal. A 10-karat gold ring has 10 parts gold and 14 parts other metal.

Figure 2 Lead (Pb) with antimony (Sb) makes glass yellow, while cobalt (Co) makes glass a deep blue.

DIG DEEPER

Did You Know?

Lead Poisoning

Lead was once used in paints to make a pure-white base that does not dull the colours of other paints. The famous Dutch painter Vincent van Gogh may have suffered from lead poisoning. Lead poisoning can cause the retinas of the eyes to swell, making light appear in rings around objects, the way it does in van Gogh's painting *The Starry Night* (**Figure 4**).

Hidden Hazards

Metals are very useful, but they can also cause problems.

- Nickel (Ni) is often used to make the posts and wires of jewellery worn in pierced ears. Many people are allergic to nickel and may experience a painful rash that can lead to infection if the rash is scratched.
- The term "heavy metal" is not just the name for a type of music. Some metals, such as mercury (Hg) and lead (Pb) are very dense, or "heavy." Mercury was once widely used in mining and industrial processes, and frequently found its way into the environment. When foods containing mercury or lead are eaten, internal organs, including the brain, can be damaged.
- Lead is often mixed with other elements to provide strength and durability. For example, leaded glass breaks less easily than plain glass. Leaded crystal is composed of up to 24 % lead. Some of the lead can dissolve in acidic drinks such as fruit juice because metals react with acids.
- The plumbing in many older buildings is made of copper pipes held together with solder that contains lead. Both copper and lead are toxic, and both dissolve more readily in hot water than in cold water. That is why hot water from the faucet should never be used to cook food.

Figure 4 Van Gogh painted *The Starry Night* while in an asylum at Saint-Rémy, France, in 1889. Van Gogh painted the stars with rings, which suggests he may have suffered from lead poisoning.

DIG DEEPER

Career Link

Plumbers install plumbing fixtures—bathtubs, showers, sinks, and toilets—and appliances such as dishwashers and water heaters. They also replace old pipes that can be a source of lead. To find out more about being a plumber,

GO TO NELSON SCIENCE

CITIZEN ACTION

RECYCLE A CELLPHONE AND SAVE A GORILLA

You may be surprised to learn that our society's increasing reliance on cellphones is endangering the lives of gorillas in Africa. A key part of cellphones and laptops is the metal element tantalum (Ta), shown in Figure 5. Tantalum is strong and has the unique ability to store electrical charges. These properties make tantalum ideal for making the tiny components of circuit boards in small electronic devices. Unfortunately, most of the world's tantalum comes from mines in the rainforests of central Africa. This is where the endangered lowland gorillas live (Figure 6).

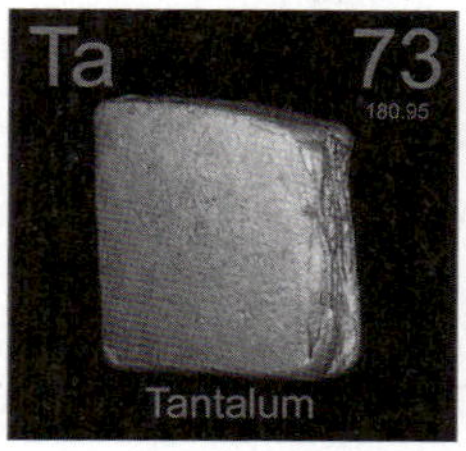

Figure 5 Tantalum is a lightweight metal used in small electronic devices.

Figure 6 Lowland gorillas in central Africa

Mining destroys habitats and threatens wildlife. Gorillas are hunted and killed as food for the miners. The demand for cellphones has increased dramatically in recent years, so many more miners have come to the area. Since this mining boom began, the population of eastern lowland gorillas has dropped by approximately 70 %.

Old cellphones can be recycled to reclaim and reuse tantalum. This reduces the need for further mining. In addition, refurbishing and reusing cellphones are ways to keep the hazardous materials in them out of the environment.

What Can You Do to Help?

- Start a cellphone recycling program at school. Get family and friends to help.
- Make and display posters about the need for cellphone recycling.
- Research the EcoCell program, which is helping restore the habitat of lowland gorillas in Africa.
- Buy a new cellphone only when necessary.

GO TO NELSON SCIENCE

6.3 Wrap Up

- Most elements are metals.
- Metal elements are usually shiny, are good conductors, and are malleable and ductile.
- Many things are made from or with metals because of their useful properties.
- Some metals are poisonous when taken in with food or drink.

CHECK YOUR LEARNING

1. What properties of metals makes them useful in these applications? K/U A
   (a) Aluminum is used to make the outside surface of some car doors.
   (b) Thin gold wires are used to connect one part of a computer chip to another.
   (c) The bottoms of some frying pans are made of copper.
2. Why is it important to take old car batteries containing lead to a hazardous waste depot for recycling? K/U A
3. Why do people in older homes often let their tap water run for a few minutes before filling their drinking glass? K/U A
4. Why are some people sensitive to certain types of jewellery? A

6.4 Non-metals

13	14	15	16	17	18
					He
B	C	N	O	F	Ne
Al	Si	P	S	Cl	Ar
Ga	Ge	As	Se	Br	Kr
In	Sn	Sb	Te	I	Xe
Tl	Pb	Bi	Po	At	Rn

Figure 1 The non-metals (in red) are located on the right of the periodic table. The only exception is hydrogen, element number 1.

Metals far outnumber non-metals on the periodic table (**Figure 1**). And yet, well over 90 % of all known substances are made only of non-metals. This means that the majority of the things around you, including your body, contain mostly non-metals (**Figure 2**).

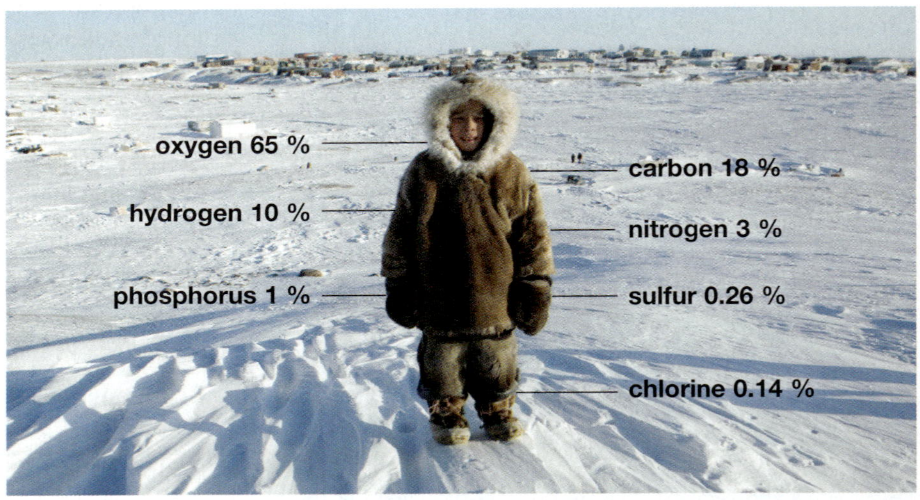

Figure 2 Over 95 % of the mass of the human body is made up of non-metals.

▶ **non-metal:** an element that is usually a gas at room temperature or a dull, brittle solid, and a poor conductor of thermal energy and electricity

Properties of Non-metal Elements

What is a **non-metal**?

- Many non-metals are gases at room temperature. A few are solids and one is a liquid.
- Solid non-metals do not have the metallic shine of metals. Most have a dull lustre. Some, such as sulfur, are brightly coloured.
- Solid non-metals are brittle, which means they break or shatter when bent or struck.
- Non-metals are usually poor conductors of electricity and thermal energy.

Examples: carbon (C), sulfur (S), oxygen (O), chlorine (Cl), iodine (I), neon (Ne)

A Solid Non-metal

Sulfur is a brittle, bright yellow solid. Sulfur burns with a blue flame to produce sulfur dioxide. In nature, sulfur is often found deposited at the edges of hot springs and geysers. It is also often found around volcanoes (**Figure 3**). Links between sulfur atoms within hair proteins are what make hair curly. Sulfur is also used to make important industrial chemicals like sulfuric acid.

Figure 3 Sulfur is one of the few non-metal elements that is a solid.

Non-metal Elements in Air

Most gases in air are non-metal elements. The composition of air is

- 78 % nitrogen
- 21 % oxygen
- 0.9 % argon
- and 0.1 % neon, helium, krypton, and compounds like water and carbon dioxide (**Figure 4**)

Oxygen is the gas all animals need to breathe. Recall from Chapter 2 that green plants release oxygen into the air during photosynthesis.

Nitrogen makes up 78 % of the air we breathe, but we do not normally notice it. However, nitrogen in the compressed air that scuba divers breathe can cause serious problems. As the diver goes deeper, the amount of dissolved nitrogen in the diver's blood increases. Surfacing too quickly causes bubbles of nitrogen to form in the diver's blood. This can result in pain, dizziness, loss of feeling, or even death.

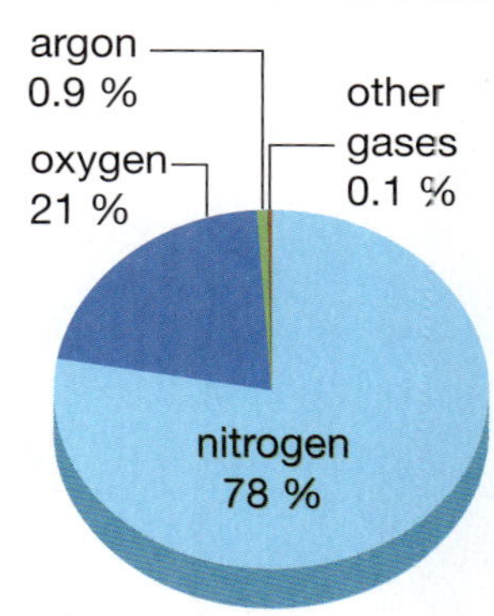

Figure 4 The composition of air

To React or Not to React

The last two groups in the periodic table are non-metals that have very different properties.

Group 17: Halogens

The halogens in Group 17 are fluorine (Fl), chlorine (Cl), bromine (Br), iodine (I), and astatine (At). All these elements are highly reactive. Fluorine reacts explosively with hydrogen to form hydrogen fluoride. In water, this substance makes hydrofluoric acid, which is used to etch glass. Pure chlorine, bromine, and iodine are poisonous to living things (**Figure 5**). Chlorine is used to disinfect water for drinking or swimming because it kills bacteria and other pathogens. Bromine has a wide range of uses, including dyes, medicines, pesticides, and photography. Iodine dissolved in alcohol is used to disinfect wounds. Campers often use iodine tablets to make water safe to drink because iodine kills bacteria. Another common use of these elements is in halogen lights. These lights are very bright, but also give off a great deal of thermal energy. This property makes them a potential fire hazard.

Figure 5 Three members of the halogen family. At room temperature, chlorine is a yellow gas, bromine is a red liquid that readily vaporizes, and iodine is a purple solid. Iodine changes to a purple gas when heated.

Group 18: Noble Gases

Unlike the halogens, the noble gases in Group 18 rarely react. But this does not mean they have no uses. Balloons filled with helium (He) rise because helium is less dense than air. Xenon (Xe) is used in the bluish, high-intensity headlights of certain vehicles. Neon (Ne) and argon (Ar) give off brightly coloured light when electricity passes through them. This makes neon and argon useful for decorative lights (**Figure 6**).

Figure 6 Each noble gas glows with its own distinctive colour when electricity passes through it.

DIG DEEPER
Web Link

For more about non-metal elements,

GO TO NELSON SCIENCE

6.4 Wrap Up

- There are fewer non-metal elements than metals.
- Most of the things around you are made up of non-metal elements.
- Non-metals are on the right side of the periodic table.
- Like metals, non-metals have distinctive properties that make them useful.

UNIT TASK
Bookmark

You can apply what you learned about non-metal elements to the Unit Task described on page 276.

CHECK YOUR LEARNING

1. List at least three properties of non-metals. K/U
2. Noble gases do not react easily. How is this property useful? K/U A
3. Nitrogen becomes a liquid at –196 °C. A rubber ball placed in liquid nitrogen becomes very brittle. What do you think would happen if you tried to bounce the ball? T/I
4. Hydrogen is less expensive and less dense than helium. Why do you think helium is used to fill balloons instead of hydrogen? K/U A
5. Why do you think hydrofluoric acid is always stored in plastic bottles? T/I
6. List three useful applications of halogens. K/U A

6.5 Metalloids: Part Metal, Part Non-metal

Only a few elements in the periodic table are classified as metalloids (**Figure 1**). They are even fewer in number than the non-metals, but they have some very useful properties.

13	14	15	16	17	18
					He
B	C	N	O	F	Ne
Al	Si	P	S	Cl	Ar
Ga	Ge	As	Se	Br	Kr
In	Sn	Sb	Te	I	Xe
Tl	Pb	Bi	Po	At	Rn

Figure 1 The green boxes are elements that are metalloids. They are along the staircase line on the right side of the periodic table.

Properties of Metalloids

- Metalloids have properties in between those of metals and non-metals.
- All metalloids are solids at room temperature.
- Some metalloids are weak conductors of electricity and thermal energy.

Examples: boron (B), silicon (Si), germanium (Ge), arsenic (As)

Semiconductors

Metalloids play an important role in the electronics industry (**Figure 2**). Some metalloids conduct electricity to some degree. This property makes silicon and germanium useful to make semiconductors. Semiconductors are used to make parts of electric circuits in electronic devices. The flow of electricity in a semiconductor can more easily be controlled than the flow of electricity in a conductor such as copper. This property makes it possible for computer chips to do complex tasks.

Figure 2 Here are a few examples of metalloid elements: (a) boron; (b) silicon; (c) germanium; (d) arsenic.

A Toxic Element

Arsenic was once the most famous metalloid. Like many of its neighbours in the periodic table, arsenic is toxic and has often been used as a poison. While poisoning people with arsenic may be the stuff of old detective novels, this element still has harmful effects today. Arsenic was used in industrial processes. It was used as a wood preservative for outdoor building materials and as an insecticide on fruit trees. As a result, it has escaped into the environment and contaminated soil, water, air, and food. Although it is no longer used as a wood preservative, many older outdoor structures, such as decks, picnic tables, and playground equipment, may still be hazardous.

RESEARCH THIS SILICON OR SILICONE?

SKILLS HANDBOOK 4.A.

SKILLS: Researching, Analyzing the issue, Communicating

The names silicon, silica, and silicone are so similar that they are often confused. Pure silicon is made up of only atoms of the metalloid silicon. Silica, however, contains silicon as well as oxygen. Silicone is a plastic-like substance that contains silicon, carbon and oxygen. Research the uses of these three related substances. Present your research in a chart that summarizes the properties and uses of these substances.

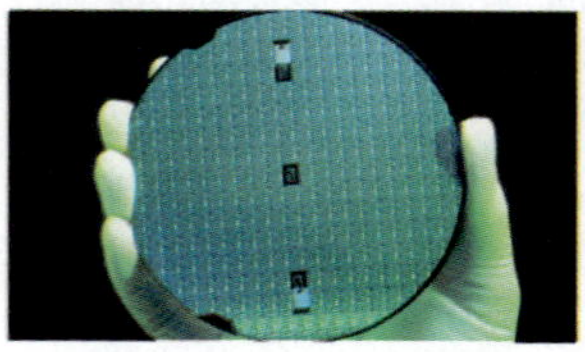

Figure 3 A thin wafer of the element silicon.

GO TO NELSON SCIENCE

6.5 Wrap Up

- A metalloid is an element that has properties in between those of metals and non-metals.
- Metalloids are used to make the semiconductors in computer chips.

CHECK YOUR LEARNING

1. What is a semiconductor? K/U
2. Describe the location of the metalloids on the periodic table. K/U
3. Why might it be a good idea for a community to have picnic tables in their parks tested for arsenic? What other items might the community also want to have tested? T/I A
4. What properties of metalloids most affect things that you use in your everyday life? K/U A
5. Use a three-column table to classify the following elements as metals, non-metals, or metalloids: carbon, arsenic, hydrogen, silver, silicon, aluminum, boron, phosphorus, iron. K/U C
6. (a) Describe the properties of metalloids that make them unsuitable for necklaces.

   (b) Describe the properties of metalloids that make them unsuitable for extension cords. K/U A
7. Place these elements in order from best conductor of electricity to worst. Explain your thinking. K/U T/I

   Zn Br Ge

6.6

Charcoal Pendants or Diamond Pencils?

What do charcoal, a pencil lead, and a diamond have in common (Figure 1)? Surprisingly, one element—carbon—makes up all three substances.

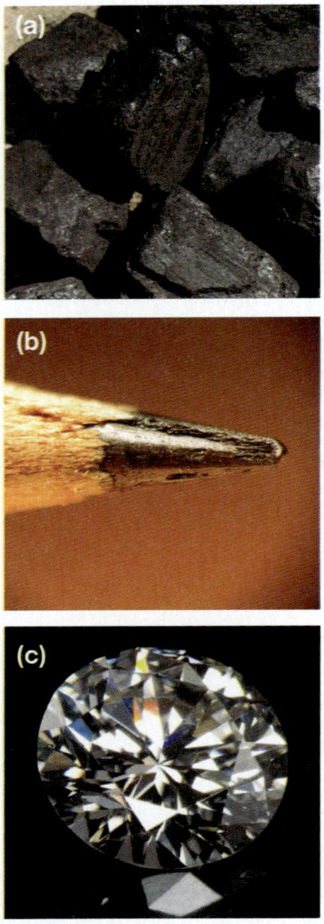
Figure 1 (a) Charcoal, (b) graphite, and (c) diamond are three forms of the element carbon.

Charcoal

Charcoal is a solid that is black, crumbles easily, and burns readily. It is made by slowly burning logs. Charcoal consists of up to 98 % carbon mixed with ash and other chemicals. We restart the burning process by lighting charcoal briquettes in our barbecues.

Graphite

Graphite is a soft form of carbon. The carbon atoms in graphite are joined together in layers. Pencil "lead" is actually an impure form of graphite. When you press a pencil lightly on paper, the graphite leaves carbon atoms on the surface of the paper. The structure of graphite makes it a good conductor, which is an unusual property for a non-metal. Products made from graphite composites include fishing rods, golf clubs, bicycle frames, and pool cues.

Diamonds

Diamonds form under extremely high temperature and pressure. They sparkle brilliantly after they are cut. They do not conduct electricity. They also rank at the top of the hardness scale—and the price scale! The carbon atoms in a diamond are connected in a pattern that gives the diamond its hardness.

TRY THIS COPPER-PLATING YOUR PENCIL

SKILLS HANDBOOK 3.B.5.–3.B.7., 3.3.9.

SKILLS: Performing, Observing, Analyzing, Communicating

Graphite is a non-metal with the unusual property of being a good conductor. You can coat the graphite tip of a pencil with copper. The method you will use is called electroplating. Many beautiful objects, such as silver trays and gold jewellery, are electroplated. To be electroplated, an object must be able to conduct electricity.

Equipment and Materials: eye protection; lab apron; 9-V battery; medium-sized beaker; graduated cylinder; 2 electrical wires with clips; 2 pencils with both ends sharpened; dilute copper(II) sulfate solution

1. Put on your lab apron and eye protection.
2. Pour about 50 mL of copper(II) sulfate into a beaker.

> **CAUTION: TOXIC**
> Copper(II) sulfate is a skin and eye irritant. Wash your hands thoroughly with soap and water after pouring the solution.

3. Connect the electrical wires to one end of each sharpened pencil.
4. Connect the electrical wires to the two terminals of the 9-V battery.
5. Place the free tips of the pencils into the copper(II) sulfate solution so that the two pencil tips do not touch (**Figure 2**). Make sure that the wire connectors are above the solution.

Figure 2 Copper-plating apparatus

6. After several minutes, disconnect the wires from the battery and remove the pencils from the solution. Rinse the pencils gently in water, and observe the tips.
7. Follow your teacher's instructions for the disposal of the materials and equipment.
8. Wash your hands thoroughly with soap and water.

A. What did you observe at the pencil tips while there was electricity flowing in the circuit? T/I

B. Where did the solid that formed on one of the pencil tips come from? What element is it? T/I

C. Is this a chemical or physical change? Explain. T/I C

D. What physical property of graphite allowed this change to occur? T/I

The Great Canadian Diamond Rush

Diamonds were first discovered in the Northwest Territories in the 1980s. Now, Canada is one of the top three diamond producers in the world. Mines in the Northwest Territories, Nunavut, and northern Ontario produce more than ten million carats of Canadian diamonds each year. That is more than 15 % of the world's supply!

Benefits and Drawbacks

The diamond rush has benefited the Canadian North in many ways. However, any form of mining disturbs the environment. The large open-pit mines in which companies dig for diamonds can scar the landscape and can lead to harmful runoff (**Figure 3**). Aboriginal peoples are also displaced by diamond mining. About 70 % of Canada's Native reserves are on or near areas that mining companies are interested in.

Figure 3 The Diavik Diamond Mine on Lac de Gras, Northwest Territories

DIG DEEPER

Reading Tip

Revise Your Inferences

Sometimes, you come across new information that conflicts with an inference you already made. For example, you read that Canada will benefit economically from diamond mining. However, after you read about the impact of diamond mining on cost of living, you may change your opinion.

DIG DEEPER

Career Link

To learn more about career opportunities in diamond mining in Canada,

GO TO NELSON SCIENCE

Diamond mining is less harmful to the environment than gold mining. In gold mining, a poisonous substance called cyanide is used to purify the gold. Still, diamond mining in northern Canada has caused lakes to be drained and streams to be buried. This means that fish habitats have been lost, and water quality is poorer.

Diamond mining also has drawbacks for society. Mining brings high-paying jobs in the mines, as well as in construction and transportation. However, this sometimes means that there are not enough workers for jobs in the neighbouring towns. Also, a shortage of available housing for all the workers that moved up north suddenly can cause rent and housing prices to skyrocket.

Some diamonds mined in parts of Africa are controversial. They are called "blood diamonds"—diamonds from mining operations that finance war and terrorism. Canadian diamonds are among those marketed as "conflict-free" diamonds. Each Canadian diamond is sold with a certificate that guarantees its source. Some are etched with a tiny polar bear trademark (**Figure 4**).

Figure 4 A tiny polar bear is the trademark for Canadian diamonds.

6.6 Wrap Up

- Charcoal, graphite, and diamond are different forms of carbon.
- Diamond mining in Canada's North has brought economic benefits, but it has also raised social and environmental concerns.

CHECK YOUR LEARNING

1. Which property of graphite is not shared by charcoal and diamond? K/U
2. Which properties of charcoal, graphite, and diamond make each of them useful? K/U
3. Why is diamond mining less harmful to the environment than gold mining? K/U
4. Suppose you are the mayor of a town near a newly discovered diamond deposit. What guarantees would you want before a mine is allowed to open? A
5. Do you think buying Canadian diamonds is better or worse for society than buying diamonds from elsewhere? Explain your thinking. C A

TECH CONNECT OSSLT

Tough as Titanium

What is three times stronger than steel but weighs half as much? What also kills germs and helps keep broken bones together? It is titanium!

Titanium (Ti) is element number 22 on the periodic table. Pure titanium does not exist in nature. Instead, it has to be extracted from the minerals found underground (**Figure 1**). Separating and refining titanium from its minerals is difficult. This is why titanium metal is so expensive.

Figure 1 Titanium occurs mainly in the mineral deposits rutile and ilmenite. These minerals are widely distributed in Earth's crust.

The fact that titanium is stronger than steel but has half its mass makes it ideal for use in airplanes. Alloys of titanium are also used to make sports equipment such as hockey sticks and tennis racquets, and jewellery (**Figure 2**). Currently, titanium is a popular choice for men's wedding rings. It is stronger than gold or platinum but costs much less.

The most common use of titanium is to make titanium oxide. This substance is the white pigment in house paint and toothpaste. Titanium oxide is ideal for these uses because it has a bright white colour and is non-toxic.

Figure 2 Titanium is lightweight and very strong, making it ideal for use in sports equipment.

Titanium also has important medical applications. Artificial joints for hips and knees are made from titanium alloys (**Figure 3**). They are lighter than steel joints and last a long time because titanium is strong and does not corrode.

Titanium has the unique ability to absorb ultraviolet light. Surgical tables made of titanium can be conveniently sterilized simply by shining ultraviolet light on them. The absorbed energy kills any bacteria on the table. Similarly, walls and equipment in the operating room can be sterilized by spraying them with a mist of titanium compounds and then shining ultraviolet light on them.

GO TO NELSON SCIENCE

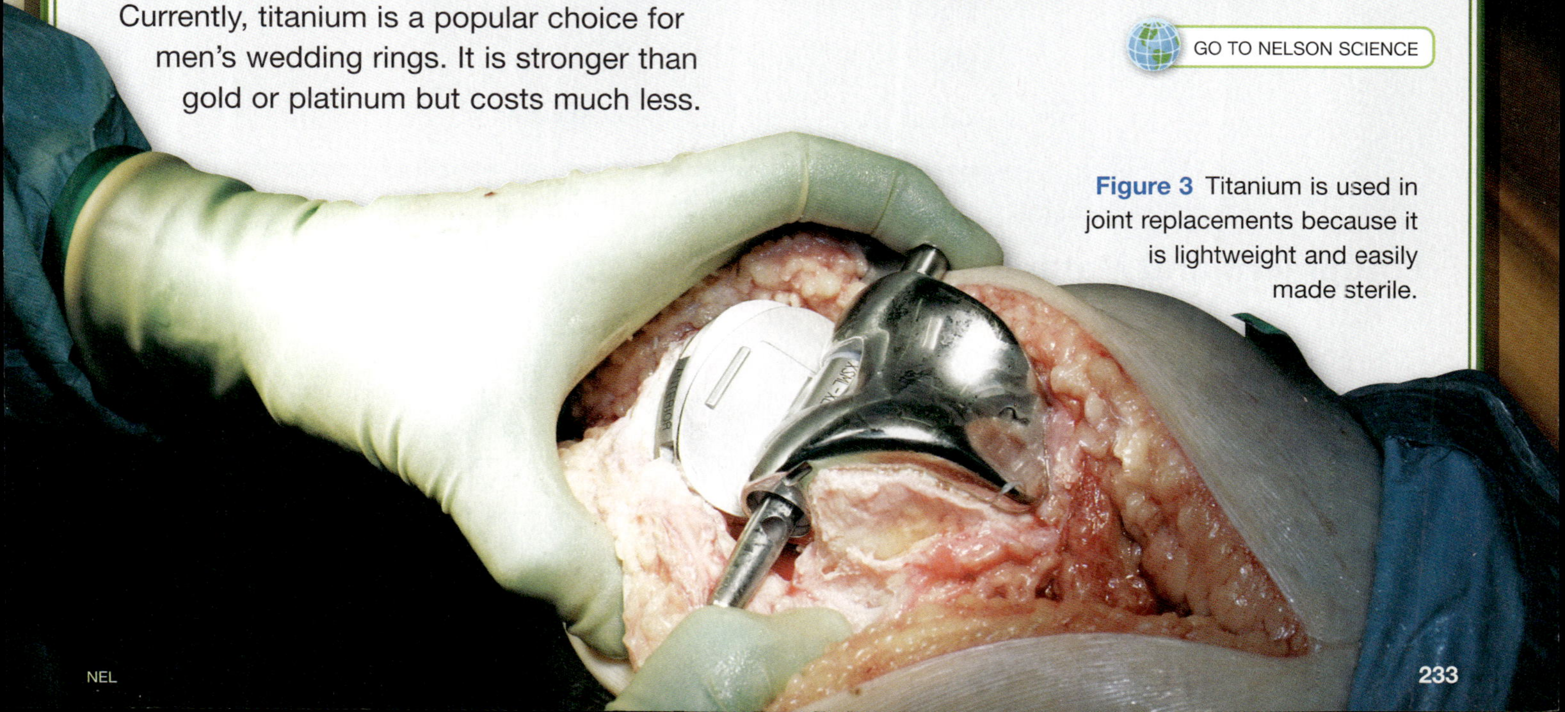

Figure 3 Titanium is used in joint replacements because it is lightweight and easily made sterile.

Family Resemblances

SKILLS MENU
- Questioning
- Hypothesizing
- Predicting
- Planning
- Controlling Variables
- Performing
- Observing
- Analyzing
- Evaluating
- Communicating

The groups in the periodic table are also called chemical families. Like any family, they share some characteristics. The elements in group 1 are called the alkali metals. The elements in group 2 are called the alkaline earth metals.

SKILLS HANDBOOK 3.B.5.–3.B.7., 3.B.9.

Purpose

To compare properties of elements in two groups of the periodic table

Procedure

1. Copy **Table 1** in your lab notebook.
2. Put on your lab apron and eye protection.
3. Observe your teacher's demonstration. Notice the appearance of freshly cut lithium (Li) and sodium (Na) and how each element behaves when placed in water. Record your observations.

Table 1 Observations

Metal	Appearance of metal	Temperature of water	Observations
Li		cold water	
Na		cold water	
Mg		cold water	
Mg		hot water	
Ca		cold water	

Equipment and Materials

lab apron

eye protection

3 small beakers

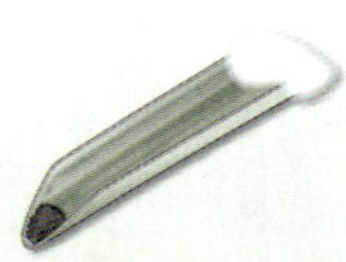
scoopula

magnesium metal

calcium metal

4. Place a small strip of magnesium (Mg) into each of two beakers.
5. Record the appearance of the magnesium.
6. Add about 10 mL of cold water to the first beaker and 10 mL of hot water to the second beaker. Allow the mixtures to sit for 1 min. Record your observations.
7. Add about 30 mL of cold water to a third beaker. Then use the scoopula to add a small piece of calcium (Ca) to the beaker. Record your observations.
8. When all samples have been observed, follow your teacher's instructions for returning or disposing of all materials and equipment.
9. Wash your hands thoroughly with soap and water.

CAUTION

Irritant

The mixture that is produced in Step 7 is an irritant. Avoid skin contact. Dispose of this mixture as directed by your teacher.

Analyze and Evaluate

(a) Compare and contrast the properties of the two alkali metals. T/I

(b) Compare and contrast the properties of the two alkaline earth metals. T/I

(c) Compare the reactivity in water of the two groups of metals. T/I

Apply and Extend

(d) Use the observed patterns to predict the physical properties and the relative reactivity of K, Rb, Cs, and Fr. T/I

(e) Use the observed patterns to predict the physical properties and the relative reactivity of Sr and Ba in water. T/I

Chapter 6 KEY CONCEPTS SUMMARY

Elements cannot be broken down into simpler substances.

- All substances are made up of combinations of elements in the periodic table.
- Elements are organized in the periodic table in order of increasing atomic number.

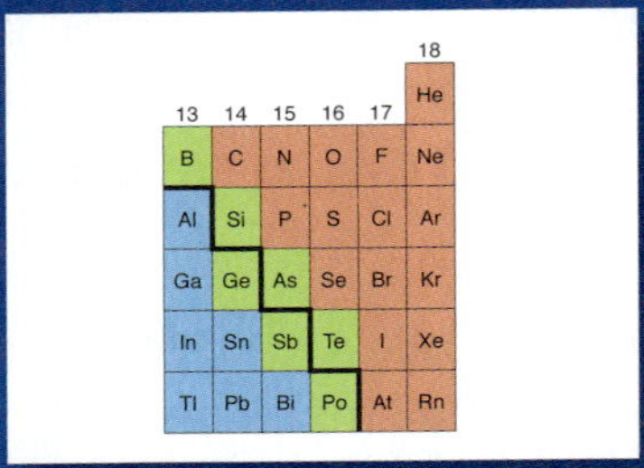

Elements of the same column (group) of the periodic table have similar properties.

- Elements in the same family have the same number of electrons in their outermost orbits.

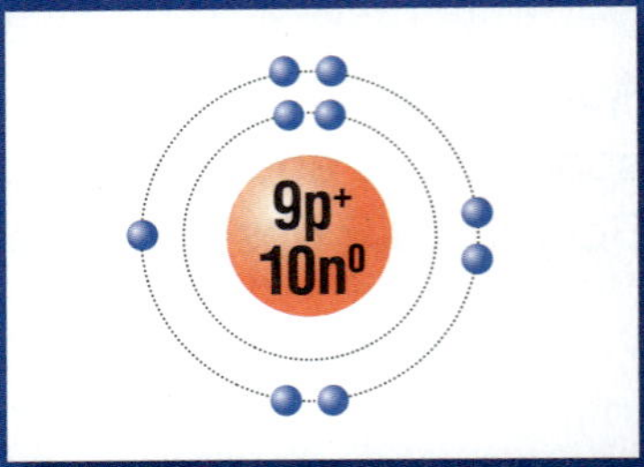

Atoms consist of protons and neutrons in a central core surrounded by electrons.

- Protons have a positive charge.
- Electrons have a negative charge.
- Neutrons have no charge.

Metalloids have properties in between those of metals and non-metals.

- Metalloids are used to make semiconductors found in computer chips.

Elements can be both beneficial and harmful.

- Some elements are harmful to humans even in very small quantities.
- Some elements are beneficial to humans, but mining these elements can be harmful to humans and the environment.

Metals and non-metals have characteristic physical properties.

- Metals are elements that have lustre and are conductors, malleable, and ductile.
- Non-metals are elements that are mostly gases or dull, powdery solids.

WHAT DO YOU THINK NOW?

You thought about the following statements at the beginning of the chapter. You may have encountered these ideas in school, at home, or in the world around you. Consider them again and decide whether you agree or disagree with each statement.

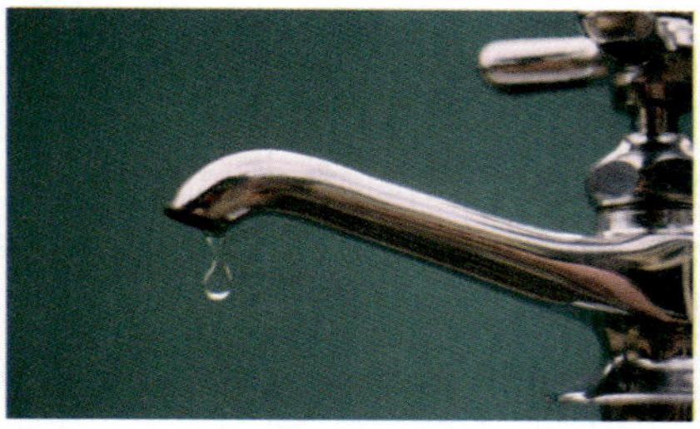

1 Water is a common element.
Agree/disagree?

2 Pure gold makes the best gold jewellery.
Agree/disagree?

3 Most of the substances around us are made from metals.
Agree/disagree?

4 Most metals are magnetic.
Agree/disagree?

5 Using hot tap water to prepare food can be bad for your health.
Agree/disagree?

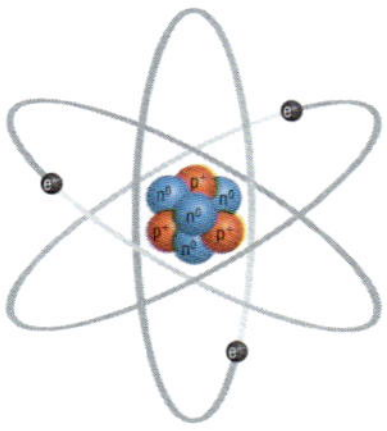

6 An atom is the smallest particle of an element.
Agree/disagree?

How have your answers changed since then?
What new understanding do you have?

BIG Ideas

- Elements and compounds have specific properties that determine their uses.
- The use of elements and compounds has both positive and negative effects on society and the environment.

DIG DEEPER

Vocabulary

proton p. 210
neutron p. 210
electron p. 210
element p. 211
periodic table p. 211
atomic number p. 212
period p. 212
group or chemical family p. 212
metal p. 220
conductivity p. 220
malleable p. 220
ductile p. 220
alloy p. 220
non-metal p. 224

CHAPTER 6 SELF-QUIZ

The following icons indicate the Achievement Chart category addressed by each question.

K/U Knowledge/Understanding
C Communication
T/I Thinking/Investigation
A Application

For each question, select the best answer from the four alternatives.

1. Which is the name for the columns in the periodic table? (6.1) K/U
   (a) periods
   (b) protons
   (c) groups
   (d) gases

2. Which phrase correctly describes what an element's atomic number is? (6.1) K/U
   (a) the element's place in a period
   (b) the number of groups the element is in
   (c) the number of protons in an atom of an element
   (d) the number of electrons in the outermost layer of the atom

3. Elements used to make semiconductors are called (6.5) K/U
   (a) non-metals
   (b) metalloids
   (c) metals
   (d) noble gases

4. Which of these elements is in the same period as calcium (Ca) in the periodic table? (6.1) K/U
   (a) beryllium (Be)
   (b) chromium (Cr)
   (c) radium (Ra)
   (d) rubidium (Rb)

5. When tested, an unknown metal is found to be strongly magnetic and mostly shiny with some dull, flaky orange patches. The unknown metal is most likely to be (6.2, 6.3) T/I
   (a) lead (Pb)
   (b) silver (Ag)
   (c) titanium (Ti)
   (d) iron (Fe)

Indicate whether each of the statements is TRUE or FALSE. If you think the statement is false, rewrite it to make it true.

6. Silicon is one of the metalloids. (6.5) K/U
7. Most of the elements are non-metals. (6.1) K/U
8. Metalloids are good conductors of thermal energy and electricity. (6.3) K/U
9. Mining can have both positive and negative effects on an area. (6.4) K/U
10. Sulfur is a bright yellow element that is malleable and ductile. (6.4) K/U

Copy each of the following statements into your notebook. Fill in the blanks with a word or phrase that correctly completes the sentence.

11. Metalloids are better conductors of electricity than __________ but not as good as __________. (6.5) K/U
12. The ability of an element to be drawn into a wire is called __________. The ability of an element to bend is called __________. (6.3) K/U
13. The core of an atom contains __________ that are positively charged and __________ that have no charge. (6.1) K/U

14. Most elements in living things are __________. (6.4) K/U

15. A(n) __________ is the smallest piece of an element with the properties of that element. Each one has a unique number called the __________ __________. (6.1) K/U

16. The three main forms of pure carbon found in nature are __________, __________, and __________. (6.6) K/U

Match each term on the left with the most appropriate description on the right.

17. (a) period	(i) ability to allow electricity and thermal energy to flow through (6.2, 6.3) K/U
(b) conductivity	(ii) one kind of matter that cannot be separated into two or more simpler types of matter (6.1) K/U
(c) ductility	(iii) a row in the periodic table (6.1) K/U
(d) group	(iv) the ability to be pulled into a wire (6.2) K/U
(e) element	(v) a column in the periodic table (6.1) K/U

Write a short answer to each of these questions.

18. Explain the difference between a group and a period in the periodic table. (6.1) K/U

19. A dull red, brittle element can be found on the side of some packs of matches. What type of element is it? Explain your reasoning. (6.2, 6.3, 6.4) T/I

20. What is a primary marketing message that is used to sell Canadian diamonds? How are the diamonds identified? (6.6) K/U C

21. Many pans used for cooking are made of metal because metals easily conduct thermal energy. What would be the advantage of making the handle of a pan out of wood? What would be a disadvantage? (6.3, 6.4) A

22. A mixture of silver and mercury called an amalgam is used by dentists to fill cavities in teeth caused by tooth decay. What are two properties you think an amalgam should have to be useful for fillings? (6.3) A

23. Compare and contrast graphite and charcoal. What do these two substances have in common? How are their properties and uses different? (6.6) K/U C

24. Why do metalloids play an important part in the electronics industry? (6.5) K/U A

25. The elements in Group 17, the halogens, and the elements in Group 18, the noble gases, are very close to each other in the periodic table. How do their chemical properties compare? (6.4) K/U

26. Your friend buys a new pair of earrings, but after wearing them she notices that her ears are sore and red. What could she conclude? (6.3) A

27. If you know that an element is a solid at room temperature, do you have enough information to classify the element as a metal, metalloid, or non-metal? Why or why not? (6.2, 6.3, 6.4, 6.5) T/I

28. Why are the halogens found in nature only in compounds? (6.4) A

29. Why do elements in a group have similar chemical properties? (6.1) K/U

30. What is "periodic" about the periodic table? (6.1) K/U

Self-Quiz

To do an online self-quiz,

GO TO NELSON SCIENCE

CHAPTER 6

REVIEW

The following icons indicate the Achievement Chart category addressed by each question.

K/U Knowledge/Understanding
T/I Thinking/Investigation
C Communication
A Application

What Do You Remember?

1. Write the chemical symbol for each element listed below. (6.1) K/U C
   (a) barium
   (b) chlorine
   (c) sodium
   (d) oxygen
   (e) mercury
   (f) hydrogen

2. Write the name for the element indicated by each symbol below. (6.1, 6.5) K/U C
   (a) C
   (b) N
   (c) P
   (d) Ti
   (e) Si
   (f) Mg

3. What is the purpose of the staircase line in the periodic table? (6.1, 6.5) K/U C

4. How are the elements grouped in a column of the periodic table alike? (6.1) K/U

What Do You Understand?

5. What characteristic determines the order of the elements in the periodic table? (6.1) K/U

6. Explain how the terms in each pair below are different. (6.1, 6.3) K/U C
   (a) an element and an atom
   (b) ductile and malleable
   (c) a proton and a neutron
   (d) a proton and an electron

7. Compare the electrical conductivity of aluminum, sulfur and phosphorus. (6.2, 6.3, 6.4, 6.5) K/U

8. Describe how you could determine whether an unknown element is a metal or non-metal. (6.2, 6.3, 6.4) K/U T/I

9. Compare and contrast a typical metalloid with typical metals and non-metals. (6.3, 6.4, 6.5) K/U

10. In your notebook, copy and complete **Table 1**. (6.1) K/U

Table 1

Element name	Element symbol	Atomic number	Number of protons	Number of electrons
arsenic				
	Zr			
		82		
			17	
				25

11. What do you think is the most useful property of metals? Use examples from everyday life to explain. (6.3) C A

12. You and a classmate are examining the periodic table. He says that since there are so many different kinds of metals, more substances must be made up of metals than of non-metals. Is your classmate right? Why or why not? (6.3, 6.4) T/I C

13. The atomic model for fluorine is shown in **Figure 1**. (6.1) K/U T/I
    (a) How many protons does it have? What is the atomic number for fluorine?
    (b) What do the blue dots represent? How does their number compare to the number of protons?
    (c) In what group of the periodic table is fluorine located?

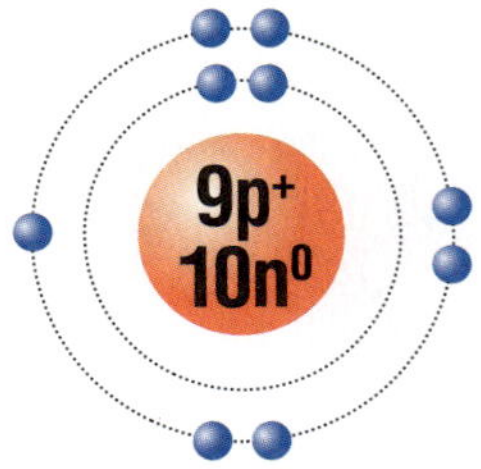

Figure 1

Solve a Problem

14. Examine the atomic models in **Figure 2**. Use the information in each model to answer the questions that follow. (6.1, 6.8) T/I
    (a) What is the element shown in model (a)?
    (b) What is the element shown in model (b)?
    (c) What is the element shown in model (c)?
    (d) Compare the number of electrons in the outer orbit of each atom's nucleus. Is this number the same for all or is it different?
    (e) To what groups in the periodic table do each of these elements belong?

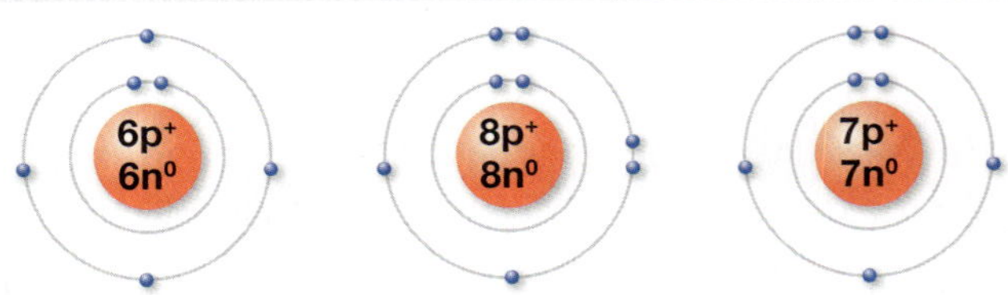

Figure 2

Create and Evaluate

15. Examine the atomic models shown in **Figure 1** and **Figure 2**. (6.1) T/I
    (a) Draw an atomic model for neon (Ne).
    (b) What makes a noble gas, such as Ne, chemically stable, or unreactive?

Reflect on Your Learning

16. Think about what you learned in this chapter. Give examples of three things you learned about elements that you found interesting or helpful. Explain your choices. C A

17. Based on what you learned in this chapter, explain why metalloids are sometimes called "the elements of the future." A C

Web Connections

18. What are some uses of titanium that were not mentioned in the Tech Connect? Why is titanium such a useful metal? What does the term "bio-compatible" mean, and how does it apply to titanium? T/I C A

19. Research current methods used to make synthetic diamonds. How successful are they? What are the environmental advantages of this technology? T/I C A

20. Why is it important to recycle metal components from computers? What are the consequences of not doing this? T/I C A

21. Microprocessors are the "brains" of many electronic devices. Research how metalloids, semiconductors and microprocessors are related. T/I C A

GO TO NELSON SCIENCE

CHAPTER 7

Chemical Compounds

KEY QUESTION

What are chemical compounds and how can they be useful, harmful, or both?

KEY CONCEPTS

- A chemical compound is a pure substance made up of two or more elements.
- A chemical formula is used to represent a pure substance, which can be an element or a compound.
- A molecule is a particle of matter made up of two or more atoms.
- Compounds can be decomposed into simpler substances.
- Many important chemical changes involve oxygen.
- Our use of compounds has both positive and negative effects on society and the environment.

New chemical compounds make high tech sports equipment possible.

Energy Drinks, Boost or Bust?

Everyone needs a boost every now and then, especially when you are about to take a test, or are going to race down a mountain at high speed.

Many people think that an energy drink is a quick and easy way to get that boost.

A Controversial Ingredient

Taurine is a controversial ingredient for two reasons. First, no one knows if it really does what the advertising claims it does. The results of the studies that have been done are unclear.

ENERGY DRINKS CLAIM TO

* **increase performance**
* **improve alertness**
* **stimulate metabolism**
* **make you "feel better"**
* **increase concentration and reaction speed**

But do energy drinks deliver on their claims? How do they work? Are there any dangers from consuming energy drinks?

Sugars and Stimulants

Most energy drinks will give you at least a short-term increase in energy and concentration. This is because they are largely made up of sugars such as glucose and sucrose, and stimulants such as caffeine, inositol, and ginseng. Energy drinks typically contain twice as much caffeine as regular soft drinks, which is as much as a strong cup of coffee. They also contain the compounds taurine and glucuronolactone, which claim to help the body remove toxins.

Second, no one knows what the effects of consuming large quantities of taurine are. No harmful effects have been found so far, but studies continue.

Too Much of a Good Thing?

As with many things we eat or drink, just because a small amount may be okay or even good for us, it does not mean that consuming a large amount is a good idea. Drinking too many energy drinks can lead to upset stomach, insomnia, nervousness, headache, and tachycardia (accelerated or rapid heartbeat).

WHAT DO YOU THINK?

Many of the ideas you will explore in this chapter are ideas that you have already encountered. You may have encountered these ideas in school, at home, or in the world around you. Some of the statements are not true. Consider each statement and decide whether you agree or disagree with it.

1 Elements can join chemically to form more complex substances.
Agree/disagree?

2 When a compound breaks down, gases are always released.
Agree/disagree?

3 The combination of oxygen with a substance always produces a flame.
Agree/disagree?

4 Most plastics completely break down in a few years.
Agree/disagree?

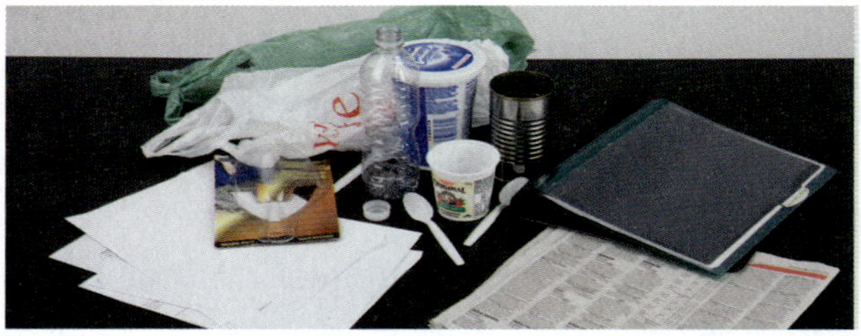

5 To protect the environment, we need to reduce the amount of materials we use, reuse what we can, and recycle when possible.
Agree/disagree?

6 All metals rust.
Agree/disagree?

FOCUS ON

WRITING

Writing a Science Report

When you write a science report, you use a specific format to organize your investigation. Use the strategies listed next to the report below to help you with the Unit Task on page 276.

DIG DEEPER

Writing Tip

As you work through this chapter, look for tips like this. They will help you develop literacy strategies.

Write a brief and accurate testable question and hypothesis.

List every item that you will need to conduct your investigation.

Write steps in the procedure in proper order.

Cleaning "Green"

Testable Question

Which cleaning product will work best to remove stains from cotton?

Hypothesis

Store-bought detergent will work better on food stains than homemade cleaners because it has more cleaning power.

Experimental Design

Identical pieces of cotton cloth will be stained with various foods and soaked in various cleaners. After 30 min, the stains will be compared.

Equipment and Materials

eye protection; lab apron; graduated cylinder; 9 pieces of cotton; ...

Procedure: Each stain will be tested with each of the 2 cleaners.

1. Put on eye protection and lab apron.
2. Label "vinegar," "detergent," and "tap water" on 3 beakers each.

Present your observations in a logical way. For example, you might present your observations in a data table. Then, evaluate how your results support your hypothesis. Does your evidence help you answer the testable question?

7.1 PERFORM AN ACTIVITY

Colour to Dye For

SKILLS MENU

- Questioning
- Hypothesizing
- Predicting
- Planning
- Controlling Variables
- Performing
- Observing
- Analyzing
- Evaluating
- Communicating

You may have seen tags on your clothing that say "Wash separately with like colours." That is a warning that the fabric contains a dye that is not permanent. This means the dye is not attached to the fabric. It might wash out in the water and stain other clothes. However, most dyes used in clothing are permanent. They attach to molecules in the fabric and form a new substance that will not wash away.

DIG DEEPER

Reading Tip

The Overall Picture

Before doing any activity, read all the steps in the procedure. List the terms you do not understand. Find out what they mean. Read the procedure again.

SKILLS HANDBOOK 3.B.5.–3.B.7., 3.B.9.

Purpose

To observe factors involved in dyeing fabrics and determine the method that produces the best colour

Procedure

1. Copy **Table 1** into your notebook. Allow enough room in each cell to write a sentence.

Equipment and Materials

eye protection

lab apron

protective gloves

permanent marker

4 beakers

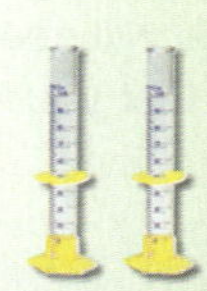
2 graduated cylinders

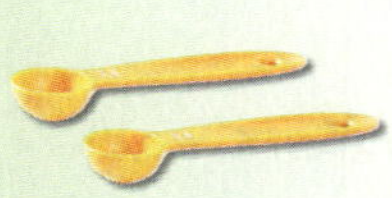
2 tablespoons

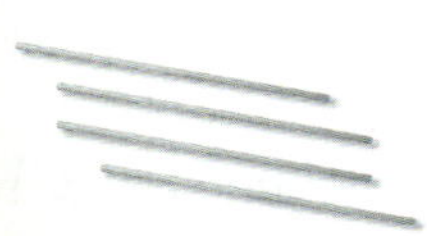
4 glass stirring rods

salt

water

paper towel

8 strips of cotton, 3 cm × 10 cm

8 strips of wool, 3 cm × 10 cm

synthetic dye

natural dye

Table 1

Dye solutions	Cotton (C)	Wool (W)
Synthetic dye, with salt		
Synthetic dye, no salt		
Natural dye, with salt		
Natural dye, no salt		

2. Use a permanent marker to write *C* on each of eight strips of cotton fabric. Then, write *W* on eight strips of wool fabric.
3. Use the marker to label each beaker: synthetic dye, with salt; synthetic dye, no salt; natural dye, with salt, natural dye, no salt.
4. Put on eye protection, a lab apron, and protective gloves.
5. Choose a synthetic dye to test. Pour 150 mL of the dye solution into each of the two clean beakers labelled "synthetic dye."
6. Add 1 tbsp salt to the beaker labelled "synthetic dye, with salt." Stir the solution with a stirring rod.
7. Repeat Steps 5 and 6 for the natural dye solution you choose.
8. Place two cotton strips and two wool strips in the solution in each beaker. Note the time you placed the fabric strips into each dye solution.
9. Gently stir the solutions every 5 min for 30 min. Dispose of the dye solutions according to your teacher's instructions.
10. Describe the colour of the fabric strips. Record your observations. Then, rinse the strips in the sink for 2 min.
11. Place a few paper towels in a stack. Lay the fabric strips on the paper towel stack, beside the beakers in which they were dyed. Allow the strips to dry overnight.

CAUTION

Irritant/Stains

Dyes might irritate skin or stain skin or clothing. Wear eye, skin, and clothing protection, and immediately wash any skin or clothing that comes in contact with a dye.

Analyze and Evaluate

(a) What did the strips look like when they came out of each dye? How does this new colour compare to what the strips looked like after rinsing and drying them? In **Table 1**, write a sentence that describes each fabric strip and its dye solution. T/I C

(b) Did adding salt have any effect on how well each dye coloured the strips? Explain your thinking. T/I

7.2 Putting Atoms Together

Figure 1 Graphite is made entirely of carbon (C) atoms.

In Chapter 6, you read that all substances are made of the elements of the periodic table. The way that the atoms in a substance are arranged determines the substance's properties. For example, the carbon atoms in graphite are arranged in layers that are only loosely held together (Figure 1). That is why graphite is soft and brittle. These properties also allow some of the graphite to slide off of your pencil and onto paper as you write.

Molecules

Most substances are not made up of individual atoms. Instead, they are made up of molecules. A **molecule** is a group of atoms that are chemically joined together. For example, the air you breathe contains many kinds of molecules. These molecules include oxygen, carbon dioxide, and water vapour.

- **molecule:** two or more atoms that are chemically joined
- **subscript:** the small number that follows the symbol for an element in a chemical formula and shows how many atoms of the element are present

Oxygen and Ozone

Oxygen molecules make up 21 % of air. Each oxygen molecule is made up of two atoms of oxygen joined together (Figure 2(a)). A chemical formula is used to describe the number and type of atom in each substance. The chemical formula for an oxygen molecule is O_2. The **subscript**, 2, tells you that there are two atoms of oxygen in each molecule of oxygen.

Oxygen can also be found as O_3 molecules. This type of oxygen has a special name: ozone (Figure 2(b)). Because O_2 and O_3 molecules are put together differently, these substances have different chemical and physical properties. Your body breathes in O_2 molecules to survive. O_3, on the other hand, is toxic and highly reactive. Inhaling ozone damages lung tissue.

DIG DEEPER

Did You Know?

The Good and Bad of Ozone

Ozone high in the atmosphere protects Earth by absorbing harmful UV radiation from the Sun. Near Earth's surface, ozone is an irritating air pollutant in smog.

Figure 2 (a) An oxygen molecule (O_2) consists of two oxygen atoms joined together. (b) An ozone molecule (O_3) consists of three oxygen atoms.

Molecular Elements

Like an oxygen atom, a molecule of oxygen is also an element. Oxygen is an element that occurs as molecules. There are two different kinds of oxygen molecules, O_2 and O_3. Hydrogen (H_2) and phosphorus (P_4) are other examples of elements that occur as molecules (**Figure 3**).

There are seven elements that form molecules consisting of two atoms: H_2, N_2, O_2, F_2, Cl_2, Br_2, and I_2. A molecule that contains two atoms is sometimes called "diatomic." The prefix *di-* means "two."

Figure 3 This diver is using a torch that burns hydrogen gas (H_2) with a flame so hot that the torch can be used effectively underwater.

Compounds

Other than oxygen (O_2) and nitrogen (N_2) in air, it is rare to find elements on their own. Most of the pure substances we come across contain more than one element.

Water, for example, is made up of two different elements—hydrogen (H) and oxygen (O). A pure substance that contains two or more elements is called a **compound**. The chemical formula for the compound water is H_2O. That means that each molecule of water has two H atoms and one O atom. The formula for water is read H-2-O. Note that the subscript 1 is not written. **Figure 4** shows models of three compounds found in the air.

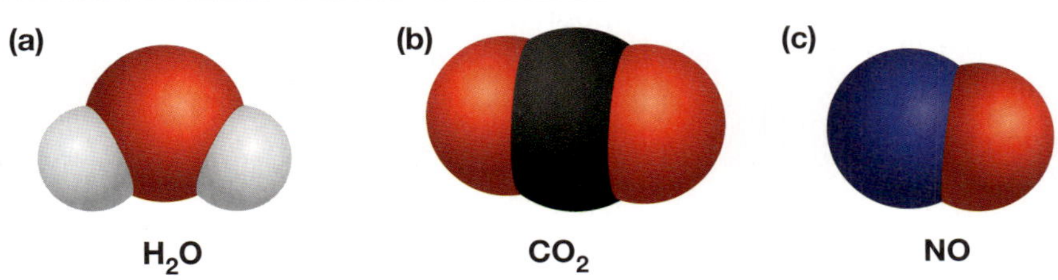

Figure 4 (a) Water (H_2O) is made of two hydrogen (H) atoms and one oxygen (O) atom. (b) Carbon dioxide (CO_2) is made of one carbon (C) atom and two oxygen (O) atoms. (c) Nitric oxide (NO) is made of one nitrogen (N) atom and one oxygen (O) atom.

DIG DEEPER

Learning Tip

Understanding Molecular Models

In these molecular models, the white spheres represent hydrogen atoms, the black spheres represent carbon atoms, the red spheres represent oxygen atoms, and the blue spheres represent nitrogen atoms.

▶ **compound:** a pure substance made up of two or more elements that are chemically joined

Interpreting Formulas

A chemical formula tells you important information about a pure substance.

- The symbols in the formula tell you the elements in the substance.
- If there is only one symbol, the substance is an element.
- If there is more than one symbol, the substance is a compound.
- The subscripts tell you how many atoms of each element are in each molecule of the substance.

SAMPLE PROBLEM: Interpreting Chemical Formulas

Interpret the chemical formula for sodium carbonate (Na_2CO_3).

Step 1: Determine how many different elements are in the formula.

- Sodium carbonate contains 3 elements: Na, C, and O.

Step 2: Determine how many atoms of each element are in the formula.

- Na_2CO_3 is made up of Na_2, C, and O_3, so Na_2CO_3 has 2 atoms of sodium (Na) 1 atom of carbon (C) and 3 atoms of oxygen (O).

Step 3: Determine the total number of atoms in the chemical formula.

- 2 Na atoms + 1 C atom + 3 O atoms = 6 atoms
  There are 6 atoms in the formula for sodium carbonate.

Practice: Interpret the formulas of these common compounds:

(a) ammonia (NH_3)

(b) butane (C_4H_{10})

RESEARCH THIS WHAT DIFFERENCE DOES A SUBSCRIPT MAKE?

SKILLS HANDBOOK 4.C.2., 4.C.6., 4.C.7.

SKILLS: Researching, Communicating, Evaluating

You have learned how to interpret chemical formulas. Sometimes, two compounds may contain the same elements but have different numbers of atoms. In this activity, you will find out how important the chemical symbols and subscripts in a chemical formula are.

1. Research the chemical names and properties of the following pairs of substances:
   (a) CH_4 and C_2H_6
   (b) H_2O and H_2O_2
   (c) CO and CO_2
2. Compare and contrast the chemical properties and uses of each pair of compounds.

A. Do compounds with the same elements always have similar properties or uses? T/I

7.2 Wrap Up

- Elements are pure substances that contain only one type of atom. They are represented by the chemical symbols found in the periodic table.
- Compounds are pure substances made up of two or more elements.
- Atoms can chemically join together to form larger units called molecules.
- Molecules may contain atoms of different elements or atoms of the same element.
- Chemical formulas state the type and number of atoms in each molecule of a pure substance.

CHECK YOUR LEARNING

1. In your own words, explain how a compound and a molecule are alike and how they are different. K/U C

2. Classify each of the following as an element or a compound. K/U

(a) O_3
(b) KCl
(c) NO
(d) $MgCl_2$
(e) Fe
(f) Cl_2

3. Which part of **Figure 5** represents the following? K/U

(a) a compound
(b) an element that exists as diatomic molecules
(c) individual atoms of an element

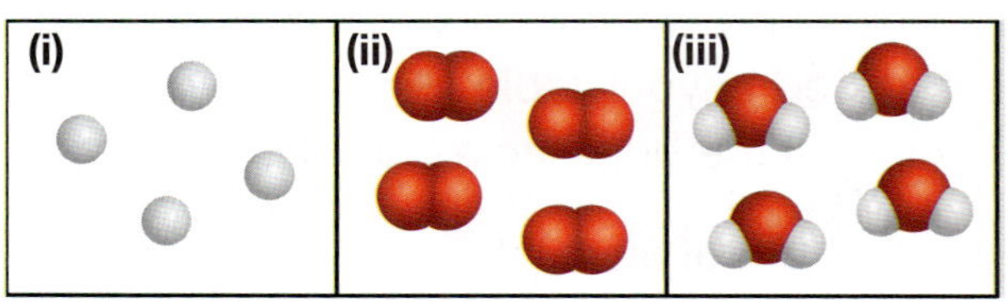

Figure 5

4. (a) What information do the subscripts in the chemical formulas O_2 and O_3 provide?

(b) How do the properties of these forms of oxygen differ? K/U

5. Interpret each of these chemical formulas. K/U

(a) F_2
(b) C_3H_8
(c) KBr
(d) $CaCO_3$
(e) HCl
(f) CH_4
(g) $AgNO_3$
(h) Fe_2O_3

6. Use a Venn diagram with three circles to show relationships among elements, molecules, and compounds. Use the elements found in compounds such as water and carbon dioxide as examples in your diagram. K/U C

7. Nitric oxide (NO) and nitrogen dioxide (NO_2) are air pollutants produced by automobiles. Nitrous oxide (N_2O) is used in dentistry to relax patients and as a power booster in racing. How are the chemical formulas of these compounds alike? How are they different? K/U A

8. What is the difference between a molecule that is an element and a molecule that is a compound? Give an example of each. K/U

7.3 PERFORM AN ACTIVITY

Matching Models and Formulas

SKILLS MENU

- Questioning
- Hypothesizing
- Predicting
- Planning
- Controlling Variables
- Performing
- Observing
- Analyzing
- Evaluating
- Communicating

The chemical formula shows the number and types of atoms in each molecule of a substance. For example, the chemical formula for water is H_2O. This tells us that each molecule of water has two hydrogen atoms and one oxygen atom. But chemical formulas do not tell us how the atoms are arranged in a molecule. In this activity, you will make models of common molecules to see how they are put together.

SKILLS HANDBOOK 3.B.5.–3.B.9.

Purpose

To build models of some common molecules

Procedure

1. Copy **Table 1** into your notebook.
2. The elements in the model kit are colour-coded. In your notebook, record the name and colour of each element you need.
3. Use the "atoms" and "bonds" provided to build a model of each combination of elements listed in your table (**Figure 1**). The molecule is complete when all the bond holes are filled.

   Note: You can use only one atom of element 1 in each molecule. However, you can use as many atoms of element 2 as are needed.

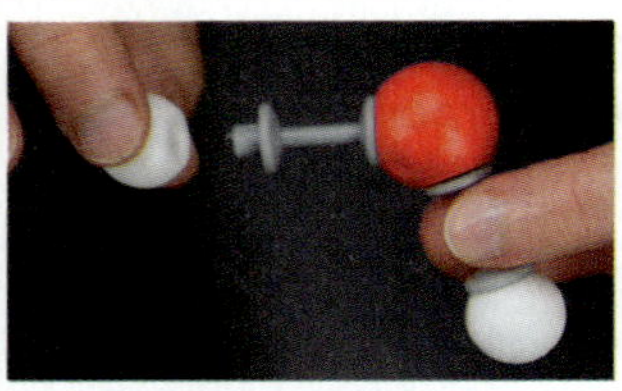

Figure 1 A water molecule is made up of two H atoms and one O atom.

Equipment and Materials

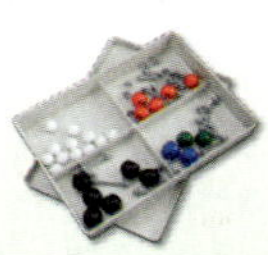

molecular model kit

coloured pencils

Table 1

Element 1 (use only one)	Element 2 (use more than one, if necessary)	Sketch of molecule	Chemical formula	Chemical name	Use
H	H			hydrogen	fuel
Cl	Cl			chlorine	pool disinfectant
H	Cl			hydrogen chloride	used to make hydrochloric acid
O	H			water	drinking
C	H			methane	fuel
N	I			nitrogen tri-iodide	contact explosive
O	O		O_2	oxygen	supports burning
C	O			carbon dioxide	puts out fire

4. Count the number of atoms of each element in your model. Sketch each molecule and write its chemical formula in your data table.

Analyze and Evaluate

(a) Separate the molecules you built into two groups: elements and compounds. T/I C

(b) Why is the word "diatomic" used to describe the molecules of the elements? T/I

(c) Compare the bonding in the Cl_2 and O_2 molecules. T/I

Apply and Extend

(d) The chemical formulas of hydrogen peroxide (H_2O_2) and acetylene (C_2H_2) are similar. Build these molecules. How are their structures different? T/I

(e) Research the chemical properties and uses of hydrogen peroxide and acetylene. Write a paragraph to summarize your findings. T/I C

(f) Propane (C_3H_8) is a fuel commonly used for barbecues. Use your kit to build a model of a propane molecule. Draw and label your model. T/I C A

Shocking Water

You read in Section 7.2 that water is a compound of the elements hydrogen and oxygen. Energy is needed to break water into its elements. One way of supplying this energy is by passing electricity through water. This process is called electrolysis. In this activity, you will use electrolysis to break apart water molecules.

SKILLS MENU
- Questioning
- Hypothesizing
- Predicting
- Planning
- Controlling Variables
- Performing
- Observing
- Analyzing
- Evaluating
- Communicating

SKILLS HANDBOOK 3.B.3., 3.B.5.–3.B.9.

Purpose

To separate water (H_2O) into its two elements

Procedure

1. Put on your eye protection and lab apron.
2. Place the battery jar containing the electrolysis apparatus on a table. Half-fill it with tap water.
3. Obtain two test tubes. Fill one test tube with water.
4. While holding a thumb over the mouth of the filled tube, invert the test tube and lower it into the jar. Remove the thumb from the test tube's mouth when it is below the surface of the water in the jar.
5. Place the test tube in the clamp. Position the test tube so that the one electrode is immediately beneath the opening.

CAUTION
Flames
This activity involves open flames. Tie back long hair and tuck in loose clothing.

Equipment and Materials

eye protection

lab apron

electrolysis apparatus

tap water

scoopula

matches

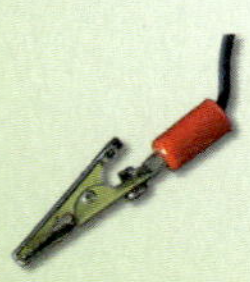
2 wires with alligator clips

9-V battery

stirring rod

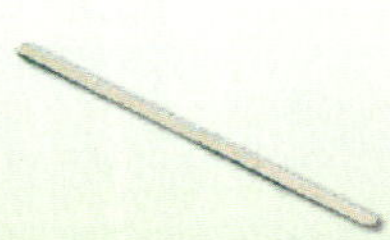
2 wooden splints

test tube clamps

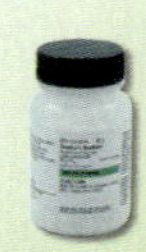
sodium sulfate

6. Repeat Steps 3 to 5 with the other test tube.
7. Add one scoopula of sodium sulfate to the water in the jar. Use a glass stirring rod to mix the sodium sulfate until it is all dissolved.
8. Attach one end of each wire to the terminal posts on the support bracket. Connect the other ends to the DC power source (either a 9-V battery or a laboratory-regulated DC power supply). Power should now be turned on.
9. Bubbles will begin to rise from both electrodes into the test tubes. When one test tube has been nearly filled with gas, disconnect the apparatus from the power source. Observe how much gas is in each test tube. Record your observations.
10. Hold the test tube containing the larger amount of gas. Release the spring clip and lift the test tube from the water, keeping it upside down.
11. Have a partner light one of the wooden splints. Move the inverted test tube's mouth over the lit end of the splint at an angle. Observe what happens. Record your observations.
12. Now remove the second test tube from the clamp. Gently lift up the test tube allowing the water that is in the test tube to drain into the battery jar. Hold the test tube horizontally to conduct the test.
13. At the same time have a partner light another wooden splint, then blow out the flame, but leave the tip glowing. Move the test tube's mouth over the glowing splint. Observe what happens. Record your observations.
14. Wash your hands thoroughly at the end of the lab.

CAUTION

Safety

Point the opening of the test tube away from yourself, your partner, and everyone else when conducting the tests with the splints.

Analyze and Evaluate

(a) What happened when the gas that almost filled a test tube was tested with a lit wooden splint? What happened when the gas in the other test tube was tested with a glowing wooden splint? T/I

(b) What was the gas in each test tube? How do you know? T/I

(c) Compare the volumes of the gases. Does this result match what you would expect from the formula for water? Explain. T/I

Apply and Extend

(d) Design an experiment to measure how quickly the gases are produced in this experiment. T/I

7.5 Taking Compounds Apart

As you know, elements cannot be broken down into simpler substances. Compounds, however, can be broken down. That is what happened when electricity was passed through water in the activity in Section 7.4.

▶ **decomposition:** the breaking down of a compound into simpler substances

▶ **electrolysis:** the process during which electricity is used to break down a compound into simpler substances

Decomposition of Compounds

Decomposition is a chemical change in which compounds are broken down into simpler substances. Many explosives work because of decomposition. For example, nitroglycerin is the explosive chemical inside sticks of dynamite. When dynamite is detonated, each nitroglycerin molecule quickly decomposes into many smaller gas molecules like nitrogen (N_2) and carbon dioxide, (CO_2). The energy released in the explosion makes these gases molecules move very quickly. These rapidly expanding gases help give dynamite its destructive power.

In general, compounds do not break down by themselves. They need some help. Energy must be added for decomposition to take place. Energy can be added in two ways:

- using electricity
- adding thermal energy to the compound

Electrolysis

During **electrolysis**, electricity is passed through a dissolved compound, causing it to break down into simpler substances.

Electrolysis of Water

Water is a very stable compound. On its own, water will not break down into its two elements. When electricity is passed through water, some of the water molecules break down. The new products formed are hydrogen gas (H_2) and oxygen gas (O_2). The formula of water (H_2O) tells us that a water molecule has twice as many hydrogen atoms as oxygen atoms. Therefore, you might predict that twice as much hydrogen as oxygen is produced. Look closely at the electrolysis apparatus shown in **Figure 1**. Notice that the amount of gas collected in each test tube shows that this is true.

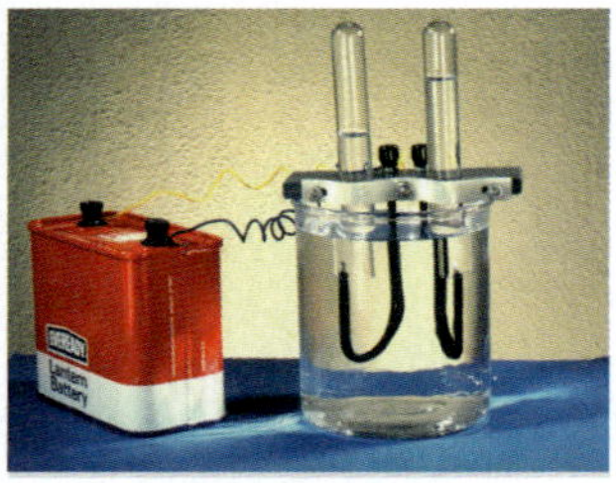

Figure 1 Some water molecules decompose into hydrogen and oxygen when electricity passes through water.

Refining Aluminum

Most metals in the periodic table are not found as elements in nature. Instead, they exist only in compounds. This means that useful metals such as aluminum and copper must first be separated from their compounds before they can be used. This can be done using electrolysis. For example, aluminum is the light-weight metal used to make foil and auto parts. Aluminum is never found in nature as an element, but minerals similar to the compound aluminum oxide (Al_2O_3) are common. Pure aluminum is made by passing electricity through liquid aluminum oxide.

Heating and Decomposition

Sometimes, heating causes compounds to break down.

Lighting a Match

Decomposition occurs when you light a match. The match head contains a mixture of chemicals including potassium chlorate ($KClO_3$). Rubbing the match head across the rough strip on the matchbox releases thermal energy. This causes potassium chlorate to break up into two simpler substances—oxygen gas (O_2) and potassium chloride (KCl). Oxygen, with the thermal energy, causes other chemicals in the match to catch fire.

Baking Muffins

A baker uses decomposition when using baking soda to bake muffins. The thermal energy of the oven causes sodium hydrogen carbonate ($NaHCO_3$), or baking soda, to decompose into three simpler substances (**Figure 2**). Bubbles of carbon dioxide produced in this process cause the muffin batter to rise, as shown in **Figure 3**.

DIG DEEPER

Did You Know?

Aluminum—More Precious than Gold

200 years ago, aluminum was rare. This made aluminum objects such as cutlery more expensive than those made of gold. Only kings were served with aluminum knives and forks at important dinner parties. Less important nobles were served with gold cutlery. Then, after scientists learned how to produce aluminum using electrolysis, its price dropped to only a few cents per gram.

DIG DEEPER

Career Link

Bakers use chemistry to make better cakes and pastries. To learn more about being a baker,

GO TO NELSON SCIENCE

Figure 2 The thermal energy of an oven decomposes baking soda into water (H_2O), carbon dioxide (CO_2), and sodium carbonate (Na_2CO_3).

Figure 3 (a) $NaHCO_3$ added to batter breaks down during heating. (b) Baking produces carbon dioxide (CO_2) bubbles, making baked goods rise.

Decomposition of Hydrogen Peroxide

Some compounds break down naturally at room temperature. One such compound is hydrogen peroxide (H_2O_2) shown in **Figure 4**. This compound does not break down into elements. It decomposes into oxygen gas (O_2) and water (H_2O).

Many factors, including light, make hydrogen peroxide break down faster. That is why hydrogen peroxide is always packaged in a dark bottle (**Figure 5**). Certain substances help speed up the breakdown of hydrogen peroxide. These substances are called catalysts. A **catalyst** speeds up a chemical change without being used up or chemically changed itself.

▶ **catalyst:** a substance that speeds up a chemical change without being consumed or chemically changed itself

Have you ever used hydrogen peroxide to treat a cut? Your blood contains catalysts, including one that speeds up the breakdown of hydrogen peroxide, called catalase. This is why you see bubbles form when hydrogen peroxide is placed on a cut. These bubbles are oxygen gas.

Figure 4 A molecular model of hydrogen peroxide (H_2O_2)

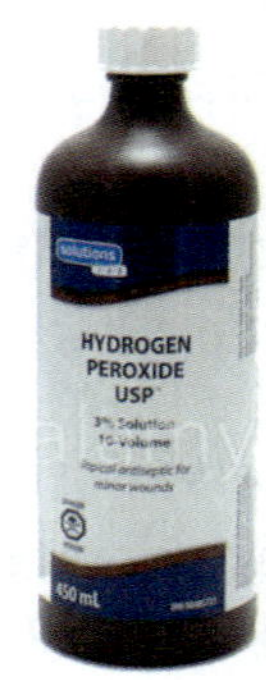

Figure 5 Hydrogen peroxide is kept in dark bottles because light can cause the molecules to decompose.

Uses of Hydrogen Peroxide

Hydrogen peroxide is a toxic chemical that kills certain bacteria. This makes it useful for disinfecting cuts and contact lenses. In addition, the bubbling action of hydrogen peroxide helps to physically remove dirt from a wound or a contact lens. Bleaches such as hydrogen peroxide react with compounds that give things their colour. The result is a new compound that has no colour. This process can make stained teeth white (**Figure 6**) and dark hair lighter in colour.

Figure 6 A brighter smile is the result of a chemical process between hydrogen peroxide and stains on your teeth.

TRY THIS ELEPHANT TOOTHPASTE

SKILLS HANDBOOK 3.B.3., 3.B.5.–3.B.9.

SKILLS: Controlling Variables, Performing, Observing, Analyzing, Communicating

Many beauty supply stores sell two different hydrogen peroxide solutions: 3 % hydrogen peroxide and 6 % hydrogen peroxide. Because both solutions contain hydrogen peroxide, they should undergo the same processes. But what effect does the difference in concentration have on how they react?

Equipment and Materials: eye protection; lab apron; 2 large narrow test tubes; 1 beaker; 2 stirring rods; scoopula; 3 % hydrogen peroxide solution; 6 % hydrogen peroxide solution; liquid dish soap; dry yeast

CAUTION: IRRITANT
Hydrogen peroxide is an irritant. Avoid skin contact. In case of skin contact, wash the affected area with lots of cool water.

1. Put on your eye protection and lab apron.
2. Pour 3 % hydrogen peroxide into one test tube to a depth of about 3 cm. Pour the same amount of 6 % hydrogen peroxide into the other test tube. Label each test tube appropriately.
3. Add 3 drops of liquid dish soap to each test tube.
4. Use a different stirring rod to mix the contents of each test tube.
5. Place the test tubes in the beaker.
6. Place the beaker and its contents in the sink to avoid spilling.
7. Add a small amount of yeast (enough to cover the tip of the scoopula) to each test tube. Mix the contents of each with its stirring rod, and observe.

A. What evidence is there that a chemical change occurred in this activity? T/I

B. Describe a test that you can perform to confirm that the gas produced is oxygen. T/I

C. What was the role of yeast in this activity? T/I

D. Which concentration of hydrogen peroxide do you think is more effective at removing hair colour? Why? A

7.5 Wrap Up

- Compounds can break down, or decompose, into simpler substances.
- Energy is needed to decompose a compound. Common sources of energy are heating and electricity.
- Some substances can speed up chemical changes without being used up. These substances are known as catalysts.

UNIT TASK Bookmark

You can apply what you learned about the effects of decomposing certain chemicals to the Unit Task described on page 276.

CHECK YOUR LEARNING

1. (a) Compare and contrast the chemical formulas of water and hydrogen peroxide.
   (b) How do the products of their decomposition compare? K/U
2. (a) What property of hydrogen peroxide makes it suitable for use in changing hair colour?
   (b) Is this a chemical or physical property? How do you know? K/U
3. Find two examples of household substances that are stored in opaque plastic bottles. For each substance, describe its purpose and usefulness. What are the hazards associated with its handling and disposal? A
4. Consumer products such as hydrogen peroxide have a limited shelf-life. T/I A
   (a) What do you think the term "shelf-life" means?
   (b) Why do school labs often store hydrogen peroxide in dark bottles and in the refrigerator?
5. Carbonated soft drinks contain carbonic acid (H_2CO_3). When this compound decomposes, it releases carbon dioxide gas and liquid water. What would you see in a soft drink as this process happens? How could you make this process occur faster? K/U C A

SCIENCE WORKS OSSLT

Extreme Close-Up of a Diaper

"I've got to pee."

These frantic words were uttered by the astronaut Alan Shepard in 1961, while waiting to blast off. NASA engineers were in a panic. Do they let him pee in the suit? Since urine contains dissolved ionic compounds, it conducts electricity. Released urine might damage electrical equipment. Finally, NASA gave him relief and said, "Okay, Alan. Go to it."

To avoid this problem for future missions, NASA developed a high-tech astronaut "diaper." This garment used a white crystalline powder called sodium polyacrylate as its absorbent material. Crystals of this compound absorb up to 400 times their mass in water. As they do, they grow larger and larger. This increase in size is a physical change. The water is held so strongly the crystals stay dry to the touch. Sodium polyacrylate is now used in infant and adult diapers and feminine hygiene products.

The absorbency of sodium polyacrylate is useful in other consumer products. For example, you can buy a small toy animal that grows to a life-sized pet in a water-filled bathtub (**Figure 2**). Gardeners sometimes mix sodium polyacrylate with soil. When water is added to the soil, the crystals absorb large amounts of water. The crystals can supply an unattended plant with water for a few weeks! Sodium polyacrylate is also used to make artificial snow, which is used in indoor snowboarding parks.

Diapers were once made from cloth. Over the years, these diapers have mostly been replaced by disposable ones (**Figure 1**). A typical disposable diaper is made of synthetic materials, mostly plastics made from chemicals found in fossil fuels. Each part of the diaper is designed with properties specific for its particular use.

Figure 1 Many different materials, each with a specific purpose, make up a disposable diaper.

Figure 2 This toy lizard is made of sodium polyacrylate. When placed in water, the lizard on the left grows to several times its original size (right).

Diaper Dissection

See for yourself how the sodium polyacrylate mentioned in **Science Works** absorbs large amounts of water.

SKILLS HANDBOOK 1.A., 3.B.5–3.B.9.

Purpose

To observe the water-absorbing property of sodium polyacrylate

Procedure

1. Put on your lab apron, protective mask, eye protection, and disposable gloves.
2. Cut a disposable diaper in half lengthwise. Hold the diaper over a large tray. Use your fingers to shred the inside filling, and shake the crystals it contains into the tray.
3. Slowly add water to the large tray, and observe what happens.
4. Use a scoopula to move a few of the crystals into a small tray.
5. Use another scoopula to add a scoop of salt to these crystals, and stir. Observe what happens.

Analyze and Evaluate

(a) Describe what happened when you added water to the crystals in the large tray. T/I

(b) What happened when you added salt to the water-filled crystals? Is this a physical or chemical change? Explain. T/I

SKILLS MENU

- Questioning
- Hypothesizing
- Predicting
- Planning
- Controlling Variables
- Performing
- Observing
- Analyzing
- Evaluating
- Communicating

CAUTION

Irritant

Do not eat the crystals or touch your face or eyes if you come into contact with them. The crystals can irritate skin and cause dehydration. Wash your hands thoroughly after the activity.

Equipment and Materials

eye protection

lab apron

disposable gloves

protective mask

scissors

large tray

small tray

2 scoopulas

disposable diaper

table salt

water

7.7 EXPLORE AN ISSUE CRITICALLY

Reducing Wastes from Plastics

SKILLS MENU

- Defining the Issue
- Researching
- Identifying Alternatives
- Analyzing the Issue
- Defending a Decision
- Communicating
- Evaluating

How often do you walk down the street and see a candy wrapper or a plastic water bottle that someone has dropped? Maybe you also see leaves or something like a half-eaten apple in the gutter. If you walk down the same street a month later, you might not see the leaves or the apple. They will probably have decomposed. However, the candy wrapper and plastic water bottle will likely still be there. They could still be there in a year or in 10 years.

Bacteria and fungi feed on natural compounds such as leaves and apples, causing them to decompose into simpler compounds. Other materials take longer to break down. Most plastics resist decomposing (**Figure 1**). Plastics contain very large molecules made up mostly of carbon and hydrogen atoms. Often, they do not react readily with other substances. Because plastics are synthetic, nature has no way of breaking their molecules down. Therefore, the atoms contained in most plastics are "trapped" and cannot be naturally recycled into new substances.

Figure 1 Leaves and other natural items that fall to the ground decompose and become part of the soil. Materials such as plastics and metals do not decompose readily and remain in the environment.

The Issue

Many items we use every day are made of plastics, from milk jugs and grocery bags to packaging and car parts. What happens to these items when we are finished with them? What can we do to reduce the amount of plastic we use? How can we reuse more plastic? Which plastics can be recycled?

Role: You are a member of the Envirothon team in your school. Your team's project this year is to develop plans for how each household can reduce, reuse, or recycle in order to decrease the amount of plastic waste they produce.

Audience: You will present the results of your research to the entire team.

Goal

To develop a plan to reduce the wasteful consumption of synthetic materials and environmental pollution.

Gather Information

Look around your home. List ways you see plastic items being used. Look for ways to reduce, reuse, or recycle plastic materials. Research organizations or local businesses and services that focus on doing these three things.

Identify Solutions

Consider the following questions to help you identify solutions:

- How can you reduce the amount of packaging you bring home from stores? What alternatives are there to plastic bags? What other materials can be used instead of plastic?
- What are some ways you can reuse plastic items?
- Are there any compostable plastics?
- Most plastic items are imprinted with a small triangle. Although all plastics could be recycled, recycling centres usually accept only plastics with certain numbers. Check with a local recycling centre and find out what plastics they recycle. Why do they recycle only certain plastics?

Make a Decision

Decide what steps the households in your community can take to reduce their use of plastic materials, and how they can reuse existing plastic items. Also decide on ways to raise awareness of the need to recycle, and ways recycling programs may be improved.

DIG DEEPER
Web Link

To learn more about reducing, reusing, and recycling,

GO TO NELSON SCIENCE

UNIT TASK
Bookmark

You can apply what you learned about environmentally sensitive products to the Unit Task described on page 276.

7.8

Oxygen: Properties, Uses, and Impacts

Have you ever thought about what makes Earth different from other planets? For one thing, Earth's atmosphere is about 21 % oxygen. Oxygen atoms combine easily with many substances.

Quick Chemical Changes with Oxygen

▶ **combustion:** a rapid chemical change involving oxygen and a fuel that releases large amounts of energy

Combustion, or burning, is a rapid chemical change that involves oxygen and a fuel. Gasoline, propane, and natural gas are common fuels. Combustion releases energy that is stored in the fuel molecules as thermal energy and light. We use the energy from combustion to heat our homes and to cook. Sometimes combustion gets out of control. A simple campfire, for example, can easily result in a forest fire if it is not tended carefully (**Figure 1**).

Figure 1 The thermal energy released in a forest fire is a result of the rapid combination of oxygen and the trees, which act as fuel.

Slow Chemical Changes with Oxygen

Not all of the chemical changes that use oxygen are rapid. Slow, controlled changes that use oxygen take place all the time in our bodies' cells. You read about the process called cellular respiration in Chapter 2. During cellular respiration, our cells use oxygen to extract energy stored in nutrient molecules like sugar. The energy released is used for important cell functions like movement, growth, and reproduction. Some of the waste energy of cellular respiration is responsible for keeping our body temperatures at 37 °C.

Rusting is a slow chemical change that converts iron into a reddish-brown compound called iron oxide. Alloys that contain iron, such as steel, also can rust. Both oxygen and water are needed for rusting to occur, and the presence of salt makes it happen faster. Rusting weakens metallic objects. Millions of dollars are spent in Canada each year replacing metal objects damaged by rusting. Painting the metal or covering it with oil protects it from contact with oxygen and can slow down this chemical change (**Figure 2**).

▶ **rusting:** a slow chemical change of iron reacting with oxygen and water

Figure 2 Removing rust and painting steel towers is dangerous but necessary work.

RESEARCH THIS PEDAL TO THE NON-METAL?

SKILLS: Researching, Analyzing the Issue, Communicating

SKILLS HANDBOOK 4.C.

Compare the advantages and disadvantages of using non-metallic materials to make car bodies.

Manufacturers are using more non-metallic materials such as plastics and fibreglass to reduce the weight of cars. These materials have other advantages as well. For example, they do not corrode (**Figure 3**). Research three different materials used to make car bodies. List the advantages and disadvantages of each material. Choose what you think is the best material, and give reasons for your choice.

Present your research in the form of an advertisement poster, video, brochure, or magazine ad.

Figure 3 The metal in this car rusted. Fibreglass and plastic would not rust.

GO TO NELSON SCIENCE

7.8 Wrap Up

- Chemical changes involving oxygen can be slow or fast.
- Combustion is a rapid chemical change involving fuel and oxygen that releases thermal energy and light.
- Rusting is a slow chemical change involving iron, oxygen, and water.
- During cellular respiration, oxygen combines with food to release the energy we need.

CHECK YOUR LEARNING

1. Compare and contrast combustion and rusting using a T-chart. K/U C
2. Why can you classify burning and respiration as similar reactions? T/I C
3. Name a way you could stop iron objects such as a basketball hoop or other playground equipment from rusting. K/U
4. Give two examples of each of the following: K/U
   (a) positive effects of chemical changes with oxygen
   (b) negative effects of chemical changes with oxygen
5. Name a place in Canada where rusting might occur:
   (a) very quickly
   (b) moderately
   (c) very slowly K/U A

Put Out That Fire!

Firefighters are trained in the science of fire, which can be explained using the fire triangle (**Figure 1**). To burn, fire needs three things: fuel, thermal energy, and oxygen. Removing any one of these requirements will put out a fire. In this activity, you will use your knowledge of chemistry to make your own fire extinguisher.

SKILLS MENU

- Questioning
- Hypothesizing
- Predicting
- Planning
- Controlling Variables
- Performing
- Observing
- Analyzing
- Evaluating
- Communicating

Figure 1 Firefighters know that you can put out a fire by removing any part of the fire triangle.

CAUTION

Irritant

Chemicals you might choose to use could be irritants. Wash your hands after using these chemicals.

SKILLS HANDBOOK 3.A., 3.B.

Purpose

To design and test a fire extinguisher that will produce a gas that puts out a small fire

Procedure

Day 1

1. Work in a small group to design a fire extinguisher.
2. As you design your extinguisher, keep the following in mind:
   - A fire extinguisher must remove one or more of the requirements for a fire.
   - The gas in a fire extinguisher should not burn or cause other things to burn. It should be denser than air so that it keeps oxygen away from the fire.
   - A fire extinguisher has to work immediately when needed. If your extinguisher uses a chemical change to produce the desired gas, the materials that react must be kept apart, and then they must be easily mixed when needed.

Equipment and Materials

eye protection

lab apron

small candle

matches

ingredients for fire extinguisher

- You should be able to point the fire-extinguishing ingredient toward the fire.
- All materials you use must be non-toxic.

3. As you design your extinguisher, consider the following questions:
   - Which gas will you produce to put out the fire?
   - What common materials will you use to produce the gas?
   - What will you use to store the ingredients, and how will you keep them separated inside the extinguisher?
   - How will the ingredients mix when you need to use them?
   - How will you aim your extinguisher toward the fire?
4. Sketch and label your design. Organize the decisions you made in a chart to explain how your extinguisher works.
5. Show your teacher your design and chart for approval.
6. Make a list of the materials you will need. Decide who will bring materials to class.

Day 2

1. Put on your eye protection and lab apron.
2. Using the materials your group selected, assemble your fire extinguisher.
3. Light the candle.
4. Try your fire extinguisher to see if it puts out the flame.
5. If it does not, adjust your design until your model works successfully.
6. Set up the extinguisher again and demonstrate it to the class.

DIG DEEPER

Writing Tip

Check and Revise

When revising a design, check that the procedure is described as clearly as possible. Read each step aloud. Have a partner mime the action described. If your partner has difficulty visualizing what to do, change the wording to make the description easier to understand.

CAUTION

Flames

This activity involves open flames. Tie back long hair and tuck in loose clothing.

Analyze and Evaluate

(a) Which requirement for a fire did your extinguisher eliminate: fuel, thermal energy, or oxygen? Explain your choice. T/I

(b) How did your extinguisher eliminate this requirement? T/I

(c) Use a flow chart or a diagram to summarize how your extinguisher worked. C

Apply and Extend

(d) Would your extinguisher put out a grease fire? Explain your answer. T/I

(e) Compare your design to that of another group in your class. What features of other extinguishers do you think could improve your design? T/I

Chapter 7 KEY CONCEPTS SUMMARY

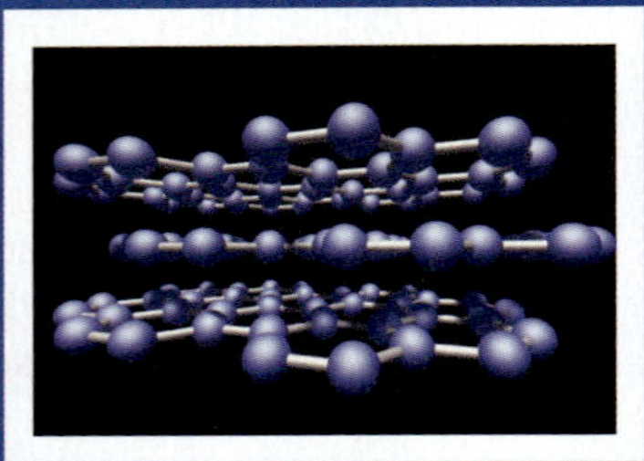

A chemical compound is a pure substance made up of two or more elements.

- The properties of a compound differ from the properties of the elements it contains.
- Compounds can be formed by combining elements or other compounds.

A chemical formula is used to represent a pure substance, which can be an element or a compound.

- The letters show the elements that are in a molecule of the substance.
- Subscripts give the number of atoms of each element.

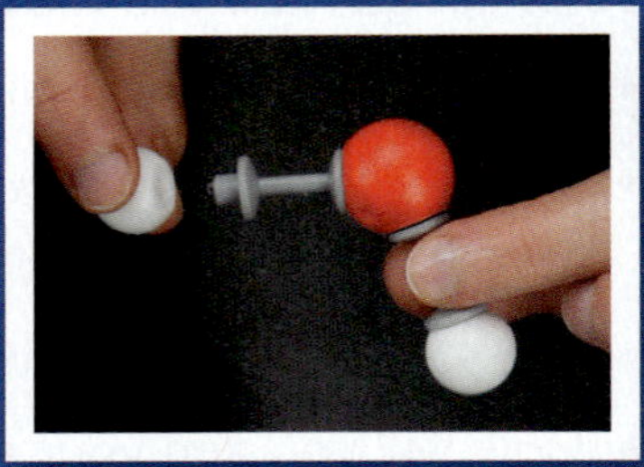

A molecule is a particle of matter made up of two or more atoms.

- Molecular models are used to represent molecules.

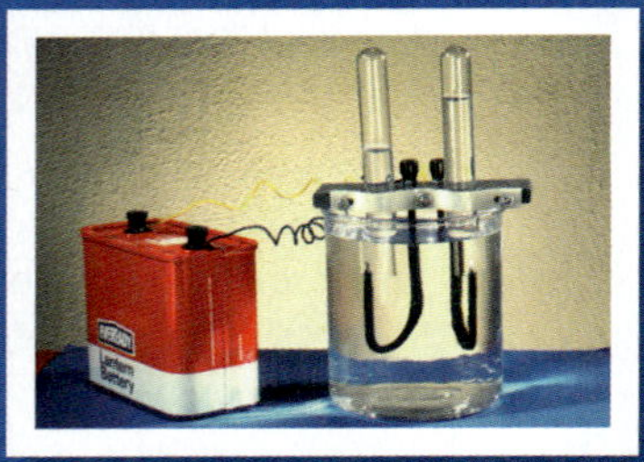

Compounds can be decomposed into simpler substances.

- decomposition produces elements or simpler compounds
- energy is required for decomposition to occur

Many important chemical changes involve oxygen.

- Burning and rusting are chemical changes that use oxygen from the air.
- Lab tests can be used to identify gases formed by chemical changes.

Our use of compounds has both positive and negative effects on society and the environment.

- An understanding of the properties of compounds helps us to better reuse, recycle or dispose of them appropriately.

WHAT DO YOU

THINK NOW?

You thought about the following statements at the beginning of the chapter. You may have encountered these ideas in school, at home, or in the world around you. Consider them again and decide whether you agree or disagree with each statement.

1 Elements can join chemically to form more complex substances. **Agree/disagree?**

4 Most plastics completely break down in a few years. **Agree/disagree?**

2 When a compound breaks down, gases are always released. **Agree/disagree?**

5 To protect the environment, we need to reduce the amount of materials we use, reuse what we can, and recycle when possible. **Agree/disagree?**

3 The combination of oxygen with a substance always produces a flame. **Agree/disagree?**

6 All metals rust. **Agree/disagree?**

How have your answers changed since then? What new understanding do you have?

BIG Ideas

- Elements and compounds have specific properties that determine their uses.
- The use of elements and compounds has both positive and negative effects on society and the environment.

DIG DEEPER

Vocabulary

molecule p. 248
subscript p. 248
compound p. 249
decomposition p. 256
electrolysis p. 256
catalyst p. 258
combustion p. 264
rusting p. 265

CHAPTER 7

SELF-QUIZ

The following icons indicate the Achievement Chart category addressed by each question. | K/U Knowledge/Understanding | T/I Thinking/Investigation | C Communication | A Application

For each question, select the best answer from the four alternatives.

1. Which is the simplest type of substance? (7.2) K/U
   (a) a solution
   (b) an element
   (c) a mixture
   (d) a compound

2. Which of these elements exists in nature as a diatomic molecule? (7.2) K/U
   (a) Fe
   (b) O
   (c) K
   (d) Na

3. Glucose, a type of sugar, has the formula $C_6H_{12}O_6$. What is the total number of atoms in a molecule of glucose? (7.2) K/U
   (a) 6
   (b) 12
   (c) 24
   (d) 3

4. Which of these chemical formulas represents an element? (7.2) K/U
   (a) H_2
   (b) CO
   (c) CH_4
   (d) CCl_4

5. What does the formula N_2 represent? (7.2) T/I
   (a) an atom and an element
   (b) an element and a molecule
   (c) a molecule and a compound
   (d) a compound and an element

6. A powder inside disposable diapers absorbs up to 400 times its mass in water. What type of change is happening as the powder absorbs water? (7.6) K/U
   (a) chemical
   (b) physical
   (c) electrical
   (d) mechanical

7. A fire has several requirements. When a forest burns, which requirement does a tree provide? (7.9) K/U
   (a) fuel
   (b) oxygen
   (c) carbon dioxide
   (d) thermal energy

Indicate whether each of the statements is TRUE or FALSE. If you think the statement is false, rewrite it to make it true.

8. Oxygen in the air is made up of single atoms. (7.2) K/U

9. Each molecule of carbon dioxide contains one carbon atom and two oxygen atoms, so the chemical formula for this compound is CO_2. (7.2) K/U

10. Ozone (O_3) is a diatomic molecule. (7.2) K/U

Copy each of the statements into your notebook. Fill in the blanks with a word or phrase that correctly completes the sentence.

11. Pure substances include both elements and _________. (7.2) K/U

12. The smallest unit of an element is a(n) _________. (7.2) K/U

13. In the chemical formula C_2H_6, the numbers 2 and 6 are called _________. (7.2) K/U

Match the term on the left with the appropriate term on the right. Terms on the right may be used more than once.

14. (a) He
    (b) Cl_2
    (c) CH_4
    (d) NO

    (i) atom
    (ii) molecule

    (7.2) A

15. (a) P_4 (b) HCl (c) Cu (d) Br_2

(i) element
(ii) compound
(7.2) A

Write a short answer to each of these questions.

16. Some dyes remain permanently in hair and fabric, and some wash out over time. Explain why some dyes are permanent and some are not. (7.1) C K/U

17. Explain why all molecules are not considered compounds. (7.2) K/U

18. Why does a glowing splint burst into flames around oxygen? (7.8) K/U A

19. Sulfuric acid is a chemical that is widely used in industry. In addition to being used on its own, as an acid, it is used to make other products, such as fertilizer and rubber tires. Interpret the chemical formula for sulfuric acid (H_2SO_4). (7.2) T/I

20. (a) What effect does the high temperatures of an oven have on baking soda?

    (b) How does this make a cake rise? (7.5) K/U

21. What kinds of atoms and how many of each kind are represented by these formulas? (7.2) K/U C

    (a) NH_4NO_3

    (b) HNO_3

    (c) $C_{12}H_{22}O_{11}$

22. What is formed when hydrogen peroxide decomposes?

23. Glucose (simple sugar) ferments to ethyl alcohol and carbon dioxide. The sugar in grapes or from grain ferments with yeast to make the alcohol and carbon dioxide. The carbon dioxide is the gas that bubbles out of beer or champagne. $C_6H_{12}O_6$ (glucose) reacts to form 2 C_2H_5OH (ethyl alcohol) and 2 CO_2.
    Although fermentation is a complex process, what kind of chemical reaction describes the overall result? Explain your thinking. (7.4, 7.5) K/U A C

24. What do disposable diapers and potting soil that stores water have in common? (7.6) A

25. Explain how ozone (O_3) can be both helpful and harmful. (7.2) K/U

26. Why is it necessary to store substances such as hydrogen peroxide and certain medicines in dark bottles? (7.4) K/U

27. During an electrolysis of water experiment, 20 mL of hydrogen is produced. What volume of oxygen is produced? Explain your answer. (7.4) (7.5) A T/I

28. Certain plastics and plant tissues both contain large amounts of carbon, hydrogen, and oxygen. Why, then, do discarded plastics cause more problems for the environment than dead plants? (7.7) A C

29. Burning and rusting both involve a chemical change involving oxygen and another substance. What is the key difference between burning and rusting? (7.8) K/U

Self-Quiz

To do an online self-quiz,

GO TO NELSON SCIENCE

CHAPTER 7

REVIEW

The following icons indicate the Achievement Chart category addressed by each question.

K/U Knowledge/Understanding | T/I Thinking/Investigation | C Communication | A Application

What Do You Remember?

1. Which of these chemical formulas represents an element? (7.2) K/U
   (a) Br_2
   (b) NO
   (c) H_2O
   (d) SO_2
2. Sucrose, or table sugar, has the formula $C_{12}H_{22}O_{11}$. What is the total number of atoms in a molecule of sucrose? (7.2) K/U
   (a) 3
   (b) 12
   (c) 22
   (d) 45
3. Which of these elements exists as a diatomic molecule found in air? (7.2) K/U
   (a) N
   (b) C
   (c) Rn
   (d) Ca
4. A fire extinguisher prevents which of these from supporting a fire? (7.9) K/U
   (a) the fuel
   (b) oxygen gas
   (c) carbon dioxide gas
   (d) the catalyst
5. The compound manganese dioxide is sometimes used to make chemical processes go faster. It is not changed by the process. Manganese dioxide is an example of (7.5) K/U
   (a) an element
   (b) a mixture
   (c) a catalyst
   (d) a type of energy
6. Hydrogen peroxide bubbles when it is placed on a cut. What gas forms these bubbles? (7.8) K/U
   (a) O_2
   (b) H_2
   (c) H_2O
   (d) CO_2
7. If two atoms are chemically joined, do they always form a compound? Explain. (7.2) K/U
8. In your own words, describe how water is decomposed. (7.5) K/U C
9. Describe the role of salt (NaCl) in the process of dyeing fabrics. (7.1) K/U
10. Does each of the following formulas represent an atom or a molecule? (7.2) K/U
    (a) Ar
    (b) P_2O_5
    (c) I_2
    (d) Ag
11. Does each of the following formulas represent an element or a compound? (7.2) K/U
    (a) S_8
    (b) HBr
    (c) Cr
    (d) H_2

What Do You Understand?

12. Oxygen exists in a form with molecules of O_3, which is known as ______. (7.2) K/U
13. A ______ must be added to many substances to increase the speed at which decomposition happens. (7.5) K/U
14. A cake rises because of the ______ bubbles that form when baking soda breaks down. (7.5) K/U
15. When discarded materials are ______, they are used to make new products. (7.7) K/U
16. Carbon tetrachloride is a compound that was once used for dry cleaning. Interpret the chemical formula for carbon tetrachloride (CCl_4). (7.2) K/U

17. Explain what occurs when a steel beam rusts. (7.4) K/U

18. Copy and complete **Table 1** in your notebook with the words Yes or No. (7.2) K/U C

Table 1

Formula	Element	Compound	Consists of only atoms	Consists of only molecules
H_2O_2				
Fe				
N_2				
CH_4				
O_3				

19. Carbonated soft drinks contain carbonic acid (H_2CO_3). When you first open the bottle, some carbonic acid is converted into water (H_2O) and carbon dioxide (CO_2). Why is this an example of decomposition? (7.5) K/U A

Solve a Problem

20. Describe two different ways you can reuse or recycle items made of plastic. (7.7) K/U A

21. The underside of cars that are left outside during the winter can begin to rust. What can car owners do to prevent this problem? (7.8) K/U A

Create and Evaluate

22. When electricity is passed through water, two colourless gases form. When hydrogen gas burns, water is produced. Explain how each of these chemical changes shows that water is a compound, not an element. (7.4) (7.5) T/I

23. Many people incorrectly believe that the numbers on the bottom of a plastic container describe the quality of the plastic. What do these numbers really mean? Present this information in an informative display. T/I C A

Reflect on Your Learning

24. Consider a typical day in your school week. Estimate the number of plastic or other non-biodegradable items that you use. Compare this estimate to the number of biodegradable or reusable items that you use. Why would it be good for the environment to change this ratio? How can you change your daily habits to make these changes? A C

25. In this chapter, you learned how chemical changes can have both positive and negative effects on society and the environment. How did your understanding of chemical compounds change after reading the chapter? T/I C

Web Connections

26. Usually, engineers do everything they can to prevent rust. However, a type of steel called Core-Ten™ is designed to rust. Research how this type of steel was developed and how it is used. T/I A

27. Research the latest developments in plastics and other materials that have been created from non-petroleum-based sources and are designed to decompose quickly. For example, some plastic shopping bags now include plant starch that is biodegradable. T/I A

28. Research the advantages and disadvantages of using plastic pop bottles as opposed to aluminum pop cans. Write a paragraph explaining which you think people should choose to use and explain why. T/I C A

GO TO NELSON SCIENCE

Exploring Matter

CHAPTER 5

Properties of Matter

KEY CONCEPTS

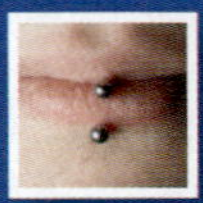

A substance can be identified by is physical or chemical properties.

Understanding the physical and chemical properties of a substance helps us identify its uses.

The use of substances can have both positive and negative effects in our daily lives.

Changes in substances can be either physical or chemical.

CHAPTER 6

Elements and the Periodic Table

KEY CONCEPTS

Elements cannot be broken down into simpler substances.

Elements of the same column (group) of the periodic table have similar properties.

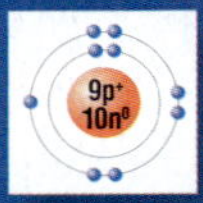

Atoms consist of protons and neutrons in a central core surrounded by electrons.

Metals and non-metals have characteristic physical properties.

Metalloids have properties in between those of metals and non-metals.

Elements can be both beneficial and harmful.

CHAPTER 7

Chemical Compounds

KEY CONCEPTS

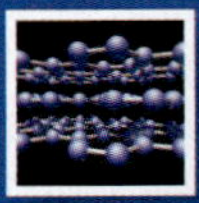

A chemical compound is a pure substance made up of two or more elements.

A chemical formula is used to represent a pure substance, which can be an element or a compound.

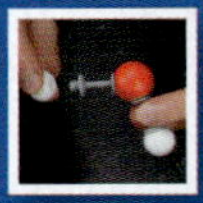

A molecule is a particle of matter made up of two or more atoms.

Compounds can be decomposed into simpler substances.

Many important chemical changes involve oxygen.

Our use of compounds has both positive and negative effects on society and the environment.

MAKE A SUMMARY

Consider the substances you learned about in this unit. All of them have uses that have positive effects and negative effects. For example, think about the effects and uses of sodium chloride (NaCl). Do you think this substance is a Super Hero? Or, do you think it is more of a Super Villain?

1. Choose any substance that interests you. Tell your teacher your choice: each student in the class should choose a different substance. Here are some possibilities:
    - copper (Cu)
    - iron (Fe)
    - ozone (O_3)
    - water (H_2O)
    - methane (CH_4)
    - hydrogen peroxide (H_2O_2)
    - carbon dioxide (CO_2)
    - sulfuric acid (H_2SO_4)
2. Research the substance's physical and chemical properties and its uses. Pay close attention to any societal or environmental benefits or hazards associated with it.
3. Display what you found out in a creative way, portraying your substance as either a Super Hero or a Super Villain, or both. You might make a trading card, a movie poster (starring your Super Hero or Super Villain), a comic strip, an ad for a theme park, or a storyboard for a TV commercial. Or think of an idea of your own!
4. Create a class display of your work.

Questions

1. What were your three favourite Super Heroes or Super Villains from the display? Explain why you chose them. T/I
2. Choose a Super Villain from the display. How might it have been portrayed as a Super Hero? C T/I

CAREER LINKS

List the careers mentioned in this unit. Choose two of the careers that interest you, or choose two other careers that relate to Exploring Matter. For each of these careers, research the following information:

- educational requirements (secondary and post-secondary)
- skill/personality/aptitude requirements
- potential employers
- salary
- duties/responsibilities

Assemble the information you have gathered into a brochure. Your brochure should compare and contrast your two chosen careers and explain how they connect to the information in this unit.

Figure 1 A firefighter

GO TO NELSON SCIENCE

Cleaning "Green"

There are dozens of cleaners to choose from at the store. Many store-bought cleaners contain toxic chemicals that can be harmful to us if used incorrectly. They can also be harmful to the environment even when they are used correctly.

SKILLS MENU

- Questioning
- Hypothesizing
- Predicting
- Planning
- Controlling Variables
- Performing
- Observing
- Analyzing
- Evaluating
- Communicating

SKILLS HANDBOOK 3.B.

Is the solution to stop buying store-bought brands and to start mixing together various household ingredients to make our own cleaners? Some people think these cleaners are cheaper, "greener" or more environmentally friendly, and just as effective. Are they right?

Testable Question

Is a homemade cleaner as effective at cleaning different types of dirty surfaces as a store-bought brand?

Prediction

Predict which cleaner will be the most effective for cleaning each type of surface you decide to test.

Experimental Design

In this investigation, you will develop a recipe for a homemade "green" cleaner. You will then design a procedure to test your cleaner against store-bought cleaners by cleaning similarly soiled surfaces.

CAUTION

Safety

Follow all safety precautions for using glassware and chemicals. Mix chemicals in a well-ventilated area.

Do not mix bleach and ammonia.

Equipment and Materials

eye protection

lab apron

protective gloves

scoopula

water

graduated cylinder

cleaner ingredients

plastic spray bottles

store-bought cleaner

surfaces to clean

Procedure

1. Use the Internet to research different recipes for cleaners that can be made with common household items. Recipes should be environmentally friendly. This means that the ingredients come from natural sources, are degradable, and are not harmful when properly handled. Select a recipe to use in this investigation.
2. Show the recipe to your teacher for approval. Then assemble your ingredients and place them on your worktable.
3. Design a procedure that allows you to fairly test and compare the cleaning ability of your cleaner against that of a store-bought brand.

   As you design your procedure think about the following:
   - What kinds of dirt and stains will you clean?
   - How will you avoid any bias in interpreting your test results?
   - How many surfaces will you need to prepare to test both cleaners on at least two different types of dirt and stains?
   - How much of each cleaner will you use in each test?
   - How will you decide when a surface is clean?
4. Show your procedure to your teacher for approval.
5. Put on your lab apron, eye protection, and gloves. Use your recipe to make your cleaner and prepare the dirty surfaces.
6. Follow your procedure to test each sample. Record your observations in a table.
7. Once your tests are complete, follow your teacher's instructions for returning or disposing of all materials and equipment.

Analyze and Evaluate

(a) Which cleaner was the most effective? Explain. T/I

(b) Compare the ingredients in your cleaner against the listed ingredients in the store-bought brand. T/I

(c) Do you think cleaning power is the only important consideration? Which cleaner do you think was safer to use? Which was better for the environment? Which was least costly? T/I A

Apply and Extend

(d) Based on this investigation, would you change the recipe of your cleaner? If so, how? Explain your reasoning. T/I C

(e) Prepare a presentation in which you apply the results of your investigation to convince others to use your cleaner. C A

CHECKLIST

Assessment

Your completed Performance Task will be assessed according to the following criteria:

KNOWLEDGE/ UNDERSTANDING

- Know the properties of various elements and compounds.
- Know that mixing chemicals can result in other chemicals being formed.

THINKING/ INVESTIGATION

- Plan the design, manufacturing, and testing of a product.
- Evaluate a substance in terms of its effect on society and the environment.
- Evaluate a substance in terms of its safety.

COMMUNICATION

- Prepare a report that communicates different aspects of a project.

APPLICATION

- Relate the properties of a finished product to the properties of its components.
- Apply basic concepts of science to real-world applications.

UNIT C

SELF-QUIZ

The following icons indicate the Achievement Chart category addressed by each question.

K/U Knowledge/Understanding | T/I Thinking/Investigation | C Communication | A Application

For each question, select the best answer from the four alternatives.

1. At the boiling point, a substance changes rapidly from (5.7) K/U
   (a) a solid to a liquid
   (b) a liquid to a gas
   (c) a liquid to a solid
   (d) a gas to a solid
2. What is the property of metals that allows them to be hammered into thin sheets? (6.2, 6.3) K/U
   (a) lustre
   (b) viscosity
   (c) ductility
   (d) malleability
3. What forms when atoms of different elements chemically join? (7.2) K/U
   (a) a mixture
   (b) a compound
   (c) a different element
   (d) a diatomic element
4. Which is true about a proton? (6.1) K/U
   (a) carries a positive charge and is located in the nucleus
   (b) carries a negative charge and has much less mass than the electron
   (c) carries a positive charge and has much less mass than the electron
   (d) carries no charge and orbits the nucleus
5. Aspirin has the formula $C_9H_8O_4$. What is the total number of atoms in a molecule of aspirin? (7.2) K/U
   (a) 12 (c) 17
   (b) 13 (d) 21
6. Which of these chemical formulas represents an element? (7.2) K/U
   (a) K_2S
   (b) C_2H_6
   (c) P_4
   (d) CaO

Indicate whether each of the statements is TRUE or FALSE. If you think the statement is false, rewrite it to make it true.

7. Hardness is an example of a physical property. (5.2) K/U
8. Salt melts ice on roads because it lowers the freezing point of the ice/salt mixture. (5.7) K/U
9. Most of the elements in the periodic table are non-metals. (6.1, 6.3, 6.4) K/U
10. The molecule with chemical formula H_2S contains three atoms. (7.2) K/U

Copy each of the following statements into your notebook. Fill in the blanks with a word or phrase that correctly completes the sentence.

11. In the chemical formula C_4H_{10}, the numbers 4 and 10 are called ________. (7.2) K/U
12. Stating that something is shiny is the same thing as saying it has ________. (5.2) K/U
13. Combustibility is an example of a ________ property. (5.3) K/U
14. ________ are commonly used in semiconductors. (6.5) K/U

Match each term on the left with the appropriate term on the right. Terms on the right may be used more than once.

15. (a) Ne — (i) atom
    (b) Br_2 — (ii) molecule
    (c) C
    (d) P_2O_5

    (7.2) K/U

Write a short answer to each question.

16. Classify each of the following properties of carbon dioxide as physical or chemical. (5.2, 5.3) K/U
    (a) colourless
    (b) forms carbonic acid in water
    (c) will not burn
    (d) freezes at –78.5 °C
    (e) odourless

17. Explain why the bottom of a clothes iron is made of metal, but the handle is made of a non-metallic substance, such as plastic. (6.3) T/I A

18. Describe how you could use a piece of paper, scissors, and a match to explain and demonstrate the difference between physical and chemical changes to someone. (5.4) T/I C

19. Dry ice is frozen carbon dioxide. Dry ice changes directly from the solid state to the gaseous state in a process called sublimation. Is sublimation a physical change or a chemical change? Explain. (5.4) K/U

20. Two different compounds can form when hydrogen combines with oxygen. Both compounds in **Figure 1** are made with hydrogen and oxygen. Write a chemical formula for each compound. (7.2) K/U

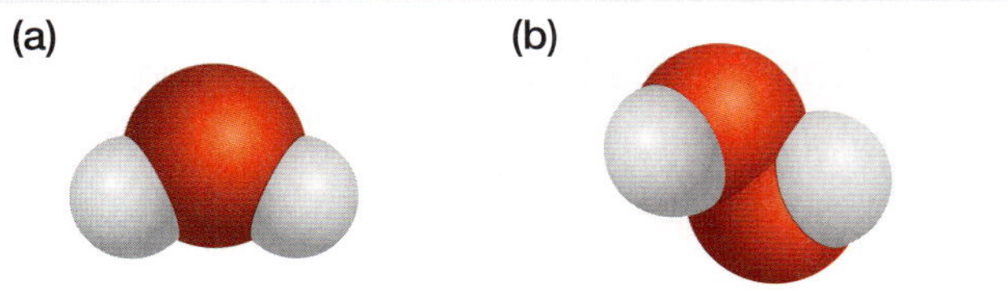

Figure 1

21. Hydrogen peroxide is commonly used to whiten teeth and lighten hair. Is the action of hydrogen peroxide a physical or chemical change? Explain. (7.4) K/U

22. Classify each of the following elements as a metal, a non-metal, or a metalloid. (6.3, 6.4, 6.5) K/U
    (a) phosphorus
    (b) calcium
    (c) nitrogen
    (d) potassium
    (e) oxygen
    (f) silicon

23. In your own words, explain why the term "periodic" is used to describe the periodic table of elements. (6.1) K/U C

24. Density is a physical property that can be calculated by dividing a substance's mass by its volume. Pyrite "fool's gold" has a density of 5 g/cm^3. Gold has a density of 19.3 g/cm^3. If you pick up equal-sized samples of each substance, which one would feel heavier? Explain your answer. (5.7) T/I

25. Akeem observed the following properties of an element: brittle, shiny, melting point of 3410 °C, grey colour, temperature of 27 °C. Which of these properties is a characteristic property? (5.7) K/U

26. An atom of potassium contains 19 protons. There is one electron in the outer orbit. (6.1) K/U
    (a) What is the atomic number of potassium?
    (b) How many electrons are in an atom of potassium?
    (c) Potassium is one of the alkali metals. What happens to the outermost electron when potassium reacts with other substances?

UNIT C

REVIEW

The following icons indicate the Achievement Chart category addressed by each question.

K/U Knowledge/Understanding T/I Thinking/Investigation C Communication A Application

What Do You Remember?

1. Which of the following is a list of characteristic physical properties? (5.7) K/U
   (a) acidity, lustre, and viscosity
   (b) colour, boiling point, and reactivity
   (c) density, melting point, and freezing point
   (d) malleability, conductivity, and combustibility

2. Which of the following is an element that consists of molecules? (7.2) K/U
   (a) CO_2 (c) O_3
   (b) Ca (d) Rh

3. Which is true about an electron? (6.1) K/U
   (a) carries a negative charge and is located in the nucleus
   (b) carries a negative charge and has much less mass than the proton
   (c) carries a positive charge and has much less mass than the neutron
   (d) carries no charge and orbits the nucleus

4. Which of these elements exists as a diatomic molecule? (7.2) K/U
   (a) Mn (c) Cl
   (b) He (d) Mg

5. Propane, a common fuel, has the chemical formula C_3H_8. What is the total number of atoms in a molecule of propane? (7.2) K/U
   (a) 3
   (b) 5
   (c) 8
   (d) 11

6. Which of these chemical formulas represents an element? (7.2) K/U
   (a) NaI
   (b) H_2
   (c) NO_2
   (d) CF_2Cl_2

7. The decomposition of hydrogen peroxide (H_2O_2) is a common laboratory process. What are the products of the decomposition of hydrogen peroxide? (7.5) K/U
   (a) two elements
   (b) three elements
   (c) two compounds
   (d) one element and one compound

8. Which physical property describes the ability of a metal to bend? (5.3) K/U
   (a) lustre (c) ductility
   (b) malleability (d) hardness

9. Which is evidence that a physical change has taken place? (5.4) K/U
   (a) A gas is produced.
   (b) A precipitate forms.
   (c) There is a change in state.
   (d) There is a change in colour.

10. Which of these is a non-metal? (6.4) K/U
    (a) iron (c) lead
    (b) cobalt (d) chlorine

11. How can you tell burning sugar is a chemical change? (5.4) K/U
    (a) The sugar melted.
    (b) It darkens.
    (c) The sugar changed shape.
    (d) The sugar became warmer.

12. Which statement about non-metal elements is true? (6.3, 6.4) K/U
    (a) All non-metals are ductile.
    (b) Most non-metals are good conductors.
    (c) Most non-metals are gases at room temperature.
    (d) Non-metals bend when hammered.

13. Fructose, a common sugar, has the formula $C_6H_{12}O_6$. How many hydrogen atoms are in one molecule of fructose? (7.2) K/U
(a) 6
(b) 12
(c) 18
(d) 24

14. What does the formula N_2 represent? (7.2) K/U
(a) an atom and an element
(b) an element and a molecule
(c) a molecule and a compound
(d) a compound and an element

15. Rust is formed when iron combines with which chemical? (7.8) K/U
(a) H_2
(b) O_2
(c) CO_2
(d) NaCl

16. A small amount of sodium chloride (NaCl), or table salt, disappears when poured into water, but a new substance is not formed. This is an example of what physical property? (5.2) K/U

17. An odour is produced when you mix two liquids. This is evidence that what kind of change has occurred? (5.4) K/U

18. Why is a test for magnetism not always useful to identify metals? (6.2) K/U

19. Galvanized steel is steel that has been dipped into melted zinc (**Figure 1**). After being dipped, a thin coating of zinc sticks to the steel. What do you think is the purpose of galvanizing steel? (7.8) T/I A

Figure 1 Nails made from galvanized steel

20. What is electrolysis? Explain how it is used to decompose water. (7.4, 7.5) K/U

21. You mix eggs, sugar, flour, and other ingredients and placed the batter in a hot oven. 30 min later, delicious cake smells bring you back to the kitchen. List as many pieces of evidence as you can to show that a cake baking involves chemical change. (5.4) T/I A

22. Identify each of the following as a physical change or a chemical change, and give reasons for your answer. (5.4) K/U A
(a) water droplets forming on the bathroom mirror
(b) grinding wheat into flour
(c) toasting marshmallows
(d) making a peanut butter and jelly sandwich
(e) lighting a Bunsen burner
(f) dissolving sugar in tea
(g) beating a raw egg until it is creamy

23. Copy and complete **Table 1** in your notebook with the words Yes or No. (7.2) K/U T/I

Table 1

Formula	Element	Compound	Consists of only atoms	Consists of only molecules
H_2O				
Cu				
F_2				
C_2H_6				
P_4				

24. If you leave your bicycle outside, it might develop rusty spots. Describe two different ways you can help prevent rust from forming on your bicycle. (7.8) K/U A

25. The element helium has the atomic number 2. The element neon has the atomic number 10. List two things these elements have in common. (6.1) K/U

The following icons indicate the Achievement Chart category addressed by each question.

K/U Knowledge/Understanding T/I Thinking/Investigation C Communication A Application

What Do You Understand?

26. Name two compounds that you might find in your kitchen. (7.2) A

27. (a) What is the difference between a qualitative property and a quantitative property?

    (b) Choose an element or a compound. List two of its qualitative properties and two of its quantitative properties.

    (c) How can a property such as temperature be either a quantitative or a qualitative property? (5.2, 6.1, 7.2) K/U T/I

28. Copy and complete **Table 2** by placing an X in the correct column in each row to show the types of elements that form each compound. (6.3, 6.4, 7.2) K/U

Table 2

Compound	Two metals	Two non-metals	One metal and one non-metal
NaCl			
CO_2			
KF			
SO_2			
CaO			

29. Your friend says that matter is "anything you can see and touch." What is wrong with this definition? (5.2) T/I

30. What feature on the periodic table separates the metals from the non-metals? (6.3, 6.4) K/U

31. What is the same about the following processes: combustion and rusting? What is different about them? (7.8) T/I

32. How does pouring water onto a fire put it out? (7.8, 7.9) K/U A

Solve a Problem

33. You are given a sample of an unidentified element. (5.2, 5.6, 5.7, 6.2) T/I

    (a) Describe a series of tests that would help you determine if the substance is a metal.

    (b) After determining if the substance is metallic, you wish to identify the specific element in the sample. Describe how you could use a characteristic physical property to determine the element's identity.

34. Some metals, such as platinum and gold, are found near the centre of the periodic table and they are found in nature in pure form. What can you conclude about the reactivity of these metals? (5.3, 6.3) T/I A

35. You want to build one molecular model for carbon monoxide (CO) and one for carbon dioxide (CO_2). (7.2, 7.3) K/U

    (a) How many model atoms of carbon do you need in total?

    (b) How many model atoms of oxygen do you need in total?

36. You accidentally lose the cap of the hydrogen peroxide bottle and decide to pour it into a clear glass jar with a lid. Weeks later you cut your finger and go to the medicine cabinet to get the jar of hydrogen peroxide. When you pour the hydrogen peroxide on your cut you do not see any of the normal fizzing and bubbling. What do you think has happened? (7.5) K/U T/I

37. Carbonated soft drinks contain carbonic acid (H_2CO_3). When this compound decomposes, it releases carbon dioxide gas and liquid water. What would you see in a soft drink as this process happens? How could you make this process occur faster? (7.5) K/U A

Create and Evaluate

38. Northern Ontario has vast regions of forests and lakes. It is also rich in many precious metals. The Sudbury basin produces 9 % of the world's nickel. Considerable amounts of copper, silver, and zinc are also mined there. Lac des Iles, Canada's only platinum mine, is ranked among the top 10 in the world. Create a table with two columns. List the benefits in one column and in the other list the drawbacks of having a successful mining operation in the region (**Figure 1**). (6.3) K/U A

Figure 1

Reflect on Your Learning

39. How does studying the physical and chemical properties of elements and compounds help you to make more informed decisions as a consumer and as a citizen? C A

40. In this unit, you learned that chemistry is not confined to a laboratory, but affects our everyday lives in many ways. What was the most exciting everyday application of chemistry you learned in this unit? A

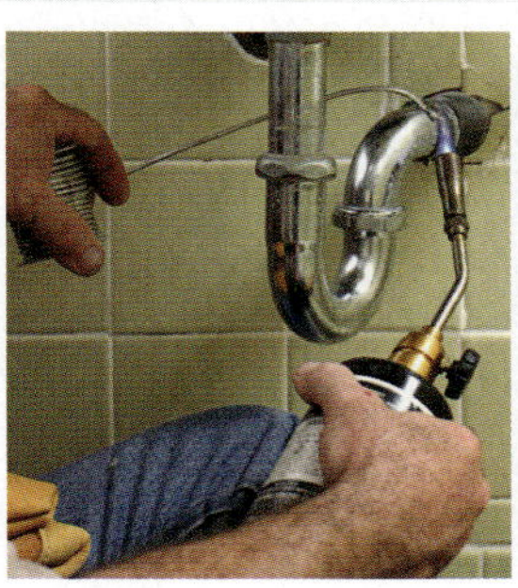

Figure 2

Figure 3

Web Connections

41. Research and report on the symptoms and long-term effects of mercury poisoning in living things. What are the main sources of mercury in the environment today? What can be done to reduce or eliminate these sources? T/I C

42. Solder is a soft metal mixture that plumbers use to join pipes (**Figure 2**). One type of solder is made by mixing tin (Sn) and bismuth (Bi). When a length of solder wire is attached to a heavy mass at one end and suspended from the ceiling at the other end, it slowly increases in length. T/I C A
    (a) What physical property of solder does this observation demonstrate?
    (b) Research the properties of this mixture of tin and bismuth and the properties of these individual elements. Does the mixture share some of the properties of the elements that it includes? Explain your answer.
    (c) Now consider the properties of the compound table salt, or sodium chloride (NaCl). Are some of the properties of this compound the same as the properties of the elements it contains, Na and Cl? Explain.

43. Artists rely on various materials and chemical processes to create paintings, sculptures, and other works of art. Research to find the most unusual material or process used to make an artistic piece. Compare your findings with a classmate's. Whose discovery was more unusual or surprising? C A

44. The Boeing 787 "Dreamliner" jet made its first flight in December 2009 (**Figure 3**). Research the materials that were used to construct this jet. How did these materials help make the 787 the most fuel-efficient airliner of its time? T/I A

GO TO NELSON SCIENCE

UNIT

D Space Exploration

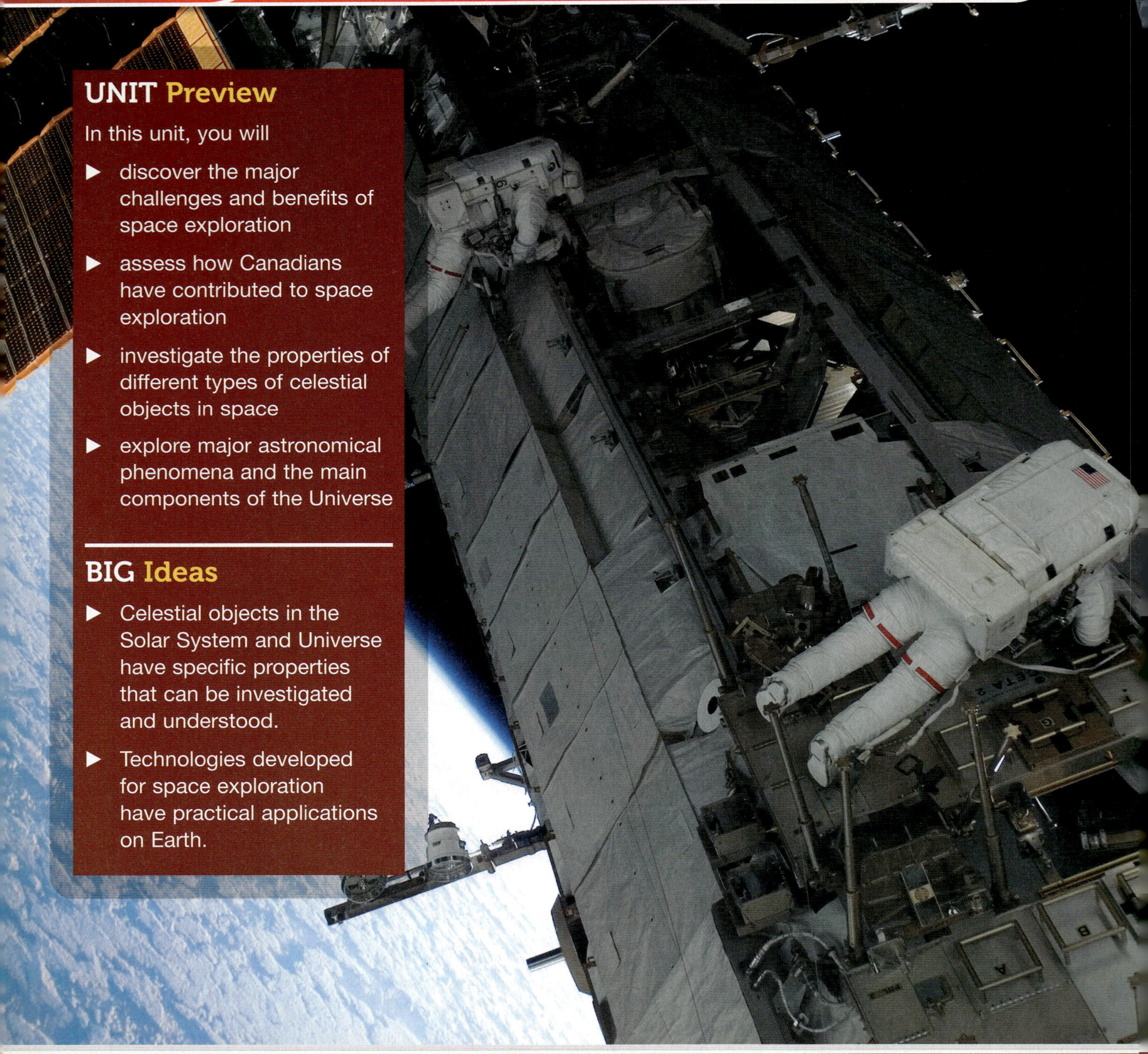

UNIT Preview

In this unit, you will

- discover the major challenges and benefits of space exploration
- assess how Canadians have contributed to space exploration
- investigate the properties of different types of celestial objects in space
- explore major astronomical phenomena and the main components of the Universe

BIG Ideas

- Celestial objects in the Solar System and Universe have specific properties that can be investigated and understood.
- Technologies developed for space exploration have practical applications on Earth.

Focus on STSE

Space Spinoffs

Space exploration has expanded our knowledge of the Universe, and given us technology that we use everyday.

- Invisible braces are made of a ceramic material originally developed for spacecraft.
- Ski boots have folds similar to those on spacesuits. This allows the boot to support a skier's foot but stay flexible so the skier can move around.
- Joystick controllers used in computer games and on vehicles for people with disabilities were originally developed to control the Apollo lunar rover.

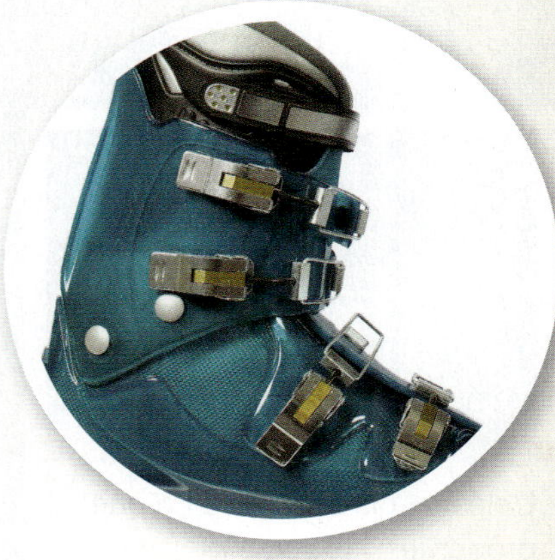

There are many other examples of technologies used on Earth that were first developed for space exploration. In medicine, examples include the pacemakers used to treat heart conditions, and a device that can monitor intensive care patients from their homes. These devices were first developed to monitor astronauts and spacecraft at a distance. Another example is the ear thermometer. The infrared sensor that can give an accurate temperature in seconds was initially developed to measure the thermal energy of distant stars.

Space research has given us many new technologies that have an impact on the way we live here on Earth.

Think/Pair/Share

1. Space missions are costly and dangerous. Do you think humans should continue to explore space? Write a brief paragraph to explain your thinking. C
2. Working in pairs, decide which of the space spinoffs described above you think is most important. Present your choice to the class with your reasons. C

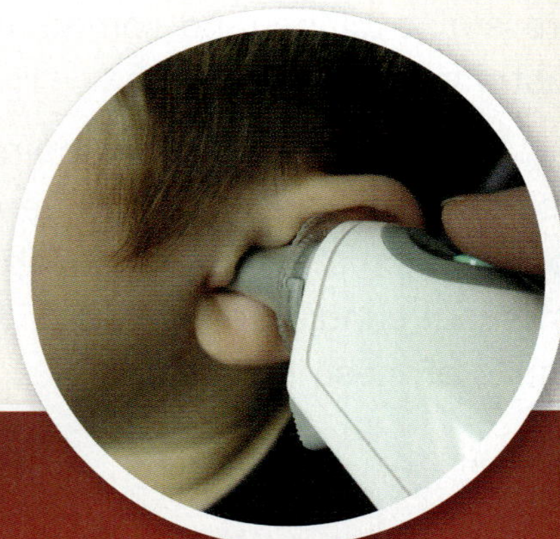

Space Exploration

CHAPTER 8

Touring the Night Sky

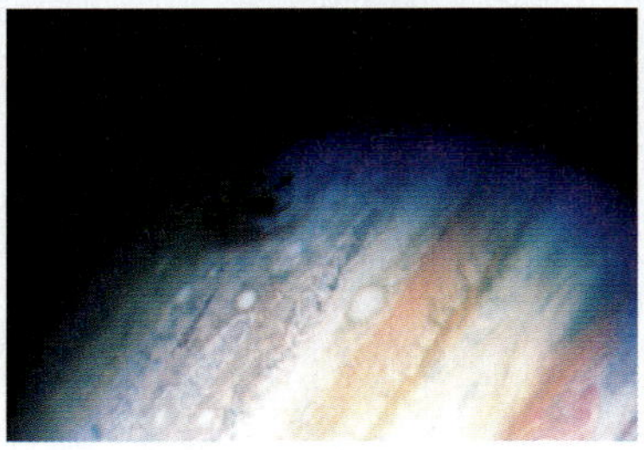

Celestial objects have unique properties.

CHAPTER 9

Our Place in Space

Our Universe is in a state of continuous change.

CHAPTER 10

Space Research and Exploration

Space exploration is beneficial to humankind.

Unit Task Preview

Astronaut Training Test

Being an astronaut is an incredibly challenging job. It requires great precision and composure under stressful, exciting, and sometimes frustrating conditions.

In this Unit Task, you will complete a series of astronaut training tests. You will then recommend suitable astronaut candidates based on the results of these tests.

UNIT TASK

Assessment

You will be assessed on:

- demonstrating an understanding of space travel and training
- conducting each test and recording your results
- summarizing test results and explaining how each test would be useful for selecting successful astronaut candidates

UNIT TASK

Bookmark

The Unit Task is described in detail on page 404. As you work through this unit, look for this bookmark and see how the section relates to the Unit Task.

Getting Started: Useful Concepts and Skills

Concepts Review

1. What features of Earth visible in **Figure 1** help keep us alive? T/I

Figure 1

2. Copy **Table 1** into your notebook. Categorize the following objects into the correct column: Mars, the Moon, the Sun, Earth, stars. K/U

Table 1

Emits light	Reflects light

3. Using your own words, explain why the Sun appears to move across the sky. K/U

4. Select two items from the list of technological tools and devices below and write a brief paragraph summarizing its importance to the study of astronomy. C A

    (a) spectroscope

    (b) spacecraft

    (c) life-support systems

Skills Review

5. (a) Explain how you could build a model of the Earth and Sun in order to investigate their motions. T/I

    (b) How might your model best be used to explain the reason for day and night? T/I

6. (a) Name the device shown in **Figure 2**. K/U

Figure 2

    (b) Why is this device important in astronomy?

    (c) Explain how this device can be used to investigate celestial objects. K/U A

7. Briefly explain what safety precautions you should take when testing a sundial (**Figure 3**). T/I

Figure 3 A sundial

CHAPTER 8

Touring the Night Sky

KEY QUESTION

What can we detect in the Universe?

KEY CONCEPTS

- The Solar System includes the planets and other celestial objects that orbit the Sun.
- The Universe includes stars, planets, moons, and galaxies.
- The Moon is the closest celestial object to Earth.
- A star is a giant mass of gases that emit light and huge amounts of energy.
- The Sun is our star.
- Galaxies contain billions of stars, planets, and other celestial objects.

When comet Shoemaker–Levy 9 collided with Jupiter on July 21, 1994, it created disturbances that caused the dark smudges visible on the upper part of the planet in this photograph.

Search

Home

SHOEMAKER-LEVY 9 SLAMS INTO JUPITER

July 23, 1994 Last night, the scientific community witnessed the last of a series of dramatic collisions in space. A comet slammed into the largest planet in the Solar System—Jupiter.

Comets are icy objects believed to be left over from the formation of the Solar System. This comet had been caught by Jupiter's strong gravitational pull.

In March 1993, American astronomers Carolyn Shoemaker, Eugene Shoemaker, and Canadian David Levy were examining photographs taken through a powerful telescope. They were looking for objects in space that might someday collide with Earth. Instead, they found something that no one had ever seen before—a comet travelling around a planet. Most comets travel around the Sun.

Shoemaker, Shoemaker, and Levy's observations got the attention of other astronomers. They all started to watch this comet. They found that the comet was being pulled apart by Jupiter's gravity.

The fragmented comet had now stretched out over a million kilometres. This long stretch included chunks of rock and ice ranging from a few metres to 2 km in diameter. They were all travelling around Jupiter.

Scientists believed that the comet, now named Shoemaker–Levy 9, would crash into the surface of Jupiter this July. It was the first time such an impact had been predicted.

Excitement grew, and on July 16, telescopes from around the world were pointed at Jupiter. Just after 10:00 p.m. on the night of July 16, the wait was over. Professional astronomers and amateurs alike finally got to see the comet–planet collision.

The impacts were difficult to see. Jupiter was just below the horizon. But when the planet rotated a few days later, scientists could see the impact. Large dark spots—some bigger than Earth—could be seen. Satellite images showed fireballs more than 3000 km tall.

Shoemaker, Shoemaker, and Levy will go down in history. This highly skilled team made one of the most dramatic discoveries in the history of astronomy.

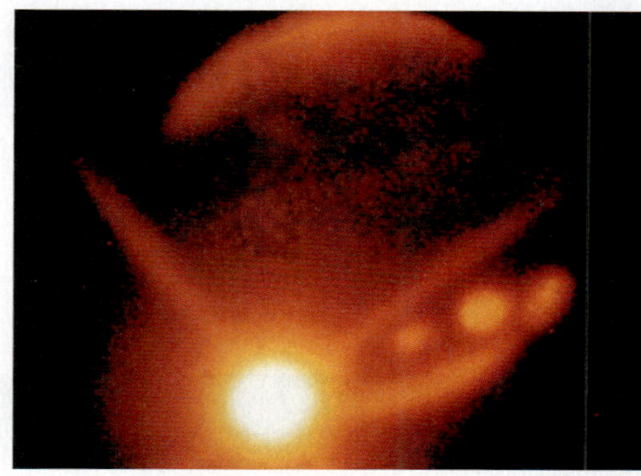

An enormous fireball, larger than Earth, was created from the comet's impact with Jupiter.

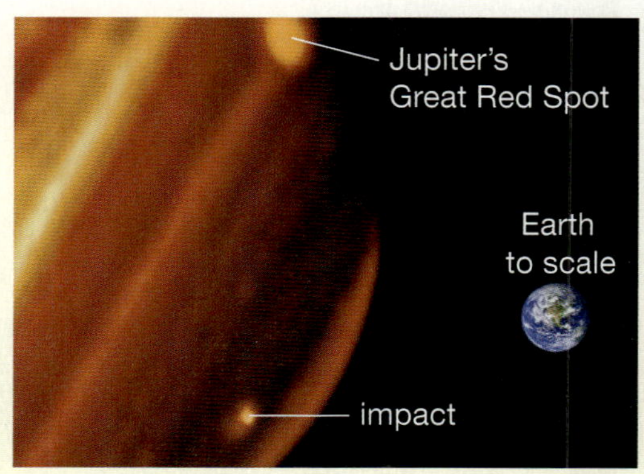

Shoemaker–Levy 9 was thought to be a once-in-a-lifetime event. However, on July 20, 2009, there was another impact scar near Jupiter's south pole. The impact crater looks like a bright spot in the infrared image shown here. It is about the size of the Pacific Ocean.

WHAT DO YOU THINK?

Many of the ideas you will explore in this chapter are ideas that you have already encountered. You may have encountered these ideas in school, at home, or in the world around you. Some of the statements are not true. Consider each statement and decide whether you agree or disagree with it.

1 The Universe contains more stars than you can count in a lifetime.
Agree/disagree?

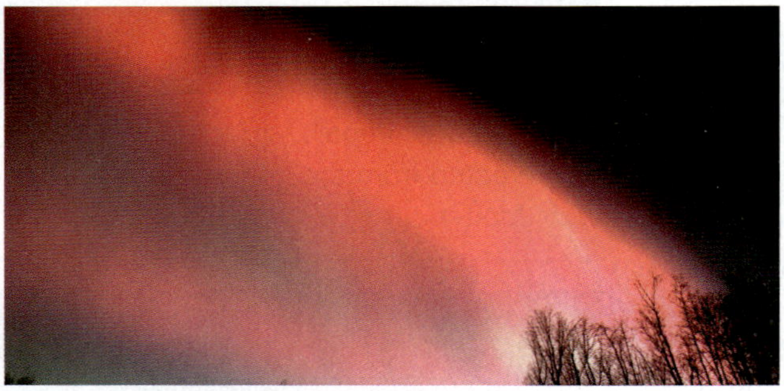

2 Particles from the Sun cause the aurora borealis.
Agree/disagree?

3 A star looks like the Moon when viewed through a powerful telescope.
Agree/disagree?

4 The Moon is the only celestial object that orbits Earth.
Agree/disagree?

5 Jupiter is very similar to Earth.
Agree/disagree?

6 Stars give off different colours of light.
Agree/disagree?

FOCUS ON
READING

Finding the Main Idea

The main idea narrows the general topic to a specific point or opinion. Use the following strategies to find the main idea in the text excerpt below from Section 8.3:

- use clues in the title and headings
- check the first, second, and last sentences
- look for clues in the figures and captions
- look for signal words (therefore, in conclusion, in summary)
- look for repeated words and bolded words

> **A Nearly Solid Interior**
>
> The Moon started out as hot liquid (molten) rock, but it is now mostly solid. We know this because of experiments done by NASA in the 1970s. Astronauts placed vibration-measuring instruments on the Moon. Later, an old moon rocket was deliberately crashed into the surface of the Moon. The effect was like hitting a bell with a hammer. The Moon started to vibrate from the crash, and the instruments measured the vibrations. If this experiment had been done on Earth, which has liquid rock in its core, the vibrations would have stopped after a few minutes. However, on the Moon, the vibrations continued for more than three hours. Scientists concluded that the Moon was mostly solid because solid rock vibrates longer than liquid rock.
>
> NEL 8.3 The Moon 301

DIG DEEPER

Reading Tip

As you work through the chapter, look for tips like this. They will help you develop literacy strategies.

Finding the Main Idea *in Action*

The main idea states a clear point or opinion. The facts, data, and examples in a reading support the main idea. Clues in the text can be used to figure out the main idea. Below is an example of how one student used clues to find the main idea in the above paragraph.

Clues	Information from the text
Use titles and headings	The heading says the interior is almost solid.
Check the first and last sentences	The first sentence repeats the idea in the heading. The last sentence repeats the idea in the first sentence.
Look for signal words	The word "concluded" suggests the main idea is coming up.
Look for repeated words	The words "mostly solid" are used in the first and last sentences.
Main Idea: Under its thin crust the moon is almost solid.	

8.1

The Night Sky

If you live in a city, the night sky may not seem very exciting. If you go into a rural area or out on a large lake or ocean, the night sky dazzles. On a clear night, you will see more stars than you can count.

- **celestial object:** any object that exists in space
- **star:** a massive collection of gases in space that emits large amounts of energy
- **luminous:** producing and giving off light; shining

Stars

Any object in space, such as the Sun or the Moon, is a **celestial object**. Most of the celestial objects we see at night are stars. A **star** is a massive collection of gases. Stars produce and emit (give off) their own light, which makes them **luminous**. Since they are so far away from Earth, they look like tiny pinpoints of light. It is amazing how many stars we can see on a clear night with the unaided eye. Even more amazing is the fact that we can only see a very small percentage of the stars in space (**Figure 1**). There are many more stars that can only be seen with powerful telescopes.

Our Star, the Sun

The Sun is a star. Compared to other stars, the Sun is medium-sized. However, compared to Earth it is enormous. The Sun has a mass almost 340 000 times that of Earth and a volume that is 1.3 million times the volume of Earth!

Figure 1 Many stars are too dim or too far away to see.

From Earth, the Sun appears to be much larger than other stars, but this is only because it is closer to us than any other star. The Sun is located about 150 million kilometres from Earth. The next-closest star to Earth is about 300 000 times farther from us than the Sun.

Many stars are brighter than the Sun and many others are much dimmer. Again, it is because the other stars are much farther away from us that they appear tiny and dim.

Life would not be possible on Earth without energy from the Sun. The size, brightness, and distance of the Sun from Earth all contribute to what makes our relationship with the Sun unique. You will read more about this relationship later in this chapter.

DIG DEEPER

Learning Tip

The Order of the Planets

To remember the order of the planets in the Solar System, repeat this sentence: **M**y **V**ery **E**xcellent **M**other **J**ust **S**ent **U**s **N**achos. The first letter of each word will help you remember Mercury, Venus, Earth, Mars, Jupiter, Saturn, Uranus, Neptune.

Planets

A **planet** is a large, round celestial object that travels around a star in a closed or continuous path called an **orbit**. To be classified as a planet, a celestial object must

1. orbit a star (such as the Sun) and no other celestial object,
2. contain enough mass so that its gravity pulls it into a round or spherical shape, and
3. be able to clear other celestial objects out of its orbit.

There are eight planets travelling around the Sun: Mercury, Venus, Earth, Mars, Jupiter, Saturn, Uranus, and Neptune. The planets, the Sun, and countless smaller rocky and icy objects that orbit the Sun make up the **Solar System**.

- **planet:** a large, round celestial object that travels around a star
- **orbit:** the closed path that a celestial object takes around another object; elliptical or circular in shape
- **Solar System:** the Sun and all the celestial objects that travel around it

Two Types of Planets

Mercury, Venus, Earth, and Mars are all made of solid material. For this reason, they are called the "terrestrial planets." Jupiter, Saturn, Uranus, and Neptune are made up mostly of gases, and possibly have small rocky cores. They are also much larger than the terrestrial planets. As a result of these properties, they are called "gas giants." Earth is unique from the other planets. Of all the planets in the Solar System, Earth is the only planet that contains liquid water and living things.

From Earth, planets appear bright, like stars. However, unlike stars, planets are not luminous. We are only able to see planets because they reflect the radiant energy, or light, of the Sun. Five planets are visible from Earth with the unaided eye: Mercury, Venus, Mars, Jupiter, and Saturn (Figure 2).

Figure 2 When visible, Venus (top right) is one of the brightest objects in the night sky.

Moons

Many planets have smaller celestial objects travelling in orbit around them. Any celestial object travelling in orbit around a larger celestial object is called a **natural satellite**. The planets are natural satellites of the Sun, and moons are natural satellites of the planets.

▶ **natural satellite:** a celestial object that orbits a larger celestial object

Like planets, moons are not luminous. Instead, they reflect the light of the Sun. The Moon shines brightly in the night sky because it reflects light from the Sun.

Planets can have different numbers of moons. Mercury and Venus do not have any moons, while Jupiter and Saturn each have more than 60 moons.

Galaxies

Some of the objects you may observe in the night sky are gigantic collections of gas, dust, stars, and planets known as **galaxies**. Galaxies contain billions of individual stars! Galaxies are huge, but are very far away, and can only be seen in detail with a telescope. All of the millions of galaxies together, indeed everything that exists, are part of the **Universe** (**Figure 3**).

▶ **galaxies:** huge, rotating collections of gas, dust, stars, planets, and other celestial objects

▶ **Universe:** everything that exists, including all matter, energy, and space

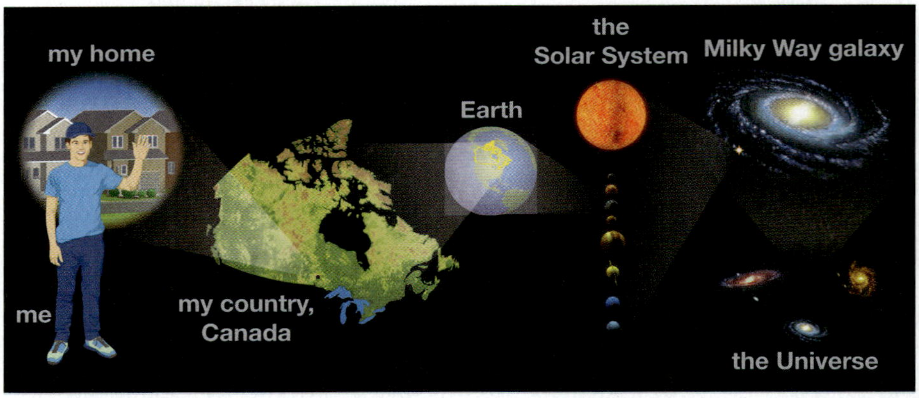

Figure 3 The Universe contains everything that exists, including you!

Our Galaxy, the Milky Way

Earth is part of the Milky Way galaxy. On very clear nights, we can see the Milky Way as a band of light spread across the sky. Some ancient observers thought it looked like spilled milk, so they named it the Milky Way. Other cultures have different names for our galaxy. The Chinese call it the Silver River.

DIG DEEPER
Web Link
To learn more about the different names for our galaxy,
GO TO NELSON SCIENCE

Movement in the Sky

You may have noticed that celestial objects appear to slowly move across the sky. One movement you have likely witnessed is the rising and setting of the Sun. The Moon also rises and travels across the sky. If you observe the stars carefully over several hours, you will see that groups of stars also appear to move across the night sky.

It Only Appears to Move

Most of the movement we observe is not the celestial objects moving, but Earth spinning in space and orbiting the Sun. The Sun appears to rise and set because our side of Earth is facing the Sun during the day, but is turned away from the Sun at night. Suppose you sit in a rotating chair in your classroom and someone spins the chair around. From your position, it *seems* like the walls and desks of the classroom are moving, but it is really you who is moving! We call this optical illusion of movement "apparent motion."

When we observe celestial objects during the day or at night, we must always remember that we are on a spinning planet that orbits the Sun. Thus, the Sun, stars, and the other planets appear to move across the sky. The apparent motion of the Sun and stars across the sky is predictable. If you were to stand high above the North Pole and look down, Earth rotates beneath you in a counterclockwise direction, from west to east. So, when you are standing on Earth, the Sun and stars appear to travel from east to west.

The apparent motion of celestial objects appears relatively slow to us. You need to observe a celestial object for many minutes to notice its apparent motion. In Chapter 9, you will be able to observe the relative motion of celestial objects.

Polaris Unmoving

If you observe the night sky regularly, you will notice one celestial object in the night sky that does not appear to move. It is Polaris, the North Star. This star is located almost directly above the North Pole. Consider a figure skater who looks at a point above her head on the ceiling as she spins in place. Even though she is spinning, the point on the ceiling does not appear to move because it is directly above her (**Figure 4**). In the same way, Polaris appears stationary as we observe it from a spinning Earth.

Figure 4 Polaris looks stationary to an observer on Earth because Earth's rotation is like the motion of this spinning skater.

TRY THIS CREATE A HORIZON DIAGRAM

SKILLS HANDBOOK 1.B.6., 2.A., 3.B.

SKILLS: Observing, Communicating

In this activity, you will create a diagram that shows the positions of celestial objects in the night sky.

Equipment and Materials: flashlight; red cellophane; plain paper; clipboard; pencil; ruler

Part A: Prepare

1. Cover the end of the flashlight with red cellophane.
2. Draw a line at the bottom of your paper to represent the horizon, the place where sky and ground meet.

Part B: Observe and Record

3. With an adult, go to an outdoor location and observe the night sky.
4. On your diagram, sketch any large trees or buildings you observe along the horizon. Use these objects as landmarks.
5. Carefully sketch the brightest objects you see in the night sky.
6. Record the date and exact time of your observations. Describe the position from which you made the observations and the appearance of the clouds in the sky.

A. When you get back to the classroom, compare diagrams with a classmate. Why do you think some objects were seen differently? T/I

8.1 Wrap Up

- You can learn a lot about space by simply observing the night sky.
- Stars that are actually larger and brighter than our Sun still look like tiny pinpoints of light because they are vastly farther away.
- The Solar System is made up of the Sun, along with the planets and other celestial objects that orbit it.
- A galaxy is a huge, rotating collection of gas, dust, and billions of stars, planets, and other celestial objects.
- The Universe is made up of all the matter and space that exists.
- The apparent motion of celestial objects in the night sky occurs because Earth is spinning on its axis as well as orbiting the Sun.

CHECK YOUR LEARNING

1. How do you think early peoples learned about objects in space before the invention of the telescope? K/U
2. Arrange the following celestial objects in order of relative size from largest to smallest: planet, Solar System, star, galaxy, Earth's Moon, Universe. K/U
3. What causes the apparent motion of the Sun in the sky as seen by an observer on Earth? K/U
4. Explain why Polaris seems to stay in one place. K/U
5. Label each column in **Table 1**. K/U

Table 1 Types of Planets

Column 1	Column 2
Mercury	Jupiter
Venus	Saturn
Earth	Uranus
Mars	Neptune

Tools That Help Us See Celestial Objects

Astronomers cannot conduct experiments on celestial objects, they can only observe them at a distance. However, today's technology allows us to see farther into outer space than ever before. Our current knowledge of the Universe has been greatly improved by the development of the telescope. For example, when Carolyn Shoemaker, Eugene Shoemaker, and David Levy discovered a comet orbiting Jupiter, they were looking at images produced by a powerful telescope.

DIG DEEPER

Did You Know?

Amateur Astronomy

Many celestial objects have been discovered by amateur astronomers. Canadian amateur astronomer David Levy is a good example. Along with co-discovering Shoemaker-Levy 9, he has discovered 21 other comets and 41 asteroids.

Telescopes

A telescope is an instrument that allows us to see objects that are very far away. Telescopes use mirrors and lenses to collect beams of light from a source and produce a magnified image. A telescope can collect up to 1 million times more light than the human eye can. **Figure 1** shows two basic kinds of telescopes: a refracting telescope and a reflecting telescope.

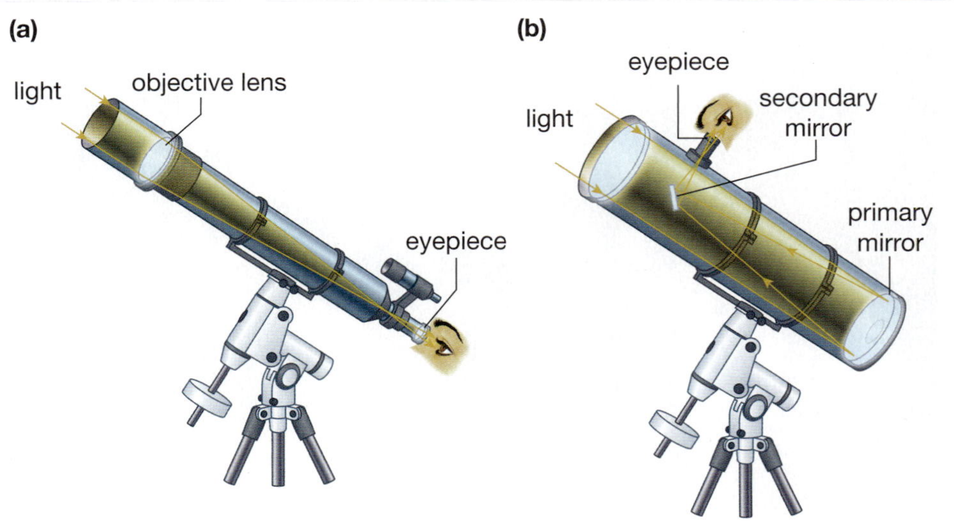

Figure 1 (a) A refracting telescope uses lenses to magnify images. (b) A reflecting telescope uses a series of mirrors.

DIG DEEPER

Career Link

Telescope operators gather data about celestial objects. They work in observatories and control the telescopes using computers. To learn more about becoming a telescope operator,

GO TO NELSON SCIENCE

Refracting Telescopes

A **refracting telescope** is a simple tube with a lens (piece of curved glass) at each end. The lenses refract, or bend, light. The observer can view objects through the eyepiece or attach a camera to the telescope to record an image.

Refracting telescopes are known for providing sharp images. The larger the lens, the more light it can gather and the sharper the image. However, a refracting telescope with a very large lens is expensive and difficult to build.

▶ **refracting telescope:** a telescope that uses lenses to gather and focus light

Reflecting Telescopes

The size limit for lenses in refracting telescopes means that they cannot always be used to observe celestial objects that are very distant. The scientist Isaac Newton solved this problem. He replaced the lenses with curved mirrors. A telescope that uses mirrors instead of lenses is called a **reflecting telescope**.

▶ **reflecting telescope:** a telescope that uses mirrors to gather and focus light

Telescopes in Space

Astronomers realized that Earth's atmosphere prevented images from being as clear as they could be. One solution was to place telescopes on high mountaintops. Another solution was to place telescopes above the atmosphere, in space. In 1990, the Hubble Space Telescope (HST) was launched into orbit around Earth. The largest mirror in the HST is 2.4 m in diameter. The HST produces clearer images than telescopes on Earth four times its size (**Figure 2**).

Figure 2 The Hubble Space Telescope was the first telescope to provide deep, clear images of the Universe.

DIG DEEPER

Did You Know?

The End of the HST?

The James Webb Space Telescope, successor to the Hubble Space Telescope, is expected to launch in 2014. This telescope has a mirror that is 6.5 m in diameter.

CITIZEN ACTION

SKILLS HANDBOOK 4.A.

Light Pollution

Light pollution is excess light produced by cities and towns at night. The wasted light costs millions of dollars a year, wastes energy, and makes astronomical observations more difficult. The excess light overpowers the dim light coming from celestial objects (**Figure 3**).

Use the Internet and other electronic and print resources to investigate the causes and effects of light pollution and ways that it can be minimized.

A. Compile a list of different ways that you, as an individual, could reduce light pollution. T/I C

B. Write an article for your school newspaper suggesting ways to reduce light pollution. How could you inform others in your school or community about the light pollution problem? T/I C

Figure 3 Light pollution in a large city

GO TO NELSON SCIENCE

8.2 Wrap Up

- Telescopes use lenses or mirrors to collect light and focus it as an image.
- Telescopes are often placed on high mountains or in space to get a clearer view of the sky.

CHECK YOUR LEARNING

1. Much of our current knowledge about celestial objects was gained using telescopes. What makes the telescope such an important tool for astronomical research? T/I
2. What is the difference between a reflecting telescope and a refracting telescope? K/U
3. (a) What were some of the limitations of telescopes that astronomers encountered?
   (b) Give two examples of how astronomers were able to overcome these limitations. K/U
4. Where could you go if you wanted to get the clearest observation possible of the night sky? Explain why. C A

TECH CONNECT OSSLT

The James Webb Space Telescope

We may not be able to travel back in time, but telescopes are the closest thing we have to a time machine. The farther away an object is in space, the longer its light takes to reach us. Today, the Hubble Space Telescope (HST) offers astronomers amazing views of the farthest reaches of the Universe. However, as time passes, Hubble requires more extensive maintenance and repairs to keep working. For this reason, Hubble will be replaced by an even bigger telescope called the James Webb Space Telescope (JWST), scheduled for launch in 2014 (**Figure 1**). Astronomers believe this successor to Hubble will be able to see farther than ever before. The JWST is designed to detect the first galaxies that formed in the early Universe.

Named after the NASA chief who worked during the Apollo moon missions, JWST will be a telescope unlike any other. The James Webb Space Telescope will have a very large mirror—6.5 m in diameter. This is about 3 times as wide as Hubble's mirror. The mirror will be made from one of the lightest metals known, beryllium, making it even lighter than that of the HST. The large, honeycomb-shaped mirror will be made up of 18 hexagonal segments. The JWST's giant mirror will sit on top of a plastic sunshield the size of a tennis court. The sunshield, which acts as an umbrella, will keep the telescope in the shadow of the Sun. This allows the JWST to stay very cold, around –225 °C. The telescope needs to be kept at these low temperatures because it is designed to detect thermal energy emitted from distant objects. The large mirror and sunshield are too large to fit into a rocket, so they will be folded up. Once JWST is in orbit, 1.5 million km from Earth, the mirror and sunshield will unfold itself and begin its task.

Astronomers are excited about what this next-generation telescope will reveal. In a few years, we may be able to see images from the earliest moments of the Universe, when the very first stars began to shine!

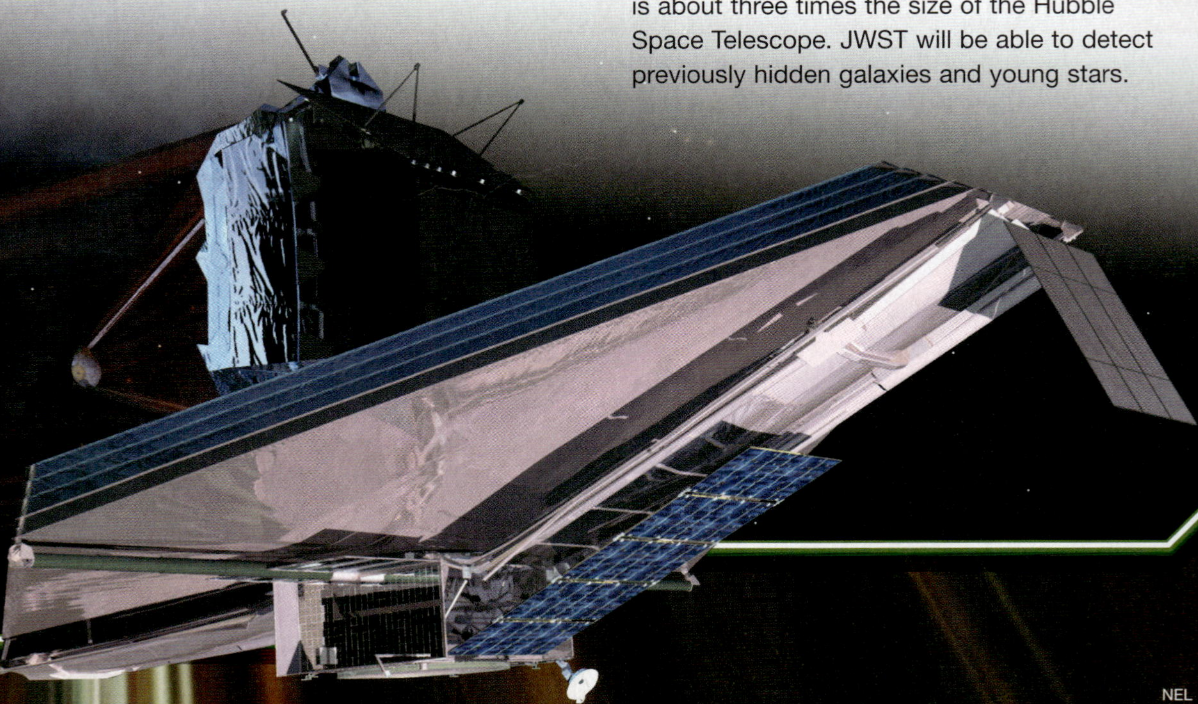

Figure 1 The James Webb Telescope (JWST) is about three times the size of the Hubble Space Telescope. JWST will be able to detect previously hidden galaxies and young stars.

The Moon

The Moon appears as the largest, brightest object in the night sky because it is the closest celestial object to Earth.

Formation of the Moon

Scientists believe that 4.6 billion years ago, Earth was struck by an object the size of Mars. This collision caused molten rock from Earth to fly out into space, forming the Moon (**Figure 1**).

Figure 1 It is believed that the Moon formed from the collision of a celestial object with Earth.

A Nearly Solid Interior

The Moon started out as hot liquid (molten) rock, but it is now mostly solid. We know this because of experiments done by NASA in the 1970s. Astronauts placed vibration-measuring instruments on the Moon. Later, an old moon rocket was deliberately crashed into the surface of the Moon. The effect was like hitting a bell with a hammer. The Moon started to vibrate from the crash, and the instruments measured the vibrations. If this experiment had been done on Earth, which has liquid rock in its core, the vibrations would have stopped after a few minutes. However, on the Moon, the vibrations continued for more than three hours. Scientists concluded that the Moon was mostly solid because solid rock vibrates longer than liquid rock.

Molten Rock That Cooled

As the molten rock that formed the Moon cooled, heavy matter sank to the centre of the ball, and lighter materials floated to the surface, forming three layers (**Figure 2**).

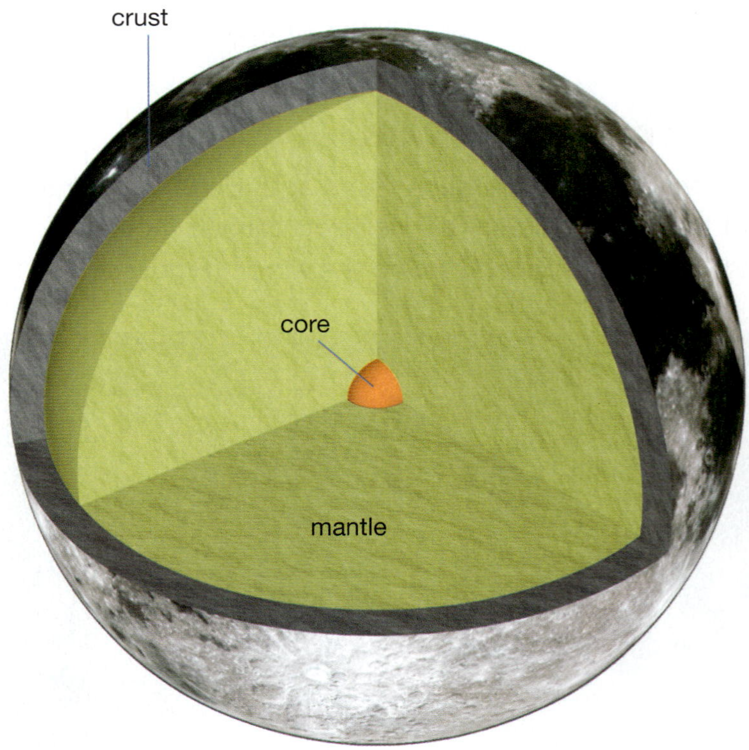

Figure 2 The Moon is made up of a thin crust, a thick mantle, and a small core.

Figure 3 The surface of the Moon is covered with craters.

"Seas" and Mountains

Later in the Moon's history, volcanic eruptions took place, spilling molten rock onto the surface and leaving large areas of dark rock. We call these areas maria, or "seas," because early scientists thought the dark areas on the Moon were bodies of water.

Mountains are also found on the edges of many of these maria. Many of these mountains were formed as a result of asteroid impacts (**Figure 3**).

An Uneven Surface

Small, rocky celestial objects in orbit around the Sun are called **asteroids**. Asteroids have collided with the Moon's surface, creating bowl-shaped holes in the rocky surface called **craters**. The Moon's surface has become very uneven because of these collisions.

- **asteroids:** small, rocky, irregularly shaped celestial objects that orbit the Sun
- **craters:** circular or bowl-shaped holes in surfaces caused by the impact of a celestial object

Learning about the Moon

The Moon is the only celestial object on which humans have landed. Between 1969 and 1972, six Moon landings were made and 12 astronauts were able to explore the Moon's surface. The information about the Moon gathered on these missions is extremely valuable. Many of the space technology spinoffs described earlier were the results of information gathered on these Moon missions.

Moon Rocks

One of the most important achievements of the Moon missions was the collection of moon rocks (**Figure 4**). The rocks on the Moon are extremely old—even older than the rocks on Earth. These ancient rocks gave scientists valuable information about what the early Solar System was like and how the Moon formed.

Working on the Moon

Astronauts working on the Moon had to wear bulky spacesuits to protect them from the harsh conditions. There is no atmosphere on the Moon, so there is no air. Surface temperatures range from 100 °C in direct sunlight to –100 °C in the shade.

Figure 4 Astronaut James B. Irwin, lunar module pilot, uses a scoop to make a trench in the Moon's soil during one of the Apollo missions. Mount Hadley is visible in the background.

DIG DEEPER

Reading Tip

Finding the Main Idea

The first sentence of a paragraph is called the topic sentence and this is where the main idea is usually stated. For example, "One of the most important achievements of the Moon missions was the collection of moon rocks."

DIG DEEPER

Did You Know?

Moon Rocks

NASA astronauts have collected more than 2400 samples of rock, dust, sand, and pebbles from the Moon.

Figure 5 A Moon landing site shows a lunar rover.

Exploring the Moon's Surface

With its fine layer of dust, deep craters, and tall mountains, the Moon's surface is uneven. This can make it a dangerous place to move around. To make moving around easier, astronauts on Moon missions drove the lunar rover, an electric jeep-like vehicle (**Figure 5**).

There may someday be permanent facilities on the Moon to provide a base for future exploration of the Solar System. If humans return to the Moon, new technology may be needed to deal with its rough surface conditions. You will learn more about the challenges of space exploration in Chapter 10.

Water on the Moon

In 1998, a robot probe called *Prospector* landed on the Moon and found evidence of ice hidden in craters in the Moon's polar regions. These pockets of ice might someday be melted to supply a future Moon base with water. They might also be broken down into oxygen (for astronauts to breath) or hydrogen (for rocket fuel).

UNIT TASK

Bookmark

You can apply what you learned about what it is like to explore, collect samples, and perform experiments on other worlds to the Unit Task described on page 404.

8.3 Wrap Up

- Scientists believe the Moon formed when an object the size of Mars struck Earth, sending molten rock into space.
- The Moon's surface includes mountains, craters, maria, and dust.
- The Moon is made up of a thin crust, a thick mantle, and a small partially liquid core.
- Someday permanent bases may be built on the Moon.

CHECK YOUR LEARNING

1. Create a comic strip to show what scientists think happened during the "Big Whack." K/U C
2. How did scientists determine that the Moon is mostly made of solid rock? K/U
3. Describe how the surface of the Moon got its craters. K/U
4. What conditions on the Moon's surface do you think engineers had to consider when they designed the lunar rover? K/U
5. Why do you think the Moon's surface is so different from Earth's? K/U
6. Do you think explorers from Earth should ever return to the Moon? Explain your answer. A C

8.4

Planets

After the Sun, the largest objects in the Solar System are planets. Astronomers recognize eight planets: the four inner terrestrial planets and the four outer gas giants.

Planet, or Something Else?

For hundreds of years, the word "planet" did not have a good scientific definition. Pluto was declared the ninth planet when it was discovered in 1930. However, it is different from the other planets.

- It is smaller than Earth's Moon.
- It has a flattened orbit that crosses the orbit of Neptune so that sometimes it is closer to the Sun than Neptune (**Figure 1**).
- Its moon is almost as large as itself.

Then, in the late twentieth century, other small, planet-like objects were discovered farther from the Sun. Some were larger than Pluto. Astronomers did not know what to call these objects. It was obvious that a clearer definition was needed for the term "planet." In 2006, scientists decided on the three-part definition on page 293.

Pluto does not satisfy the third rule. It is not able to clear other objects out of its orbit. Thus, Pluto is no longer considered a planet. It is now considered to be a dwarf planet. A **dwarf planet** is a celestial object that orbits the Sun and has a spherical shape, but which does not clear its orbit.

There are five recognized dwarf planets: Ceres, Haumea, Makemake, Eris, and Pluto. However, astronomers estimate that there are at least 70 000 icy objects beyond Neptune that orbit the Sun. Some dwarf planets, such as Eris, are much larger than Pluto and take nearly twice as long to orbit the Sun.

DIG DEEPER

Did You Know?

Other Celestial Objects

Astronomers estimate that there are at least 70 000 icy objects in orbit around the Sun beyond Neptune with about the same composition as Pluto. The dwarf planet Eris is 27 % larger than Pluto and takes nearly twice as long to orbit the Sun.

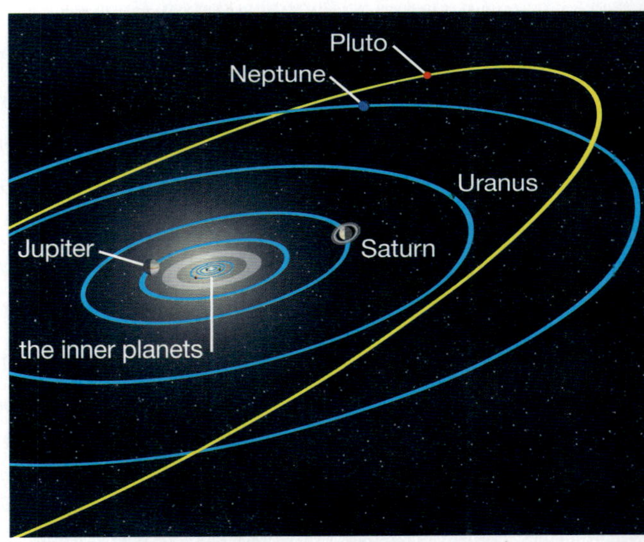

Figure 1 Pluto's tilted orbit crosses the path of Neptune's orbit.

▶ **dwarf planet:** a celestial object that orbits the Sun and has a spherical or round shape, but does not clear its orbit

The Inner Planets

Mercury, Venus, Earth, and Mars are the planets closest to the Sun. The inner planets are called terrestrial planets because they have a solid surface made up of rock. Earth is the largest of the inner planets and the only planet known to support life. Mercury was named after the speedy Roman messenger god because it travels around the Sun very quickly. From Earth, Venus appears brighter than any planet or star (except the Sun) and can be seen with the unaided eye on a clear night. Mars may once have had liquid water, but now it is so cold that it has huge amounts of ice beneath its surface.

DIG DEEPER

Did You Know?

Europa

Europa is one of the 60 moons of Jupiter. Scientists think that there is water more than 100 km below Europa's surface. It is possible that Europa's liquid water and relatively warm temperatures might support life of some kind.

Outer Planets

The outer planets are known as gas giants, because they are made up mostly of gas. Jupiter is the largest of all the planets, and it has more than 60 moons. Ganymede, Jupiter's largest moon, is bigger than the planet Mercury. Saturn is famous for its rings, which are made up of ice crystals and can only be seen through a telescope (**Figure 2**). Jupiter, Neptune, and Uranus also have rings, but they are much more faint. Uranus is mostly gas, but it may have liquid water and a solid core. One of Neptune's moons, Triton, may once have been a ball of ice and dust—a comet—that was captured by Neptune's gravity when it passed too close to the planet.

The distances between the inner planets and the outer planets are too large to be shown to scale in a diagram. **Figure 3** shows the relative sizes of the eight planets in the Solar System.

Figure 2 Saturn's rings are composed of small pieces of ice and a bit of dust. This makes them reflect the Sun's light very effectively.

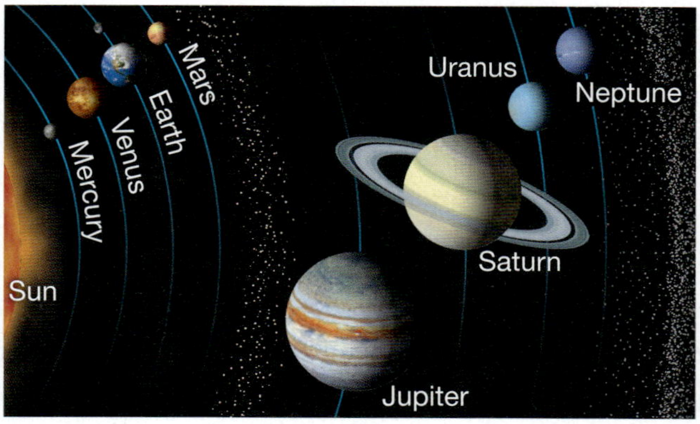

Figure 3 The planets in the Solar System—Mercury, Venus, Earth, Mars, Jupiter, Saturn, Uranus, and Neptune. (Note: This diagram is not to scale.)

RESEARCH THIS WHAT IS IT LIKE ON OTHER PLANETS?

SKILLS HANDBOOK 4.A., 4.B.

SKILLS: Researching, Analyzing the Issue, Communicating, Evaluating

You are familiar with Earth's weather, climate, and landforms. But, what are the weather conditions like on other planets?

1. Research more information about the conditions on each planet, such as the composition of the atmosphere, extreme temperatures, storm systems, surface features, and gravity.

A. Choose one planet and describe what it would be like to visit it. How would you deal with the conditions on the planet? Would living there be possible? Why or why not? T/I C

B. Choose a format to present your research, such as an illustrated poster, a written report, or a video presentation. T/I C

GO TO NELSON SCIENCE

8.4 Wrap Up

- A dwarf planet orbits only the Sun and is shaped like a sphere, but (unlike a planet) orbits among many other objects.
- The four inner terrestrial planets are made of solid matter.
- The four outer gas giant planets are huge and made mostly of gases around a solid core.

CHECK YOUR LEARNING

1. Which of the planets can be viewed without a telescope? K/U
2. What made scientists decide that a new definition of the term "planet" was needed? K/U
3. Only five objects are currently classified as dwarf planets. They are Ceres, Pluto, Haumea, Makemake, and Eris. Do you think other dwarf planets will be found? Why or why not? T/I
4. Imagine what it would be like to take a trip around the Solar System. What unique sights might you see on the other planets or moons compared to what you see on Earth? Write a brief postcard message to a friend or draw a comic strip illustrating such a journey. K/U C
5. Make a diagram showing the location of all the planets in relation to the Sun. Label each planet and include one significant fact that would help someone identify it. K/U C

8.5 The Stars

DIG DEEPER

Did You Know?

Distance to the Sun

The Sun is about 0.000 016 ly from Earth. This means that light from the Sun takes about 0.000 016 years (or 8 min) to reach us.

On a clear night, countless stars can be seen in the sky. Stars vary greatly in size, brightness, temperature, and colour. We have learned a great deal about stars by studying these properties.

▶ **light year:** the distance that light travels in a year; the unit is ly

Measuring Distances

It would be impractical to use kilometres to measure the large distances between stars. Instead, we use a much larger unit—the light year. One **light year** (1 ly) is the distance that light travels through space in one year. In space, light travels at a constant speed of 300 000 km/s. That is fast enough to circle Earth 7.5 times in one second!

▶ **luminosity:** the amount of energy released by a star each second

Star Brightness

The brightness of a star depends on how much energy it produces and gives off. **Luminosity** is the amount of energy produced by a star each second. A star's luminosity is measured on a numbered scale where the Sun is assigned a luminosity of 1. For example, the very bright star Sirius has a luminosity of 22. This means that Sirius is 22 times brighter than the Sun. When viewed from Earth, however, Sirius appears smaller and less bright than the Sun (**Figure 1**). This is because the Sun is closer to Earth. Stars that are farther away appear dimmer than the Sun, even if they are actually brighter.

Figure 1 Although it is the brightest star in the night sky, Sirius still appears smaller and dimmer than the Sun.

Apparent Magnitude

Apparent magnitude is a measurement of how bright a star appears from Earth. Apparent magnitude depends on both luminosity and distance. It is measured on a scale, where brightness decreases as the numbers increase (**Figure 2**). Very bright stars have an apparent magnitude of 1 or lower. The apparent magnitude of the Sun is –26.7 and the apparent magnitude of Sirius is –1.5. This means that the Sun appears to be much brighter than Sirius when viewed from Earth. The dimmest celestial objects that can be seen with the unaided eye have an apparent magnitude of 6. Each step on the scale represents an increase or decrease in brightness by a factor of 2.5.

► **apparent magnitude:** a measurement of how bright an object is when viewed from Earth; it depends on luminosity and distance

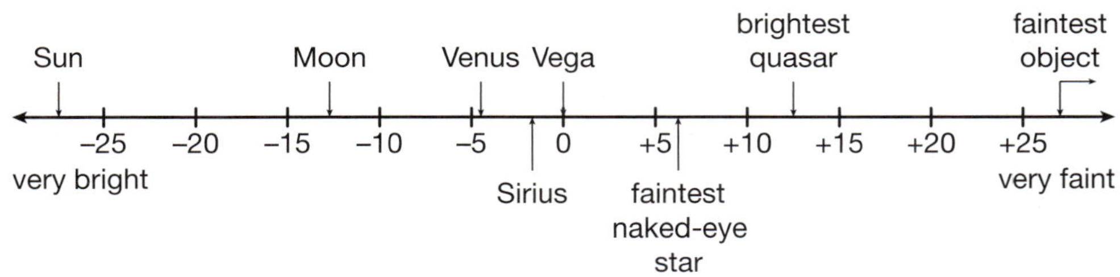

Figure 2 Apparent magnitude of some celestial objects

Absolute Magnitude

To better compare the brightness of stars, scientists compare how bright two stars would be if they were the same distance from Earth. **Absolute magnitude** is the measure of how bright a celestial object would appear if it were 33 ly from Earth. As with the apparent magnitude scale, the brightness of a celestial object decreases as the numbers increase.

Using the absolute magnitude scale, the absolute magnitude of Sirius is 1.4 and the absolute magnitude of the Sun is 4.8. This means that Sirius is actually a much brighter star than the Sun.

► **absolute magnitude:** a measurement of how bright an object would be if it were 33 ly from Earth

Star Colour and Temperature

The colour of stars depends on their temperature. Reddish stars are the coolest, and bluish–white stars are the hottest. For example, Betelgeuse is a red star and Rigel is a blue–white star (**Figure 3**). Our Sun appears yellowish–white and falls midway between bluish stars and reddish stars.

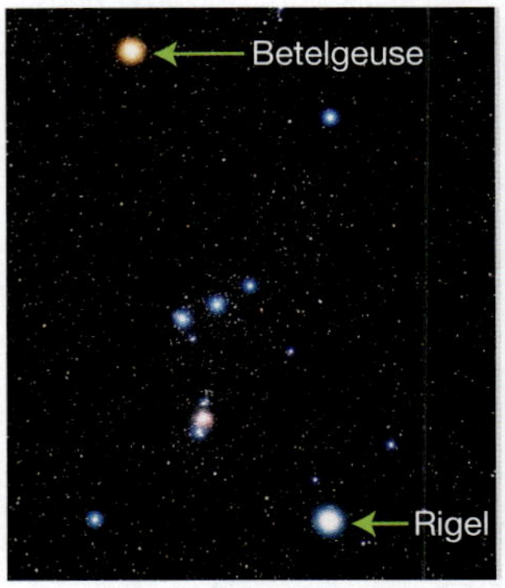

Figure 3 Betelgeuse, with a surface temperature of 3000 °C, appears reddish. Rigel, with a surface temperature of nearly 11 000 °C, appears bluish.

DIG DEEPER

Reading Tip

Finding the Main Idea

Use the title and headings to get a sense of what a passage is about. For example, "Plotting the Properties of Stars" suggests that the properties of stars can be plotted graphically to reveal a relationship.

Plotting the Properties of Stars

Astronomers plot the properties of stars on a special type of diagram that compares absolute magnitude (brightness) with surface temperature and colour (**Figure 4**). When stars are plotted according to these three properties, patterns emerge that indicate that these properties are related. For example, as the temperature of a star increases, its colour becomes more blue than red. Similarly, the hotter a star is, the brighter it is.

Most stars fall into a band on the diagram called the main sequence. Our Sun, which is yellow in colour, falls somewhere in the middle according to this system.

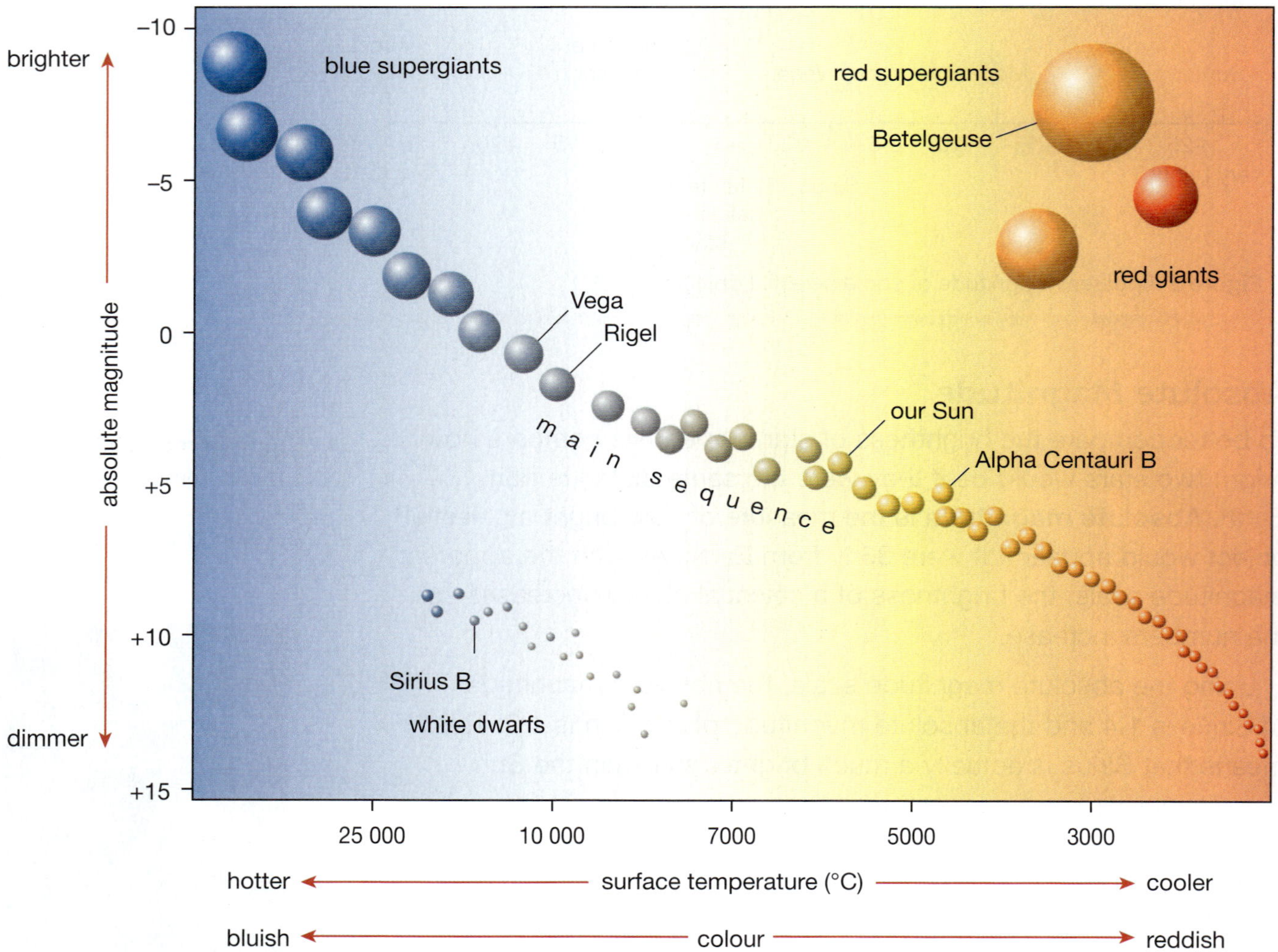

Figure 4 This diagram arranges stars into groups based on their brightness, colour, and surface temperature.

Grouping Stars

Grouping stars helps scientists to see patterns. For example, stars that are bright, reddish, and cool appear in the same areas of the diagram. These stars are called red giants. Other types of stars include blue supergiants, red supergiants, and white dwarfs.

8.5 Wrap Up

- The brightness of stars can be measured by their luminosity, apparent magnitude, and absolute magnitude.
- Luminosity is the amount of energy a star produces.
- Apparent magnitude is how bright a star appears from Earth.
- Absolute magnitude is how bright a star would appear if viewed at a set distance from Earth.
- There is a relationship between the colour of a star and its temperature. Cooler stars are reddish-white and hotter stars are bluish-white.

CHECK YOUR LEARNING

1. Explain why astronomers needed a different measurement unit to express distances between stars. What unit do they use? K/U
2. Which do you think is more useful for identifying a star's actual brightness: apparent magnitude or absolute magnitude? Explain. K/U
3. Deneb has an apparent magnitude of 1.3. Sirius has an apparent magnitude of –1.5. Is Deneb brighter or dimmer than Sirius? T/I
4. Write a short paragraph to explain how the distance of a star from Earth affects its apparent magnitude (brightness, as viewed from Earth). Use an example to illustrate your point. K/U C
5. Look at the diagram astronomers use to classify star properties (**Figure 4**). T/I C
   (a) Which has a higher surface temperature, Alpha Centauri B or the Sun? Explain.
   (b) How do the properties of Vega compare with the Sun?
6. Compare a red giant to a white dwarf, by referring to some of the properties in **Figure 4**.
7. **Table 1** lists data on the colour of several stars. K/U T/I A
   (a) Using what you read about star colour and temperature, arrange the stars in order from hottest to least hot.
   (b) Use **Figure 4** to explain how an astronomer might categorize these stars.

Table 1

Example(s)	Colour
Zeta Orionis	blue
Rigel, Spica	bluish-white
Vega, Sirius B	white
Polaris, Procyon	yellowish-white
Sun, Alpha Centauri B	yellow
Arcturus, Aldebaran	orange
Betelgeuse, Antares	red

8.6 The Sun

The Sun is the star at the centre of the Solar System. The Sun warms our planet and provides light and thermal energy. Processes, or reactions, occurring inside our Sun provide this energy. This energy is very important to life on Earth. Recall from Chapter 2 that plants use sunlight to produce the food needed by most organisms.

Layers of the Sun

Early astronomers thought the Sun was featureless—that it was a white smooth, glowing ball in the sky. Today, we know that the Sun is actually composed of many different layers, with each level performing different activities. The layers of the Sun are composed of gases (**Figure 1**). The Sun has no solid surface like Earth, but it does have a definite structure.

The Core

▶ **nuclear reactions:** reactions in which the nuclei of atoms change; these types of reactions produce large amounts of energy

In the core of the Sun, **nuclear reactions** occur that change the composition and energy of atomic nuclei. Atoms of hydrogen (H) fuse, or join, to form atoms of helium (He). These nuclear reactions release immense amounts of energy, causing the Sun's core to reach temperatures of about 15 million °C.

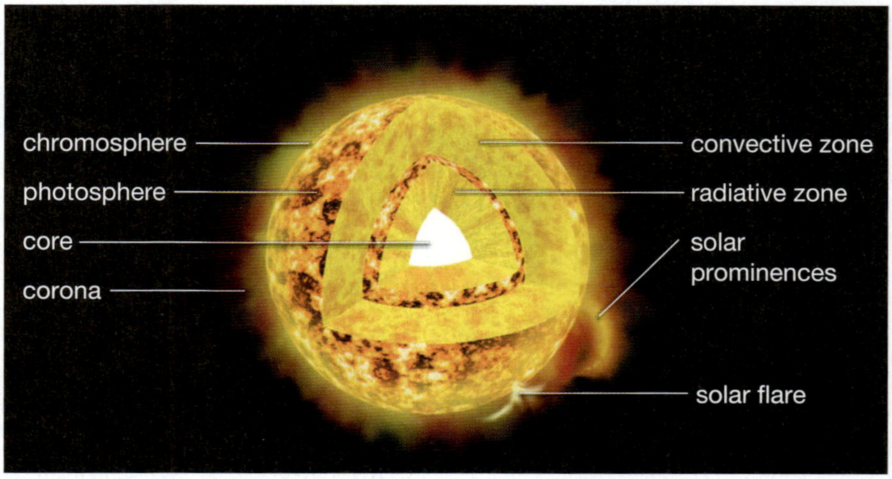

Figure 1 The Sun's layers include a core, several zones, the surface, and the atmosphere.

From the Core to the Atmosphere

Energy produced in the Sun's core moves outward into the next layer, the radiative zone. The energy moves outward into the convective zone, where cool matter sinks and hot matter rises. Energy continues to move outward until it reaches the Sun's surface, the photosphere, where energy escapes as light.

The region above the photosphere is called the solar atmosphere. It is divided into the chromosphere and the corona. The chromosphere makes up the inner atmosphere, and is very hot—about 60 000 °C hotter than the photosphere! The thin outer layer, called the corona, glows brilliantly. It extends millions of kilometres into outer space.

The Sun's Surface Features

Today, with the aid of telescopes, we know that the Sun's surface is not smooth and featureless. There are two distinct features of the photosphere—sunspots and solar flares.

Sunspots

Galileo first studied sunspots about 400 years ago. A **sunspot** is a dark region of cool matter on the photosphere, caused by twists in the Sun's magnetic field. Sunspots vary in both size and regularity. Often, sunspots appear in pairs of opposite polarity. For example, one will be magnetic north, the other will be magnetic south (**Figure 2**).

A **solar flare** is a huge explosion of gas and charged particles (**Figure 3**). Often found near sunspots, solar flares release huge amounts of energy and can reach temperatures of tens of millions of degrees Celsius.

- **sunspot:** a dark area on the Sun's surface that is cooler than the area surrounding it
- **solar flare:** gases and charged particles that are blasted above an active region of the Sun's photosphere

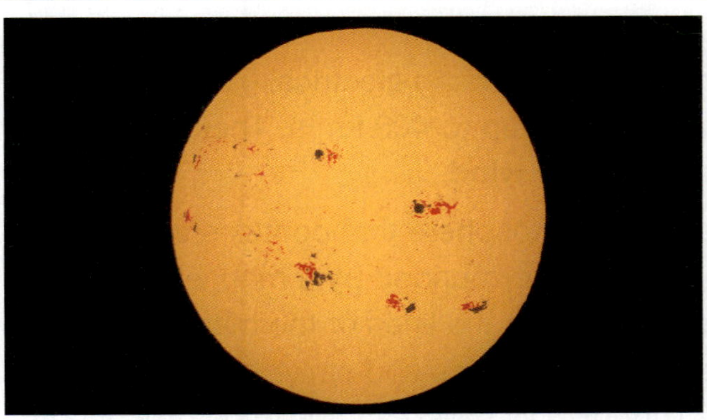

Figure 2 In this image, red indicates "north pole" and blue indicates "south pole".

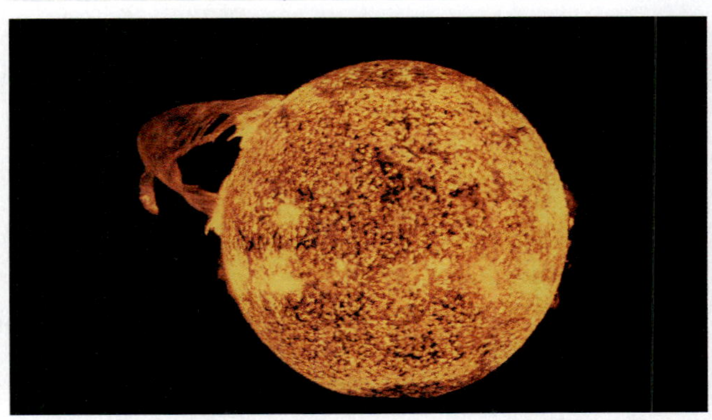

Figure 3 The large structure at the top left corner is a gigantic solar flare.

Energy from the Sun

Every second, the Sun releases enough energy to meet the needs of every person on Earth for the next 500 000 years. The Sun is the driving force behind Earth's climate and weather and also provides the energy needed for life to exist on Earth (**Figure 4**). The list below shows how, either directly or indirectly, almost all of the energy we use on Earth comes from the Sun.

- Plants use sunlight to grow and to produce food.
- Animals eat plants that are grown in sunlight.
- Coal was formed from the remains of prehistoric plants, so it is a form of stored energy from the Sun.
- Oil, including gasoline, is formed from the remains of prehistoric organisms, which obtained energy from plants, and hence, the Sun.

Figure 4 Almost all of the energy on Earth comes from the Sun.

The Auroras

The Sun's corona is constantly releasing electrically charged particles into space at speeds of up to 400 km/s. These travelling particles are known as the **solar wind**.

Earth is surrounded by an atmosphere that contains atoms of different gases. Earth is also surrounded by a magnetic field. Particles of solar wind spiral along the lines of the magnetic field. As the particles move toward the poles and strike the gases in Earth's atmosphere, a light display called an **aurora** is produced. Auroras occur at altitudes as high as 400 km above Earth's surface.

- **solar wind:** electrically charged particles that flow from the Sun's corona out into space
- **aurora:** coloured lights in the sky produced when the solar wind comes in contact with Earth's magnetic field and the atmosphere

Auroras are usually seen at night at the North and South Poles. Over the North Pole, the display is called the aurora borealis, or northern lights (**Figure 5**). Over the South Pole, it is called the aurora australis, or southern lights. The aurora appears green when the particles strike oxygen and red when they strike nitrogen.

Figure 5 The northern lights are featured in many Aboriginal legends. The Cree people, for example, call the aurora "the dance of the spirits."

8.6 Wrap Up

- The interior of the Sun is made up of several layers.
- The Sun's energy comes from nuclear reactions in its core. Energy radiates out of the core through the upper layers.
- The Sun's surface looks like boiling liquid. It has dark spots called sunspots and huge plumes of gas called solar flares.
- The Sun produces enormous amounts of energy. Most of the energy we use on Earth comes from the Sun.
- Auroras are light displays produced when the solar wind interacts with Earth's magnetic field and the atmosphere at the North and South Poles.

CHECK YOUR LEARNING

1. Where do the Sun's nuclear reactions take place? K/U
2. Create a graphic organizer describing light's journey from the core of the Sun to Earth. K/U C
3. How are solar flares different from sunspots? K/U
4. Describe how the Sun indirectly powers gasoline-powered automobiles. K/U
5. The Sun produces more than enough energy to meet our needs. Suggest ways we could use the Sun's energy more effectively. A
6. How would you explain to a friend how an aurora forms? K/U C

8.7

EXPLORE AN ISSUE CRITICALLY

SETI

SKILLS MENU

- Defining the Issue
- Researching
- Identifying Alternatives
- Analyzing the Issue
- Defending a Decision
- Communicating
- Evaluating

"SETI" stands for the Search for Extra-Terrestrial Intelligence. "Extra-terrestrial" means outside of Earth, so SETI is the search for life in outer space. Searching for alien life may sound more like science fiction than science, but some scientists believe it is worthwhile (Figure 1). So far, there is no evidence of extra-terrestrial life. However, some scientists argue that the development of life on Earth suggests that other forms of life may have developed somewhere else as well. They believe that making contact with an alien civilization would give us access to important new information and technologies.

On the other hand, many scientists think that SETI is a waste of time. They wonder why people would continue looking for extra-terrestrial beings when years of searching has produced no evidence. These scientists believe that the time, money, personnel, and equipment used for SETI would be better used on other projects. Others point out that announcing our presence to a potentially advanced alien civilization could be extremely dangerous. We have no idea if aliens would be friendly or not.

Figure 1 The gigantic Arecibo Radio Telescope in Puerto Rico is used by SETI researchers to search for faint alien radio signals.

SKILLS HANDBOOK 4.A., 4.B., 4.C.

The Issue

Some people argue that making contact with an alien civilization would benefit humanity. Others argue that the search for alien life is pointless, wasteful, or potentially dangerous.

You and a partner are presenting at a conference where it will be decided whether it is worth continuing to fund SETI projects.

Goal

To research and evaluate the benefits and drawbacks of SETI; to decide whether or not SETI should continue to receive funding; and convince a group of students, teachers, and parents of your particular point of view

Gather Information

Work with a partner to research the following:

- history of SETI
- methods used for SETI exploration
- passive SETI versus active SETI
- arguments for and against SETI (costs, research findings, etc.)
- what makes life possible on Earth so that we could recognize it elsewhere in the Universe

Make a Decision

Decide which side of the SETI issue you and your partner are on. If you are on different sides, compromise. Pick a side of the argument and represent that viewpoint.

Communicate

Prepare a short presentation of your findings and opinion in the form of a commercial message, a play, a PowerPoint presentation, a graphic organizer, or a video. Create visual materials such as diagrams and illustrations to explain difficult concepts.

Be sure to present general information about SETI first, then state your opinion and give specific reasons that support your position. Remember you are scientists, so your presentation has to include scientific facts and/or theories. Include a bibliography of the sources you used in your research.

DIG DEEPER
Web Link
For more information about SETI,
GO TO NELSON SCIENCE

8.8 Galaxies

As you read in Section 8.1, a galaxy is a huge grouping of stars, planets, and dust clouds. Galaxies vary in size, shape, and brightness.

Looking Back in Time

You have read that the distances in space are so huge that they are measured in terms of light years. As a result, the deeper into outer space that we direct our telescopes, the farther back in time we can see. For example, the Andromeda Galaxy is the closest galaxy to the Milky Way, and it is more than 2 million light years away. So, the light we see today from the Andromeda Galaxy left that galaxy more than 2 million years ago.

DIG DEEPER

Learning Tip

Large Numbers

A trillion is a million million. That is 1 000 000 000 000.

Types of Galaxies

Galaxies are huge. They contain between a billion and a trillion stars. Galaxies vary in size from a few thousand to half a million light years across.

Astronomers classify galaxies into four basic shapes: elliptical, spiral, lenticular, and irregular. Most galaxies are elliptical. **Table 1** shows the four types of galaxies.

Table 1 **Types of Galaxies**

Elliptical galaxy	Spiral galaxy	Lenticular galaxy	Irregular galaxy

Elliptical galaxies are shaped like spheres or ovals.	Spiral galaxies are flat and circular, with curving arms.	Lenticular galaxies are disc-shaped with a bulge in the middle.	Irregular galaxies get their name from their appearance.

The Milky Way

The Milky Way is a spiral galaxy with curved arms containing about 200 billion stars that rotate around its centre (**Figure 1**). It is about 100 000 ly across. The Sun is located about 30 000 ly from the centre.

Clusters of Galaxies and Stars

Galaxies are not spread out evenly across space. Some are bunched together in small groups. The Milky Way and its nearest neighbours, including the Andromeda Galaxy, are grouped together in a cluster of about 30 galaxies called the Local Group.

Stars are not spread out evenly within galaxies. A **star cluster** is a group of stars held close together by gravity. The Pleiades star cluster is one example (**Figure 2**). These clusters can contain anywhere from two or three stars to millions of stars, relatively close together in space.

▶ **star cluster:** a dense group of stars held together by gravity

Figure 1 Our galaxy, the Milky Way, appears as a band of light in the sky.

Figure 2 Also known as the "Seven Sisters," the Pleiades star cluster is probably the most famous star cluster.

8.8 Wrap Up

- Galaxies vary in size, shape, and brightness. There are four main types of galaxies: elliptical, spiral, lenticular, and irregular.
- The Milky Way is a spiral galaxy and is part of a galaxy cluster called the Local Group.
- Star clusters are dense groups of stars within a galaxy.

CHECK YOUR LEARNING

1. Of the four types of galaxies described in this section, which two are most similar in shape? Explain why. T/I C
2. Why are you "looking back in time" when you see light from distant galaxies? K/U
3. (a) Describe the Milky Way galaxy in terms of its size and shape.
   (b) Where is our Solar System located within the Milky Way? K/U C
4. Why do you think we cannot see the individual stars in other galaxies with the unaided eye? T/I

Chapter 8 KEY CONCEPTS SUMMARY

The Solar System includes the planets and other celestial objects that orbit the Sun.

- The four planets closest to the Sun are the terrestrial planets: Mercury, Venus, Earth, and Mars.
- The four planets farthest from the Sun are the gas giants: Jupiter, Saturn, Uranus, and Neptune.

The Universe includes stars, planets, moons, and galaxies.

- Some celestial objects, like stars, are luminous. Others, such as the Moon and planets, are non-luminous.
- Much of what we have learned about celestial objects has come from observations using telescopes.

The Moon is the closest celestial object to Earth.

- Scientists believe the Moon formed when an object the size of Mars struck Earth.
- The Moon is made up of a thin crust, a thick mantle, and a partly liquid core.
- The Moon's surface includes craters, maria, mountains, and dust.

A star is a giant mass of gases that emit light and huge amounts of energy.

- Stars vary greatly in size, brightness, temperature, and colour.
- A star's brightness can be measured by its luminosity, absolute magnitude, and apparent magnitude.

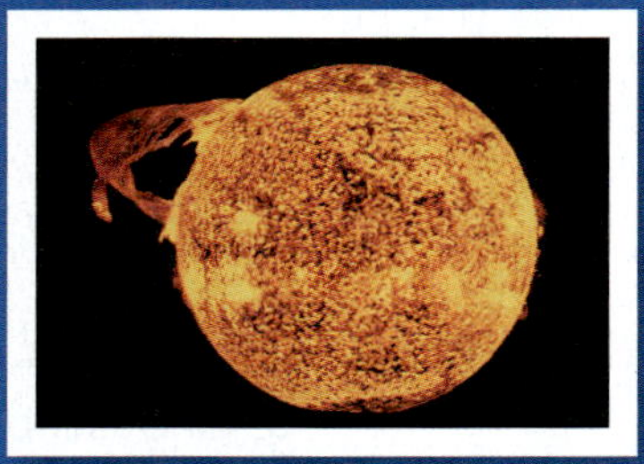

The Sun is our star.

- The Sun is vital to life on Earth. Almost all the energy we use on Earth comes from the Sun.
- The interior of the Sun is made up of several layers.
- Sunspots and solar flares are features of the Sun's surface.

Galaxies contain billions of stars, planets, and other celestial objects.

- Galaxies vary in size, shape, and brightness.
- Four main types of galaxies include elliptical, spiral, lenticular, and irregular.
- Our galaxy, the Milky Way, is a spiral galaxy.

WHAT DO YOU THINK NOW?

BIG Idea

- Celestial objects in the Solar System and Universe have specific properties that can be investigated and understood.

You thought about the following statements at the beginning of the chapter. You may have encountered these ideas in school, at home, or in the world around you. Consider them again and decide whether you agree or disagree with each statement.

1 The Universe contains more stars than you can count in a lifetime.
Agree/disagree?

2 Particles from the Sun cause the aurora borealis.
Agree/disagree?

3 A star looks like the Moon when viewed through a powerful telescope.
Agree/disagree?

4 The Moon is the only celestial object that orbits Earth.
Agree/disagree?

5 Jupiter is very similar to Earth.
Agree/disagree?

6 Stars give off different colours of light.
Agree/disagree?

How have your answers changed since then? What new understanding do you have?

DIG DEEPER

Vocabulary

celestial object p. 292
star p. 292
luminous p. 292
planet p. 293
orbit p. 293
Solar System p. 293
natural satellite p. 294
galaxies p. 294
Universe p. 294
refracting telescope p. 298
reflecting telescope p. 298
asteroids p. 302
craters p. 302
dwarf planet p. 305
light year p. 308
luminosity p. 308
apparent magnitude p. 309
absolute magnitude p. 309
nuclear reactions p. 312
sunspot p. 313
solar flare p. 313
solar wind p. 314
aurora p. 314
star cluster p. 319

CHAPTER 8

SELF-QUIZ

The following icons indicate the Achievement Chart category addressed by each question.

K/U Knowledge/Understanding T/I Thinking/Investigation
C Communication A Application

For each question, select the best answer from the four alternatives.

1. Which of these planets is farthest away from the Sun? (8.4) K/U
   (a) Mars
   (b) Saturn
   (c) Jupiter
   (d) Neptune

2. Which of the following does a refracting telescope use to gather light and focus images? (8.2) K/U
   (a) mirrors
   (b) lenses
   (c) sensor
   (d) film

3. What colour are the hottest stars? (8.5) K/U
   (a) whitish
   (b) reddish
   (c) yellowish
   (d) bluish-white

4. What type of celestial object orbits the Sun, has enough mass to form a sphere, and is large enough to clear its orbit? (8.1, 8.4) K/U
   (a) dwarf planet
   (b) planet
   (c) asteroid
   (d) star

Indicate whether each of the statements is TRUE or FALSE. If you think the statement is false, rewrite it to make it true.

5. One light year is the time it takes for light to travel through space in a year. (8.5) K/U

6. The amount of energy given off by a star is called its apparent magnitude. (8.5) K/U

7. A dwarf planet orbits another planet. (8.4) K/U

8. Stars are evenly spread out within galaxies. (8.8) K/U

9. The surface of the Moon is covered with bowl-shaped depressions called craters. (8.3) K/U

10. Not all objects in the night sky can be seen with the unaided eye. (8.1, 8.2) K/U

Copy each of the following statements into your notebook. Fill in the blanks with a word or phrase that correctly completes the sentence.

11. A(n) _________ galaxy has curving arms. (8.8) K/U

12. The electrically charged particles from the Sun's corona that strike gases in Earth's atmosphere to produce auroras are called _________. (8.6) K/U

13. The Sun's energy is produced by _________. (8.6) K/U

14. _________ is a measure of how bright an object would appear if it were located 33 ly away from Earth. (8.5) K/U

15. _________ is excess light produced by cities and towns at night. (8.2) K/U

16. The Milky Way is a _________-shaped galaxy. (8.8) K/U

17. Stars are _________, which means that they produce and emit their own light. (8.1) K/U

18. The _________ is at the centre of the Solar System. (8.6) K/U

19. A _________ telescope uses mirrors to collect and focus light form objects in space. (8.2) K/U

20. A celestial object that orbits the Sun but which does not clear its orbit is called a _________. (8.4) K/U

Match each term on the left with the most appropriate description on the right.

21. (a) satellite — (i) the brightness of a star as if it were located 33 ly from Earth

(b) colour — (ii) indicates a star's surface temperature

(c) absolute magnitude — (iii) a small, rocky celestial object that orbits the Sun

(d) asteroid — (iv) a celestial object that orbits another, larger, celestial object (8.1, 8.3, 8.5) K/U

Write a short answer to each of these questions.

22. Explain the strategies that astronomers used to overcome the limitations of telescopes when observing celestial objects. (8.2) K/U C

23. Why has it become more difficult in many areas near cities or towns to see stars in the night sky with the unaided eye? (8.2) K/U

24. What can scientists learn about the history of the Solar System by studying rocks from Earth's moon? (8.3) K/U

25. Imagine that you see the aurora borealis for the first time while on vacation. Write a postcard to a friend describing the event, explaining the natural conditions that cause the borealis to occur. (8.6) K/U C

26. Sketch a diagram of the Sun and include the following labels: photosphere, chromosphere, corona, solar flare. (8.6) C

27. Look at the diagram of the properties of some common stars (**Figure 1**). Compare the temperature and absolute magnitude of Vega with that of Betelgeuse. (8.5) T/I

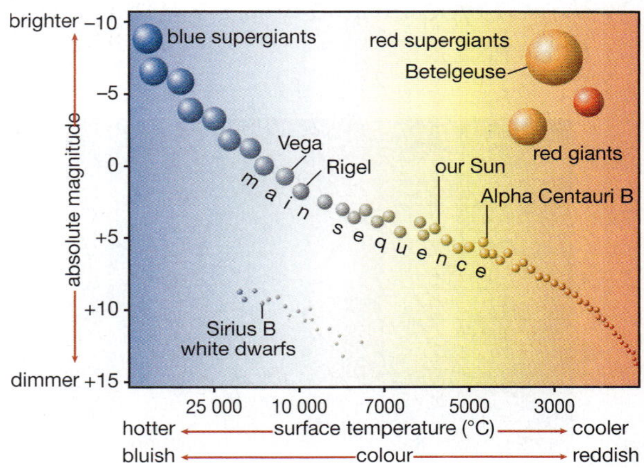

Figure 1

28. The official definition of a "planet" was recently changed. Why do you think astronomers decided the term needed a better definition? (8.4) K/U

29. Draw a series of labelled pictures showing how scientists think the Moon was formed. (8.3) K/U C

30. Describe which activity on the Sun has to do with the northern lights or auroras that form in Earth's atmosphere. (8.6) K/U

31. How is the planet Jupiter different from Earth? Give at least 3 specific examples. (8.1, 8.4) K/U

32. Using a diagram, explain how stars are organized within galaxies and star clusters. (8.8) K/U C

Self-Quiz

To do an online self-quiz, GO TO NELSON SCIENCE

CHAPTER 8

REVIEW

The following icons indicate the Achievement Chart category addressed by each question.

K/U Knowledge/Understanding
T/I Thinking/Investigation
C Communication
A Application

What Do You Remember?

1. Which type of galaxy is shown in **Figure 1**? (8.8) K/U

Figure 1

2. Describe the characteristics of a star. (8.1, 8.5) K/U

3. (a) In which galaxy is our solar system located?
   (b) How did our galaxy get its name? (8.1, 8.8) K/U

4. How are dwarf planets different from planets? (8.4) K/U

5. What causes the aurora australis and aurora borealis? (8.6) K/U

6. Describe the shape of a spiral galaxy. (8.8) K/U

7. How do astronomers know that the inside of the Moon is mostly solid rock? (8.3) K/U

8. Define a light year. (8.5, 8.8) K/U

9. What reaction in the Sun creates thermal energy? (8.6) K/U

What Do You Understand?

10. (a) What is an asteroid?
    (b) Explain how asteroids may have contributed to the formation of the Moon. (8.3) K/U

11. Write a short paragraph explaining how electricity generation can be indirectly powered by the Sun. (8.6) K/U C

12. Neptune and Pluto are both spherical objects orbiting the Sun. Why is Neptune considered a planet while Pluto is only a dwarf planet? (8.4) K/U

13. Explain why astronomers place land-based telescopes on mountaintops or above the atmosphere in space. (8.2) K/U

14. Describe or draw a picture to show how energy travels from the centre of the Sun to reach Earth. (8.6) K/U C

15. Many craters are visible on the Moon, but not on Earth. Why do you think this is? (8.3) K/U C

16. Suppose someone said to you that the Sun is a featureless ball of glowing gas. How would you describe some of the active features on the Sun's surface? (8.6) K/U T/I

17. "New theories often develop to replace old ones after new discoveries are made." Write a brief paragraph explaining how the reclassification of Pluto as a dwarf planet reflects this statement. (8.4) A C

18. Betelgeuse is a cool, reddish star, while Rigel is a bluish–white star. Refer to **Table 1**. Which star is closer to Earth? How do you account for the fact that their apparent magnitudes are so similar? (8.5) T/I

Table 1 Properties of Some Stars

Star	Distance from Earth (ly)	Apparent magnitude
Rigel	900	0.12
Aldebaran	65	0.85
Betelgeuse	1500	0.41

19. Compare the planets of the inner Solar System with those in the outer Solar System. Use a graphic organizer. (8.1, 8.4) T/I C

20. **Table 2** shows the apparent magnitude and absolute magnitude for several different stars. Which star gives off the most light? How do you know? (8.5) T/I

Table 2 Apparent and Absolute Magnitude of Some Stars

Star	Apparent magnitude	Absolute magnitude
Sun	–26.8	4.83
Sirius	–1.45	1.5
Vega	0.04	0.5
Betelgeuse	0.41	–5.6
Deneb	1.99	–7.5

Solve a Problem

21. A star with a luminosity of 3 is located 2 ly from Earth. Another star with a luminosity of 4 is located 100 ly from Earth. Which do you think will have greater apparent magnitude? Explain. (8.5) T/I A

Create and Evaluate

22. Imagine you are the head of a large government space agency. The agency has decided to set up a colony on a planet or natural satellite in the Solar System. It is your job to choose which one. Consider what you know about the conditions on each planet or satellite and their locations in the Solar System. Then write a speech or brief opinion piece that you might present to the public to convince them your choice is the best one. (8.4) C A

Reflect on Your Learning

23. Consider what happened when scientists decided they needed a new definition of "planet." Also think about why astronomers decided they needed a new model or theory to explain the movements of the planets. What do these events tell you about the way scientific theories change over time? A

Web Connections

24. In light of the recent discovery of water on the Moon, humans may be sending new missions there to further study its properties. Research proposals for future missions, and write a news report presenting your findings. T/I C

25. The Kuiper Belt is a spherical region of space located beyond Neptune. It contains icy celestial objects that orbit the Sun. The Oort cloud extends almost 3 ly from our Sun and is believed to be the region of space where long-period comets come from. Research the objects found in these distant regions of the Solar System. Summarize your findings in a news report. T/I C

26. The night sky has been a source of curiosity worldwide historically. A
    (a) Research explanations other cultures have of astronomical phenomena.
    (b) Explain how observations of the natural world lead to science.

27. Research and report on important discoveries made with the help of the Hubble Space Telescope. A

GO TO NELSON SCIENCE

CHAPTER 9 Our Place in Space

KEY QUESTION

How does the motion of celestial objects affect life on Earth?

KEY CONCEPTS

- The Solar System consists of all the objects orbiting the Sun.
- The motion of Earth causes day, night, and the seasons.
- Earth's motion affects our view of the night sky.
- Celestial objects can be used to predict annual events and to navigate.
- The positions of the Sun, Earth, and the Moon result in the lunar cycle and eclipses.
- Scientists have strong evidence that the Universe began around 14 billion years ago.

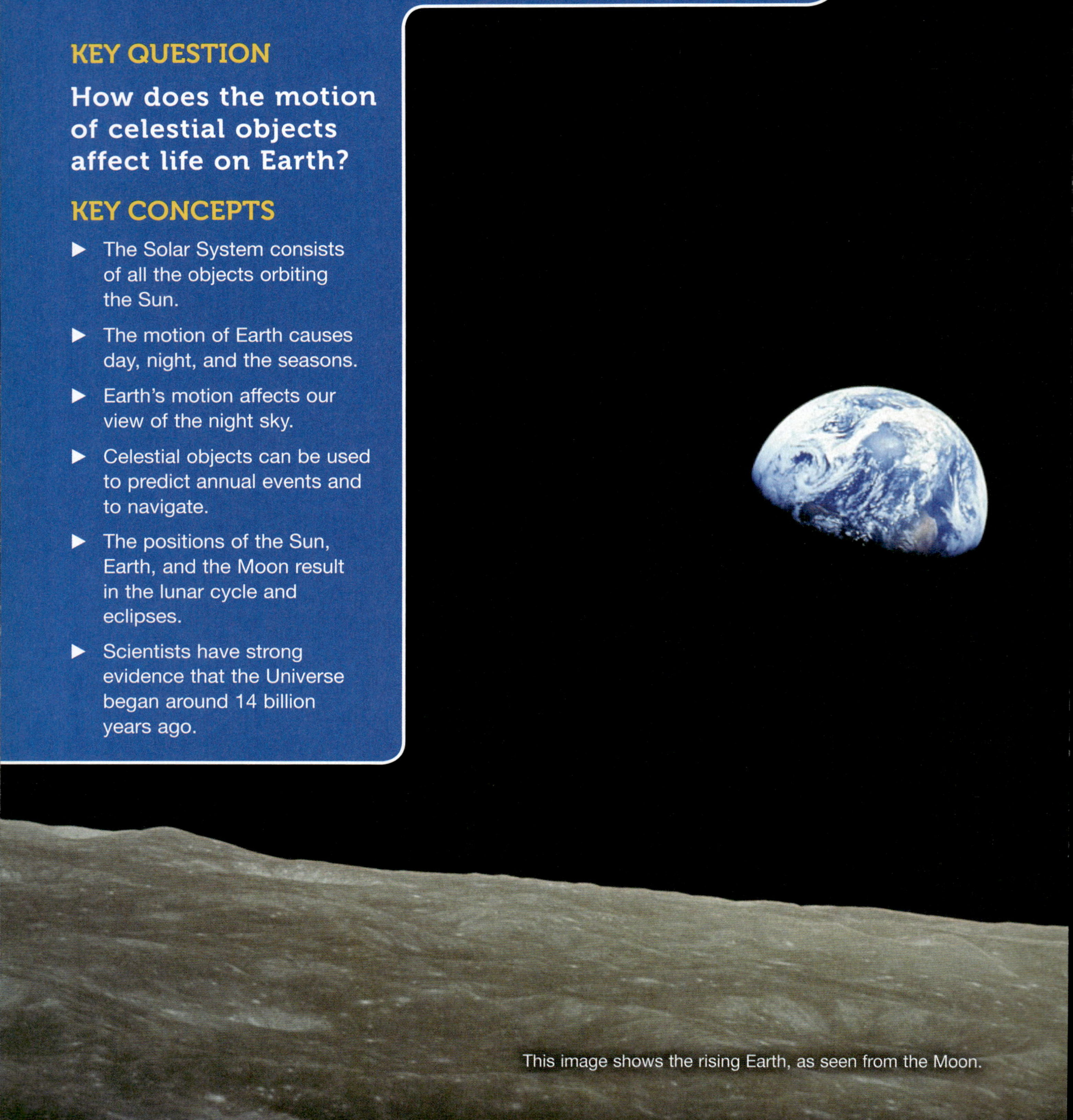

This image shows the rising Earth, as seen from the Moon.

THE IMPORTANCE OF EARTH'S LOCATION

Many people know that Earth is the third planet from the Sun and that the Moon revolves around Earth. What they may not have thought about is how important the Sun, the Moon, and the planets are to the existence of life on Earth.

The Sun, in particular, plays an important role in the Solar System. The force of gravity between the Sun and Earth keeps Earth in its orbit. Light, or radiant energy, from the Sun allows plants to grow and reproduce. This energy also creates Earth's weather systems and keeps Earth from freezing. It also provides the light that allows us and other animals to see.

If Earth were closer to the Sun, it would be too hot for plants and animals to survive. If Earth were farther away, it would be too cold for life to exist. There would not be enough sunlight for plants to grow or animals to stay warm. Earth is in the best possible position within the Solar System for life to be sustained.

WHAT DO YOU THINK?

Many of the ideas you will explore in this chapter are ideas that you have already encountered. You may have encountered these ideas in school, at home, or in the world around you. Some of the statements are not true. Consider each statement and decide whether you agree or disagree with it.

1 An eclipse of the Sun occurs when Earth is between the Sun and the Moon.
Agree/disagree?

2 The Sun, the Moon, and planets orbit Earth.
Agree/disagree?

3 The size of the Universe is changing.
Agree/disagree?

4 We see different stars at different times of the year because Earth is going around the Sun.
Agree/disagree?

5 We always see the same side of the Moon.
Agree/disagree?

6 When it is summer in the northern hemisphere, it is winter in the southern hemisphere.
Agree/disagree?

FOCUS ON
READING

Summarizing

DIG DEEPER

Reading Tip

As you work through the chapter, look for tips like this. They will help you develop literacy strategies.

You summarize a text by making it shorter. Focus on the main idea and key points, leaving everything else out.

What to look for	Where to look/What to look for	What to do
clues about what the main idea and key points are	title, headings, topic sentences, bolded words	think of other words you could use to communicate the main idea and key points
information you do not need to understand the main idea	the body of a paragraph	leave unimportant information out
text patterns	signal words such as "because," "however," and "generally"	organize your summary in the same way

Seasons on Other Planets

Other planets in the Solar System also experience seasons.

Any planet that has a tilted axis also has seasons. Mars, Saturn, and Neptune are much farther from the Sun than Earth, so it takes longer for them to make a complete revolution around the Sun. As a result, their seasons are much longer and, in general, much colder. Uranus, on the other hand, has an axis that is tilted so much that the planet is practically lying on its side. Its poles face the Sun for long periods of time, producing extreme seasons.

No Tilt, No Seasons

Planets that do not have a tilted axis, do not have seasons. Mercury, Venus, and Jupiter fall into this category.

Summarizing *in Action*

First, decide what is important and what is not important. The main idea is very important. Facts, examples, and descriptions are not as important for a summary. Second, use words that you understand to restate the main idea. Here is how one student summarized the text selection above, from Section 9.3.

Clues	Information not needed	Text pattern
The heading suggests planets that don't tilt don't have seasons.	a planet's distance from the Sun	Compare/Contrast (organizes information by how things are like and not like each other)
Earth, Mars, Saturn, Neptune, and Uranus have seasons.	how long the seasons are how cold a planet is	
Mercury, Venus, and Jupiter don't have seasons.	how much a planet tilts	

Summary: Planets on a slant have seasons. Planets that stay straight up don't have seasons.

9.1 Motion in the Solar System

Early astronomers noticed that the motions of celestial objects seemed to follow certain patterns. From this observation, they developed a theory about the structure of the Universe.

- **geocentric model:** model of the Universe in which Earth is at the centre
- **heliocentric model:** model of the Universe in which the Sun is at the centre

Explaining Motion in the Solar System

Ancient Greek philosophers believed that the Sun and the planets orbited Earth, which remained motionless at the centre of the Universe (**Figure 1**). The view that the Universe was "Earth-centred" became known as the **geocentric model**. It was the leading model of the Universe for more than 2000 years.

A New Idea: Heliocentric Model

The geocentric model of the Universe did not explain everything. When viewed from Earth, some of the planets, such as Mars, seemed to move in one direction, stop, and then move in the opposite direction. This observation could not be explained if all the planets circled Earth.

In the 16th century, Nicolaus Copernicus presented a model called the **heliocentric model**, in which the Sun was at the centre of the Universe (**Figure 2**). The heliocentric model was much better at explaining the motion of the planets than the geocentric model. In this new model, the Sun is at the centre of the Universe.

DIG DEEPER

Learning Tip

Word Origins

Geo means "Earth" and "centric" means "centre," so the term *geocentric* means "Earth-centred." Similarly, *helios* means "Sun," so *heliocentric* means "Sun-centred."

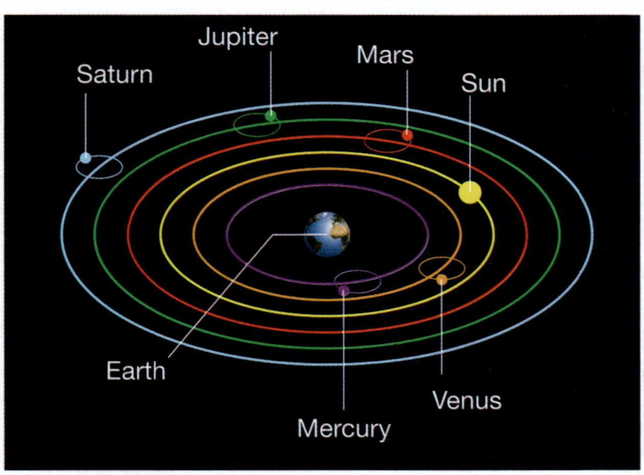

Figure 1 The geocentric model

Figure 2 The heliocentric model

Distances in the Solar System

Distances within the Solar System are too large to be measured in kilometres but are not large enough to be measured in light years. For this reason, astronomers use a more convenient unit to measure them, called the astronomical unit. One **astronomical unit (AU)** is equal to about 150 000 000 km, which is the average distance between Earth and the Sun. The AU provides a more manageable way to measure large astronomical distances. For example, using AUs, the distance from the Sun to Neptune can be written as 30 AU, rather than as 4 500 000 000 km. The astronomical unit is also useful for comparing the sizes of celestial objects.

▶ **astronomical unit (AU):** the average distance between Earth and the Sun; about 150 million km

DIG DEEPER

Learning Tip

Scale Models

Scientists often use scale models when dealing with very large distances like those in the Solar System.

TRY THIS DISTANCES IN THE SOLAR SYSTEM

SKILLS HANDBOOK 3.B.

SKILLS: Observing, Communicating

In this activity, you will create a scale drawing to represent distances in the Solar System.

Equipment and Materials: calculator; metre stick; tape (or paperweight); ruler; coloured pencils; roll of art paper

1. Copy **Table 1** into your notebook. It shows the average distance from the Sun to each planet.
2. In your drawing, let 20 cm = 1 AU. The distances for the first three planets have been calculated for you. Complete the calculations for the other planets in the third column of your table. (Tip: To convert distances using this scale, simply cross-multiply the average distance value in AU by 20 to obtain the scale distance value in centimetres.)
3. Once you have completed **Table 1**, work in small groups to construct your Solar System drawing.
4. Using a metre stick, measure out 8 m of paper from the roll. Secure the paper with tape or paperweights so it is flat. You may have to take the paper into the hall or outside to stretch it to the full length.
5. Draw the Sun on one edge of the paper.
6. Using the scaled distances that you calculated, draw each planet in a long row on the paper.

A. Some of the planets in your model are very far apart. What would the effect on your model be if you increased the scale by 10? T/I

B. Would it be practical to construct a model of the Solar System using the same scale for both planetary diameters and distances? Explain your reasoning. T/I

Table 1 Planetary Distances from the Sun

Planet	Average distance from Sun (AU)	Scale distance from Sun (cm)
Mercury	0.4	8
Venus	0.7	14
Earth	1.0	20
Mars	1.5	
Jupiter	5.2	
Saturn	9.5	
Uranus	19.2	
Neptune	30.0	

Other Objects in the Solar System

Besides the Sun and planets, there are millions of smaller objects in the Solar System. Most of these can only be seen with a telescope.

Asteroids

Recall from Chapter 8 that an asteroid is a small celestial object made mostly of rock and metal. Asteroids often have irregular shapes and range in diameter from 10 m across to more than 900 km across. There are millions of asteroids in the Solar System. Most are located in the asteroid belt between Mars and Jupiter.

Meteoroids

- **meteoroid:** a small celestial object made of rock or metal that is smaller than an asteroid
- **meteor:** a meteoroid that burns up as it travels through Earth's atmosphere; it forms a bright streak of light in the night sky
- **comet:** a large chunk of ice, rock, and dust that orbits the Sun

A **meteoroid** is a piece of rock or metal that is smaller than an asteroid. A meteoroid can range from the size of a dust particle, to up to several kilometres across. When a meteoroid enters Earth's atmosphere, friction causes the meteoroid to burn up in a bright streak of light across the sky, called a **meteor**. This bright streak is what we commonly call a "shooting star." Most meteors are so small that they completely burn up before they reach Earth's surface. A meteor that survives Earth's atmosphere and reaches the surface is called a meteorite. If a large meteorite strikes Earth, it can leave a huge crater (**Figure 3**).

Comets

A **comet** is a large chunk of ice, dust, and rock that orbits the Sun. Comets range from 1 km to 10 km in diameter. When a comet gets close enough to the Sun, its frozen gases are evaporated by the solar wind. The comet develops a glowing head and a long tail, often bright enough to be seen in Earth's sky (**Figure 4**).

Figure 3 Manicouagan Crater in northern Québec was formed by the impact of a 5 km-wide meteorite.

Figure 4 Comet Hale-Bopp

Planetary Motion

All celestial objects undergo some type of motion. Planetary motion can be divided into two types of motion: revolution and rotation.

▶ **revolve:** to orbit a central point or object

Revolution

The eight planets in the Solar System **revolve**, or travel around the Sun. Planetary orbits are elliptical, or oval-shaped. Earth completes one revolution around the Sun in about 365 days, which is equal to one Earth year. The farther a planet is from the Sun, the slower it moves along its orbit and the longer it takes to orbit the Sun (**Figure 5**).

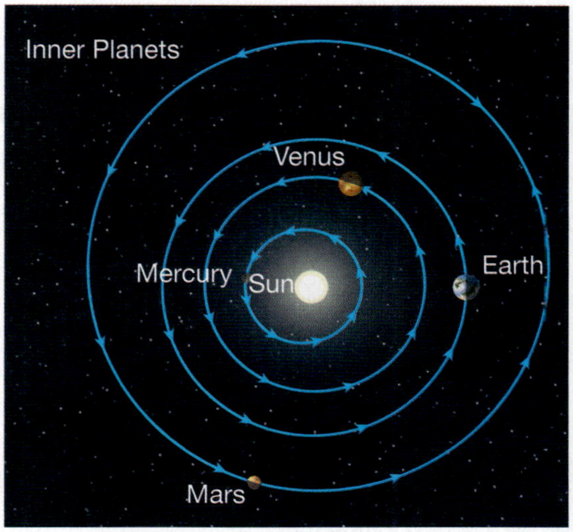

Figure 5 Earth completes one revolution around the Sun in 365.25 days.

Rotation

You have probably noticed that the Sun appears to move across the sky, rising in the east and setting in the west. This apparent motion of the Sun is due to Earth's rotation. As Earth revolves around the Sun, it also **rotates**, or turns, on its axis. An **axis** is an imaginary line around which an object rotates. Earth's axis runs from pole to pole (**Figure 6**). Earth, along with most other planets, rotates in a west-to-east direction. Venus is unique in that it rotates in the opposite direction from the other planets.

▶ **rotates:** moves, or turns, around a centre point

▶ **axis:** an imaginary line around which an object rotates

We may not notice the speed of Earth's rotation, but we do notice its effect: day and night. As Earth rotates, the portion of Earth that faces the Sun experiences day, while the portion of Earth facing away from the Sun experiences night (**Figure 7**).

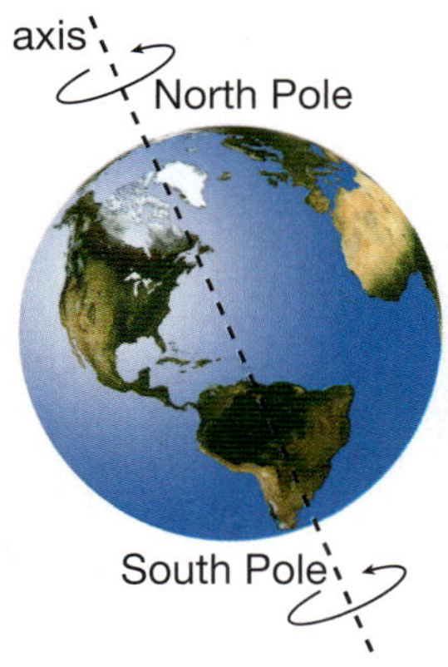

Figure 6 Earth rotates on its axis once every 24 hours.

Figure 7 We experience night and day because of Earth's rotation on its axis.

The amount of time it takes a planet to complete one rotation on its axis is equal to the length of one day on that planet. For example, Earth completes one rotation on its axis every 24 hours, which is equal to one Earth day.

DIG DEEPER

Did You Know?

Measuring Time

In the past, people built sundials to measure time by the Sun. Mechanical clocks eventually replaced sundials, but clock time is still based on the Sun's position in the sky. This is why when you travel, the time on your watch has to be changed to match the local time zone.

Effects of Earth's Motion in Our Daily Lives

Due to Earth's motion, sunlight does not strike everywhere on Earth at the same time. This affects how we live our daily lives. One example of this is timekeeping.

Time Zones

The time of day when the Sun is directly overhead is known as "solar noon." Solar noon in Toronto occurs 4 min earlier than in Kitchener because Kitchener is about 100 km west of Toronto. Four minutes may not sound like much, but if you were trying to catch a train, it would make all the difference. Over the centuries, many countries created their own systems for measuring time instead of relying on the Sun's position. These systems did not always agree with one another and resulted in confusion. The problem was solved in 1879 when Sir Sandford Fleming, a Canadian scientist, proposed a worldwide system in which Earth is divided into regions that each have a standard time, called a **time zone**. Today's time zones are based on his system. Canada spans six time zones (**Figure 8**). Within a time zone, all the clocks are set to the same time.

▶ **time zone:** a vertical section of Earth's surface that has a uniform time

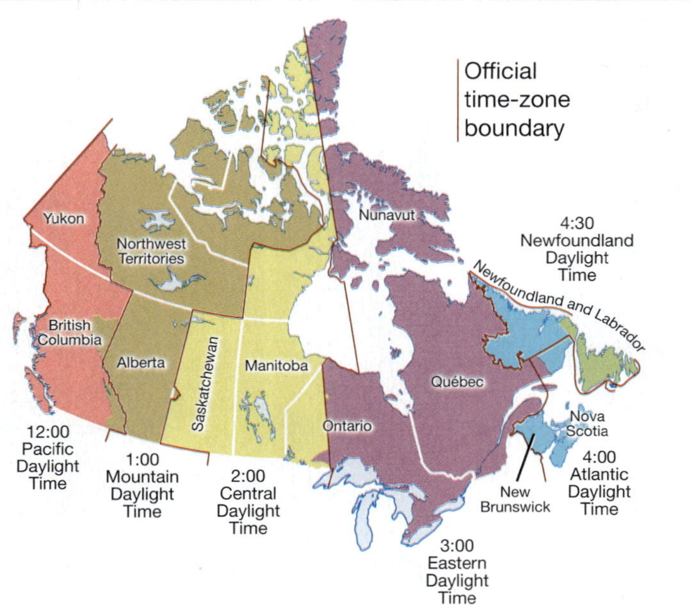

Figure 8 Most of Ontario is in the Eastern Time zone.

DIG DEEPER

Reading Tip

Restate the Idea

After reading a text, it is a good idea to restate what you have read in your own words. Then ask yourself what the main idea of your passage is and compare it to what is written in the text to see if you have captured the main idea.

Imagine that you are looking at Earth from the North Pole. Now imagine that Earth is divided into 24 equal zones. In a perfect world, each time zone would be equal, for a total of 24 time zones around the globe. In reality, the shapes of some time zones are modified so that a time zone change does not occur in the middle of a city. It would be very confusing if one city had two different time zones! In some cases, extra times zones with only a half-hour time difference have been created.

DIG DEEPER

Web Link

For more information about time zones,

GO TO NELSON SCIENCE

9.1 Wrap Up

- People once believed that Earth was the centre of the Universe. This model of the Universe is called the geocentric model.
- People later realized that the Sun was the centre of the Solar System. This model of the Universe is called the heliocentric model.
- Distances within the Solar System are measured in astronomical units (AU). One AU equals the average distance between Earth and the Sun.
- In addition to planets and moons, the Solar System also contains asteroids, meteoroids, and comets.
- The planets revolve around the Sun in elliptical orbits. Each planet also rotates around its axis.
- Earth rotates, or makes a full turn (360°), on its axis once every 24 hours.
- At any given time, half of Earth experiences daylight while the other half experiences darkness.
- Time zones are divisions of Earth's regions, where each region has the same time.

CHECK YOUR LEARNING

1. How would a day on Mercury be similar to a day on Earth? How would it be different? K/U
2. Explain to a classmate the difference between a comet and an asteroid. K/U C
3. (a) What observations may have caused people to believe in the geocentric model of the Universe?
   (b) What observation helped cast doubt on the geocentric model? K/U T/I
4. How many times does Earth rotate on its axis in one year? K/U
5. What factor determines the length of a day on Earth? K/U
6. Why are there different time zones on Earth? K/U

Chicxulub Crater

Have you ever seen a "shooting star"? People often feel lucky when they see one, and sometimes make a wish. A shooting star is more accurately called a meteor, or—if it reaches Earth's surface—a meteorite.

Meteorites

Most meteorites are quite small and do little or no damage when they hit the ground. However, there have been several large impacts during the history of Earth.

The Extinction of Dinosaurs

Since the 1980s, scientists have suspected that a very large meteorite may have led to the mass extinction of the dinosaurs and other species alive at the time. The evidence was a thin layer of the element iridium found in rocks all over the world. Layers of rock can be used to measure time. Older layers are deeper in the ground and younger layers are closer to the surface. The iridium was always found in the same layer—the layer corresponding to the time when dinosaurs last roamed Earth.

The Evidence

Iridium is a very rare element on Earth, but it is common in meteors. Geologists hypothesized that a meteorite 10 km across struck Earth and exploded, spreading iridium across the globe. The impact would have released a dust cloud that would have blocked out the Sun's light for years. The climate change it created may have killed off many of the world's plants, disrupting food chains and leading to mass extinctions.

The Crater?

No impact crater had been found that matched such a huge meteorite. The crater from this impact would have been about 200 km in diameter. Craters this large are hard to find because they can only be seen from the air and, over time, plants grow over them and erosion wears down their shape.

Then, in the early 1990s, geologists were studying a strange rock formation located about a kilometre underwater along the coast of Mexico. It was round, 180 km in diameter, and filled with rock that had been exposed to extreme heat. This matched a crater that a 10 km meteorite would produce (**Figure 1**). Today, this feature is known as the Chicxulub Crater, and many scientists believe it was created by the meteorite believed to be responsible for the extinction of dinosaurs.

Figure 1 A meteorite 10 km across would produce devastating effects.

Seasons

As Earth revolves around the Sun, the northern and southern hemispheres experience the seasons—spring, summer, autumn, and winter. Many people mistakenly believe that seasons occur because Earth moves closer to the Sun in the summer and farther away in the winter. In fact, seasons change because of Earth's tilt.

Earth's Tilt

You may have noticed that it gets dark earlier in January and February than in summertime, when the Sun sets much later in the evening. This is a result of Earth's tilt.

Earth's axis is tilted 23.5° from the vertical when compared to its orbit around the Sun (**Figure 1**). This causes each of Earth's hemispheres to point away from the Sun at one point in Earth's orbit and toward the Sun at another point (six months later). When a hemisphere is tilted toward the Sun, it experiences more hours of sunlight each day. Due to this, the amount of sunlight an area receives each day is more extreme at the poles.

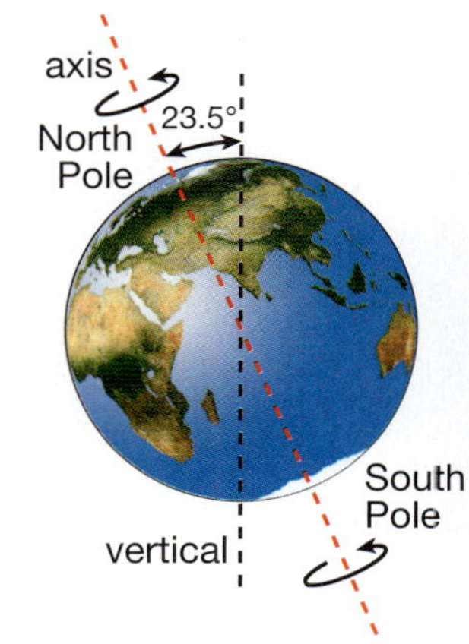

Figure 1 Earth's axis is tilted 23.5° from the vertical.

Seasons and the Tilt of Earth

Earth's tilt is the main cause of the seasons. When the northern hemisphere is tilted toward the Sun, sunlight spreads over a relatively small area of Earth's surface. This causes the temperature to increase, and we experience summer (**Figure 2(a)**). When the northern hemisphere is tilted away from the Sun, sunlight spreads over a larger area of Earth's surface. This causes the temperature to decrease, and we experience winter (**Figure 2(b)**). When it is winter in the northern hemisphere, it is summer in the southern hemisphere.

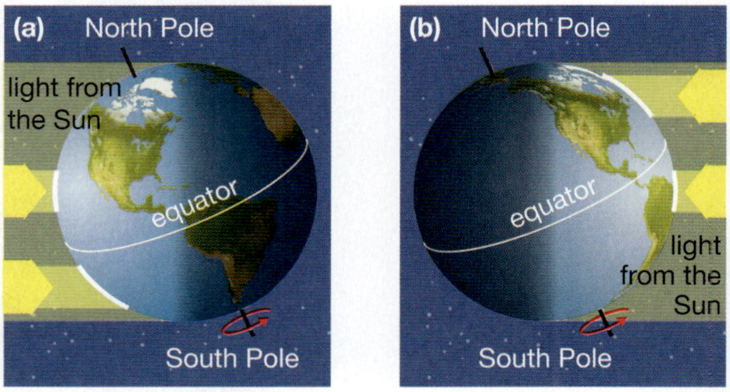

Figure 2 (a) When a hemisphere is tilted toward the Sun, it experiences more hours of sunlight each day at a stronger intensity (due to the angle of sunlight). (b) The opposite effect occurs when it is tilted away from the Sun.

What about the seasons at the Equator? The intensity of the Sun's rays at the Equator changes very little over the year, so the temperature remains fairly constant and warm (**Figure 3**). The equatorial regions do not have four seasons. Instead, they have two—the rainy season and the dry season—that depend on tropical rains.

DIG DEEPER
Web Link
To learn more about the effects of Earth's tilt,
GO TO NELSON SCIENCE

Figure 3 (a) Countries in the northern and southern hemispheres experience the four seasons. (b) Countries near the equator experience rainy and dry seasons, with a more constant climate.

Solstices and Equinoxes

When Earth's axis is most inclined toward or away from the Sun, we call it a **solstice**. Solstices occur twice each year. In the northern hemisphere, this occurs around June 21 (summer solstice) and December 21 (winter solstice). Between the solstices are two days when day and night are of equal length, called **equinoxes**. In the northern hemisphere, the spring equinox occurs around March 21 and the autumn equinox occurs around September 21 (**Figure 4**).

- **solstice:** a day when Earth's axis is most inclined toward, or away from, the Sun
- **equinoxes:** days when the hours of daylight equal the hours of darkness

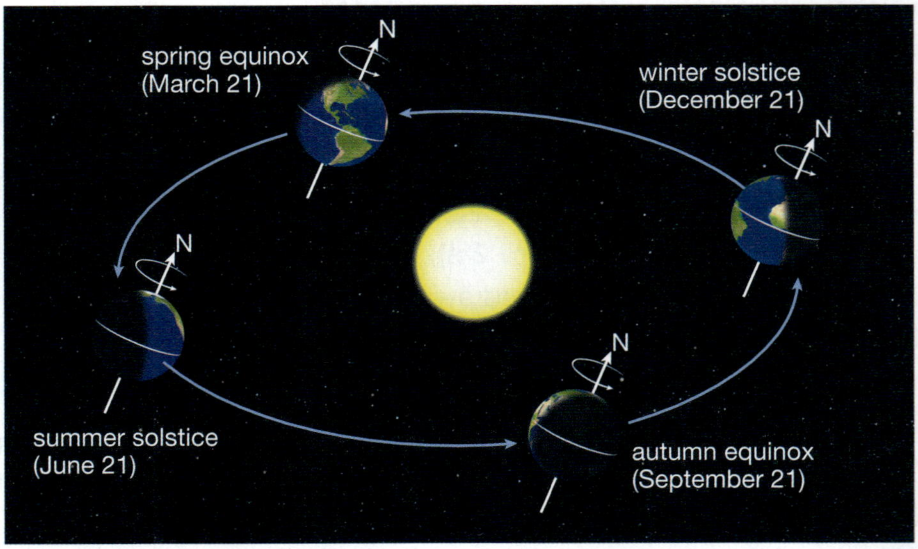

Figure 4 During winter in the northern hemisphere, the Sun's rays do not reach the North Pole, which is tilted away from the Sun.

Seasons on Other Planets

Other planets in the Solar System also experience seasons.

Any planet that has a tilted axis also has seasons. Mars, Saturn, and Neptune are much farther from the Sun than Earth, so it takes longer for them to make a complete revolution around the Sun. As a result, their seasons are much longer and, in general, much colder. Uranus, on the other hand, has an axis that is tilted so much that the planet is practically lying on its side. Its poles face the Sun for long periods of time, producing extreme seasons.

No Tilt, No Seasons

Planets that do not have a tilted axis, do not have seasons. Mercury, Venus, and Jupiter fall into this category.

9.2 Wrap Up

- Earth revolves around the Sun in an orbit once each year.
- The tilt of Earth's axis affects the amount of daylight in an area.
- The seasons in the northern and southern hemispheres occur because Earth remains tilted as it orbits the Sun, and the tilt of Earth causes different amounts of solar energy to reach Earth's surface.
- When it is winter in the northern hemisphere, it is summer in the southern hemisphere.
- Other planets with tilted axes also have seasons.

CHECK YOUR LEARNING

1. What causes the seasons? K/U
2. Use diagrams to explain why it is winter in the northern hemisphere when it is summer in the southern hemisphere. K/U C
3. Explain why the South Pole experiences 24 hours of sunlight when the North Pole experiences 24 hours of darkness. K/U
4. What is the difference between a solstice and an equinox? K/U
5. If scientists discovered another planet revolving around the Sun that was not tilted, what would you know about the seasons on that planet? A

9.3 Star Patterns

If you have ever looked up at the night sky in winter, you probably would recognize the pattern of stars shown in **Figure 1**. The Ancient Greeks named this pattern of stars Orion, after the giant hunter that Zeus placed among the stars. Today, Canadians and Americans still call it Orion. Others, like Australian Aborigines, call it "Julpan," which means canoe.

Figure 1 The three bright stars in a row are commonly known as Orion's Belt.

Constellations and Asterisms

A **constellation** is a large region of stars in the night sky. The stars in a constellation may be many light years from one another, but from Earth, they appear to be grouped closely together. A pattern of stars observed in the night sky is called an **asterism**. A constellation may contain several asterisms, or may itself be an asterism. The Big Dipper (**Figure 2**) is an asterism within the constellation called Ursa Major, or Great Bear (**Figure 3**).

- **constellation:** a region of stars in the night sky
- **asterism:** a pattern of stars in Earth's night sky

Think of a constellation as a county and stars within it as the towns and villages. There are 88 known constellations. Well-known constellations that can be seen from the northern hemisphere include Leo (the Lion) and Scorpio (the Scorpion).

Figure 2 The Big Dipper is a familiar asterism.

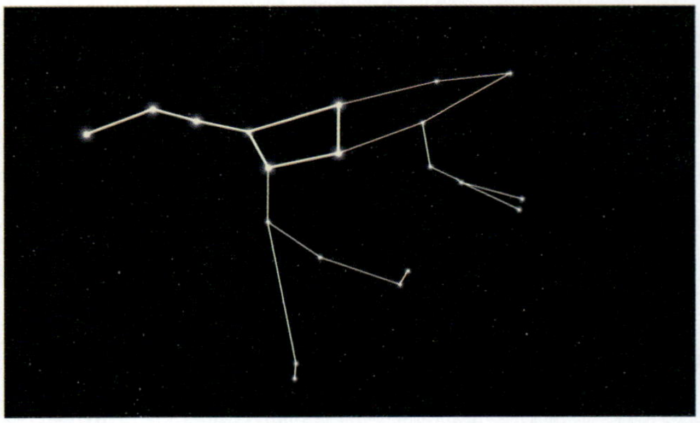

Figure 3 The handle of the Big Dipper is the tail of the Great Bear.

Changes in the Night Sky

Ancient astronomers also wondered why different constellations were visible at different times of the year. Earth's revolution around the Sun changes our view of the night sky with each passing season. In the winter sky, we see stars such as Orion that are opposite the Sun. Other constellations are still there, but they are in our sky during the day. We cannot see them because the Sun's light is too bright. In the summer sky, we see constellations such as Lyra in the night sky. Orion is invisible to us because it is in the sky during the day.

Polaris, the North Star, is different from most stars—it is always visible above the northern horizon. This means that we can see Polaris all year long in the northern hemisphere. The same is true for some constellations. A star or constellation that is always visible in the sky is said to be **circumpolar**. Circumpolar constellations in the northern hemisphere include Cassiopeia, Cepheus, Ursa Major, Ursa Minor, Camelopardalis, and Draco (**Figure 4**).

DIG DEEPER
Web Link
To learn more about constellations,
GO TO NELSON SCIENCE

► **circumpolar:** a star, or constellation, that never sets below the horizon

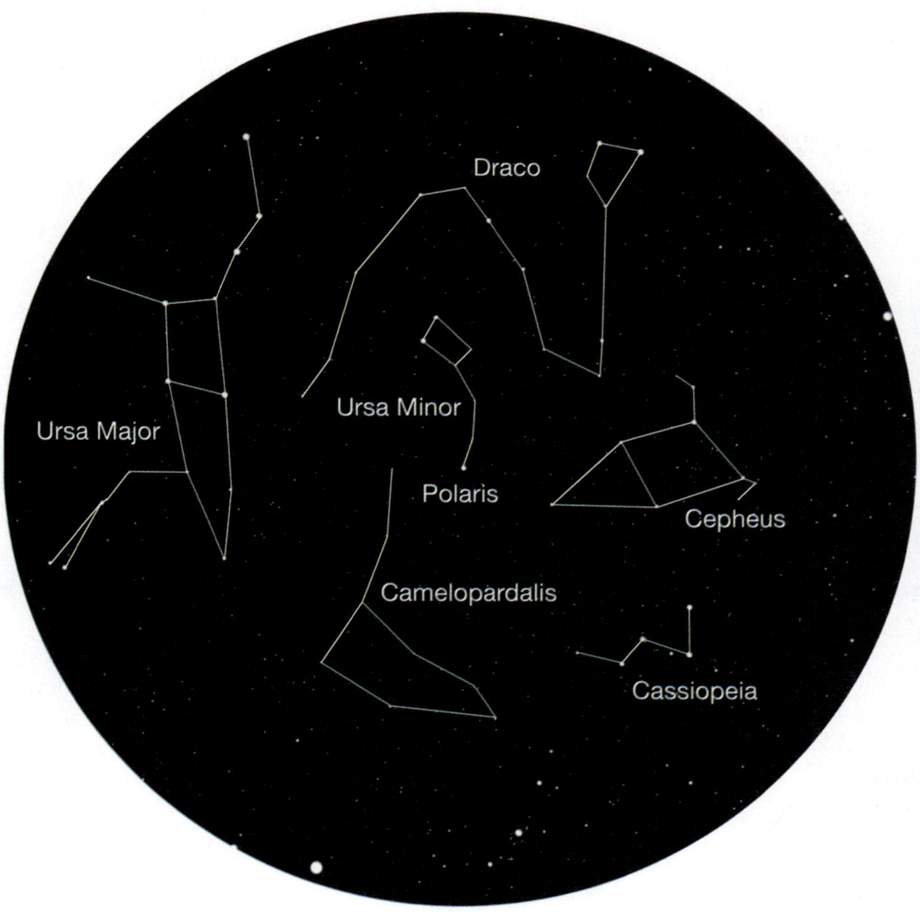

Figure 4 Circumpolar stars and constellations surround Polaris.

TRY THIS MAPPING CONSTELLATIONS

SKILLS HANDBOOK 3.B., 4.B.

SKILLS: Observing, Analyzing, Communicating

If you view the night sky at different times of the year, you see different constellations. In this activity, you will model the changes in the night sky in your area.

Equipment and Materials: paper

1. Conduct research on the Internet to find a seasonal star map for your area.
2. Draw the constellations you would see in the sky during each season. Use one sheet of paper for each constellation.
3. Hang the pictures in order of the seasons on different walls in your classroom.
4. Stand in the centre of the classroom. Orbit the classroom with your back to the centre of the room, so you are facing the wall. Record your observations.

A. What do you think you would find if you were to observe the same constellation rise and set each night for two weeks? T/I

B. Explain why some constellations are only seen at certain times of the year. T/I

GO TO NELSON SCIENCE

9.3 Wrap Up

- A constellation is a region of stars as viewed in Earth's night sky.
- An asterism is a pattern of stars as seen from Earth.
- Earth's rotation makes it look like constellations are moving across the sky from east to west.
- Earth's revolution around the Sun makes us see different constellations with each passing season.

CHECK YOUR LEARNING

1. What is the difference between an asterism and a constellation? K/U
2. Why do most constellations appear to move from east to west across the sky each night? K/U
3. Why do we see different constellations in the night sky during different seasons? K/U
4. What is a circumpolar constellation? K/U

How People Use the Stars

Many civilizations, both past and present, use the apparent motion of celestial objects as tools. For example, many cultures track the cyclical movements of the Sun, the Moon, and the stars to determine when to plant and harvest crops. Stars may be used to navigate while sailing in the oceans.

Using the Stars on Land

Ancient Egyptians noticed that the star Sirius always appeared in the summer sky (**Figure 1**). Its appearance coincided with the annual flooding of the Nile River. The Egyptians used their observations of the star Sirius to predict the annual floods of the Nile River. The flooding provided rich soil for growing crops. Therefore, the appearance of Sirius in the sky signalled the time to plant.

Many of the most spectacular buildings in ancient cultures were built to keep track of important celestial events, such as the solstices and equinoxes. In Britain, the ancient Celts used Stonehenge to predict the timing of the seasons, which was important for their agriculture, and also for their religion.

In Mexico, the ancient Mayans built the giant pyramid at Chichén Itzá to mark the equinoxes. On the equinox, the shadow of the serpent Kukulcan appears along the stairway (**Figure 2**).

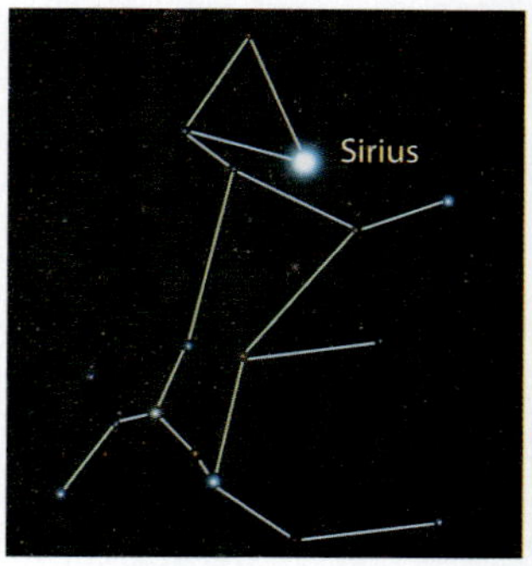

Figure 1 Sirius is in the Canis Major (Big Dog) constellation. Sirius is the dog's nose in that constellation. It is known as the "Dog Star."

DIG DEEPER
Web Link

To learn more about the role stars play in different cultures,

GO TO NELSON SCIENCE

Figure 2 During the equinoxes, the setting Sun casts a shadow of the serpent Kukulcan on the northern steps of the pyramid.

Using the Stars at Sea

Ancient sailors used the stars to find their way at sea. For example, sailors used their knowledge of the movement of specific stars, and where the stars would rise and set on the horizon, to determine where their ships were located during the journey (**Figure 3**).

Figure 3 Ancient sailors relied on the position of the Sun, Moon, and stars to help them navigate on the ocean.

Azimuth and Altitude

In geography, latitude and longitude are used to pinpoint a location on Earth's surface. In astronomy, two similar measurements are used to pinpoint the location of a celestial object, as viewed from Earth.

Azimuth is the distance measured from north along the horizon to a point directly below the celestial object. North has an azimuth of 0°, east has an azimuth of 90°, south has an azimuth of 180°, and west has an azimuth of 270°. **Altitude** is the angular height of a celestial object, measured from the horizon. **Figure 4** shows how azimuth and altitude are measured.

- **azimuth:** the horizontal angular distance from north measured (in °) eastward along the horizon to a point directly below a celestial object
- **altitude:** the angular height of a celestial object above the horizon; measured vertically (in °) from the horizon

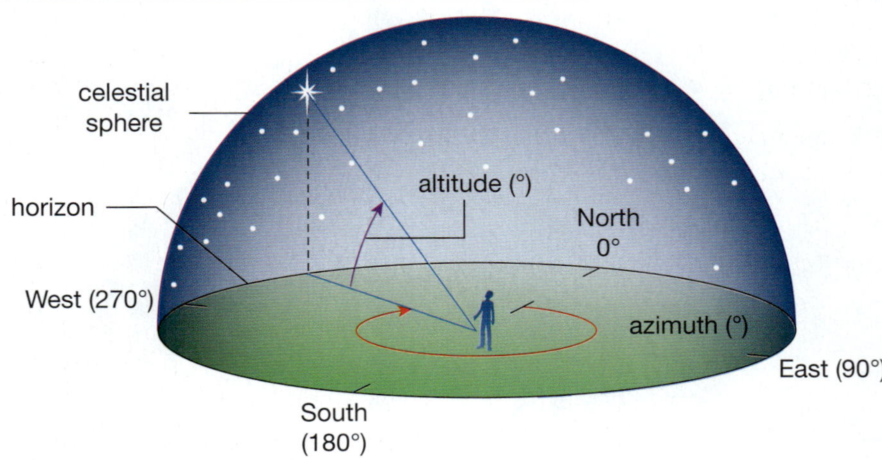

Figure 4 Azimuth and altitude are measured with reference to the horizon.

You can use your hands to determine the position of celestial objects using azimuth and altitude. If you hold your hand out at arm's length, the width of your index finger is about equal to 1° (**Figure 5**). The width of a fist at arm's length is about equal to 10° (**Figure 6**). If you extend the fingers of your hand, the width from the tip of your pinky finger to the tip of your thumb is about equal to 20° (**Figure 7**). Using this method, you can estimate the azimuth and altitude coordinates of any celestial object in the sky relative to your position on Earth.

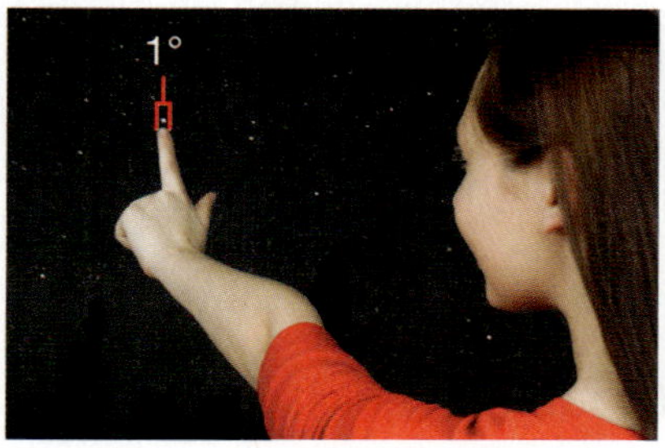

Figure 5 One finger measures 1°.

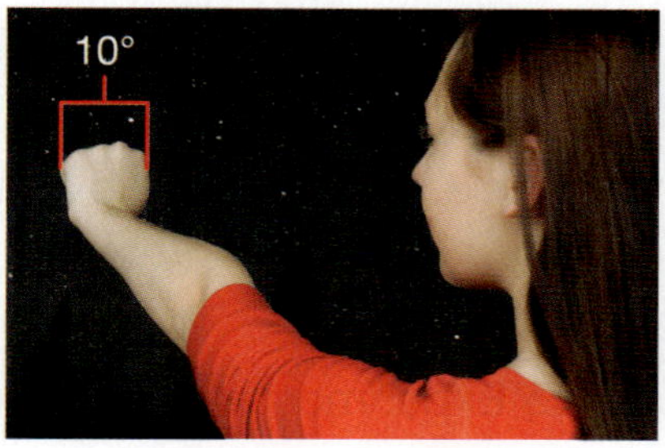

Figure 6 A fist measures 10°.

Figure 7 An outstretched hand measures 20°.

9.4 Wrap Up

- Many ancient and modern-day cultures have studied the movement of celestial objects for agriculture, religious beliefs, and cultural events.
- Historically, stars have been used to navigate at sea.
- Azimuth and altitude can be used to describe the positions of celestial objects.

CHECK YOUR LEARNING

1. What was the link between agriculture and astronomy in ancient cultures? K/U
2. Why did ancient cultures build large structures such as Stonehenge? K/U
3. Explain how sailors were able to find their way at sea before the invention of navigation devices. K/U C
4. Explain to a classmate how you can measure the azimuth of a celestial object. K/U C

9.5

CONDUCT AN INVESTIGATION

What Can You See Tonight?

SKILLS MENU

- Questioning
- Hypothesizing
- Predicting
- Planning
- Controlling Variables
- Performing
- Observing
- Analyzing
- Evaluating
- Communicating

In this activity, you will go outside and observe the motion of celestial objects.

SKILLS HANDBOOK 3.B., 4.A., 4.B.

Testable Question

How do celestial objects move across the night sky?

Prediction

Read the Experimental Design and Procedure. Predict an answer to the Testable Question.

Experimental Design

Accompanied by an adult, you will go outside at night and record three observations of what you see in the night sky. Observations will be made an hour apart. You will repeat the procedure after several hours, and then again after 3 days.

Procedure

Part A: Observe and Record

1. Find a place near your home where you have a clear view of the southern sky. Use the compass to determine which direction is south.
2. On a piece of paper, sketch trees, or buildings that are on the southern horizon. You will use these as landmarks to draw the objects you see in the night sky.

Equipment and Materials

compass

paper

pencil

3. Draw the brightest objects you see in the night sky. Be sure to identify the location, date, and time on your horizon diagram.
4. Using your fist, measure the azimuth and altitude of the brightest objects you saw. Write down the azimuth (Z) and altitude (A) measurements beside the objects on your horizon diagram (**Figure 1**).
5. After one hour has passed, repeat your observations and create a new horizon diagram. Record the azimuth and altitude in your diagram.
6. Wait another hour, and then repeat your observations and create a new horizon diagram. Record the azimuth and altitude in your diagram.

Part B: Repeat and Predict

7. Analyze your observations. How did the objects you observed move over the course of the evening?
8. After 3 days, return to the same location at the same time, and repeat Steps 1 to 6.

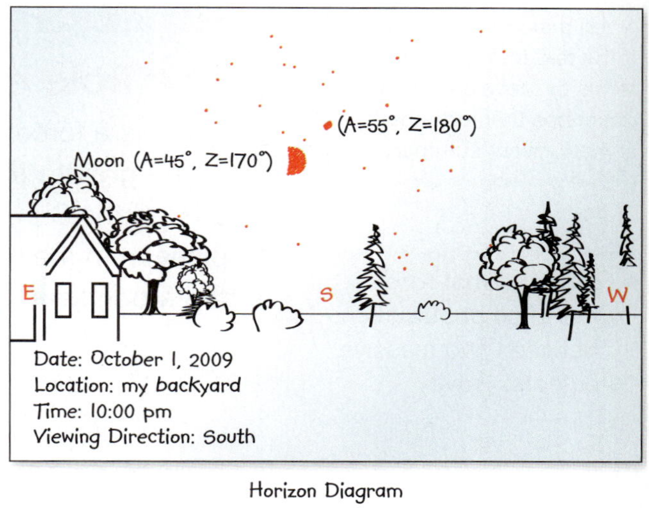

Figure 1 Step 3

Analyze and Evaluate

(a) Answer the Testable Question. T/I

(b) How did your observations compare with your predictions? T/I

(c) In Part A, how did the position of the celestial objects you observed change over time? T/I

(d) How did your second set of observations (Part B) compare with the first (Part A)? If you were to make another set of observations a few days later, what do you think would be the result? T/I

Apply and Extend

(e) Research which stars and constellations should be visible from your location at this time of year. Make another observation, and sketch any constellations you are able to observe. T/I A

9.6

Phases of the Moon

DIG DEEPER

Reading Tip

Organization of Text

Follow the same organization as the text when making a summary. If the text introduces new terms or ideas and then describes them, do the same for your summary.

The constellations change from season to season, but the Moon appears to change over the course of a month. Have you ever wondered what keeps the Moon in orbit around Earth?

The Role of Gravity

There is a force of attraction between all objects in the Universe called **gravitational force**, or gravity. The planets are kept in orbit around the Sun by the gravitational force between the Sun and the planets. In the same way, the gravitational forces of both Earth and the Moon keep the Moon in orbit around Earth (**Figure 1**).

▶ **gravitational force:** the force of attraction between two massive objects; gravity

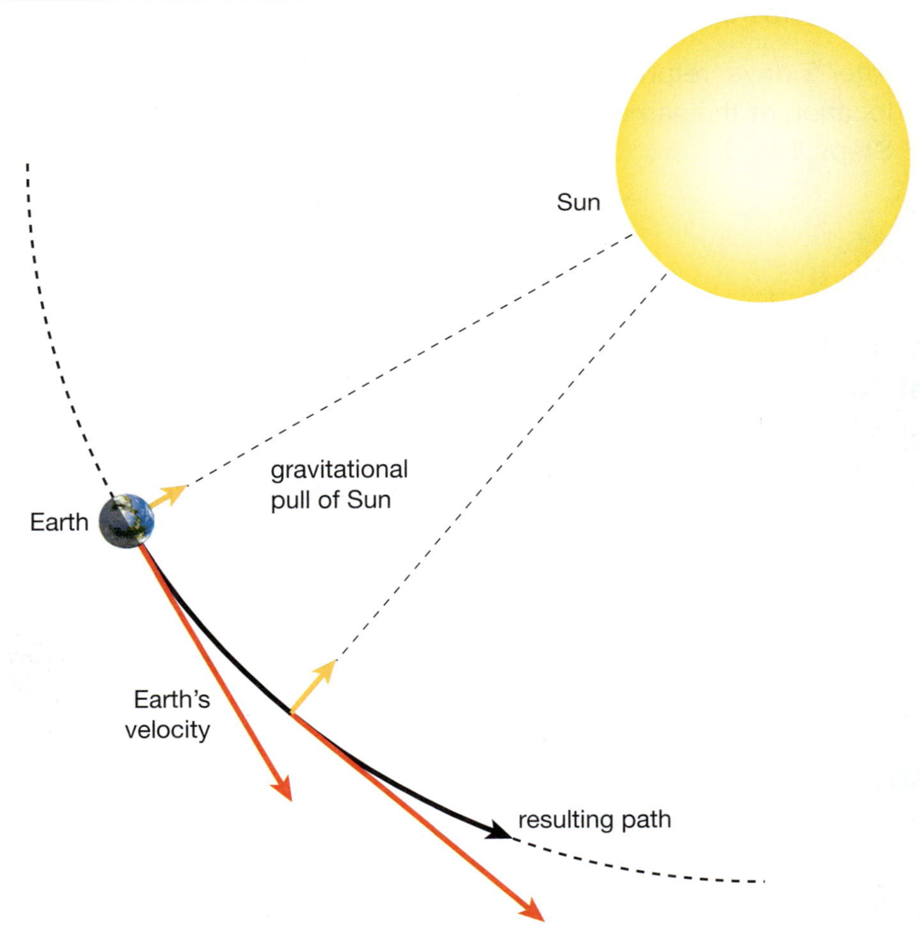

Figure 1 Earth stays in orbit because of the gravitational forces between the Sun and Earth. The Moon stays in orbit around Earth because of similar forces.

The Lunar Cycle

The Moon takes about 27 days to complete one revolution around Earth. In addition, the Moon takes about 27 days to complete one rotation on its axis. This is why the same side of the Moon faces Earth at all times.

Although it seems to glow, the Moon does not emit its own light. Instead, it reflects light from the Sun. As the Moon orbits Earth, we see different amounts of the sunlit side of the Moon each night of the month, which we refer to as the **lunar cycle** (Figure 2). The changes in appearance of the Moon are called the **phases of the Moon**.

The lunar cycle begins with the new Moon, when the sunlit side of the Moon is facing away from Earth. When the Moon is new, we see no Moon at all. As the Moon continues to revolve and rotate, we continue to see more of its sunlit side—the visible surface of the Moon is increasing, or **waxing**.

The Moon continues to wax until we can see all of its sunlit side during a full Moon. After the full Moon, the part of the Moon we see begins decreasing, or **waning**. We see less and less of the sunlit side until we are back to the new Moon.

DIG DEEPER

Did You Know?

Phases of Planets

Some planets have phases visible from Earth. Venus is a famous example.

- **lunar cycle:** the sequence of phases made by the Moon during one orbit around Earth
- **phases of the Moon:** the changing amount of illuminated Moon surface visible from Earth
- **waxing:** increasing in size; a period during which the visible surface of the Moon increases
- **waning:** decreasing in size; a period during which the visible surface of the Moon decreases

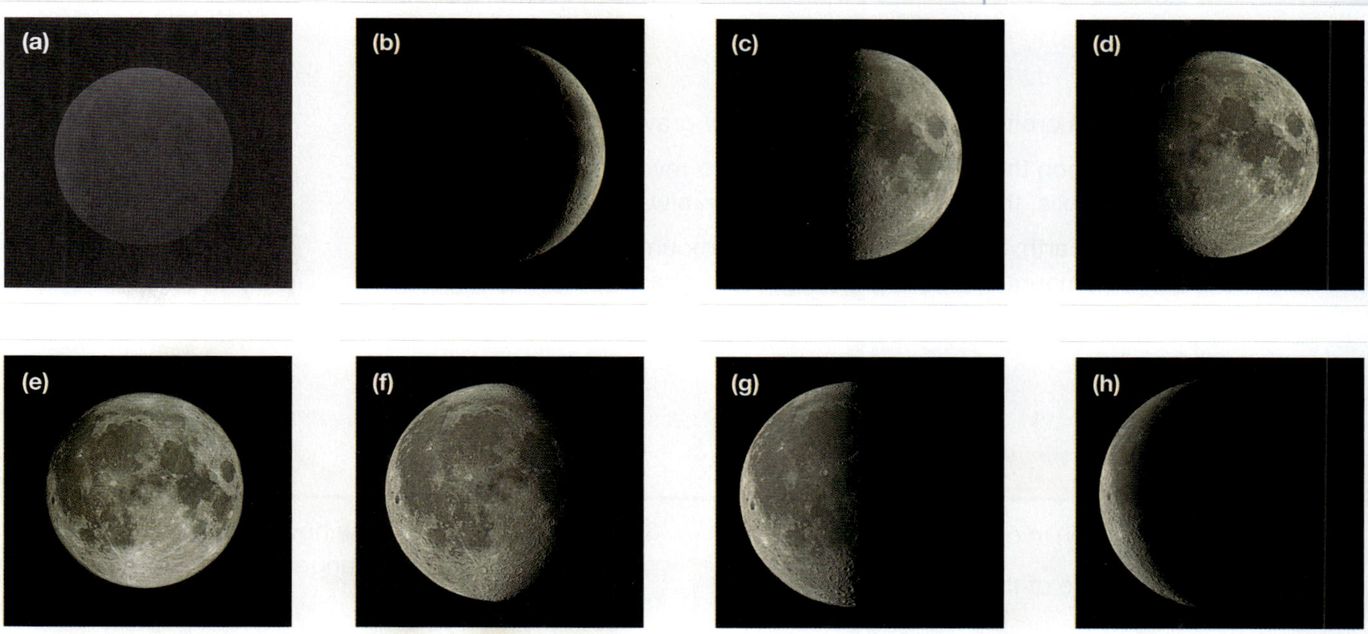

Figure 2 The phases of the Moon as seen from Earth: (a) new Moon (image darkened to represent the new Moon because we cannot see it); (b) to (d) waxing phases; (e) full Moon; (f) to (h) waning phases

TRY THIS MODELLING THE LUNAR PHASES

SKILLS HANDBOOK 3.B.

SKILLS: Observing, Communicating, Analyzing

You will work in a group to model the lunar phases.

Equipment and Materials: a lamp with a 40- to 60-watt incandescent light bulb (lampshade removed); polystyrene ball; pencil (**Figure 3**)

Figure 3

1. Choose one person in your group to represent Earth.
2. Have another member of your group stick the pencil into the polystyrene ball. This is the Moon.
3. Turn out the classroom lights and turn on the lamp, which represents the Sun.
4. Ask Earth to stand several metres away from the Sun.
5. Have the Moon revolve around Earth in a counterclockwise direction, while Earth watches the lighted part of the Moon and determines the phases visible as the Moon continues in its orbit.
6. Have each member of your group take a turn being Earth and observing the Moon go through its phases.

A. Starting with the new Moon phase, use the model to describe the main phases of the Moon during one lunar cycle. T/I

B. How could the activity be changed to make the phases of the Moon appear more distinct? T/I

9.6 Wrap Up

- The Moon remains in orbit around Earth because of gravity.
- Since it takes the Moon the same amount of time to revolve around Earth and to rotate on its axis, the same side of the Moon always faces Earth.
- As the moon orbits Earth, we see changes in the amount of illumination on the near side. The amount of illumination is called a phase of the moon.

CHECK YOUR LEARNING

1. What holds the Moon in orbit around Earth? K/U
2. Why is the same side of the Moon always facing Earth? K/U
3. Draw a diagram to show how the size of the sunlit portion of the Moon changes during a month. K/U C
4. Explain the difference between the waxing phases and the waning phases of the Moon. K/U C

9.7 Eclipses

When we look at the Sun and the Moon in the sky, they appear to be about the same size. Yet the Sun's diameter is actually 400 times that of the Moon. The Sun is also 400 times more distant from Earth than the Moon. Since the Sun is so much farther away from Earth, in the sky it appears to be about the same size as the Moon.

Solar Eclipse

When the Moon is directly between Earth and the Sun, the Moon can completely block the light from the Sun and cast a shadow on Earth. This produces a **solar eclipse** (Figure 1). A total eclipse occurs somewhere on Earth once every two years. If you are lucky enough to be somewhere along the path, you will see the spectacular sight of a total solar eclipse.

When the Moon appears to cover the Sun, only the outermost layer of the Sun, the corona, is visible (Figure 2). We do not normally see the corona of the Sun because the rest of the Sun is so bright. During a total eclipse, astronomers can safely study the corona. The path of the Moon's shadow on Earth is usually about 100 km wide.

The chance to observe a total solar eclipse is very rare. The orbit of the Moon around Earth is tilted in respect to the orbit of Earth around the Sun. This means that most of the time the Moon passes just above or just below the Sun. Partial solar eclipses occur when the Moon blocks only some of the light from the Sun.

▶ **solar eclipse:** when the Moon passes between the Sun and Earth and blocks some or all of the Sun's light

CAUTION

Solar Eclipses

Always use a suitable solar filter or other protection when looking at the Sun, especially during a solar eclipse. The Sun's powerful rays can cause permanent damage to your eyes at any time, yet some people are tempted to stare at the Sun during eclipses. Never look directly into the Sun.

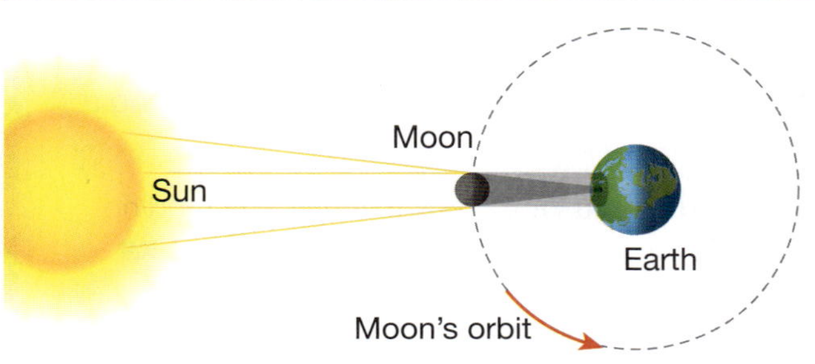

Figure 1 This diagram, looking down at the North Pole, shows the shadow of the Moon falling on Earth during a solar eclipse.

Figure 2 When a total solar eclipse occurs, the Sun's corona is most visible. This total eclipse on July 22, 2009, was visible from India to parts of China.

▶ **lunar eclipse:** when the Moon passes through Earth's shadow and does not reflect as much sunlight as usual

Lunar Eclipse

A **lunar eclipse** occurs when the Moon passes behind Earth in such a way that Earth blocks the Sun's light from hitting the Moon (**Figure 3**). The Moon cannot reflect sunlight as it normally would. If only part of the Moon passes through Earth's shadow, a partial lunar eclipse occurs. A lunar eclipse can be seen anywhere on Earth where the Moon is visible above the horizon. This is why it is much more common to see a lunar eclipse than a solar eclipse. The colour of the Moon during a total lunar eclipse depends on the amount of dust and clouds in Earth's atmosphere (**Figure 4**).

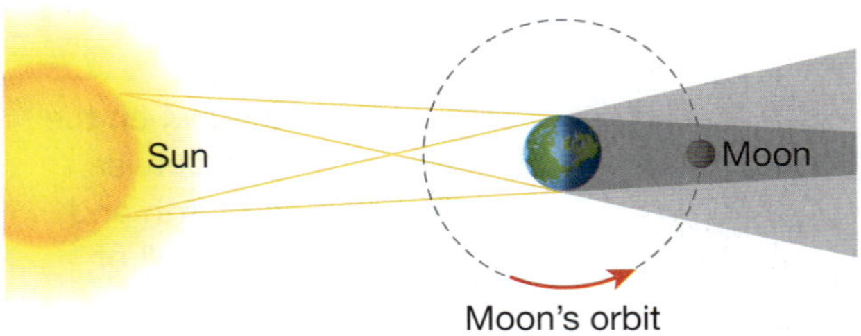

Figure 3 This diagram, looking down from the North Pole, shows the positions of the Sun, the Moon, and Earth during a total lunar eclipse.

Figure 4 The red Moon during a lunar eclipse is caused by sunlight passing through Earth's atmosphere.

9.7 Wrap Up

- A solar eclipse occurs when the Moon passes between the Sun and Earth. You must be in the path of the Moon's shadow to see a total solar eclipse.
- A lunar eclipse occurs when the Moon passes through Earth's shadow.

CHECK YOUR LEARNING

1. Why is an eclipse a good time for scientists to observe the outer atmosphere of the Sun? K/U
2. Draw a diagram to explain what happens during a solar eclipse. K/U C
3. Draw a diagram to explain what happens during a lunar eclipse. K/U C
4. Predict what you would see if you were standing on the Moon during a total lunar eclipse. T/I A

9.8

Expansion of the Universe

People have been wondering about the Universe for as long as there have been people. When we look into a clear night sky and see the Milky Way (Figure 1), we may have some questions: Where did it all come from? How big is the Universe? What will happen to the Universe over time?

Figure 1 The Milky Way's brightest section is 26 000 light years from Earth.

The Universe

How big is the Universe? It has only been in the past century that astronomers have developed technology to help us answer this question. Until the 1920s, astronomers thought that our galaxy, the Milky Way, *was* the entire Universe. Then, an astronomer named Edwin Hubble made one of the most important discoveries in science. Using the 2.5 m telescope at Mount Wilson Observatory in California, Hubble found stars that were outside of the Milky Way (Figure 2). This observation proved the existence of other galaxies, such as the Andromeda Galaxy (Figure 3). From his observations and analysis, he realized that the Andromeda Galaxy contained billions of stars and was fantastically far away.

Hubble was also the first to describe the four types of galaxies that you read about in Chapter 8. Overnight, the scale of the Universe expanded millions of times!

DIG DEEPER
Web Link

To learn more about the work of Edwin Hubble,

GO TO NELSON SCIENCE

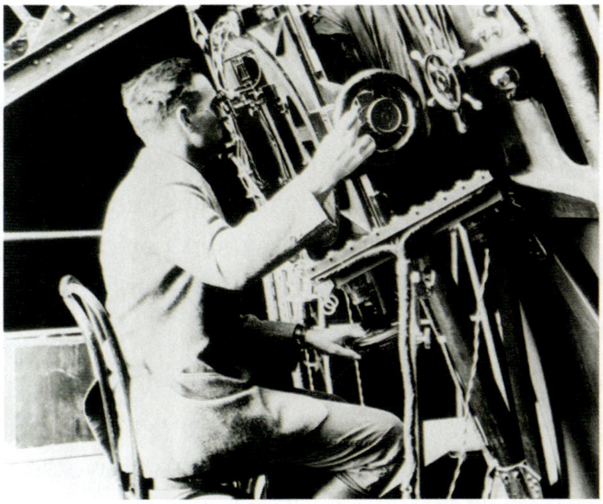

Figure 2 Edwin Hubble changed our understanding of the Universe with his discoveries.

Figure 3 Hubble discovered stars beyond the Milky Way galaxy, in the Andromeda Galaxy.

Red Shift

Hubble made a second discovery during the late 1920s by studying the patterns of light emitted from galaxies. His studies revealed that each galaxy emitted a band of colours of light that was unique, called a spectrum (**Figure 4(a)**). Hubble noticed that the visible spectrum of distant galaxies seemed to be shifting toward the red end of the spectrum, which he called a **red shift** (**Figure 4(b)**). This meant that the galaxies were moving away from the Milky Way and away from each other. The farther away the galaxies were, the greater the red shift. The greater the red shift, the faster the galaxies were moving. Hubble realized that the movement of the galaxies meant that the Universe is expanding.

▶ **red shift:** light from a galaxy shifting toward the red end of the visible spectrum

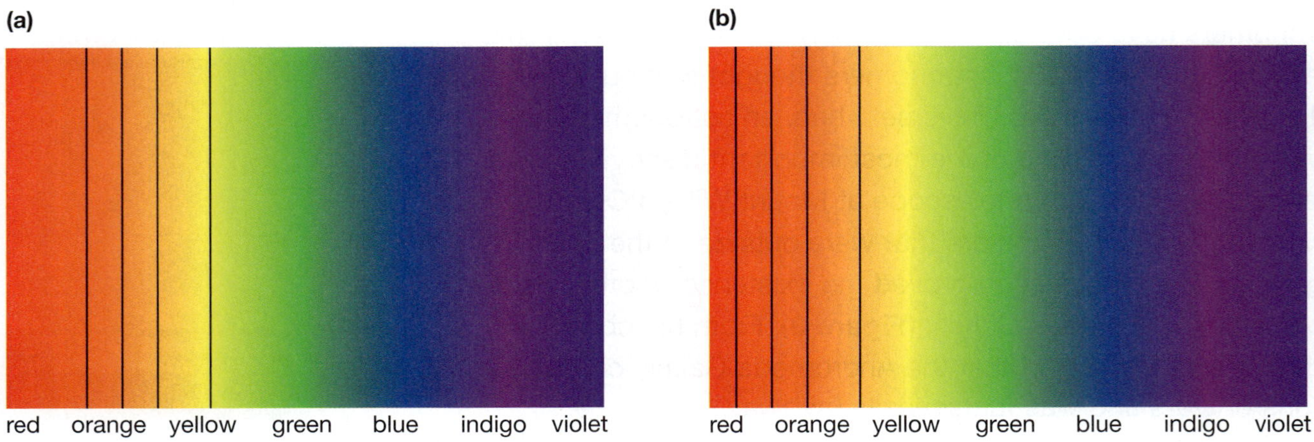

Figure 4 (a) The spectrum of a stationary galaxy; (b) The same spectrum showing red shift, indicating that the galaxy is moving away from Earth. The black lines, indicating light absorption, shift toward the red end of the spectrum.

The Big Bang

Where did the Universe come from? **Cosmologists**, scientists who study the material Universe and its physical origins, are trying to answer this question. Cosmologists build on existing evidence and make new discoveries to try to determine what the Universe is made of, how old it is, and how it began.

From their observations, cosmologists developed a theory about the origin of the Universe. The theory, called the **Big Bang theory**, states that around 14 billion years ago, everything in the Universe, including time and space, existed in a single, fixed mass that was hot and dense. Then, the mass suddenly expanded. A billion years later, the first galaxies formed.

▶ **cosmologists:** scientists who study the makeup and origins of the Universe

▶ **Big Bang theory:** the theory that the Universe began in an incredibly hot, dense expansion about 14 billion years ago

TRY THIS MODEL THE EXPANDING UNIVERSE

SKILLS HANDBOOK 3.B.

SKILLS: Performing, Observing, Analyzing, Communicating

The Universe is expanding. In this activity, you will investigate the expansion of the Universe and its effect on the movement of galaxies.

Equipment and Materials: marker; tape measure; round balloon

1. Fill your balloon with air until it is the size of a large apple. Then twist the end and hold it closed with one hand. Do not tie it shut.
2. Use the marker to draw four dots on the balloon in a line. Leave at least 1 cm of space between each dot (**Figure 5**). Each dot represents a galaxy.
3. Label the dots A, B, C, and D.
4. Now inflate the balloon to the size of a volleyball and tie it.
5. Measure the distances between the galaxies and write down the measurements.

A. What happened to the distances between galaxies as you blew up the balloon? T/I

B. Imagine that you are standing on galaxy A while the balloon is expanding. Which galaxy would appear to move away from you more quickly? T/I

C. Which one would appear to move away from you more slowly? T/I

Figure 5 Step 2

9.8 Wrap Up

- Red shift in the spectrums of distant galaxies indicates that they are moving away from us, and from one another.
- The motion of the galaxies suggests that the Universe is expanding.
- The Big Bang theory states that the Universe began in a giant expansion about 14 billion years ago.

CHECK YOUR LEARNING

1. How does red shift help astronomers determine the movement of distant galaxies? K/U
2. Summarize the conclusion that Edwin Hubble made about the Universe and the evidence he had to support his claim. K/U C
3. Were stars and galaxies formed immediately after the Big Bang? Explain. K/U C
4. What do cosmologists do? K/U

9.9 EXPLORE AN ISSUE CRITICALLY

Who Owns the Moon?

SKILLS MENU

- Defining the Issue
- Researching
- Identifying Alternatives
- Analyzing the Issue
- Defending a Decision
- Communicating
- Evaluating

In the near future, several countries and private companies plan to expand to the Moon. This leads to the question, "Who owns the Moon?" In 1967, the Outer Space Treaty was put into effect. This treaty has been signed by more than 100 countries. The treaty states that the Moon and other celestial objects belong to all people on Earth.

When placing satellites and other objects into orbit, countries and companies can apply for locations for their satellites to orbit. These locations in space are assigned on a first-come, first-served basis. Governments are investigating and trying to decide how best to share ownership of the more than 4 million hectares on the Moon's surface (**Figure 1**).

The American astronauts who walked on the Moon planted an American flag. However, this flag did not claim any part of the Moon for the United States. The flag indicated that the landing had been accomplished by the collective efforts of Americans.

Figure 1 Should anyone be able to claim land on the Moon?

SKILLS HANDBOOK 4.A., 4.B., 4.C.

The Issue

As lunar missions are planned around the world, the property rights issue needs to be solved. Your team has been asked to make a recommendation to the National Research Council on how property ownership should be decided on the Moon. You might consider the model of Antarctica, which has no ownership either and can be visited only for research purposes.

Goal

To prepare a report for the National Research Council offering a solution to the division of property ownership on the Moon

Gather Information

Work in pairs or in small groups:

- Learn more about the Outer Space Treaty and how it relates to ownership of property in space.
- Research what has been discussed about this issue, and possible solutions.

Identify Solutions

You may wish to consider the following factors as you work:

- How is ownership different from property rights? For example, a tenant has rights to an apartment but does not own it.
- How can the Outer Space Treaty be applied to this issue?
- What would be a fair way to handle this issue, keeping in mind the needs and wants of different countries and corporations?
- How would property ownership affect the distribution of resources on the Moon or other celestial bodies?
- How would we settle disputes and enforce the rules if space were to be occupied by more than a few scientists?

Make a Decision

Make a final list of recommendations concerning the way to handle property ownership in space.

Communicate

Present your recommendations at a model meeting of the National Research Council. Your presentation can be written, oral, or electronic.

DIG DEEPER

Web Link

For more information on space exploration and ownership, including the National Research Council's website,

GO TO NELSON SCIENCE

DIG DEEPER

Did You Know?

Loopholes in the Treaty

While the Outer Space Treaty forbids governments and corporations to claim ownership in space, it does not forbid individuals from owning pieces of space. People around the world have taken advantage of this omission and claimed that they own parcels of land on the Moon and have "sold" them to other citizens for profit.

Chapter 9 KEY CONCEPTS SUMMARY

The Solar System consists of all the objects orbiting the Sun.

- The Solar System is heliocentric.
- Distances within the Solar System are measured in astronomical units (AU).
- The planets revolve around the Sun in slightly elliptical orbits.

The motion of Earth causes day, night, and the seasons.

- Earth's rotation causes day and night.
- Earth's tilt and its revolution around the Sun cause the seasons.
- Countries along the equator do not experience four seasons.

Earth's motion affects our view of the night sky.

- A constellation is a region of stars in the sky.
- An asterism is a pattern of stars observed from Earth.
- Our view of constellations changes over time because of Earth's motion.

Celestial objects can be used to predict annual events and to navigate.

- Many civilizations use celestial objects to predict and plan for agricultural, cultural, and religious events.
- Sailors can use the stars to navigate the seas.

The positions of the Sun, Earth, and the Moon result in the lunar cycle and eclipses.

- The phases of the Moon are the amount of the Moon's sunlit side we see from Earth.
- A solar eclipse occurs when the Moon passes between Earth and the Sun.
- A lunar eclipse occurs when the Moon enters Earth's shadow.

Scientists have strong evidence that the Universe began around 14 billion years ago.

- Red shift provides evidence that galaxies in the Universe are moving apart rapidly.
- The Universe began with the Big Bang.

WHAT DO YOU
THINK NOW?

You thought about the following statements at the beginning of the chapter. You may have encountered these ideas in school, at home, or in the world around you. Consider them again and decide whether you agree or disagree with each statement.

1 An eclipse of the Sun occurs when Earth is between the Sun and the Moon.
Agree/disagree?

2 The Sun, the Moon, and planets orbit Earth.
Agree/disagree?

3 The size of the Universe is changing.
Agree/disagree?

4 We see different stars at different times of the year because Earth is going around the Sun.
Agree/disagree?

5 We always see the same side of the Moon.
Agree/disagree?

6 When it is summer in the northern hemisphere, it is winter in the southern hemisphere.
Agree/disagree?

How have your answers changed since then? What new understanding do you have?

BIG Idea

- Celestial objects in the Solar System and Universe have specific properties that can be investigated and understood.

DIG DEEPER

Vocabulary

SELF-QUIZ

The following icons indicate the Achievement Chart category addressed by each question. K/U Knowledge/Understanding T/I Thinking/Investigation C Communication A Application

For each question, select the best answer from the four alternatives.

1. Which statement best explains why the amount of daylight changes throughout the year? (9.2) K/U
   (a) Earth is closer to the Sun at different times of year.
   (b) The tilt of Earth in relation to the Sun changes as Earth revolves around the Sun.
   (c) The rotation of Earth slows at certain times of the year.
   (d) The revolution of Earth speeds up at certain times of the year.

2. Which term best describes the model of the Solar System shown in **Figure 1**? (9.1) K/U

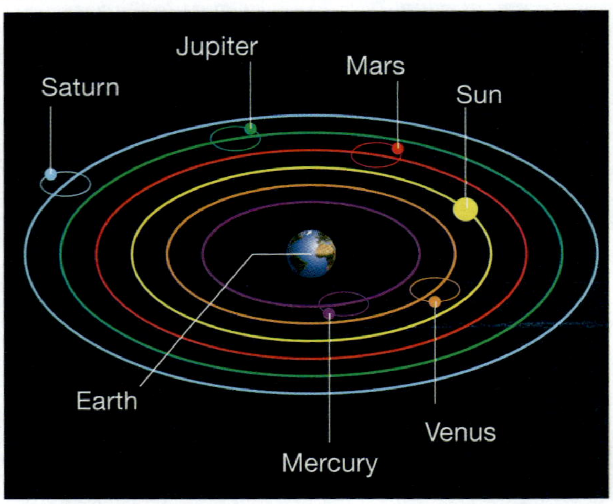

Figure 1

   (a) Moon-centred
   (b) geocentric
   (c) Sun-centred
   (d) heliocentric

3. When can we observe a solar eclipse? (9.7) K/U
   (a) when the Moon passes between Earth and the Sun
   (b) when the Sun passes between Earth and the Moon
   (c) when Earth passes between the Sun and the Moon
   (d) when the Moon is brighter than the Sun

4. Which statement best explains why people on Earth always see the same side of the Moon? (9.6) K/U
   (a) A day is the same length on Earth as it is on the Moon.
   (b) It takes 365 days for the Moon to complete one revolution around Earth.
   (c) Earth revolves around the Sun more slowly than the Moon revolves around Earth.
   (d) The Moon rotates on its axis and revolves around Earth in exactly the same amount of time.

Indicate whether each of the statements is TRUE or FALSE. If you think the statement is false, rewrite it to make it true.

5. All of the stars in the Universe are in the Milky Way galaxy. (9.8) K/U

6. The northern hemisphere is farthest from the Sun on June 21. (9.2) K/U

7. Ancient sailors used their knowledge of the stars to navigate. (9.4) K/U

8. Total solar eclipses occur somewhere on Earth during every full Moon. (9.7) K/U

9. In your own words, explain how you might use azimuth and altitude to determine the position of a celestial object in the night sky. (9.4) K/U C

10. (a) What is a scale model?
    (b) How are scale models useful in astronomy? (9.1) K/U T/I

Copy each of the following statements into your notebook. Fill in the blanks with a word or phrase that correctly completes the sentence.

11. During our ________ season, Earth's northern hemisphere is tilted toward the Sun to a greater degree than it is at other times of the year. (9.2) K/U

12. It takes Earth one year to ________ around the Sun. (9.1) K/U

13. According to the ________ theory, the Universe began about 14 billion years ago then rapidly expanded. (9.8) K/U

14. A(n) ________ is an imaginary line running through a planet from pole to pole on which it rotates, producing day and night. (9.1) K/U

Match each term on the left with the most appropriate description on the right.

15. (a) circumpolar — (i) different amounts of the surface of the Moon lit by the Sun as viewed from Earth
    (b) phases — (ii) a pattern of stars in the sky
    (c) gravity — (iii) a star or constellation that does not set below the horizon
    (d) asterism — (iv) the force of attraction between two massive objects, such as the Sun and Earth (9.3, 9.6, 9.7) K/U

Write a short answer to each of these questions.

16. What force keeps the Moon in orbit around the Sun? (9.6) K/U

17. Canada has several time zones. When it is noon in Ottawa, it is 9 a.m. in Vancouver. If noon is the time when the Sun is highest in the sky, explain why Canada needs more than one time zone. (9.1) T/I C

18. Why is it safer to view a lunar eclipse with the unaided eye than it is to view a solar eclipse? (9.7) K/U

19. An astronomer observes that the spectrum from the light of a distant galaxy has shifted toward the red end of the spectrum. What does this indicate about the galaxy? (9.8) K/U

20. Explain why it is colder in the winter in the northern hemisphere than it is in the summer. (9.2) K/U C

21. Why do we always see the same side of the Moon from Earth? (9.6) K/U

22. Imagine that you see a very red lunar eclipse while on vacation. Write a letter to a friend describing what you see and what caused it. (9.7) K/U C

23. Explain the difference between an asteroid and a comet. (9.1) K/U

24. If people in your community were to name the constellations today, do you think they would name them after wild animals, mythical figures, and common tools? What might they name constellations after? (9.3) C A

25. You're going to New Zealand in March for winter break. What kind of clothing should you pack? (9.2) T/I

Self-Quiz

To do an online self-quiz, GO TO NELSON SCIENCE

REVIEW

The following icons indicate the Achievement Chart category addressed by each question.

K/U Knowledge/Understanding
T/I Thinking/Investigation
C Communication
A Application

What Do You Remember?

1. Explain why the constellations appear to move across the sky. (9.3) K/U

2. Write a brief paragraph to explain how the red shift in the spectrums of light from distant galaxies means they are moving slowly toward us. (9.8) K/U

3. Explain what the annual climate would be like on a planet with an axis that was not tilted. (9.2) K/U

4. Why is it rare to see a solar eclipse? (9.7) K/U

5. In ancient times, sailors could not tell which way they were going at night. Explain how they were able to navigate using celestial objects. (9.4) K/U

6. (a) In your own words, explain the terms azimuth and altitude (9.5) K/U
   (b) What use do azimuth and altitude have in astronomy?(9.5) K/U

7. Briefly define the terms "rotation" and "revolution." (9.1) K/U

8. When the Moon waxes, its sunlit portion gets
   (a) smaller
   (b) brighter
   (c) larger
   (d) darker (9.6) K/U

9. When is the length of day and the length of night equal? (9.2) K/U
   (a) on the winter solstice
   (b) on the equinox
   (c) on the summer solstice
   (d) at noon

10. What did Edwin Hubble discover? (9.2) K/U

11. Write a brief paragraph explaining the relationship between a constellation and an asterism. (9.3) K/U C

12. (a) When is Earth's axis most inclined toward the Sun?
    (b) When is Earth's axis most inclined away from the Sun? (9.2) K/U

13. Explain to a friend how they might use their hands to measure the azimuth and altitude of a celestial object, as viewed from Earth. (9.4) K/U C

14. Explain why Polaris is called the North Star. (9.4) K/U

15. What portion of the Sun is visible during a solar eclipse? (9.7) K/U

What Do You Understand?

16. Draw a diagram to show the phases of the Moon. (9.6) K/U C

17. Draw a diagram to show the positions of Earth, the Moon, and the Sun during a lunar eclipse. (9.7) K/U C

18. Draw a diagram to show the positions of Earth, the Moon, and the Sun during a solar eclipse. (9.7) K/U C

19. Draw a heliocentric model of the Solar System. Include all eight planets. (9.1) K/U C

20. Why does the Moon appear reddish during a lunar eclipse? (9.8) K/U

21. Explain why Jupiter does not have seasons. (9.2) K/U

22. Explain why some planets have longer years than others. (9.1) K/U C

Solve a Problem

23. Your friend tells you that Earth is hot in summer and cold in winter because the Sun is hotter in summer and cooler in winter. Send an email to your friend or have a discussion explaining the true causes of the seasons. (9.2) K/U C

24. How do the rotation and revolution of Earth affect the apparent motion of celestial objects in the night sky? (9.2) T/I

25. Compare the model that people used to explain the Universe 2000 years ago to the one we use today. (9.1) K/U T/I

26. Your friend tells you that Ontario is hot in summer and cold in winter because Earth is closer to the Sun in summer and farther in winter. Write a letter to your friend explaining the true causes of the seasons. (9.2) K/U C

27. Suppose you are lost in the woods on a camping trip in northern Ontario. (9.4) T/I
    (a) What star could you use as a navigational guide?
    (b) How could you find that star among the thousands of stars that are visible on a clear night?

Create and Evaluate

28. More than 2200 years ago, Aristarchus of Samos, an ancient Greek astronomer, suggested that the Sun was at the centre of the Solar System. It took more than 1800 years for this theory to be revived by Copernicus. Why do you think it took so long for people to accept that the Sun was at the centre of the Solar System? (9.1) T/I

Reflect on Your Learning

29. What was the most interesting thing you learned in this chapter? C
    (a) Why was this topic interesting?
    (b) If you wanted to learn more about it, where could you look?

30. Identify one way you could you use your new knowledge of the positions of the objects in the night sky. A

31. Identify one topic covered in this chapter that you found difficult to understand.
    (a) Why was this topic difficult for you to understand? C
    (b) What did you do to help overcome your difficulty?

32. Work with a partner. Find a page in your respective notebooks and where you both took notes on the same topic on the same day. C
    (a) Make a list of the ways in which your notes are similar and different.
    (b) What can you conclude from this about your different learning styles?
    (c) Why is it important to keep good notes?

Web Connections

33. Find five asterisms or celestial objects that can be seen in the sky from Ontario. T/I

34. Research some of the ways that Aboriginal peoples tracked the movements of the Sun, the Moon, and stars. Compare their methods to the methods used by people in other cultures such as the ancient Egyptians. Can you see any similarities in the stories? T/I C

35. Research and write several paragraphs about how tides affect the lives of people who live and work on or near the sea.

GO TO NELSON SCIENCE

CHAPTER 10

Space Research and Exploration

KEY QUESTION

How has space exploration benefited humans?

KEY CONCEPTS

- Getting into space and staying in orbit are major challenges of space travel.
- Robotic space probes have explored the Solar System and beyond.
- Space can be a hostile environment for humans.
- Living and working in space presents challenges.
- Space exploration is technically challenging and expensive.
- Space technologies have benefited our lives.

SPACE BOOTS FOR BASKETBALL?

Ask an older person what kind of shoes they wore for sports when they were young. Many will have had shoes with rubber soles that wore out quickly. Often, the shoes came in only two colours—black or white!

Today, athletes can choose from a wide variety of athletic footwear. Many improvements in athletic shoes came from space technology.

- **Shoe Soles** Spacesuits worn by astronauts have to be firm enough to be pressurized. They also have to be bendable enough so the astronauts can move. The accordion-like joints in spacesuits allow astronauts to move more freely. Athletic shoe companies have used ideas from spacesuit design to improve their athletic shoes. Some modern-day shoes have a pressurized shell for the sole with joints for cushioning.
- **Shoe Insoles** NASA developed a moulding process to make space helmets impact-resistant. Athletic shoe companies use this moulding process to mould the insole of the shoe. They fill the mould with the same kind of foam used in space boots. This type of foam makes the shoes very light and durable. Today, shoe companies are still improving athletic shoes. They especially keep an eye on new space technology for innovations that can make shoe designs even better.

WHAT DO YOU THINK?

Many of the ideas you will explore in this chapter are ideas that you have already encountered. You may have encountered these ideas in school, at home, or in the world around you. Some of the statements are not true. Consider each statement and decide whether you agree or disagree with it.

1 Humans have landed on all the planets.
Agree/disagree?

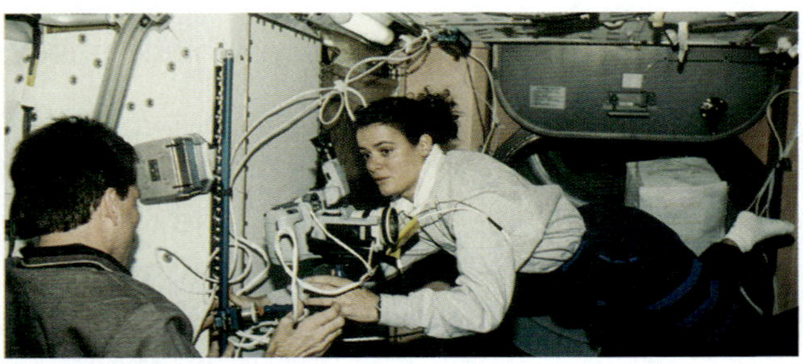

2 Astronauts float in space because there is no gravity.
Agree/disagree?

3 Computers control the launch of spacecraft.
Agree/disagree?

4 Astronauts in orbit can get swollen faces.
Agree/disagree?

5 Space travel poses hazards for astronauts.
Agree/disagree?

6 Canada has contributed to space exploration.
Agree/disagree?

FOCUS ON

READING

Synthesizing

When you synthesize, you join ideas together. You combine ideas in the text with ideas you already know and with ideas in other texts to gain new insights, make inferences, or draw conclusions. Use the strategies below to synthesize the text excerpt from Section 10.3.

DIG DEEPER

Reading Tip

As you work through the chapter, look for tips like this. They will help you develop literacy strategies.

What to look for	What to do
title or heading	Think about what you know about the topic.
information or idea in the text	Make a connection with something you know.
information or idea in the text	Compare it with similar information or a similar idea in another text.
graphics and captions in the text	Think about how they help you understand the main idea.
information or idea in the text	Ask yourself how they changed your understanding of the topic.

Features of a Spacesuit and Helmet

The main purpose of the spacesuit and helmet is to protect astronauts while they are outside the spacecraft. Temperatures change very quickly between the sunlit side of a celestial object and the side in shade. The temperatures outside a spacecraft vary between 120 °C in sunlight and –100 °C in the shade. A spacesuit is insulated to protect astronauts from these extreme temperatures. A spacesuit must also be flexible so that an astronaut can perform tasks in a space environment (**Figure 7**). Imagine trying to perform a simple task like tightening a screw while wearing hockey gloves!

Figure 7 Spacesuit gloves must be flexible so that astronauts are able to carry out their tasks during missions.

Synthesizing *in Action*

First, combine ideas in the text with ideas you already know. Then ask yourself how your understanding of the topic has changed. Below is an example of how one student used the strategies to synthesize the text excerpt about the features of a spacesuit.

Information in the text	What I already know	What I think now
The temperature range outside the spacecraft is 120 °C to –100 °C.	On Earth 35 °C feels very hot and –20 °C feels very cold.	The spacesuit must have built-in heating and air conditioning.
Spacesuit gloves need to be flexible so astronauts can use tools.	I can't put my key in the door lock with my winter gloves on.	Spacesuit gloves must be made of special materials.
The photo shows the flexible spacesuit gloves.	Hockey gloves have a lot of padding but are stiff.	The spacesuit gloves have soft padding for protection and folds so astronauts can bend their fingers.

10.1 Getting into Space

Astronauts are pioneers venturing into uncharted territory. The vehicles used to get them into space are complex and use powerful rockets. Space vehicles need to be durable as they leave and then re-enter Earth's atmosphere. Also, space vehicles have limited space. During space missions, astronauts have to take with them everything they will need.

Space Exploration

Technological developments in the past five decades have helped humans in their quest to explore outer space. We have expanded our presence from Earth to the edge of the Solar System, and beyond. We have built scientific laboratories in space. Vehicles designed for space travel, called **spacecraft**, have landed on planets, moons, and asteroids. We have also sent instruments, such as the Hubble Space Telescope, into space to gather information we cannot get from Earth. We have sent probes to planets that are not yet fit to live on, or too far away for humans to safely make the trip. Future missions are planned to send even more technologically advanced spacecraft into space (**Figure 1**).

▶ **spacecraft:** a robotic or human-occupied vehicle used to explore space and celestial objects

Figure 1 The James Webb Space Telescope (JWST), an international collaboration, will be a large infrared telescope with a 6.5 m mirror. The JWST will replace the Hubble Space Telescope when it is launched in 2014.

Types of Spacecraft

There are different types of spacecraft that serve different purposes in space exploration. Some spacecraft, such as the space shuttle in **Figure 2**, travel in low orbits around Earth on short-term missions. Other spacecraft, such as weather and communications satellites, are in long-term orbits around Earth. The **International Space Station (ISS)** is a space research laboratory that has been in low orbit around Earth since 1998 (**Figure 3**).

▶ **International Space Station (ISS):** a research laboratory in low orbit around Earth

All spacecraft have similar components. A launch vehicle is a rocket that carries the spacecraft from Earth's surface into outer space. On most human-occupied missions, astronauts sit in one section of the launch vehicle to control the spacecraft. Another section of a spacecraft, called a payload, carries cargo. Cargo can include scientific equipment and supplies. Cargo can also be other spacecraft, such as planetary land rovers (**Figure 4**). Specialized satellites may also be carried as cargo and released into space once the spacecraft is beyond Earth's atmosphere (**Figure 5**).

Figure 2 The Space Shuttle *Endeavour* lands after a mission to the International Space Station.

Figure 3 The ISS travels at an average speed of 27 700 km/h. It orbits Earth 15.7 times each day.

Figure 4 The Apollo 17 mission in 1972 carried the lunar roving vehicle to the Moon.

Figure 5 The XMM-Newton is an orbiting X-ray observatory belonging to the European Space Agency.

Space Travel

One of the biggest challenges of space travel is the very first step—getting up and out of Earth's atmosphere. Earth's atmosphere exerts air resistance that works opposite the direction of the motion of the spacecraft. This reduces the speed of spacecraft. A spacecraft needs a lot of force, or **thrust**, to lift off from Earth's surface and overcome this air resistance. To attain orbit, a spacecraft must be able to travel about 7.5 km/s. That means a spacecraft could travel the distance from London, Ontario, to Kingston, Ontario, in just under one minute!

▶ **thrust:** the force of a rocket engine

It is essential for spacecraft to travel a gravitational orbit (usually around Earth). Spacecraft must be equipped with powerful engines that will allow them to reach the speed required to place them into Earth's orbit. Only a small amount of fuel can be brought into space with the craft, and the fuel needs to be used wisely. Once in orbit, the gravitational pull of Earth keeps the spacecraft in motion. Spacecraft must also be constructed to withstand travel in space and to safely transport humans into space and then back to Earth again.

Liftoff

Many spacecraft use rocket boosters for liftoff. A **rocket booster** is a large powerful rocket engine containing fuel that attaches to a spacecraft. Rocket boosters provide most of the liftoff thrust for a spacecraft. Once the spacecraft has gained a certain altitude, the empty rocket boosters break away from the spacecraft and fall. This reduces the mass of the spacecraft and allows the spacecraft to use less fuel for the rest of its journey.

▶ **rocket booster:** a powerful rocket engine that contains fuel and attaches to a spacecraft for use during launch

A Model of Liftoff

Think about a balloon model of a rocket (**Figure 6**). If you inflate a balloon and then release it, the escaping air propels the balloon upward. It is this same principle that launches a rocket at liftoff. In a rocket, a chemical change occurs that produces hot gases. The gases are expelled through the bottom of the rocket at great speed, causing the rocket to move in the opposite direction.

Figure 6 A balloon model of a rocket liftoff.

The acceleration and speed provided by a rocket booster have to be carefully calculated based on the overall mass of the spacecraft, including the fuel. The greater the mass, the more thrust needed for liftoff. Once in orbit, a spacecraft coasts along, conserving energy. The spacecraft can do this because it is above the atmosphere and there is no air to slow it down.

Fuel Supplies

A spacecraft requires massive amounts of fuel to launch it into space (**Figure 7**). The fuel is heavy, and enormous amounts of it are burned to overcome the pull of gravity during liftoff.

Once the vehicle is in space, there is not much fuel left over for the rest of the journey. Small amounts of fuel are burned in bursts for correcting or changing direction. The spacecraft coasts for long periods between bursts in the desired direction. **Figure 8** shows the components of a Space Shuttle.

Figure 7 Rocket boosters carry more than 500 000 kg of fuel. Almost all of this fuel is used in the first 2 min after launch.

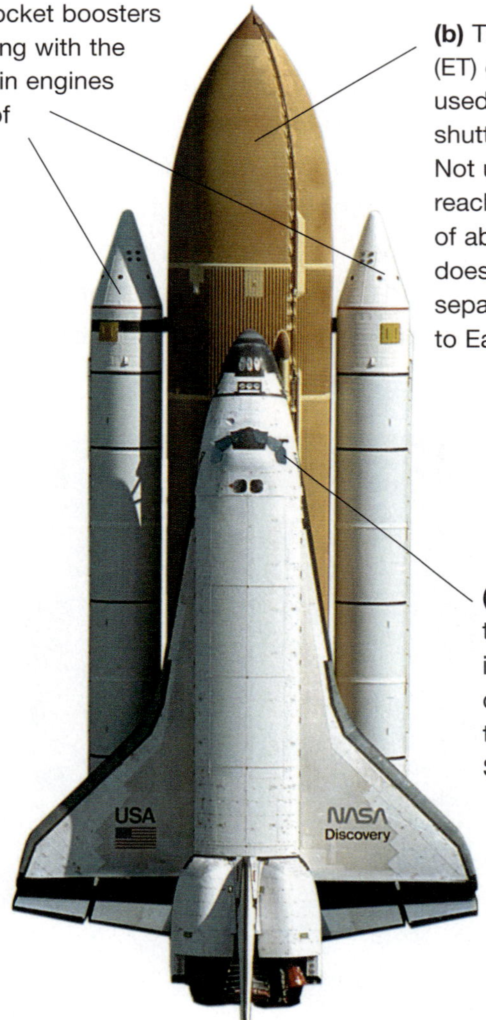

Figure 8 The space shuttle is made of three main parts: (a) two solid rocket boosters, (b) a massive external tank, and (c) an orbiter.

DIG DEEPER

Did You Know?

First Canadian in Space

Marc Garneau became the first Canadian in space in 1984. Since then, Canadian astronauts have flown on more than a dozen space shuttle missions.

DIG DEEPER

Reading Tip

Using Visuals

Graphs, tables, and images can help you to visualize the main idea of a text. For example, the diagram showing the forces that combine to keep the ISS in orbit can help you synthesize what you are reading with what you have already learned about space travel. This synthesis will help you understand how spacecraft are launched and how they attain orbit.

Orbiting Earth

A satellite in orbit moves faster when it is closer to the celestial object it is orbiting. Similarly, it moves slower when it is farther away. For example, the International Space Station orbits 380 km above Earth and travels at a speed of about 8 km/s. Global positioning satellites, on the other hand, orbit 20 000 km above Earth and travel at a speed of about 4 km/s—much more slowly. Earth's Moon is also a satellite. It is about 385 000 km from Earth and orbits at a speed of about 1 km/s. Once in orbit, a satellite moves at a constant speed.

Staying in Orbit

An object in space travels forward in a straight line unless a force acts on it. Objects launched into space from Earth are pulled back toward Earth by the force of gravity. If an object does not have enough travelling speed when it is launched, it will be pulled back to Earth's surface. If an object is travelling too fast, it will continue to move away from Earth. The combination of the forward motion of a spacecraft and the pull of Earth's gravity keeps the spacecraft in orbit (**Figure 9**).

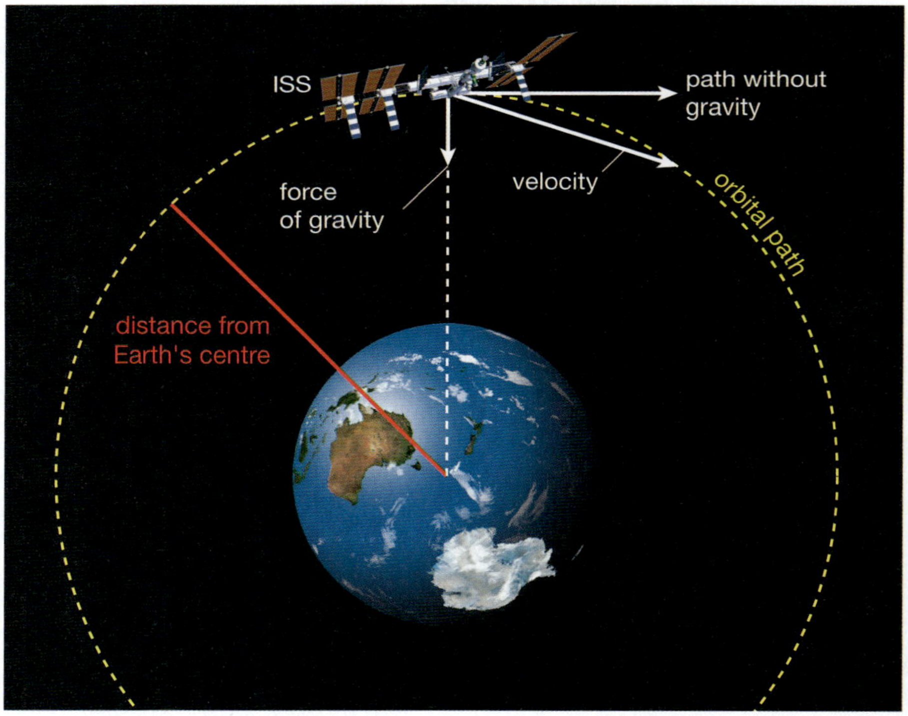

Figure 9 This diagram shows the forces that combine to keep the International Space Station in orbit around Earth.

10.1 Wrap Up

- It takes an enormous amount of energy to launch a spacecraft into space.
- Rocket boosters must have enough thrust to launch a vehicle into space.
- Most of a spacecraft's fuel is used up during the launch into space.
- Gases escaping from the rear of the rocket boosters propel the spacecraft in the opposite direction of Earth's gravity.
- The pull of Earth's gravity and the forward motion of a spacecraft keep it in orbit around Earth.

CHECK YOUR LEARNING

1. (a) Explain how spacecraft are launched into orbit.
   (b) What are some of the challenges that must be overcome? K/U
2. What must a rocket be able to do to lift off and escape Earth's atmosphere? K/U
3. Why do the empty rocket boosters separate from the space shuttle soon after launch (**Figure 10**)? K/U
4. Draw a diagram to describe what happens to rocket fuel as a spacecraft lifts off and leaves Earth's atmosphere. Label your diagram. K/U C
5. Describe how a spacecraft is able to stay in orbit around Earth. K/U
6. What will happen when the girl in **Figure 11** releases the blown-up balloon? What does this balloon have in common with rocket boosters? T/I

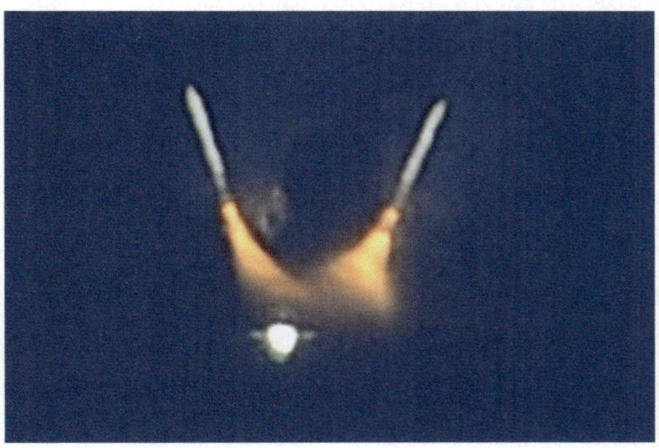

Figure 10

Figure 11

10.2 The Future of Space Exploration

Humans have landed on the Moon, sent robotic probes to distant planets, and have lived and worked in orbit aboard space stations. What do you think will be the next steps for space exploration? Read on to find out what space travel may accomplish in the near future.

DIG DEEPER

Did You Know?

Where Space Begins

The internationally recognized boundary between Earth's atmosphere and space is 100 km from Earth's surface at sea level—the Kármán line. Aeronautical scientist Theodore Kármán calculated that above 100 km, an aircraft would have to attain orbit to stay up.

Limitations of Space Travel

Over the last 50 years, engineers and scientists have overcome many of the challenges of space travel. What are some of these challenges and how have they been overcome?

Human Survival

One of the main concerns of long-term space travel is keeping the crew alive. There are many physical and mental challenges that astronauts must overcome on their missions. The following list outlines some of these challenges:

- **Basic Needs:** Some of the problems with long-term space travel include how to bring along enough air, food, and water, and what to do with waste.
- **Emotional and Mental Well-Being:** The mental health of the crew is also an issue. Crew members might get home-sick —they might miss family and Earth. They might suffer from claustrophobia—a fear of being in an enclosed space.
- **Physical Well-Being:** Physical problems, such as weakened bones and muscles due to prolonged periods in space, are another concern.
- **Boredom:** Long space voyages can be boring for astronauts.

A space mission to the Moon was possible because spacecraft can reach the Moon in less than four days. Travelling to a planet such as Mars would take seven to ten months. Sending equipment and people to Mars is far more time-consuming, expensive, and risky than to the Moon (**Figure 1**).

Figure 1 Mars' distance from Earth makes it a difficult space destination for humans.

Low Orbit Missions in Space

There are two basic locations in space where humans have lived and worked. One is in the spacecraft of the Apollo voyages to the Moon. The other is in low Earth orbit, between 200 km and 2000 km above Earth's surface. Both the space shuttle missions and the International Space Station are examples of low orbit missions.

To the Moon and Back

The Apollo missions that sent humans to the Moon used a *Saturn V* launch rocket. These rockets carried the human crew, the lunar lander, and the **command module**, the section of the spacecraft that carried the crew, toward the Moon. In all, there were 11 Apollo missions.

Apollo 13's planned Moon landing was aborted due to an oxygen tank explosion. There were six Moon landings. Of these, Apollo 15 was the first of the missions in which astronauts spent long periods of time on the Moon's surface.

▶ **command module:** the section of a spacecraft in which astronauts live, communicate with Earth, and operate controls during a flight

Apollo 15

Apollo 15 was the fourth mission to land on the Moon. In the lunar module *Falcon*, astronauts David Scott and James Irwin spent 67 hours on the Moon's surface and explored the surface of the Moon for a total of 18.5 hours. It was the first time astronauts used a motorized lunar roving vehicle (LRV) to drive around on the Moon's surface (**Figure 2**). The mission lasted a little over 12 days before the astronauts returned to Earth in the command module (**Figure 3**). On this mission, astronauts found a rock called the "Genesis Rock." It is thought that this rock is part of the original lunar crust when the Moon was formed about 4.6 billion years ago (**Figure 4**).

DIG DEEPER

Did You Know?

Walking on the Moon

On July 20, 1969, Neil Armstrong was the first person to walk on the Moon. On December 6, 1972, Eugene Cernan was the last person to walk on the Moon.

Figure 2 The lunar roving vehicle during the Apollo 15 mission.

Figure 3 The Apollo 15 command module, *Endeavour,* orbits the Moon.

Figure 4 The Genesis Rock is a sample of original lunar crust.

DIG DEEPER

Did You Know?

Canadians in Space

In July 2009, Julie Payette flew on Space Mission STS-127, her second space mission to the ISS. As a flight engineer, Julie Payette was part of the cockpit crew that operated and docked the shuttle to the ISS. At the time of STS-127, Canadian astronaut Robert Thirsk was on a six-month stay on the ISS. This was the first time two Canadians were in space at the same time.

The Space Shuttle Program

In the early 1970s, after the Apollo program ended, the United States built a mostly reusable spacecraft called a space shuttle. The Space Transportation System (STS), also called the Space Shuttle program, launched with the Space Shuttle *Columbia* (STS-1) in April 1981. Shuttles launch from Cape Canaveral, on Florida's east coast.

The International Space Station

The International Space Station (ISS) is the ninth space research facility to be built in orbit. Since its launch in 1998, more than 16 countries have participated in its construction and use (**Figure 5**). The ISS weighs 245 735 kg, and it is the largest human-made object to orbit Earth.

Figure 5 Astronaut Dave Williams represents the Canadian Space Agency during a spacewalk to continue the construction of the ISS in 2007. Williams is anchored to the foot restraint on the Canadarm2 robotic arm.

DIG DEEPER

Reading Tip

Drawing Conclusions

After reading a text, you might need to do some research to help you synthesize and draw conclusions. For example, if you wonder how much a space shuttle weighs, you might find the information and calculate the total weight to be 2050 metric tonnes. You might then conclude that future spacecraft carrying humans to Mars will need to be launched from space.

Future Space Missions

Presently, human space exploration is limited to low orbit missions. Future space missions may include further trips to the Moon, as well as building lunar bases.

Mission planners are also discussing a long-term mission to Mars. This mission would be one of the greatest challenges in space exploration and full of potential hazards. Despite the challenges, the exploration of Mars will provide us with more clues to further our understanding of the Universe, as it is believed that Mars is the most Earth-like planet. It will also give us opportunities to develop new technologies and satisfy our desire to explore. Canada's plans include designing and producing new lunar and Martian rovers for use in these upcoming missions.

RESEARCH THIS TRIP TO THE MOON

SKILLS HANDBOOK 3.B.

SKILLS: Analyzing, Communicating

You are part of a group planning a trip to the Moon. Your team has four members. You will stay on the Moon for three weeks.

1. Investigate the things you must consider before the trip. For example, how much food will you need? Are there restrictions on the types of food you can bring? Explain the decisions you make.
2. Create a checklist of considerations.

A. What obstacles, if any, might you need to overcome? Explain. T/I

B. During the Apollo missions, a *Saturn V* rocket was needed to place a lunar module on the Moon. On this trip, you are staying longer and bringing far more material with you. What technologies will be necessary to launch a spacecraft with much more cargo? T/I

10.2 Wrap Up

- There are many limitations to space travel, such as those related to food, air, and waste management.
- Human space travel is presently limited to low Earth orbit missions.
- The Apollo program sent six missions that landed on the Moon.
- A space shuttle is a reusable spacecraft that carries astronauts and supplies to and from the International Space Station (ISS).
- The ISS is an orbiting laboratory where astronauts from 16 nations work and do research.

CHECK YOUR LEARNING

1. Explain how the distance of a celestial object affects the ability of a human to travel to that celestial object. K/U
2. How might a future mission to Mars better explain our understanding of our own planet? Explain your reasoning. K/U T/I
3. Why do you think low orbit missions are the most common type of space missions? Provide at least two reasons for your thinking. T/I A
4. List some benefits of future human missions to space. A
5. There are many challenges to space exploration. Rank the following criteria from most important (1) to least important (6). Explain your reasoning. T/I
   - constant supply of water
   - protection from harsh conditions
   - constant supply of oxygen
   - constant supply of food
   - dealing with feelings of claustrophobia
   - dealing with feelings of isolation
6. (a) Would you like to become an astronaut? Explain why or why not.
   (b) What do you think an astronaut's training is like? C A

10.3 Humans in Space

One of the biggest challenges of space travel is the feeling of weightlessness that astronauts experience. In space, astronauts may experience a range of health issues. They also face hazards that do not exist anywhere on Earth.

A Feeling of Weightlessness

Have you ever been on a roller coaster and at one point in the ride felt that you were off your seat? For a moment, you may have experienced the feeling of weightlessness.

Some people think that weightlessness in orbit is caused by lack of gravity. That is not true—there is gravity. The force of gravity 100 km from Earth's surface is 97 % of the force of gravity at Earth's surface. The feeling of weightlessness occurs because orbiting objects are constantly falling toward Earth. Astronauts feel weightless because they are falling toward Earth at the same rate as their spacecraft. This explains why astronauts in the space shuttle or space station appear to be floating. Both the astronauts and the spacecraft are travelling forward, but they are also falling at the same time (**Figure 1**).

The feeling of weightlessness in spacecraft affects the human body in several ways. Together, the effects are known as Space Adaptation Syndrome (SAS).

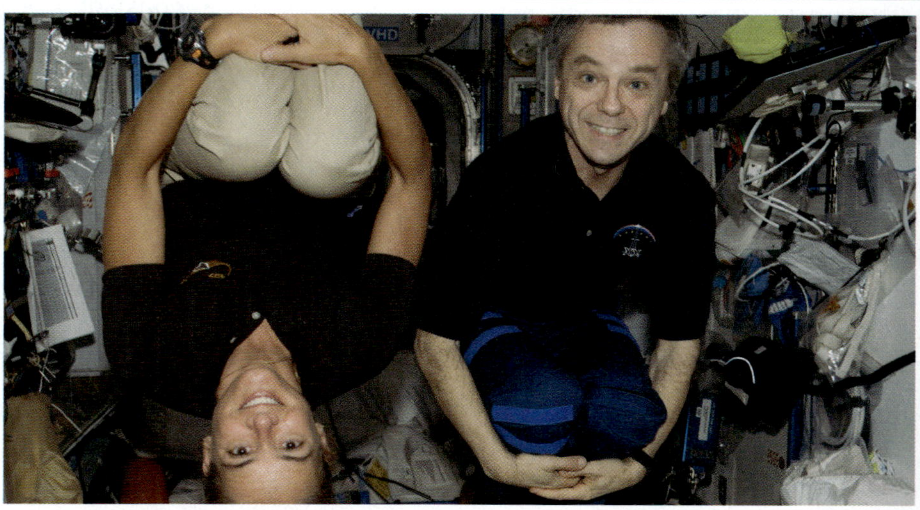

Figure 1 Julie Payette and Robert Thirsk float in the Destiny laboratory of the International Space Station.

Dizziness

About two-thirds of all astronauts experience dizziness, disorientation, and nausea during their first few days in space because they are constantly falling. The important tasks during a space mission, such as spacewalks and linking up with the ISS, often do not take place during the first few days in space. This way, the astronauts have time to adapt to space conditions.

Puffy Face and Skinny Legs

On Earth, blood and other body fluids tend to distribute themselves evenly throughout the body. When astronauts are in orbit, these fluids collect in the upper part of their bodies. This occurs because of the constant falling toward Earth the body experiences when it is in orbit. This causes puffy face, stuffy nose, and headaches. It also causes their legs to become skinnier than their upper bodies, as there is less blood pooled there (**Figure 2**). Space travel also affects the body's ability to use water properly and can cause dehydration.

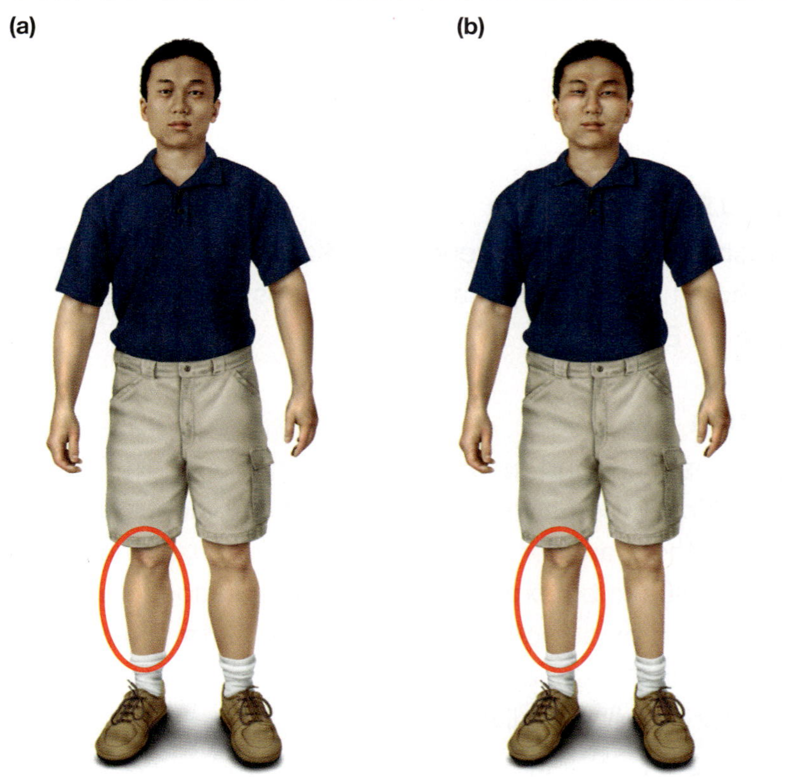

Figure 2 (a) Person on Earth (b) The same person in space. This effect is sometimes called the "puffy-face, bird-leg look."

DIG DEEPER

Did You Know?

Seasickness in Space

Researchers hypothesize that balance organs in the inner ear do not function properly in space where there is less exposure to the force of gravity. This produces a sensation similar to seasickness on Earth.

Figure 3 Canadian astronaut Robert Thirsk stayed on the ISS for 6 months, from May 27, 2009, to December 1, 2009. Thirsk, age 55, was monitored closely so that scientists could learn more about the endurance of the human body in space.

Bone and Muscle Loss

Astronauts can lose up to 2 % of the mass of their bones for each month they spend in space. This is because in an apparently weightless environment, their skeletal systems do not have to hold up their bodies. As a result, calcium and phosphorus, which are usually used for this purpose, are excreted instead of being retained in bones. This is not a very big problem on short missions of a week or two. However, astronauts who spend months at a time on the ISS have to be monitored carefully (**Figure 3**). The effect is similar to a bone loss condition known as osteoporosis.

While in space, muscles also weaken. To counter both bone loss and weakened muscles, astronauts must use a treadmill onboard the space station at least 2 h a day. Apparent weightlessness also causes astronauts to grow taller by 2 cm to 4 cm while in orbit. This is because gravity in space does not compress the backbone as much. Some astronauts (typically taller men) experience back pain when their backbone expands. Readapting to Earth takes about as long as the astronaut was in space. Some bone mass is probably lost for good.

Hazards in Space

While in space, dangerous conditions exist both inside and outside the spacecraft. Astronauts may face hazards that do not exist anywhere on Earth.

Radiation

Earth's atmosphere and magnetic field help protect us from the Sun's radiation. Low Earth orbit is within the magnetic field, so missions there are somewhat protected from radiation. But astronauts still can be exposed to harmful radiation. Shields on spacecraft provide some protection from harmful radiation. Spacesuits are also specially designed to protect astronauts from radiation during spacewalks.

Astronauts on missions outside of Earth's magnetic field—such as those to the Moon or farther—face even greater risks from the Sun's radiation. Mission planners must be able to monitor radiation levels and may have to consider cancelling missions when radiation levels are too high. If a base on the Moon is constructed, it would have to be almost completely shielded against radiation.

Equipment Failure

A spacecraft is an artificial environment that protects astronauts from the harsh conditions of space. This environment is operated by machines and computers. There are usually backup systems in case these fail. Equipment failure can cause life-threatening situations for crew members. If systems fail, there may not be enough air to breathe, food to eat, or water to drink.

Environmental System Failure

- The environmental system provides breathable air to astronauts. This system is critical to their survival.
- Returning to Earth from a low orbit mission involves many difficult tasks. If the environmental system were to fail during re-entry, astronauts would not be able to respond quickly or perform their tasks.
- Backup systems provide emergency oxygen in case of environmental system failure. The orange suits that astronauts wear during launch and re-entry have built-in survival gear (**Figure 4**).

Figure 4 These suits contain a helmet with communication gear and oxygen, as well as a parachute pack, life raft, and life preserver in case of a re-entry emergency.

Computer Failure

- Computer failure may result in an inability to control the spacecraft.
- The space shuttle has five computers to control the main engines and other systems. For critical functions, the computers "vote" to make sure that they are making the correct "decision." These computers form a backup system to ensure against any system failures.

Loss of Communication

- A spacecraft cannot operate without guidance from the Mission Control Center when it is time for re-entry. Re-entry is controlled by ground computers.
- During shuttle missions, Mission Control Center (**Figure 5**) updates the crew each morning with a "go home plan." If communication with ground control on Earth is lost, then the craft must return following this plan.

Figure 5 NASA Mission Control Center in Houston. Here, the flight director manages the flight, the safety of the crew, the experiments conducted on the mission, and all the people at Mission Control.

Illness of Crew

Astronauts have had the flu, head colds, and aches and pains during a mission. They are in direct contact with Mission Control and can request a private medical conference that will allow them to speak to a NASA medical doctor. Several astronauts have also been doctors themselves. On longer missions, astronauts have regular conferences with NASA doctors to maintain good health.

Space Junk

Human-made satellites may remain in orbit once they are no longer useful, perhaps for hundreds of years. As they deteriorate or collide with other objects, this space technology becomes waste, or **space junk**. Space junk can also be tools or other materials that astronauts accidentally lose as they work in space (**Figure 6**). For example, in November 2008, an astronaut working on the ISS let go of a backpack-sized toolbag. The orbiting toolbag can still be seen in the night sky. Even small items such as nuts and bolts travelling at a few kilometres per second could heavily damage a shuttle or a working satellite.

▶ **space junk:** wreckage from artificial objects orbiting Earth

Scientists are trying to determine how to get rid of space junk. In the meantime, space agencies and astronauts are making great efforts to minimize the amount of useless material left in space.

Figure 6 This drawing shows some of the many objects—from satellites to old rocket parts—orbiting Earth. Space junk is a threat for all spacecraft.

Working in Space

The ability to work safely in space is very important. Space has no oxygen, very little air pressure, extreme temperatures, and radiation. Astronauts wear pressurized spacesuits that provide them with oxygen and protection from lack of air pressure in space. The spacesuit also protects them from temperature extremes and small meteoroids or tools that are not secured.

Features of a Spacesuit and Helmet

The main purpose of the spacesuit and helmet is to protect astronauts while they are outside the spacecraft. Temperatures change very quickly between the sunlit side of a celestial object and the side in shade. The temperatures outside a spacecraft vary between 120 °C in sunlight and –100 °C in the shade. A spacesuit is insulated to protect astronauts from these extreme temperatures. A spacesuit must also be flexible so that an astronaut can perform tasks in a space environment (**Figure 7**). Imagine trying to perform a simple task like tightening a screw while wearing hockey gloves!

The helmet is made of a material that can withstand impacts, similar to helmets worn by athletes. **Figure 8** shows some of the features and safeguards of a helmet and spacesuit.

Figure 7 Spacesuit gloves must be flexible so that astronauts are able to carry out their tasks during missions.

DIG DEEPER
Web Link
To learn more about the spacesuits of astronauts,
GO TO NELSON SCIENCE

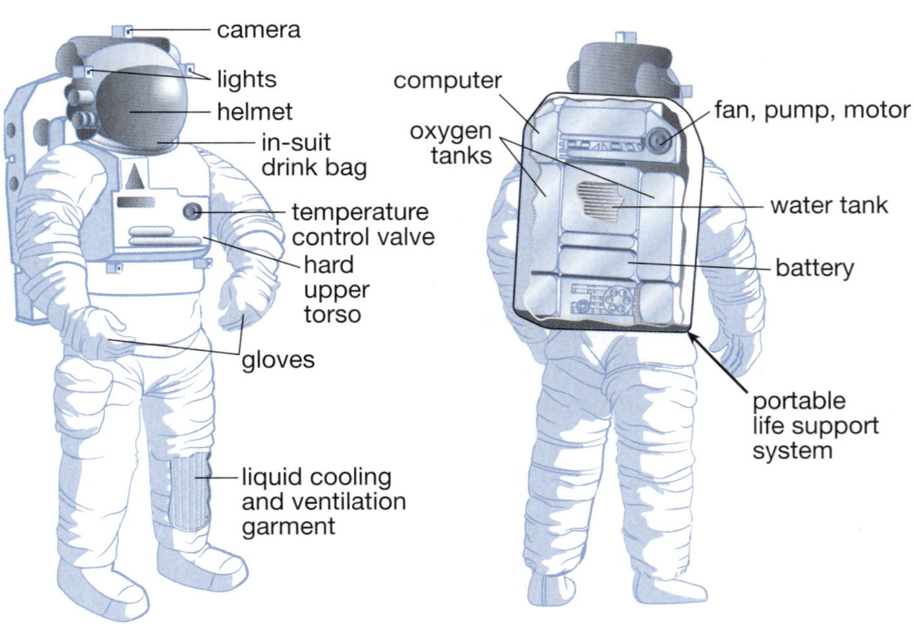

Figure 8 A spacesuit must be pressurized; supply oxygen; regulate temperature, so the astronaut does not get too hot or too cold; and offer reasonable protection against radiation.

Living in Space

Think about what you would need to pack for a camping trip. You would need to bring water, food, clothing, and something to sleep on, or in. Astronauts need to bring all of those items into space, plus one more important item—oxygen.

Life Support on the Space Station

The International Space Station's primary source of oxygen comes through electrolysis of water. In Chapter 7 you read that, in the process of electrolysis, electricity is used to break down water molecules into hydrogen molecules and oxygen molecules. On the ISS, electricity is generated using solar panels. The ISS also has solid oxygen in canisters as a backup supply. The space station recycles water in two ways:

- **Condensing water from humidity:** This is a process similar to what happens in the water cycle. The difference is that the humidity comes from the astronauts breathing and sweating.
- **Reclaiming and recycling wash water and urine:** First, urine is passed through a distillation process that separates the liquid and the gases. Then, along with other liquid wastes, the urine goes through several purification processes. It may sound surprising, but the purified water is cleaner than most of the water we drink on Earth.

Daily Living in Space

Astronauts in space have the same needs as people do on Earth. These are some of the ways astronauts adapt to life in space.

Eating

Many foods and beverages consumed on Earth are not permitted on spacecraft. For example, precautions are made to keep drinks from spilling. Spilled liquid might get into electrical equipment and damage it. Many of the foods astronauts eat are dehydrated (dried out) to preserve them. The food is then rehydrated (water is added back) and heated in microwave ovens. A typical meal could include beef stew, macaroni and cheese, pears, and a beverage.

Many of the freeze-dried foods like fruits or pasta we eat today were developed for travel in space. Take a look in your kitchen cabinets and refrigerator. You might be eating meals very similar to those astronauts eat in space!

Sleeping

Astronauts use sleeping bags that are attached to the walls, ceiling, or floor of the spacecraft to keep from floating into other areas (**Figure 9**). There is no feeling of what is up or down, so sleeping on the "ceiling" is the same as sleeping on the floor. Many crew members miss the feeling of resting their head on a pillow. The sheets and materials of the sleeping bag are not very soft, as they have to be made of a fireproof material.

Personal Hygiene

Because of the feeling of weightlessness in space, astronauts cannot use water to shower or bathe. Instead, they use rinseless shampoos that were originally developed for hospital patients. Rinseless shampoos are applied and then rubbed off, without using water. Astronauts use wipes, similar to those used for infants on Earth, to clean themselves. In space, everything floats. Special toilets work like vacuums with devices that direct urine and waste into a storage container.

Figure 9 This wall-mounted sleeping bag has arm restraints to keep arms from floating around while astronauts sleep.

TRY THIS WORKING IN SPACE

SKILLS HANDBOOK 3.B.

SKILLS: Observing, Analyzing, Evaluating, Communicating

In this activity, you will simulate what it is like to work in space.

Equipment and Materials: stopwatch; coat; helmet; gloves; a nut; a bolt

1. Work in pairs. While one person unscrews the nut from the bolt, the other person should use the stopwatch to measure how long this takes. Record the time in your notebook.
2. Have one person put on the coat, helmet, and gloves to simulate the weight of a spacesuit. Repeat Step 1, making sure to record the time in your notebook again.
3. Brainstorm three other tasks or activities that astronauts might carry out while in space. Take turns carrying out these tasks or activities with your simulated spacesuit.

A. Compare the time it took to perform each task with your bare hands with the time it took to perform the same task in your simulated spacesuit. C

B. Imagine you had to work like this for 7 hours with only water to drink, and no food. Write a journal entry about what the experience would be like. A C

C. How would you make sure that you did not lose anything such as tools or small bolts? T/I

D. How might you simulate a microgravity environment on Earth? T/I

E. How do you think the feeling of weightlessness would affect your ability to perform each task? A

10.3 Wrap Up

- The long-term effect of weightlessness in space can cause serious health problems for astronauts.
- Extreme environmental conditions in space require that astronauts take a life-sustaining environment with them.
- Equipment failure in space could cause loss of air for breathing, computer failure, and loss of communication.
- Spacesuits are constructed to protect astronauts from the extremes of the space environment and to enable them to perform their tasks.
- Astronauts must use special foods, sleeping bags, and personal hygiene items that are adapted for weightlessness.

UNIT TASK Bookmark

You can apply what you learned about working in space to the Unit Task described on p. 404.

CHECK YOUR LEARNING

1. What are three ways that space missions can affect the body? K/U
2. What are two potential hazards for astronauts during a spacewalk? K/U
3. What is space junk? Give two examples and describe why these materials pose hazards to an astronaut or spacecraft. K/U
4. Suppose you are on a two-week trip to the ISS. Write a letter to a friend that describes how you are adapting to daily routines such as eating, sleeping, and personal hygiene. A C
5. How is water recycling on the space station similar to the water cycle on Earth? K/U
6. What might humans learn by studying the health effects experienced by astronauts in space? A

10.4

Space Tools and Training

Before astronauts leave Earth on a space exploration mission, they undergo rigorous training that simulates what it is like to work in a space environment.

Apparent Weightlessness in Space

One of the biggest challenges while working in space is getting used to being in a **microgravity environment**, or in conditions where the effects of gravity are not felt, such as in orbit. The motion of the spacecraft in orbit creates a feeling of weightlessness. Astronaut crews need to train to work in a microgravity environment. To do this, astronauts perform tasks in water tanks on Earth that simulate microgravity conditions (**Figure 1**). This helps astronauts prepare for movement in both space flights and spacewalks.

▶ **microgravity environment:** an environment in which objects behave as though there is very little gravity affecting them

Figure 1 Astronauts learn how to do tasks by training in a water tank.

Space Tools

Tools used in space vary from cordless drills to robotic equipment. Tools that are used in and outside spacecraft must be tethered, or attached to something by a long cord. Tools must be very precise and suited for working in 100 °C environments with no air.

DIG DEEPER

Career Link

A specialized NASA job is to know exactly where everything is stored on the ISS. An equipment technician trains astronauts how to find what they need once they are aboard the ISS.

To learn more about equipment technicians,

GO TO NELSON SCIENCE

DIG DEEPER

Reading Tip

Making Connections

Ask yourself what you already know about a topic that relates to the information in a text. For example, you probably have some knowledge of the anatomy of the human arm. You can use this prior knowledge to help you understand the components of the Canadarm and how it works. You might draw a conclusion about how robots are designed to mimic functions that humans perform.

Canadarm

Since 1981, space shuttles have been equipped with the Canadarm (correctly called the Remote Manipulator System). Canadarm is a 15.2 m long robotic arm designed and built by Canadian engineers. It is like a construction crane attached to a shuttle. It lifts parts, captures and releases satellites, and assists astronauts in spacewalks.

In the harsh environment of space, this mechanical arm works just like a human arm. It has a "skin" that protects the internal parts of the arm from extreme temperatures. Its "nerves" are copper wiring. Its "bones" are graphite fibre. Its "muscles" are electric motors, and its "brain" is a computer. Canadarm also has six rotating joints that act like shoulders, elbows, and wrists that allow it to move loads in space.

Canadarm2

The Canadian-built Space Station Remote Manipulator System (SSRMS) is a more advanced version of Canadarm that can be anchored to the ISS. It is usually called Canadarm2. There is also a railway that moves along a length of the station that connects to the arm. This allows the arm to "walk" over the outside of the station. The arm is 18 m long and can manipulate 115 tonnes of equipment and transfer cargo from space shuttles to the ISS (**Figure 2**).

Figure 2 The astronaut shown here is anchored to the International Space Station's Canadarm2.

DEXTRE

Canada has also developed a robotic device that attaches to Canadarm2. This robot is aptly named DEXTRE (**Figure 3**). It is a two-armed robot used to do construction and repair work outside the ISS. It has movable jaws, multi-jointed arms, and video cameras. It even has a toolkit around its middle! DEXTRE can handle delicate assembly and repairs, reducing the number of spacewalks needed to repair and maintain the ISS.

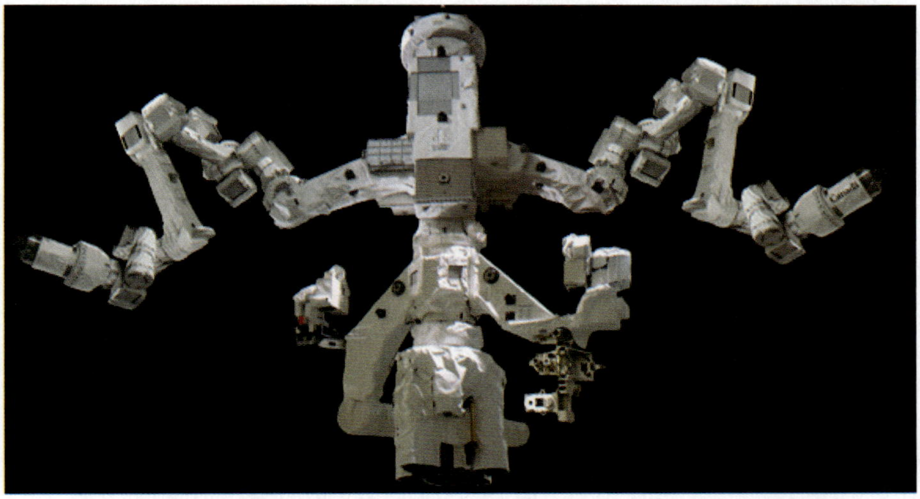

Figure 3 DEXTRE is another of Canada's robotic contributions to the ISS.

DIG DEEPER

Learning Tip

DEXTRE

DEXTRE is a nickname for "Special Purpose Dexterous Manipulator." The word "dexterous" means "careful and skillful use of the hands."

10.4 Wrap Up

- Astronauts must train to work in microgravity environments.
- Specialized space tools allow astronauts to work in space, launch satellites, and construct the ISS.
- Canadarm, Canadarm2, and DEXTRE are tools developed in Canada for space exploration.

CHECK YOUR LEARNING

1. (a) What kind of special training do astronauts need to work in space?
   (b) How is this different from training for any job on Earth? K/U
2. Describe how the Canadarm works. In which ways is the Canadarm like a human arm? K/U
3. How might robotic arms that are used in space benefit people on Earth? A
4. Describe what scientists mean when they refer to a microgravity environment. How might you experience what a microgravity environment feels like? K/U C

Microgravity Mysteries

Living things are uniquely adapted to the pull of Earth's gravity. Not surprisingly, humans are not adapted to the apparent weightless environment in space. Space crews aboard the International Space Station (ISS) often spend many months in space. In 2009, Canadian astronaut Robert Thirsk spent 186 days working aboard the ISS (**Figure 1**). In the future, astronauts may be sent on longer missions to explore more of the Solar System. It has become a priority for scientists to investigate the effects of the deep space environment on the human body. Canadian scientists are leaders in this type of medical research.

Boning Up

The apparent weightlessness in space speeds up the aging process on astronauts' bones, making them thin and brittle. When astronauts return to Earth after space missions, they often have bone conditions that mimic osteoporosis. Millennium Biologix, an Ontario-based company, along with the Canadian Space Agency, has been conducting groundbreaking experiments. One of these, known as the OSTEO mini-lab, tests the growth of cells using a synthetic bone biomaterial. Its purpose is to monitor and record the effects of space on bone loss. Spinoffs from this, and similar, projects are now being used on Earth to create bone-like material to heal fractures.

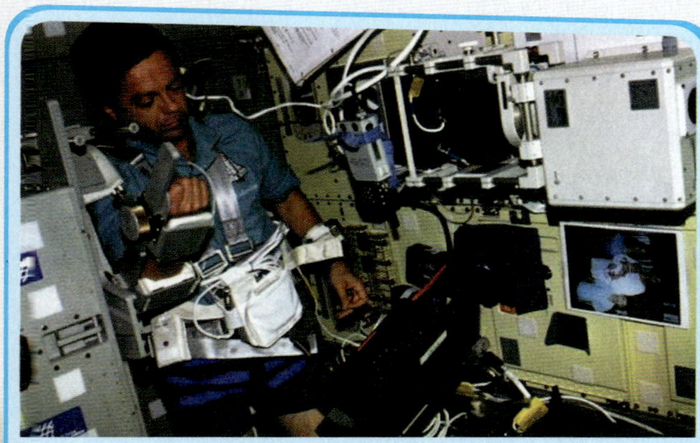

Figure 1 Astronaut Robert Thirsk measures changes in muscle forces on his bicep and tricep.

Up or Down?

Without the pull of gravity, astronauts in space have a hard time telling up from down. Being disoriented can be a matter of life and death in space—if an astronaut flips a switch the wrong way or loses their way in an emergency, the results could be fatal. A team of doctors and scientists at York University are currently testing the importance of visual signals in space. As part of these tests, astronauts aboard the ISS look at pictures on a laptop computer screen under a dark hood so that they cannot tell directions. Objects and letters on the pictures slowly begin to rotate and astronauts have to figure out which way is up or down. The results of this simple Canadian experiment will help in designing future spacecraft with better visual signs for orientation. Researchers are also working on direct medical spinoffs that they hope will help people on Earth who suffer from dizziness and balance problems.

10.5 Technology Transfer

You might be surprised to learn that there are many items developed for space technology that you encounter every day. Bar codes used to manage store items, beverage coolers, and cordless power tools were originally developed for use in space programs (**Figure 1**).

Figure 1 NASA developed a cordless drill for Apollo astronauts to drill core samples from beneath the lunar surface.

Innovative Technology

We use technologies originally developed for space exploration, or **space technology spinoffs**, in our homes, for recreation, and in the environment. These spinoffs have benefited manufacturing, medical sciences, and many other industries, including transportation, safety, and computer technology.

▶ **space technology spinoffs:** technologies originally designed for space exploration that have made their way into everyday use

Canadarm Robotic Technology

Canada's own Canadarm robotic technology has been used in the design of prosthetic, or artificial limbs. The robotic "hands" used in space technology have also been used to clean up radioactive and other harmful wastes. This space technology is also being adapted to conduct remote robotic surgery. "Telesurgery" consists of a robot that uses miniature tools to perform highly accurate surgeries (**Figure 2**).

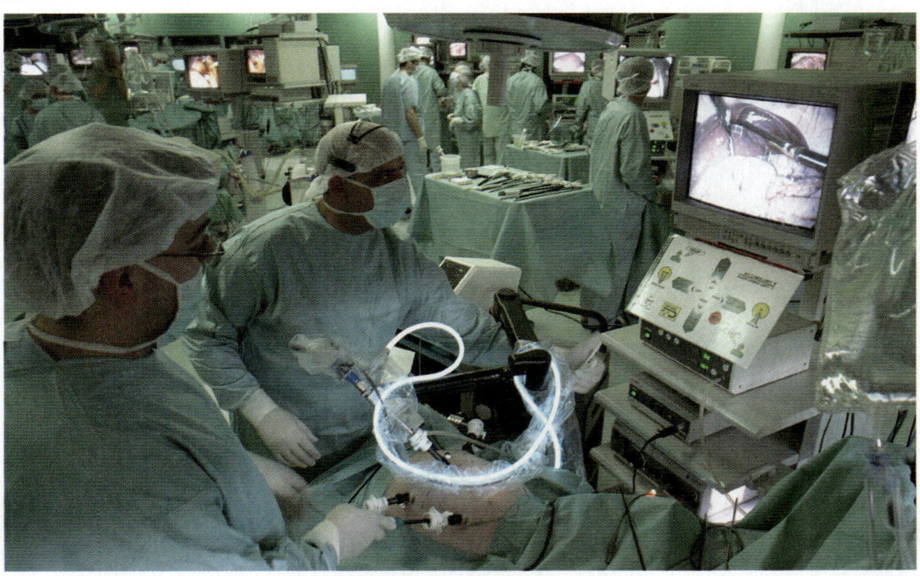

Figure 2 Doctors perform a new form of abdominal surgery on a patient where a large incision is not required. Many patients who undergo this type of surgery can go home after the procedure without spending the night at the hospital.

DIG DEEPER

Did You Know?

GPS Accuracy

Most GPS receivers can locate an object's position anywhere on Earth to within a few metres. Certain engineering and military professionals can locate objects as small as a few centimetres in diameter, or the size of a dollar coin.

From Maps to GPS Technology

In 1978, space exploration led to the development of the first global positioning system (GPS). This is a spinoff navigation system that uses satellites to determine its own location anywhere on Earth, at any time of the day. GPS is made up of three basic components: satellites that orbit Earth, monitoring stations on Earth, and GPS receivers located on Earth (**Figure 3** and **Figure 4**).

Figure 3 Handheld GPS receivers allow users to find directions anywhere on Earth.

Figure 4 Today, many automobiles come with GPS systems in the dashboard.

Today, GPS technology is available for the general public. A network of 24 GPS satellites orbits Earth twice daily. Three or more GPS satellites send a signal to a receiver on Earth, which calculates a location (**Figure 5**). GPS receivers provide directions and maps that help people locate their position on Earth.

Figure 5 The wing-like solar panels of a GPS satellite absorb radiation from the Sun, which is then stored in batteries.

TRY THIS SPINOFFS FROM A BALLOON ROCKET

SKILLS HANDBOOK 3.B.

SKILLS: Planning, Performing, Observing, Analyzing, Communicating

In this activity, you will make a balloon rocket and demonstrate how a spinoff might develop from it.

Equipment and Materials: long balloon; straw; cardboard pieces; cotton string; two chairs

1. Push the cotton string through the straw. Stretch the string across the room and tie each end to the back of a chair. Move the straw along the string so that it is close to one of the chairs. The straw represents a rocket.
2. Cut the cardboard to make 15 cm wings for your rocket. Tape the wings to the straw (Figure 6).
3. Blow up the balloon and tape it to the bottom of the straw. Keep the end of the balloon pinched closed (Figure 7).
4. When ready, release the balloon.

A. What properties does this simple rocket have in common with spacecraft you have read about in this unit? T/I

B. What are some practical uses you might consider for this rocket? Brainstorm with a partner some possible spinoffs for this rocket. T/I A

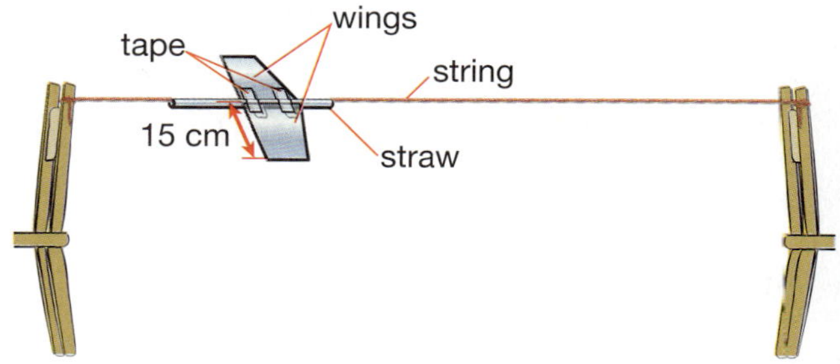

Figure 6 Step 2

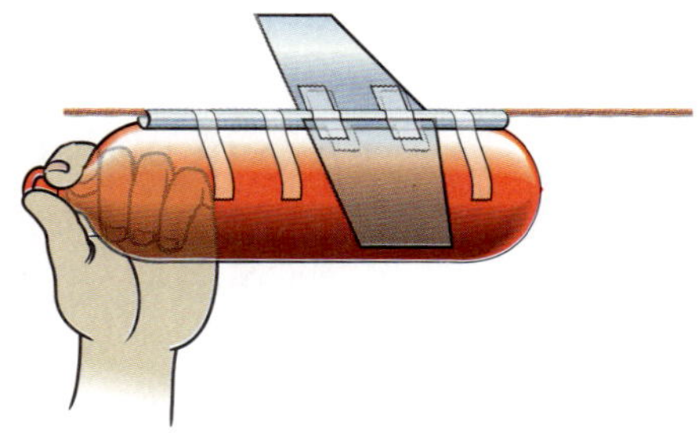

Figure 7 Step 3

10.5 Wrap Up

- Innovative space technology has many applications in industry.
- Global positioning systems (GPS) are navigation systems based on space and satellite technology.
- Robotic arm technology and surgery from remote consoles are other space technology spinoffs.

CHECK YOUR LEARNING

1. How do space spinoffs affect our daily lives? Give an example. K/U
2. How does the global positioning system (GPS) work? K/U
3. Why is GPS a valuable technology? A
4. What is the relationship between Canadarm and the new field of robotic technology? K/U

10.6 EXPLORE AN ISSUE CRITICALLY

Is Space Exploration Worth the Money?

SKILLS MENU

- Defining the Issue
- Researching
- Identifying Alternatives
- Analyzing the Issue
- Defending a Decision
- Communicating
- Evaluating

NASA's *Phoenix* Mars Lander was the sixth successful mission to Mars. The spacecraft was launched in August 2007 and landed on the red planet in May 2008. One main goal of the mission was to study the geologic history of water in order to better understand climate cycles. Another goal of the mission was to search for evidence of past or present biological life on the planet. The *Phoenix* Mars Lander was equipped with a Meteorological Station (MET), which was designed by Canadians (**Figure 1**).

Phoenix collected data on the surface of Mars for more than five months before losing power and shutting down. The MET created the first profile of an extraterrestrial atmosphere, and confirmed that Mars has ice (water), just below the surface—a sign that the planet might be able to support life.

Although the mission was successful, it was funded by taxpayers and cost Canadians more than $37 million. Some people wonder if this mission was worth the money.

DIG DEEPER

Reading Tip

Changing Your Perspective

When synthesizing a text, look for clues relating to what you already know about the topic. For example, if the text gives information about how long a spacecraft has been in operation, you may draw on your knowledge of the dangers of space travel, such as extreme temperatures and space junk, and conclude that the spacecraft has a good track record.

Figure 1 The meteorological station on the *Phoenix* Mars Lander was paid for and created by the CSA. Was it worth the money?

SKILLS HANDBOOK 4.C.

The Issue

In order to remain at the forefront of space exploration, the Canadian government must continue to spend money on research and development. Some people believe that this money would be better spent on other resources, such as health care and education.

Goal

To decide whether the Canadian government should continue to spend money on space exploration projects

Gather Information

Work in small groups. Research the ways the CSA has worked in partnership with Canadian companies, government agencies, and universities for different space projects. You may choose to focus on one of the following projects:

- SSRMS
- SCISAT-1 (**Figure 2**)
- International Space Station
- DEXTRE
- Canadarm

As you conduct your research, consider the following questions:

- How has the project benefited society in general?
- What negative effects has the project had on society? Could there be any negative effects in the future?
- How much has been spent by the CSA on this type of research? Where else could this money be spent?

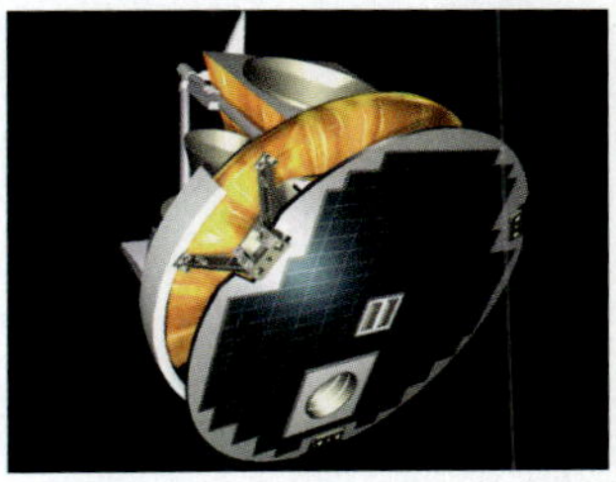

Figure 2 As it orbits Earth, SCISAT-1 measures and studies the distribution of ozone in Earth's atmosphere.

Make a Decision

Evaluate the evidence and decide what recommendation you will make to the Canadian government.

Communicate

Present your finding to the class. You may create a pamphlet, website, written report, or electronic presentation to explain the benefits and drawbacks of spending money on space exploration projects. Be sure to support your position using facts.

Chapter 10 KEY CONCEPTS SUMMARY

Getting into space and staying in orbit are major challenges of space travel.

- To escape Earth's atmosphere, a spacecraft needs lots of fuel to attain a high velocity.
- For a spacecraft to achieve orbit, its forward speed must perfectly balance the pull of gravity.

Robotic space probes have explored the Solar System and beyond.

- NASA has completed 11 missions to the Moon and has landed astronauts on the Moon 6 times.
- Missions to Mars have revealed that the planet may be able to support life.

Space can be a hostile environment for humans.

- Astronauts need a constant supply of food, water, and air while travelling in space.
- Many of the challenges to living in space have been solved by technology.

Living and working in space presents challenges.

- Special tools have been developed for space exploration.
- Due to microgravity, special arrangements must be made in space for sleeping, eating, and personal hygiene.

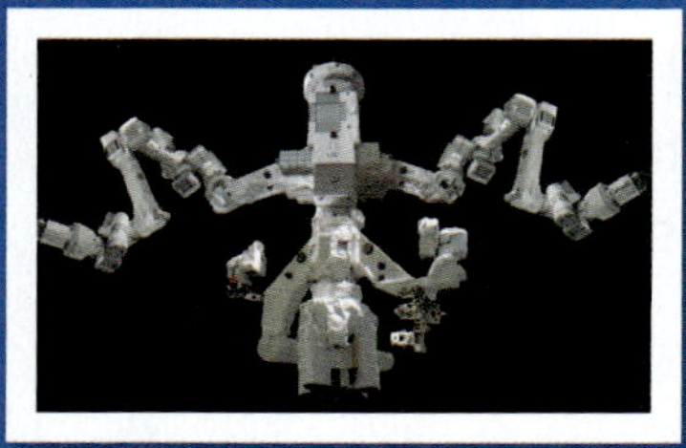

Space exploration is technically challenging and expensive.

- The development of space tools allows astronauts to perform complex tasks in space.
- NASA and the CSA have made many technological developments to further their space programs.

Space technologies have benefited our lives.

- Space technologies have led to advances in many technologies on Earth, known as space spinoffs.
- GPS technology involves low orbit satellites and receivers and monitoring stations on Earth to find locations anywhere on Earth.

WHAT DO YOU

THINK NOW?

You thought about the following statements at the beginning of the chapter. You may have encountered these ideas in school, at home, or in the world around you. Consider them again and decide whether you agree or disagree with each statement.

1 Humans have landed on all the planets.
Agree/disagree?

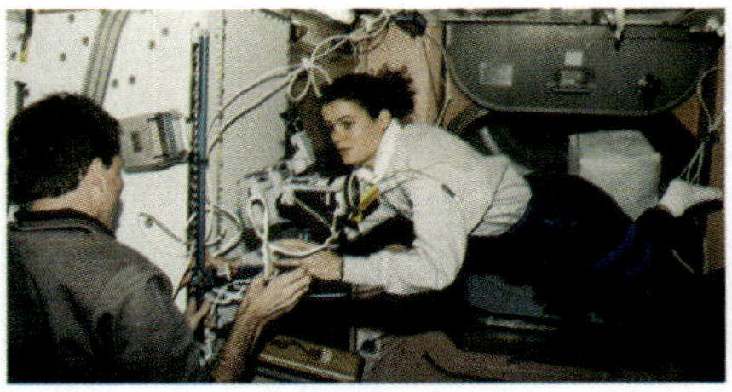

2 Astronauts float in space because there is no gravity.
Agree/disagree?

3 Computers control the launch of spacecraft.
Agree/disagree?

4 Astronauts in orbit can get swollen faces.
Agree/disagree?

5 Space travel poses hazards for astronauts.
Agree/disagree?

6 Canada has contributed to space exploration.
Agree/disagree?

How have your answers changed since then? What new understanding do you have?

BIG Idea

- Technologies developed for space exploration have practical applications on Earth.

DIG DEEPER

Vocabulary

spacecraft p. 368
International Space Station (ISS) p. 369
thrust p. 370
rocket booster p. 370
command module p. 375
space junk p. 382
microgravity environment p. 387
space technology spinoffs p. 391

CHAPTER 10

SELF-QUIZ

The following icons indicate the Achievement Chart category addressed by each question.

K/U Knowledge/Understanding
T/I Thinking/Investigation
C Communication
A Application

For each question, select the best answer from the four alternatives.

1. Which statement accurately describes how a spacecraft maintains a low Earth orbit in space? (10.1) K/U
   (a) It must travel at 7.5 km/s.
   (b) It must stay within Earth's atmosphere.
   (c) It must have a balance between its forward speed and the pull of gravity.
   (d) It must contain a balance of air, water, and oxygen.

2. Canadarm, Canadarm2, and DEXTRE are
   (a) Canadian-built space shuttles
   (b) names of lunar rovers
   (c) types of global positioning systems
   (d) robotic tools used on space shuttles and the ISS (10.4) K/U

3. Which statement accurately describes the relationship between a spacecraft and a space probe? (10.1) K/U
   (a) A spacecraft is a reusable space probe.
   (b) A spacecraft carries cargo on a space probe.
   (c) A space probe is a type of robotic spacecraft.
   (d) A space probe is a human-occupied spacecraft.

4. Which of the following pose a risk for astronauts on space missions? (10.3, 10.4) K/U
   (a) space junk
   (b) extreme temperatures
   (c) effects of a microgravity environment
   (d) all of the above

Indicate whether each of the statements is TRUE or FALSE. If you think the statement is false, rewrite it to make it true.

5. Space technologies are expensive, but have been adapted for use in everyday life. (10.5) K/U

6. Fuel used to launch the space shuttle is carried in the cargo area of the shuttle. (10.1) K/U

7. Some problems of long-term space travel involve choice of foods, the environment inside the spacecraft, and the speed of the craft. (10.2) K/U

8. The *Phoenix* Mars Lander has provided evidence as to the possible existence of life on Mars. K/U

Copy each of the following statements into your notebook. Fill in the blanks with a word or phrase that correctly completes the sentence.

9. Two places where astronauts have lived and worked in space include the Moon and ________. (10.2) K/U

10. Astronauts train in tanks of water to experience what ________ conditions will feel like in space. (10.4) K/U

11. ________ is the force needed to lift a spacecraft beyond Earth's atmosphere. (10.1) K/U

12. ________ are human-occupied or robotic vehicles used to explore space or celestial objects. (10.1) K/U

Match each term on the left with the most appropriate description on the right.

13. (a) GPS — (i) a spacecraft that orbits Earth
    (b) DEXTRE — (ii) a two-armed robotic device
    (c) SSRMS — (iii) a navigation system of satellites and Earth-based receivers
    (d) ISS — (iv) an arm that anchors itself to the International Space Station
    (10.4, 10.5) K/U

Write a short answer to each of these questions.

14. Why do astronauts on the space shuttle appear to float? (10.1, 10.3) K/U
15. How is the International Space Station an example of international co-operation regarding space travel? (10.2) K/U
16. In what ways has space exploration benefited humankind? (10.5) A
17. Describe Canada's contributions to space exploration. (10.3, 10.4, 10.5, 10.6) K/U A
18. Explain why microgravity gives many astronauts a puffy face. (10.3, 10.4) K/U T/I
19. Imagine you are an astronaut living on the International Space Station. Write a journal entry describing your activities in a typical day. (10.3, 10.4) C
20. Identify three challenges that need to be addressed before humans can build colonies on the Moon. (10.2, 10.3, 10.4) A
21. What are three benefits and three drawbacks of space exploration? (10.2, 10.3, 10.4, 10.5) A
22. What can astronauts who experience bone loss while in flight do to prevent their bones from weakening? (10.3) K/U
23. Why is it important for astronauts to exercise while on space missions (**Figure 1**)? (10.3) K/U

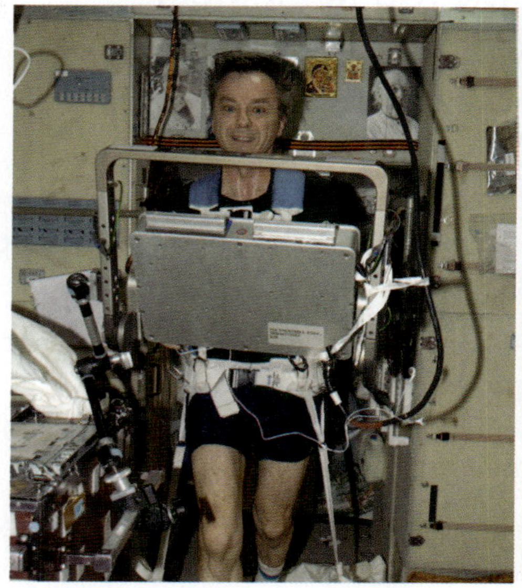

Figure 1

24. Canada is a world leader in developing technology for space exploration and research in space. In a paragraph, describe the importance of robotics in space exploration. (10.4) A C
25. What are three objects people might find in their homes that use technology originally developed for the space program? Identify each object and describe it. (10.5) K/U
26. (a) List three problems astronauts face when living in space.
    (b) Choose one problem that you have listed and describe the solution scientists have developed for the problem. (10.3) K/U

Self-Quiz

To do an online self-quiz, GO TO NELSON SCIENCE

CHAPTER 10

REVIEW

The following icons indicate the Achievement Chart category addressed by each question.

K/U Knowledge/Understanding | T/I Thinking/Investigation | C Communication | A Application

What Do You Remember?

1. (a) What is space junk?
   (b) Why is it a problem for space exploration? (10.3) K/U
2. (a) What does "GPS" stand for?
   (b) How does this technology help people? (10.5) K/U
3. (a) What was the goal of the Apollo space program in the 1960s?
   (b) In what way was it successful in meeting that goal? (10.2) K/U
4. What does a pressurized spacesuit protect astronauts against while they are working outside a spacecraft? (10.3) K/U
5. (a) What is Canadarm?
   (b) Why is it important to astronauts working on the space shuttle? (10.4) K/U
6. Describe three health problems astronauts may experience while working aboard the ISS. (10.3) K/U

What Do You Understand?

7. One of Canada's greatest contributions to research in outer space is studying the disease osteoporosis. (10.3) K/U T/I
   (a) What part of the body does osteoporosis affect?
   (b) Why is outer space a good place to study it?
8. What was the function of the *Saturn V* rocket? (10.2) K/U
9. How does a spacecraft achieve orbit, and why is it difficult? (10.1) K/U
10. Explain how an object is able to maintain a low Earth orbit. (10.1, 10.2) K/U C
11. Why is human space exploration limited to short distances in space? (10.2, 10.3) K/U
12. Why is it so important to conserve oxygen on a spacecraft? (10.2, 10.3) K/U
13. Imagine you are a scientist planning a future mission to Mars. Suggest reasons why you would want to use the Moon as a base for launching your rocket to Mars. (10.2) K/U
14. You have seen photos of small objects floating inside the International Space Station. Some people say these objects are weightless. Does this mean the objects have no weight? Explain your answer. (10.3) K/U C

Solve a Problem

15. Imagine you are hired to clean up a toxic waste spill in northern Ontario. What space technology spinoff could you use to help you get the job done? (10.5) A
16. NASA is preparing for its next mission to Mars and needs to find a place on Earth to train astronauts who are going to be working on the Martian surface. Suggest a place that might be similar to Mars, and explain your reasoning. (10.2, 10.3) C A

Create and Evaluate

17. Sending human beings into space is much more expensive than sending robots into space. Debate the pros and cons of human space travel with your classmates. Which side do you support? Why? Be sure to back up your opinion with facts. (10.1, 10.2, 10.3, 10.4, 10.5, 10.6) C A

18. Suggest an experiment to be performed aboard the ISS. Be sure to explain why you think this experiment is important and what you hope to achieve. (10.1, 10.2, 10.3) C A

19. Write a letter to one of the astronauts mentioned in this chapter asking some specific questions about what his or her space experience was like (**Figure 1**). (10.1, 10.2, 10.3, 10.4) C

Figure 1

Reflect on Your Learning

20. Write a letter to a family member as if you were an astronaut exploring the Moon during an Apollo space mission. Describe your surroundings and how being on the Moon makes you feel. K/U C

21. What information from this chapter did you already know before you studied the content? How has your knowledge of this topic changed? C

22. Name something that you learned in this chapter that changed the way you think about space or space exploration. Explain how it changed your thinking or understanding. C

23. Which topic in this chapter did you find particularly difficult to understand? Come up with two strategies to help you understand this topic before the next test or exam. C

Web Connections

24. A group of scientists supports the idea of human beings living and working in a colony on the Moon. Do you think humans should live on the Moon? Support your opinion with facts from research. T/I A

25. Research the characteristics and design of spacesuits. Find out how much they cost to develop, and describe ways that they might be used on Earth. Summarize your findings in a brochure or poster. T/I C A

26. Research Canadarm, SSRMS, and DEXTRE. Make a poster or chart that shows one of these tools, how it works, and any possible practical uses that have occurred from its use. T/I C A

GO TO NELSON SCIENCE

Space Exploration

CHAPTER 8

Touring the Night Sky

KEY CONCEPTS

The Solar System includes the planets and other celestial objects that orbit the Sun.

The Universe includes stars, planets, moons, and galaxies.

The Moon is the closest celestial object to Earth.

A star is a giant mass of gases that emit light and huge amounts of energy,

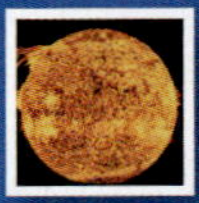
The Sun is our star.

Galaxies contain billions of stars, planets, and other celestial objects.

CHAPTER 9

Our Place in Space

KEY CONCEPTS

The Solar System consists of all the celestial objects orbiting the Sun.

The motion of Earth causes day, night, and the seasons.

Earth's motion affects our view of the night sky.

Celestial objects can be used to predict annual events and to navigate.

The positions of the Sun, Earth, and the Moon result in the phases of the Moon and eclipses.

Scientists have strong evidence that the Universe began around 14 billion years ago.

CHAPTER 10

Space Research and Exploration

KEY CONCEPTS

Getting into space and staying in orbit are major challenges of space travel.

Robotic space probes have explored the Solar System and beyond.

Space can be a hostile environment for humans.

Living and working in space presents challenges.

Space exploration is technically challenging and expensive.

Space technologies have benefited our lives.

MAKE A SUMMARY

Websites such as Wikipedia combine the knowledge of many people to create definitions and articles about different topics. However, Wikipedia can be edited by anyone, so you must cross-reference any information you find there to make sure it is accurate.

In this activity, you will use the Wikipedia model to develop definitions of concepts learned in this unit.

1. You will work in small groups. Each group will be given a coloured marker.
2. Your teacher has placed chart paper at stations around the room. A different topic is written on each piece of paper.
3. At your first station, write any information you know about that particular topic, including key words, drawings, definitions, and websites.
4. After a few minutes, move on to the next station. Take your coloured marker with you so you can recognize your contributions.
5. Read the information written by other groups. Add any new information you can. If you wish, you can even edit the existing content.
6. When you have visited each station and made comments, post the papers around the room. With your classmates, you have now summarized a unit's worth of material in useful categories.

Questions

1. (a) Look at the summaries posted around the room. Which topics did you find the most interesting?
   (b) Which topics did you find the most challenging?
2. Make a list of the top 10 words whose definitions you want to learn.
3. Which topics are best understood using pictures?

CAREER LINKS

List the careers mentioned in this unit. Choose two of the careers that interest you, or choose two other careers that relate to Space Exploration. For each of these careers, research the following information:

- educational requirements (secondary and post-secondary)
- skill/personality/aptitude requirements
- potential employers
- salary
- duties/responsibilities

Assemble the information you have discovered about each career into a brochure. Your brochure should compare and contrast your two chosen careers, and explain how they connect to space exploration.

GO TO NELSON SCIENCE

Astronaut Training Test

SKILLS MENU

- Questioning
- Hypothesizing
- Predicting
- Planning
- Controlling Variables
- Performing
- Observing
- Analyzing
- Evaluating
- Communicating

You are training to become an astronaut for a long-term trip to Mars. In this activity, you will simulate some of the stress placed on an astronaut during the long journey to Mars. You will use the results of the tests as criteria for recommending certain people as astronaut candidates.

SKILLS HANDBOOK 3.B.

Testable Question

How do tests measure the ability of astronauts to live in space?

Hypothesis

Read each astronaut training task. Then predict the outcome for each one. Explain how you made your prediction.

Procedure

Work with a partner and switch tasks for all these experiments. Use the stopwatch to time each experiment.

1. **Dexterity:** You will be given a set of 20 numbered tiles randomly organized and a numbered "control panel." Enter the numbers on the control panel, placing the tiles on the corresponding square. Repeat this activity 10 times with different sets of 20 random numbers.

Equipment and Materials

numbered "control panel"

stopwatch

mirror

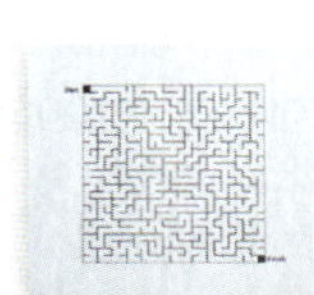

paper maze

deck of cards

2. **Disorientation Recovery:** Place a blindfold over your eyes and spin around three times. Take the blindfold off and repeat Step 1 using another set of 20 random numbers.
3. **Performing Difficult Tasks:** While looking at the image of a maze in a mirror, use a pencil to trace a route through the maze. Make sure that you cannot see your hands while you do this.
4. **Decision-Making Speed:** Shuffle a deck of cards. Deal the cards face up into a pile as fast as you can. Record the time it takes to complete the task. Continue the activity, sorting the cards as follows: by colour (red and black), by suit (hearts, clubs, spades, diamonds), by even and odd numbers (remove the face cards to do this). Record the time it takes for each task.
5. Make up your own task. Perform the task and evaluate the results.

Analyze and Evaluate

(a) How did your dexterity change each time you entered numbers on the numbered control panel? Explain your results. T/I C

(b) In Step 2, how did becoming disoriented affect the time it took to enter the numbers? C

(c) Describe what it was like to work through the maze using a mirror. Were you ever frustrated? C

(d) Look at the results of the Decision-Making Speed test. How did the decision-making affect the speed at which you dealt the cards? T/I C

(e) Analyze the results of the task you constructed in Step 5. T/I

Apply and Extend

(f) How do these tasks relate to astronauts living in space? A

(g) Who might be suitable astronaut candidates based on the results of these tests? Explain your reasoning. A C

CHECKLIST

Assessment

Your completed Performance Task will be assessed according to the following criteria:

KNOWLEDGE/ UNDERSTANDING

- Demonstrate an understanding of the requirements of space travel and training.

THINKING/ INVESTIGATION

- Conduct each test and record the results. Share your results with the rest of the class.
- Analyze the results of each classmate.

COMMUNICATION

- Prepare an oral or written report that includes a summary of the results and recommendations for astronaut candidates. Support your choices with the results of the tests.

APPLICATION

- Compare the results of the tests. What criteria did you use to determine astronaut candidates? How might these tests apply to actual events in space?

UNIT D

SELF-QUIZ

The following icons indicate the Achievement Chart category addressed by each question.

K/U Knowledge/Understanding
C Communication
T/I Thinking/Investigation
A Application

For each question, select the best answer from the four alternatives.

1. Which of the following is a function of a telescope? (8.2) K/U
   (a) magnifies an object
   (b) collects more light
   (c) increases clarity
   (d) all of the above
2. Which of the following is a distinguishing feature of the Moon? (8.3) K/U
   (a) It is a gaseous celestial body.
   (b) It has a smooth surface.
   (c) It has craters and maria.
   (d) It has an atmosphere similar to Earth's.
3. In which layer of the Sun do nuclear reactions take place, releasing immense amounts of energy? (8.5) K/U
   (a) photosphere
   (b) radiative zone
   (c) core
   (d) convection zone
4. What unit do scientists use to measure the distances between planets in the Solar System? (9.1) K/U
   (a) light year
   (b) kilometres
   (c) astronomical units
   (d) solar units
5. What does a spacecraft's rocket engine need to do in order to successfully launch into space? (10.1) K/U
   (a) travel at 7.5 km/s in the appropriate direction
   (b) disconnect when it leaves Earth
   (c) burn all the fuel in its system
   (d) maintain its speed for 2 h
6. The distance between galaxies is ________. (9.8) K/U
   (a) shrinking
   (b) expanding
   (c) constant
   (d) random
7. What is the main challenge in planning trips to Mars? (10.2, 10.3) T/I
   (a) how to practise in a Mars-like environment on Earth
   (b) how to keep the crew alive and healthy for the duration
   (c) how to escape the pull of Earth's gravity
   (d) how to communicate with Mission Control on Earth
8. What are some of the differences between working on Earth and working in a spacecraft orbiting Earth? (10.3, 10.4) K/U
   (a) In space, there is zero gravity.
   (b) In space, there is a feeling of weightlessness.
   (c) On Earth, you feel gravity less than in space.
   (d) There are no differences.

Indicate whether each of the statements is TRUE or FALSE. If you think the statement is false, rewrite it to make it true.

9. You can see some of the planets in the Solar System from Earth unaided by any devices. (8.1) K/U
10. The Moon is considered a satellite of Earth. (8.3) K/U
11. There is a distinct line between day and night, each day, everywhere on Earth. (9.1) K/U
12. Hot stars appear reddish and cool stars appear bluish. (8.5) K/U

13. The shift in light emitted by galaxies toward the red end of the spectrum led astronomers to believe that galaxies are moving away from each other. (9.8) K/U

14. The Sun produces energy through processes called nuclear reactions. (8.6) K/U

15. Robotic technology has allowed people to find locations anywhere on Earth. (10.5) K/U

16. Astronomers use the kilometre to measure distances within the Solar System. (9.1) K/U

17. Astronauts encounter risks when they are in space, including bone loss and changes in blood pressure. (10.3) K/U

18. The Moon was formed from an expansion that occurred about 14 billion years ago. (9.8) K/U

19. Pluto is no longer considered a planet because it does not orbit the Sun. (8.4) K/U

Copy each of the following statements into your notebook. Fill in the blanks with a word or phrase that correctly completes the sentence.

20. _________ are groups of stars or other celestial objects that may form a pattern within a constellation. (9.3) K/U

21. If the new Moon is directly between Earth and the Sun, you will see a(n) _________. (9.7) K/U

22. A _________ environment is an environment in which objects behave as though there is very little gravity affecting them. (10.4) K/U

23. A _________ telescope uses mirrors to collect and focus light from objects in space (8.2) K/U

24. A _________ is an enormous collection of billions of stars. (8.1) K/U

25. _________ are relatively cool regions on the Sun's surface (8.6) K/U

Write a short answer to each of these questions.

26. What causes winter to occur in the northern hemisphere on Earth? (9.2) K/U

27. Name at least one advantage and one disadvantage of placing a telescope in space (**Figure 1**). (8.2) A

Figure 1 The Hubble Space Telescope

28. What are two main differences between the inner and outer planets? (8.4) K/U

29. Define "galaxy." Give one example. (8.8) K/U

30. Why are the paths of distant planets different when seen from Earth? (8.1) K/U

31. What is the difference between Earth's rotation and its revolution? (9.1) K/U

32. Explain what Hubble discovered about the spectrum of different galaxies. (9.8) K/U C

33. Explain the Big Bang theory. (9.8) K/U C

34. Explain some of the issues that one faces if planning an extended voyage to the Moon. (10.2, 10.3) C A

The following icons indicate the Achievement Chart category addressed by each question.

K/U Knowledge/Understanding T/I Thinking/Investigation
C Communication A Application

What Do You Remember?

For each question, select the best answer from the four alternatives.

1. Which colour best describes a star that has the hottest temperature? (8.5) K/U
   (a) red
   (b) blue
   (c) yellow
   (d) green
2. Which of the following galaxies is shaped like a pinwheel? (8.8) K/U
   (a) elliptical
   (b) spiral
   (c) lenticular
   (d) irregular
3. Which of the following is a huge explosion of gas that blasts gas and particles from the Sun's surface into space? (8.6) K/U
   (a) solar flare
   (b) sunspot
   (c) solar wind
   (d) aurora
4. Which is an example of a Canadian contribution to space exploration? (10.4, 10.6) K/U
   (a) meteorological station on Mars *Phoenix*
   (b) DEXTRE
   (c) Canadarm2
   (d) International Space Station
5. Constellations in the night sky appear to move across the sky because
   (a) Earth is rotating on its axis
   (b) the stars move together through space
   (c) the Moon's gravity pulls on the stars
   (d) Earth is changing its distance from the Sun (9.1, 9.3) K/U
6. Which statement accurately describes a light year? (8.5) K/U
   (a) The distance light travels in one year.
   (b) The distance light travels between Earth and the Sun.
   (c) The time it takes for light to travel 10 trillion km.
   (d) The time it takes for light to travel from Earth to the Moon.
7. Scientists use light spectrums to determine if a celestial object is
   (a) growing larger or smaller than it appears
   (b) evolving into a different type of star
   (c) moving toward or away from an observer
   (d) none of the above (9.9) K/U

Indicate whether each of the statements is TRUE or FALSE. If you think the statement is false, rewrite it to make it true.

8. You can usually see many galaxies from Earth unaided by any devices. (8.1) K/U
9. A star's temperature and its colour are related. (8.5) K/U
10. A rocket throws high-pressure gas in one direction, propelling the rocket in the opposite direction. (10.1) K/U
11. It is summertime in Canada when Earth is closest to the Sun. (9.2) K/U
12. Astronomers often use the light year as a measure of time (8.5) K/U
13. A constellation is a pattern of stars in the sky, as viewed from Earth. (9.4) K/U
14. A star with an absolute magnitude of 4.5 gives off more light than a star with an absolute magnitude of –1.5. (8.5) K/U
15. The planets have elliptical orbits. (9.1) K/U

16. Pluto is no longer considered a planet because it does not orbit the Sun. (8.1, 8.4) K/U

17. Not all the planets in our Solar System have seasons. (9.2) K/U

18. Ancient cultures built many structures to keep track of celestial events. (9.4) K/U

19. We see the Moon because it produces and emits its own light. (9.5) K/U

20. Microgravity environments are environments in which there is zero gravity. (10.3, 10.4) K/U

21. The navigation of a spacecraft is the responsibility of the astronaut only. (10.3) K/U

22. A tiny piece of space junk could cause much damage if it hit a spacecraft. (10.4) K/U

23. Space spinoffs are technologies that are used only in space. (10.5) K/U

24. Canadians have made leading contributions to the International Space Station, including a robotic arm. (10.4) K/U

Copy each of the following statements into your notebook. Fill in the blanks with a word or phrase that correctly completes the sentence.

25. In astronomy, a _________ constellation can be seen during the entire year at Earth's poles. (9.3) K/U

26. _________ is the distance measured from north along the horizon to a point directly below a celestial object. (9.4) K/U

27. When the Moon seems fuller each night, it is said to be _________. (9.6) K/U

28. A satellite's orbit around Earth is primarily controlled by _________. (10.1) K/U

29. During a solar eclipse, the _________ is directly between the Earth and the _________. (9.7) K/U

30. _________ is a measure of how bright a star appears when viewed from Earth. (8.5) K/U

31. Mercury, Venus, Earth, and Mars are known as the _________. (8.4) K/U

Match the term on the left with the most appropriate description on the right.

32.

(a) dwarf planet	(i) keeps the planets in orbit around the Sun
(b) red shift	(ii) used during launch of a spacecraft
(c) rocket booster	(iii) provides evidence that galaxies are moving away from Earth and away from one another
(d) gravitational force	(iv) a celestial object that orbits the Sun but which does not clear its orbit

(8.4, 9.6, 9.8, 10.1)

What Do You Understand?

Write a short answer to each of these questions.

33. A time capsule preserves objects to show future generations what life was like. In what way is the Moon like a time capsule for the Solar System? (8.3) T/I C

34. (a) From your view on Earth, in what general direction do celestial bodies appear to be moving in the night sky?
    (b) Why do these bodies appear to be moving in this particular direction? (8.1) T/I

35. In what way is the expanding Universe similar to an inflating balloon? (9.8) T/I C

The following icons indicate the Achievement Chart category addressed by each question.

K/U Knowledge/Understanding T/I Thinking/Investigation
C Communication A Application

36. What is one benefit and one drawback of searching for extraterrestrial beings. (8.7) T/I A

37. What are two conditions that would make living on other celestial bodies near Earth impossible? (8.1, 8.6) T/I

38. Explain the difference between the geocentric model of the Universe and the heliocentric model. (9.1) T/I C

39. Explain how Earth's position in the Solar System makes life on Earth possible. (8.1) T/I C

40. What is the difference between a solar eclipse and a lunar eclipse? (9.7) T/I C

41. (a) In your notebook, identify each part of the telescopes shown in **Figure 1** and **Figure 2**.
    (b) What type of telescope does each figure show?
    (c) In what ways are the two telescopes similar?
    (d) In what ways are the two telescopes different? (8.2) T/I C

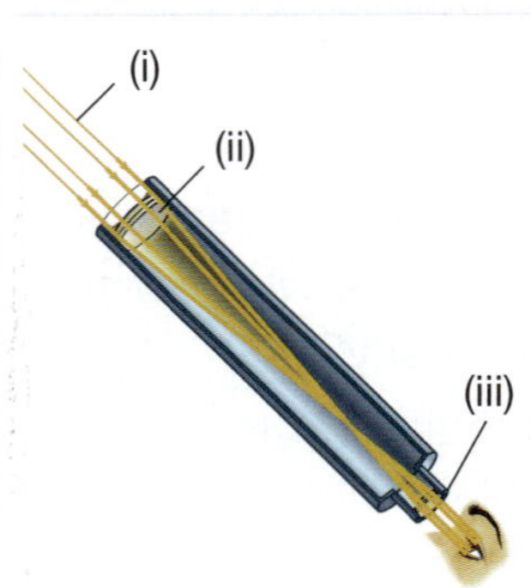

Figure 1

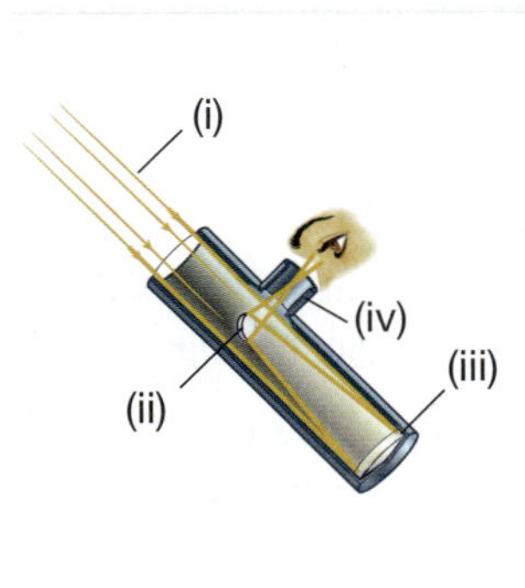

Figure 2

42. (a) Explain the difference between Earth's rotation and its revolution.
    (b) Which movement is responsible for our day and night?
    (c) Which movement is responsible for our year? K/U (9.1)

43. What evidence is there for the Big Bang theory? (9.8) K/U T/I

44. Define the following terms: meteor, meteoroid, asteroid, comet (**Figure 3**). K/U A

Figure 3

Solve a Problem

45. How is measuring the azimuth and altitude helpful when observing the positions of celestial bodies in the sky? (9.4) T/I

46. How would you convince someone that the Universe is expanding? (9.8) T/I

47. How would you explain the challenges of adapting to space? (10.2, 10.3) C A

48. You are an astronaut on the space shuttle and your communication system to Earth was down for a critical amount of time. What would you do? Explain your reasoning. (10.3) T/I C

49. An astronaut has been in space for six months, working aboard the International Space Station. (10.3) T/I
   (a) List two factors about the trip that might be hazardous for his health.
   (b) What is the cause of these dangers?

Create and Evaluate

50. The Canadian Space Agency is planning a mission to return humans to the Moon. Describe three things that you would like to research on the Moon's surface. What tools or technologies would you use to conduct this research? (8.3, 10.2) T/I C

51. You have probably heard the expression, "The Sun rises in the east and sets in the west." How could you demonstrate that this statement is not true? (8.1, 9.1) T/I

Reflect on Your Learning

52. Do you think that space tourism would be a good idea? Explain your reasoning using facts from the unit. C A

53. (a) Explain why someone might want to become an astronaut.
   (b) Explain why someone might not want to become an astronaut.
   (c) Would you want to become an astronaut? C A

54. Create a visual representation that illustrates the key concepts discussed in each chapter of this unit. C

55. After reading this unit, what questions do you still have about space exploration? C

Web Connections

56. You have read about some of the challenges and benefits of space exploration. Choose either one challenge or one benefit of space exploration and research the topic further. Write a short summary of your findings. T/I C A

57. Canadians have made many contributions to space exploration. Research one Canadian who has made a significant contribution. Write a short description of what this person has done and how his or her contributions have made a difference. T/I C A

58. Go to the Nelson Science website and download "Google Earth." Explore the Google Universe options, including the stars, planets, and satellites. How realistic is this representation? Where does Google get its data? T/I C

59. The world's largest scale model of the Solar System is the Sweden Solar System. It is modelled on a scale of 1 : 20 000 000. It uses the Ericsson Globe to represent the Sun (**Figure 4**). Research the Sweden Solar System to find out how the planets and other celestial objects are represented in the city of Stockholm, Sweden. T/I

Figure 4

GO TO NELSON SCIENCE

UNIT E Electrical Applications

UNIT Preview

In this unit, you will

- assess the costs and benefits of using renewable or non-renewable sources to produce electrical energy
- explore the properties of static and current electricity and the cost of using electrical energy
- learn how static and current electricity are alike and how they are different

BIG Ideas

- Electricity is a form of energy produced from a variety of non-renewable and renewable sources.
- The production and consumption of electrical energy has social, economic, and environmental implications.
- Static and current electricity have distinct properties that determine how they are used.

Focus on STSE

Are You an Energy Hog?

For each item below, choose the term that best describes your energy use. Be honest! K/U A

1. I turn off the lights when I am the last person to leave a room.
   (a) always (b) sometimes (c) never
2. My family uses compact fluorescent light bulbs at home.
   (a) Yes, for most or all of our lights.
   (b) We only have one or two compact fluorescent light bulbs.
   (c) No. We do not have any compact fluorescent light bulbs.
3. I open the refrigerator only long enough to take something out or put something in.
   (a) always (b) sometimes (c) never
4. When an air conditioner or furnace is on, I make sure all doors and windows are closed.
   (a) always (b) sometimes (c) never
5. I unplug my chargers when I am not using them.
   (a) always (b) sometimes (c) never

For each (a) answer, score 1 point. For each (b) answer, score 2 points. For each (c) answer, score 3 points.

5–7 points: You are an Energy Champ! Way to go! Keep looking for ways to conserve energy.

8–11 points: You are an Energy Sloth. Take more responsibility for your energy use, and you can graduate to Energy Champ.

12–15 points: You are an Energy Hog. Make a pledge to start turning off lights, unplugging chargers, and shutting doors and windows.

Electrical Applications

CHAPTER 11

Electrical Energy

Generating and using electricity affects the environment and society.

CHAPTER 12

Static and Current Electricity

Static and current electricity have distinct properties that determine their uses.

CHAPTER 13

Circuits

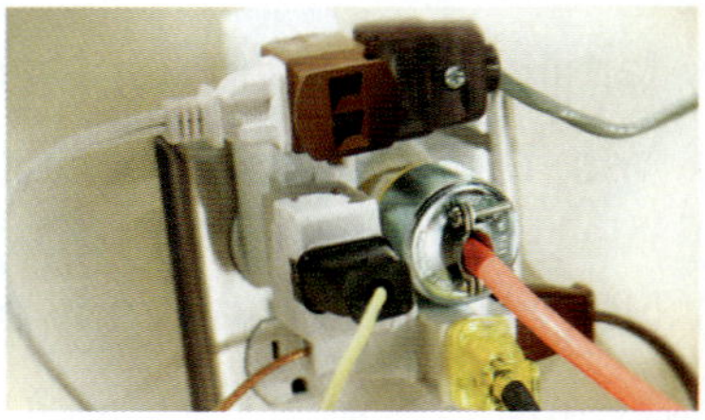

Controlling current electricity allows us to use it in a variety of ways.

Unit Task Preview

Model Home Powered by an Alternative Source of Energy

In this Unit Task, you will design and build the circuitry for a model home that will be powered by an alternative source of energy. You will research the advantages and disadvantages, including the costs of these energy sources, and present your findings to other students. Then, as a group, you will present your circuit design for the home that uses the best combination of available energy sources.

UNIT TASK
Assessment

You will be assessed on:

- researching and evaluating alternative energy sources for a home
- designing the electric circuitry for a home that uses alternative energy sources
- communicating your design to others

UNIT TASK
Bookmark

The Unit Task is described in detail on page 536. As you work through this unit, look for this bookmark and see how the section relates to the Unit Task.

Getting Started: Useful Concepts and Skills

Concepts Review

1. **Figure 1** shows two ways to generate electrical energy. K/U
   (a) Which way of producing electrical energy do you think has a bigger impact on the environment? Explain.
   (b) Does either method seem harmless? Explain.

Figure 1 (a) A nuclear plant and (b) a wind farm both produce electrical energy.

2. Can the design of a building affect how much energy it uses (**Figure 2**)? Explain. K/U

Figure 2

3. Describe the energy change that occurs in an electrical heating system. What energy is input? What energy is output? K/U

4. Decide if each action would have a positive or negative effect on the environment. Explain your thinking. K/U
   (a) hanging laundry outside to dry
   (b) using insulation in the walls of a home
   (c) burning coal
   (d) using skylights to light a home

Skills Review

5. Identify some of the costs and benefits of making your home more energy efficient. A

6. What is the person in **Figure 3** doing that is unsafe? Why is this unsafe? T/I

Figure 3

7. Compare the costs and benefits of being driven to school compared to taking public transportation. A

8. Your local power company wants to build a hydro-electric dam on a river that runs through your town. The dam will be located on part of the river shown in **Figure 4**. The company says the dam will lower the cost of electricity in your town.
   (a) Are there any other advantages, economic or environmental, to building the power plant? Explain.
   (b) Are there any disadvantages to building the power plant? Explain. A

Figure 4

CHAPTER 11

Electrical Energy

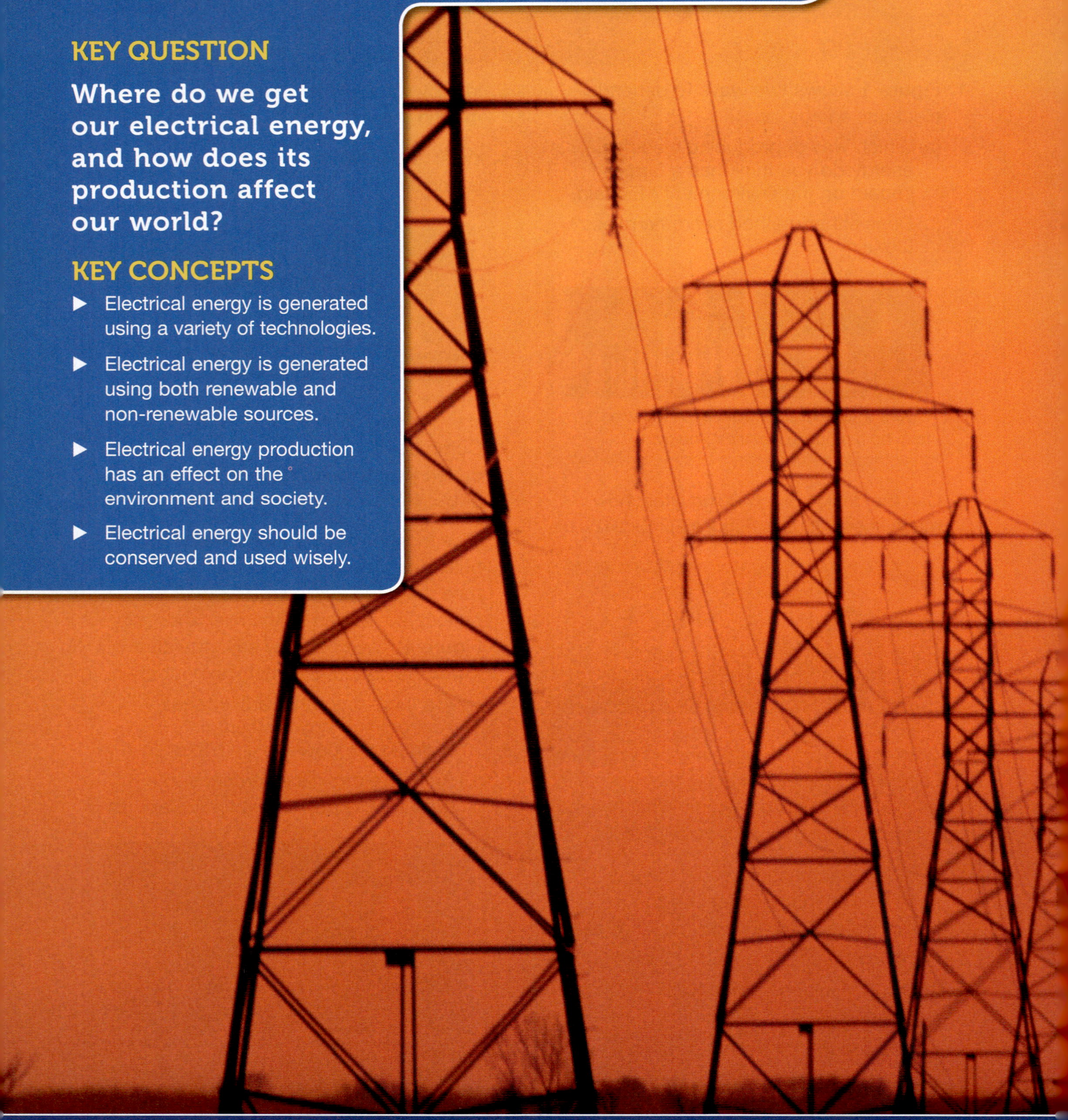

KEY QUESTION

Where do we get our electrical energy, and how does its production affect our world?

KEY CONCEPTS

- Electrical energy is generated using a variety of technologies.
- Electrical energy is generated using both renewable and non-renewable sources.
- Electrical energy production has an effect on the environment and society.
- Electrical energy should be conserved and used wisely.

UNPLUGGED

At about 4:15 p.m. on August 14, 2003, 50 million people in southern Ontario and the northeastern United States got a surprise when their lights and appliances flickered, and then stopped. This was the beginning of the largest blackout in history.

The lights went out. Subways and trains stopped, trapping commuters for hours. Traffic signals stopped working, causing traffic jams and hazardous driving conditions. On this hot and humid day, people also lost the use of their air conditioners and fans. More importantly, they lost their water supply, since the electric pumps could not work. Communication was severely limited when the cellphone network went down. Regular phones did work, but the system was overwhelmed by the number of calls going through. Television and the Internet were out. For many, the only way to learn about what was going on was by battery-operated radios.

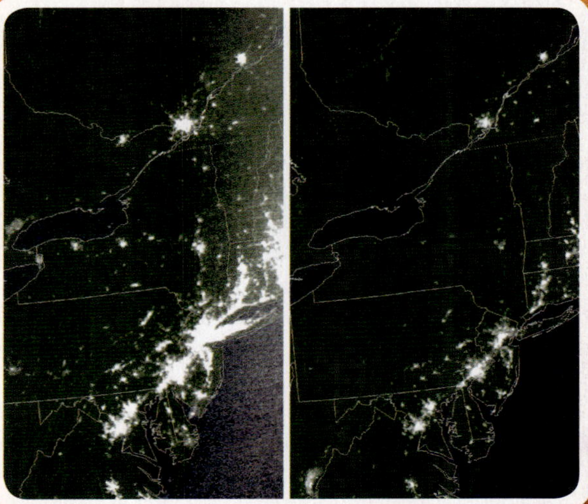

Satellite photo of eastern North America showing the difference in light levels before and during the blackout

It took days to restore electricity to southern Ontario. In Toronto, the streetcars and subways started running again four days after the blackout. What would your life be like if suddenly there was no electricity?

Skyscrapers and the CN Tower light up the downtown Toronto skyline.

The Toronto skyline at dusk during the biggest blackout in North American history

WHAT DO YOU THINK?

Many of the ideas you will explore in this chapter are ideas that you have already encountered. You may have encountered these ideas in school, at home, or in the world around you. Some of the statements are not true. Consider each statement and decide whether you agree or disagree with it.

1 Using electricity creates pollution.
Agree/disagree?

2 It is easy to produce more electricity when needed.
Agree/disagree?

3 We will run out of electricity someday.
Agree/disagree?

4 When some electrical devices are turned off, they continue to use energy.
Agree/disagree?

5 Rechargeable batteries are better for the environment than disposable batteries.
Agree/disagree?

6 All known methods of generating electricity have both advantages and disadvantages.
Agree/disagree?

FOCUS ON WRITING

Writing a Persuasive Text

Use convincing facts and organize your ideas clearly in paragraphs to persuade an audience to accept your opinion about a topic. The following is an example of a persuasive text written about energy conservation in Section 11.7. Beside it are the strategies used to write the persuasive text effectively.

DIG DEEPER

Writing Tip

As you work through this chapter, look for tips like this. They will help you develop literacy strategies.

Saving Money and Energy

State the topic in the first paragraph. → My aunt believes in conserving energy at home. She uses CFLs and has a programmable thermostat. My aunt wanted to make her

State the opinion clearly. → home even more energy efficient, but had to convince my uncle. She believed the changes would save both money and energy.

Use connecting words to link ideas. → My aunt said their furnace was old. Next, she explained the rebates the federal and provincial governments give homeowners for energy-saving upgrades. My uncle agreed to replace the furnace because the rebates would save about $2000.

My aunt decided to buy a high-efficiency gas furnace with an

State reasons to accept the opinion. → annual fuel utilization efficiency (AFUE) rating of 96.6 % (18.6 % more efficient than her old furnace). In addition, this model has a variable speed motor to power the blower. This motor uses one

Use facts to support opinion. → third of the electricity used by the motor in her old furnace. This new furnace will use a lot less natural gas and electricity.

Homeowners who want to save money can take advantage of federal and provincial rebates for home improvements that are

Restate opinion and reasons in concluding paragraph. → energy efficient. Homeowners who want to help the environment by conserving energy can switch to a high-efficiency furnace.

11.1

Generating Electricity

Nearly everywhere you go, electricity is being used. At the concert in **Figure 1**, electricity is needed for the lights, the microphones, the instruments, and the amplifiers. Electricity is also used to charge the cellphones people are using. Society depends on electricity.

Figure 1 How many different ways is electricity used at this concert?

What Is Energy?

Scientists have some difficulty defining energy because it has many forms. Some of these forms of energy include chemical energy, sound energy, and mechanical energy. When an object is able to change its environment, we say the object has energy. Energy is the ability to do a task, such as moving an object. **Table 1** lists different forms of energy and examples of each form. One familiar form of energy is electrical energy. **Electrical energy** is the energy of moving charged particles. You will learn about electric charges in Chapter 12.

▶ **electrical energy:** a type of energy that comes from the flow of charged particles; it allows electrical devices to function

Table 1 Common Forms of Energy

Form of energy	Examples
chemical	gasoline; food; fireworks
thermal	warmth from a fire
sound	voices; music
radiant (light)	sunlight; microwaves
mechanical	wind; running water
electrical	lightning; electricity for your home

Energy Transformations

There are many ways to change or transform one form of energy into another form. Your body and a car's engine change chemical energy into mechanical energy. When you bang on a drum or pull a bow across the strings of a violin, the mechanical energy is changed into sound energy (**Figure 2**). A wind or water turbine changes mechanical energy into electrical energy. Electrical appliances transform electrical energy into other forms of energy to perform specific tasks. For example, toasters and electric ovens change electrical energy into thermal energy to cook food. **Table 2** lists some common energy transformations.

Figure 2 A violinist uses the energy of motion of her arm to move the bow across the strings, producing sound energy.

Table 2 Examples of Energy Transformations

Example	Energy transformation
solar panel	light energy → electrical energy
fire	chemical energy → thermal energy and light energy
wind turbine	mechanical energy → electrical energy
car engine	chemical energy → thermal energy → mechanical energy
light bulb	electrical energy → thermal energy and light energy

DIG DEEPER

Did You Know?

Wasted Energy

Many devices that produce light also release thermal energy. The thermal energy that warms the air around a lightbulb is waste energy. An incandescent bulb transforms 95 % of electrical energy into thermal energy. A compact fluorescent bulb transforms 75 % of electrical energy into thermal energy. Therefore, an incandescent bulb is only 5 % efficient, but a CFL bulb is 25 % efficient.

TRY THIS OBSERVING ENERGY CHANGES

SKILLS HANDBOOK 3.B.6., 3 B.7.

SKILLS: Observing, Analyzing

You turn a crank for a minute or so, then flip on a switch. Suddenly you have a flashlight that is working without any batteries. What kind of energy transformations take place in one of these flashlights?

Equipment and Materials: hand-cranked flashlight

1. Turn on the flashlight.
2. Turn the handle of the flashlight crank to wind up the flashlight.
3. Describe what happens.

A. What was the source of energy for the flashlight? T/I

B. Explain which energy transformations took place to cause the flashlight to light up. T/I

C. What form of energy was produced? Was more than one kind of energy produced? How do you know? T/I

D. What energy transformations occur in a flashlight that uses batteries? A

Generating Electrical Energy

Televisions, video games, MP3 players, computers, and refrigerators all use electrical energy. Where does electrical energy come from? Understanding energy transformations is important for understanding how electrical energy is produced.

The Grid

While a tiny fraction of the electrical energy we use is provided by batteries, most of the electrical energy we use comes to us through the **electrical energy distribution grid**. This grid is commonly called "the power grid." You use the electrical energy provided by this system every time you plug the cord of an electrical appliance into a wall outlet.

- **electrical energy distribution grid:** a network that carries electricity from the energy generating station to the consumer
- **generator:** device that produces electricity by rotating a wire coil near a magnet

Generators

Most of the electrical energy distributed by the grid is produced in electrical energy generating stations. These stations use large machines called generators to produce electrical energy. A **generator** is a machine that converts mechanical energy into electrical energy. A basic generator consists of a magnet and a wire coil. As you may know, magnets have two different poles: north and south. Magnetism at these poles can cause a push or a pull. When the wire coil rotates near the magnetic poles, the negatively charged particles in the wire are forced to move. This is how electrical energy is generated (**Figure 3**). To produce the amount of electricity needed to light up a city, you have to move many large coils of wire close to many very powerful magnets. The faster the wire coils are rotated, the more electricity that is generated.

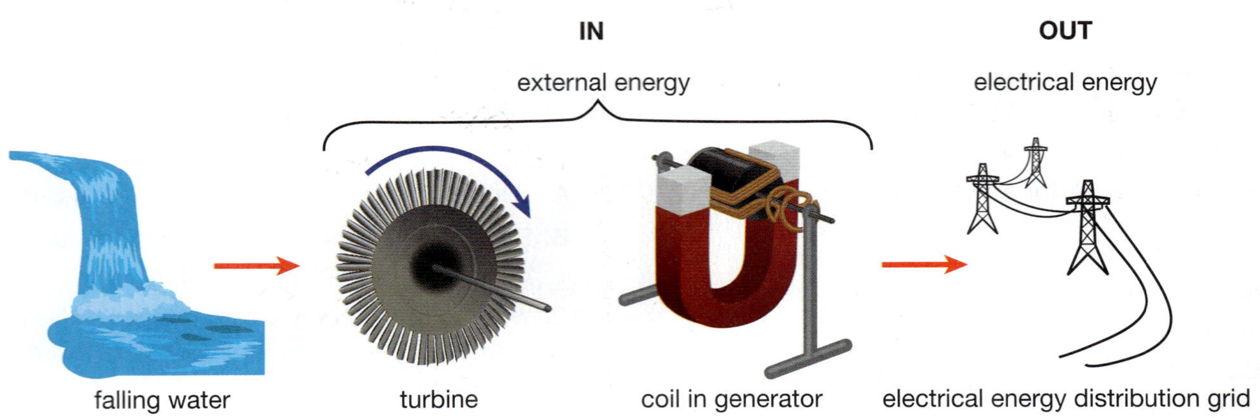

Figure 3 Falling water or another source of mechanical energy spins a turbine that, in turn, spins coil of wire between magnets to generate electrical energy.

The energy in wind, flowing water, or steam can be converted into electrical energy. The movement is used to push blades of a fan-like device called a turbine. The turbine is connected to the wire coils inside a generator. When the turbine spins, the coils of wire generate electrical energy as they spin at high speed near powerful magnets (**Figure 4**).

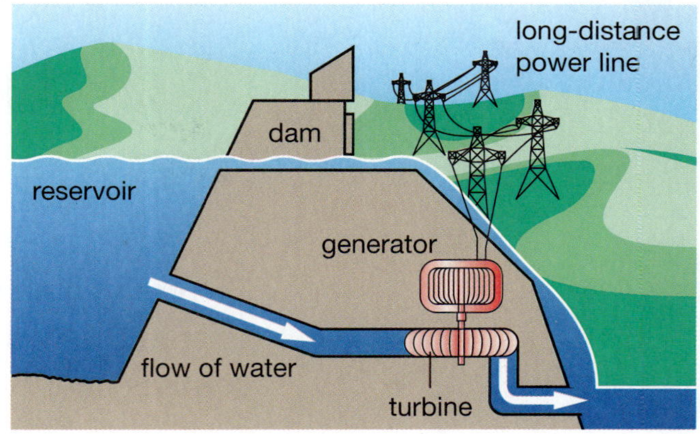

Figure 4 Some electrical energy stations use the movement of flowing water to generate electrical energy.

Measuring Electrical Energy

All energy, including electrical energy, is measured in joules (J). How quickly we use (or produce) electrical energy can be measured as the number of joules used per second. A joule per second is called a watt (W). The higher the "wattage" of a device, the more electrical energy is used per second. A 20 W compact fluorescent bulb uses 20 J of electrical energy per second. In one minute, it uses 1200 J (20 J/s × 60 s).

A joule is very small, so we often use a much larger unit of electrical energy, the kilowatt hour (kW•h), to describe energy use. A **kilowatt** is 1000 W. So, a **kilowatt hour** is equal to using 1000 W for 1 h, or 100 W for 10 h.

- **kilowatt:** 1000 watts
- **kilowatt hour:** a unit of energy equal to using 1000 W for one hour

11.1 Wrap Up

- There are various forms of energy: light, chemical, thermal, mechanical, sound, and electrical. One form of energy can be transformed into other forms of energy.
- Electrical energy comes from the flow of charged particles.
- A generator is a machine that converts mechanical energy into electrical energy.
- Wind, steam, and running water are all commonly used to supply the motion needed to produce electricity.

CHECK YOUR LEARNING

1. Name two forms of energy that are produced by a campfire. K/U
2. Explain how a generator works. K/U
3. Describe the energy transformations that occur in a clothes dryer. A
4. (a) What is the relationship between joules and watts? K/U
   (b) What is meant by the "wattage" of a 60 watt bulb? K/U
5. What is a kilowatt hour? K/U

11.2

Non-renewable Energy Sources

To produce electricity, a generator needs a source of energy to spin the wires between the magnets. Some energy sources cannot be replaced after they have been used.

Non-renewable Sources

A source of energy that is in limited supply and either cannot be renewed or takes millions of years to renew naturally is a **non-renewable energy source**.

- **non-renewable energy source:** an energy source that either cannot be renewed, or that takes millions of years to renew naturally
- **thermal generation:** a method for generating electricity in which water is heated to produce steam, which spins a turbine connected to a generator

Thermal Generation Using Fossil Fuels

Thermal generation is one way of generating electricity. In thermal generation, water is heated to produce high-pressure steam. The steam is used to spin the blades in a turbine. The turbine moves a generator, which produces electricity. Once steam has passed through the turbines it must be condensed back into liquid water before it can be reused.

A common way to heat water is by burning fossil fuels. Burning coal, oil, and natural gas currently provides more than half of the world's supply of electrical energy (**Figure 1**). Fossil fuels are classified as non-renewable sources. Fossil fuels were formed from the remains of microbes, plants, and animals that lived millions of years ago.

Figure 1 The Nanticoke generating plant uses coal as an energy source and is Canada's largest contributor to air pollution.

Table 1 lists the advantages and disadvantages of thermal generation using fossil fuels.

Table 1 Advantages and Disadvantages of Using Fossil Fuels

Advantages	• much less expensive than other sources
Disadvantages	• non-renewable
	• releases carbon dioxide and other greenhouse gases into the air, contributing to climate change
	• obtaining fossil fuels disturbs or destroys habitats (**Figure 2**)
	• refining and burning fossil fuels cause pollution, including acid precipitation
	• uses large volumes of water to cool and condense steam, which can cause thermal pollution

Figure 2 Bankhead, an abandoned coal mine in Banff National Park, Alberta. It can take decades for a used mining site to be restored to a natural state.

Thermal Generation Using Nuclear Energy

Nuclear power plants use materials such as uranium as fuel. The nucleus, or centre, of a uranium atom stores a large quantity of nuclear energy. When these atoms break apart, the atoms' nuclear energy is transformed into thermal energy. This type of reaction is called nuclear fission. A small pellet of uranium—about the size of your fingertip—produces about as much energy as burning 570 L of oil.

DIG DEEPER

Writing Tip

Connecting to the Main Idea

Conclude a persuasive text by connecting the main idea and key points. For example, a persuasive text about thermal generation using fossil fuels might conclude by saying that fossil fuels should no longer be used because the disadvantages outweigh the advantages.

DIG DEEPER

Career Link

Nuclear reactor operators maintain and control the nuclear reaction process in reactors. To learn more about a career in nuclear generation,

GO TO NELSON SCIENCE

In nuclear power plants, long, metallic tubes are filled with pellets of uranium. The tubes are then placed in a nuclear reactor. The energy released is used to heat water, producing steam. The moving steam turns a turbine in a generator. **Figure 3** shows a nuclear power plant where this process takes place.

Figure 3 More than half of Ontario's electrical energy is generated using nuclear reactors.

Radioactive waste from a nuclear reactor is stored underwater in canisters at the reactor site for more than 10 years (**Figure 4**). Then it is stored in dry storage behind concrete 1 m thick. After another 60 years, the canisters can be moved to long-term storage, but the waste is still unsafe. It will remain unsafe for tens of thousands of years.

Figure 5 The water in this pool absorbs the radiation given off by the nuclear waste stored in canisters at the reactor site.

DIG DEEPER

Did You Know?

Nuclear Plants

Ontario has three major nuclear energy plants: the Darlington, Pickering, and Bruce generating stations. These three plants currently produce about 52 % of the electrical energy used in Ontario.

Table 2 lists the advantages and disadvantages of thermal generation using nuclear fission.

Table 2 Advantages and Disadvantages of Using Nuclear Fission

Advantages	• does not cause air pollution directly
	• does not emit greenhouse gases directly
Disadvantages	• non-renewable energy source
	• produces dangerous radioactive wastes
	• mining uranium results in air, water, and land pollution
	• plants are expensive to build and maintain
	• uses large volumes of water to cool and condense steam, which can cause thermal pollution
	• some risk of serious accidents due to human error, equipment failure, or earthquakes

11.2 Wrap Up

- Sources used to generate electrical energy can be non-renewable.
- Non-renewable sources include fossil fuels and uranium.
- Producing energy from non-renewable fuels has advantages and disadvantages for society and the environment.

UNIT TASK

Bookmark

You can apply what you have learned about non-renewable energy sources to the Unit Task on page 536.

CHECK YOUR LEARNING

1. Copy **Table 3** into your notebook. List one advantage and one disadvantage of each energy source. K/U

Table 3

	Fossil fuels	Nuclear energy
advantage		
disadvantage		

2. What are the three types of fossil fuels? K/U
3. What problems are associated with storing nuclear wastes? K/U
4. Would you favour storing nuclear waste hundreds of metres below ground, as some scientists have suggested? Explain your reasoning. A
5. Would you choose nuclear energy or burning fossil fuels for generating electricity in your town? Explain your choice. T/I

11.3 Renewable Energy Sources

There are several sources of electrical energy which can be used over and over again without running out.

Renewable Sources

► **renewable energy source:** an energy source that can be replaced in a relatively short period of time

There are **renewable energy sources** that are virtually unlimited or can be replenished by natural processes in a relatively short period of time.

Hydro-electric Generation

Humans have used the energy of falling water since ancient times. For thousands of years, people built wheels near small waterfalls. The wheels were often attached to grain mills. Falling water turned the wheel and grain was ground into flour.

In a hydro-electric generating station, the energy of falling water is used to move the large blades of turbines connected to generators. Many hydro-electric generating stations use the water stored behind a dam to generate electricity. The dams control the flow of water so that it flows past a turbine connected to an electrical generator. There are 180 hydro-electric generating stations in Ontario. **Table 1** shows the advantages and disadvantages of generating electrical energy using hydro-electricity.

Figure 1 Some dams have fish ladders, which allow fish to swim around the dam.

Figure 2 In the 1980s, the James Bay Project flooded more than 12 000 km^2 of forest. As a result, the local Cree lost a portion of their hunting grounds.

Table 1 Advantages and Disadvantages of Hydro-electric Generation

Advantages	• does not release pollution into the air or water
Disadvantages	• disrupts the natural water flow pattern
	• can restrict fish movement unless fish ladders are built (**Figure 1**)
	• large land areas behind a dam are flooded, resulting in habitat loss (**Figure 2**)
	• generating stations are expensive to build
	• difficult to find suitable locations for generating stations

Wind Generation

Wind is another ancient source of energy. We use wind to move sailboats and have built windmills to pump water. A modern wind turbine is made up of large blades mounted on a tall tower (**Figure 3**). As the wind blows, the blades of a turbine turn. The turbine is connected to a generator. **Table 2** shows the advantages and disadvantages of wind generation.

Figure 3 Today, a single large wind turbine can supply electricity for about 250 Canadian homes.

Table 2 Advantages and Disadvantages of Wind Generation

Advantages	• wind is free and does not need to be transported
	• building and maintaining wind farms is less expensive than generating stations that use fossil fuels or uranium
	• can be built anywhere winds are reliable and relatively strong
Disadvantages	• objections to the way turbines change scenic views
	• need large land areas for wind farms
	• noise pollution for people living nearby
	• location issues because of migrating birds and bats

DIG DEEPER

Career Link

Solar energy technicians install, maintain, operate, and test equipment and energy systems that use solar energy. To learn more about being a solar energy technician,

GO TO NELSON SCIENCE

Photovoltaic Generation

Electrical energy can be generated without using a turbine and generator. The most abundant source of energy on Earth is sunlight. Sunlight is free, but special technology must be used to capture its energy. A **photovoltaic cell**, or solar cell, is a technology that changes light energy from the Sun directly into electrical energy (**Figure 4**). Light from the Sun excites charged particles that move, creating an electric current. **Table 3** shows the advantages and disadvantages of photovoltaic generation.

▶ **photovoltaic cell:** a device that converts light directly into electrical energy

Table 3 Advantages and Disadvantages of Photovoltaic Generation

Advantages	• free and readily available
	• can be installed anywhere that receives enough sunlight
	• does not create air or water pollution
Disadvantages	• currently expensive and inefficient
	• can only work during the day, but small systems can use batteries to store electricity for nighttime use

Figure 4 Each solar panel consists of many photovoltaic cells connected together.

Thermal Generation Using Biomass

Biomass, or biofuel, is any form of bacterial, plant, or animal matter. Biomass can also be the gases given off from decaying materials collected on farms, sewage treatment plants, and landfills.

Biomass is burned to produce thermal energy. This heats water into steam that can turn a turbine connected to a generator to produce electricity. Scientists are studying the costs and benefits of using wood and wood wastes as biofuel in Ontario. **Table 4** shows the advantages and disadvantages of thermal generation using biomass.

DIG DEEPER
Web Link
For more information about renewable energy projects in Canada,
GO TO NELSON SCIENCE

Table 4 Advantages and Disadvantages of Using Biomass

Advantages	• always renewable—trees and crops can be replanted; plant and animal waste is produced continuously; gases are continuously produced by decay
Disadvantages	• burning biomass produces greenhouse gases
	• using crops as fuel results in lower food production

Geothermal Generation

Geothermal energy plants are usually located in areas with volcanic activity. Geothermal generation uses wells drilled deep underground to access thermal energy. Hot geothermal gases are then piped to a generating station, where the thermal energy turns water into steam (**Figure 5**). The steam then turns a turbine connected to a generator to produce electricity. Geothermal energy can also be used directly to provide heating and cooling. **Table 5** shows the advantages and disadvantages of geothermal generation.

Figure 5 Geothermal generating plants use underground thermal energy to make steam to turn turbines to generate electricity.

Table 5 Advantages and Disadvantages of Using Geothermal Generation

Advantages	• geothermal energy is always present deep in Earth's crust
	• geothermal fluids can be continuously pumped back underground to be reheated and reused
	• does not produce pollution or greenhouse gases
Disadvantages	• suitable geothermal areas are rare; very few are near populated areas
	• generating stations are difficult and expensive to build

What Does the Future Hold?

In Ontario, about 26 % of all electrical energy comes from burning coal and natural gas, and 52 % of electrical energy comes from nuclear sources. This means that 78 % of Ontario's electrical energy is produced using non-renewable energy sources (**Figure 6**). Even renewable sources have consequences for society and the environment.

How can the growing demand for electrical energy be met? We need to develop ways to use less electrical energy. We also need to find ways to generate electrical energy that are effective and environmentally friendly.

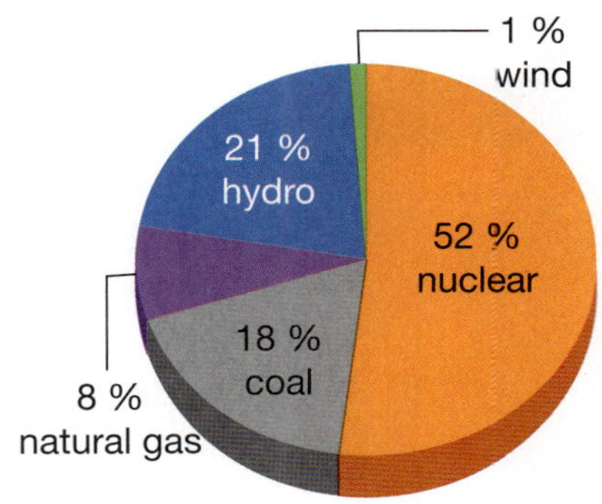

Figure 6 This circle graph shows how Ontario's electrical energy was generated in 2007.

11.3 Wrap Up

- Sources used to generate electrical energy can be renewable.
- Renewable sources used to generate electricity include hydro-electric, wind energy, solar energy, biomass, and geothermal energy.
- Producing energy using renewable sources has advantages and disadvantages for society and for the environment.

UNIT TASK

Bookmark

You can apply what you have learned about the environmental effects of electrical energy production to the Unit Task on page 536.

CHECK YOUR LEARNING

1. List one advantage and one disadvantage of each energy source in a table like **Table 6** below. K/U

Table 6

Energy source	Advantage	Disadvantage
biomass		
hydro-electric		
wind		
geothermal		
solar		

2. Do you think renewable energy sources will become more popular in the near future? Explain. A
3. Identify two technologies that you could use at home to reduce dependence on fossil fuel–generated electrical energy. A
4. If renewable sources of energy are better for the environment, why do you think we still rely mostly on non-renewable sources? T/I A
5. Which form of renewable energy would be best for your neighbourhood? Explain your reasoning. A

11.4 EXPLORE AN ISSUE CRITICALLY

Choosing a Source of Electrical Energy

SKILLS MENU

- Defining the Issue
- Researching
- Identifying Alternatives
- Analyzing the Issue
- Defending a Decision
- Communicating
- Evaluating

Canadians have become more aware of where their electrical energy comes from and the effects—economic, environmental, and social—of electrical energy generation.

SKILLS HANDBOOK 4.A., 4.C.

The Issue

Your family wants to switch from using fossil fuels to an alternative source for generating electricity.

Role: Your group must recommend a way to generate electricity. Each member of the group will research a different renewable energy source and report back on its advantages and disadvantages.

Audience: Your classmates will represent your family.

Goal

To recommend the best method of electrical energy generation

Gather Information

Gather information to answer questions such as the following:

- For what do you need electricity? How much will you need to generate? How can you find out?
- What are the costs to install and operate each system?
- What is the environmental impact of each system?
- Can unused energy be stored for future use?

Identify Solutions

Present your findings to your group. Compare the options presented by other group members.

Make a Decision

As a group, select the best option. Support your decision with facts.

Communicate

Present your recommendation to the class in a report, electronic presentation, or other method of your choosing.

UNIT TASK

Bookmark

You can apply what you learned in this activity to the Unit Task described on page 536.

FOCUS ON
MATH

Converting Units

Often, you are asked to solve a problem in which one unit of measurement is given, but another unit is asked for in the question. You have to convert one unit into another using a conversion factor.

Converting Units in Action

How many kilowatts of electrical energy does a 100 W light bulb use?

Given: wattage of light bulb = 100 W

Required: electrical energy use of light bulb in kilowatts

Analysis: 1 kW = 1000 W. Therefore, $\frac{1\ \text{kW}}{1000\ \text{W}} = 1$

To convert 100 watts to kilowatts, multiply by the conversion factor $\frac{1\ \text{kW}}{1000\ \text{W}}$. Then, cancel the units.

Solution:

$$100\ \text{W} \times \frac{1\ \text{kW}}{1000\ \text{W}}$$

$$= 100\ \cancel{\text{W}} \times \frac{1\ \text{kW}}{1000\ \cancel{\text{W}}}$$

$$= \frac{100\ \text{kW}}{1000}$$

$$= 0.1\ \text{kW}$$

Statement: A 100 W light bulb uses 0.1 kW of electrical energy.

A. A technician is testing a new light bulb. She leaves it on for three whole days. For how many hours was the light on? Use the conversion factor $\frac{24\ \text{h}}{1\ \text{day}}$.

DIG DEEPER

Learning Tip

Conversion Factors

Multiplying by a conversion factor is like multiplying by 1. It does not change the quantity, only the unit in which it is expressed.

11.5 Using Electrical Energy

We all use electrical energy. We use it at home, at work, and at play. How can we learn to use it responsibly?

Figure 1 (a) A flashlight and (b) a plasma-screen TV use very different amounts of electrical energy.

The Many Uses of Electrical Energy

Think about all the things you do that depend upon electricity. Cooking or listening to music would be much different without electricity. What about devices that use batteries? MP3 players, laptops, cellphones, and watches are just a few of the items that use energy supplied by batteries. Batteries are devices that store energy until you need to use it.

Batteries are practical and portable, but they do not solve all our electricity needs. A wireless mouse or a flashlight might run for days on two AA batteries. However, a large plasma-screen TV would need more than 200 new AA batteries each hour (**Figure 1**). That would cost more than $100 per hour. Large appliances are usually plugged into an outlet on the wall that supplies a large amount of cheaper energy from an electrical energy generating grid. The cost for running the same TV for an hour when it is plugged in is just 7¢.

Using Energy Responsibly at Home

▶ **electrical energy consumption:** the amount of electrical energy used, often measured in kilowatt hours

Using energy responsibly means reducing unnecessary energy use. **Electrical energy consumption** is how much electricity you use, measured in kilowatt hours. To measure electrical energy consumption, you need to understand how different devices use energy (**Table 1**).

Table 1 Energy Use and Costs of Household Appliances ($0.10/kW•h)

Device	Typical consumption (W)	Cost per hour ($)
heat pump or central air	15 000	1.50
water heater or clothes dryer	4 000	0.40
hair dryer	1 200	0.12
refrigerator	1 000	0.10
computer and monitor	400	0.04
light bulb	60	< 0.01

"Hidden" Consumption

Most TVs, computers, DVD players, and many other devices have a "standby" mode (**Figure 2**). This means that they continue to use energy as long as they are plugged into a wall outlet. This energy is known as "phantom load" and can account for as much as 7 % of a household's total energy use. Even if you are unsure if an electrical device uses phantom energy, unplug any electrical device when it is not being used.

Chargers for cellphones, cameras, and MP3 players are also energy consumers. They keep using energy even after a battery is fully charged. Chargers use electricity whenever they are plugged in, even if there is nothing connected to them.

Figure 2 The "standby" lights on this DVD player tell you that it is still using electricity. Use "standby" lights as a reminder to unplug your appliances.

DIG DEEPER

Writing Tip

Identify the Topic

Use the first paragraph to identify the topic and state your main idea concisely. For example, the main idea of the "Hidden" Consumption subsection is that some devices waste energy even when they appear to be turned off.

Using Energy Outside the Home

People use electrical energy to run their home appliances, televisions, and computers. They also need electrical energy outside the home. Police and fire departments, shopping malls, schools, and hospitals need electricity to communicate over phones or radio and to keep lights on and computers running. Cities and towns use electrical energy to light streetlights, supply water to homes and other buildings, and run waste treatment plants. The machinery in factories uses electrical energy. Grocery stores use electrical energy to keep food cold. A business office may use hundreds of computers, photocopiers, and other devices (**Figure 3**).

Figure 3 One computer uses up to 600 kW•h per year. In sleep mode, it will use about 20 kW•h per year. How much energy could be saved by putting the computers in sleep mode when they are not in use?

Figure 4 Electric companies use a meter attached to your house to measure how much electrical energy you use.

How We Measure Electrical Energy Use

Energy use is measured in kilowatt hours (kW•h). How do we keep track of this energy use? Electric meters in homes, schools, or businesses are owned by an energy supply company (**Figure 4**). A meter tracks how much electrical energy is used in units of kW•h. The energy company reads the meter each month to find out how much electrical energy that has been used and sends a bill to the consumer.

Efficiency

Not all electrical devices use electrical energy efficiently. **Efficiency** is how well a device converts one form of energy into another. For example, light bulbs convert electrical energy into light energy. Some of the electrical energy is wasted as thermal energy. The light bulb is not 100 % efficient. A 100 W incandescent light bulb uses four times as much energy as a 25 W compact fluorescent light bulb (CFL), but they both give off the same amount of light (**Figure 5**). The CFL is more efficient than the incandescent light bulb.

▶ **efficiency:** a measure of how much useful energy an electrical device produces compared with the amount of energy that was supplied to the device

Many small bulbs are now light-emitting diodes (LEDs). These use less energy to produce light and waste less energy than any other type of bulb, including CFLs. LEDs are used in decoration lights, computers, traffic lights, billboards, cars, and home gaming systems. LEDs also last longer than other types of lights.

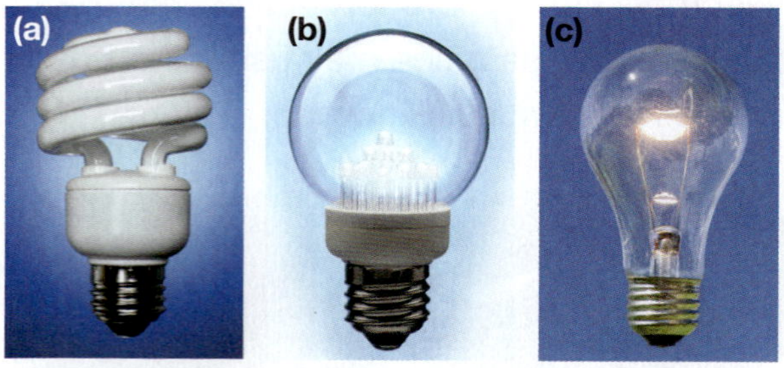

Figure 5 Compact fluorescent (a) and LED light bulbs (b) use much less electrical energy than incandescent light bulbs (c) of similar brightness.

Calculating the Cost of Electricity

To calculate the cost of using an electric device, you need to know how much energy the device uses and the price of each unit of energy. To calculate how much energy is used, you need the wattage of the device and how long the device is used. The wattage is how quickly the device consumes electrical energy. The value is given in watts. You will have to convert this value into kilowatts.

To calculate the energy used, use this equation:

$$\text{energy used (kW•h)} = \text{wattage (kW)} \times \text{time (h)}$$

To calculate the cost of operating the device, use this equation:

$$\text{cost to operate (\$)} = \text{energy used (kW•h)} \times \text{price of electricity} = (\text{\$/kW•h})$$

SAMPLE PROBLEM 1 A dishwasher has a wattage of 1200 W. A wash and air-dry cycle takes 2 h. If the cost of electricity is $0.10 per kW•h, excluding taxes and fees, how much would it cost to use the dishwasher once a day for a month (30 days)?

Given:
- The dishwasher has a wattage of 1200 W.
- The time the dishwasher is used is 2 h a day.
- The cost of electricity is $0.10 per kW•h.

Required:
- cost of using the dishwasher for 2 h
- cost of using the dishwasher every day for a month

Analysis: Convert to needed units.

The wattage is 1200 W. 1000 W = 1 kW

$$1200\text{ W} \div \frac{1000\text{ W}}{\text{kW}} = 1.2\text{ kW}$$

Solution: Calculate the amount of energy used.

$$\text{wattage (kW)} \times \text{time (h)} = \text{energy used (kW•h)}$$

$$1.2\text{ kW} \times 2\text{ h} = 2.4\text{ kW•h}$$

Calculate the cost of the energy.

$$\text{energy used} \times \text{price of electricity} = \text{cost to operate}$$

$$2.4\text{ kW•h} \times \$0.10/\text{kW•h} = \$0.24/\text{day}$$

Calculate the cost of energy for 30 days.

$$30\text{ days} \times \$0.24/\text{day} = \$7.20$$

Statement: The cost to use the dishwasher once a day for 30 days is $7.20.

Practice:

Suppose a home using a Smart Meter qualifies for a lower rate of $0.05 for energy used during non-peak hours. How much would it cost to run the dishwasher for the same amount of time if the timer on the dishwasher was set to run it at midnight (non-peak hours)?

DIG DEEPER

Did You Know?

The Cost of Energy

Energy companies charge you a fee per kW•h for the energy you use each month. The money is used to cover the cost of generating electrical energy, building and maintaining new plants and transmission lines, paying off debts, and salaries for people at the energy company.

RESEARCH THIS SMART METERS

SKILLS: Researching, Analyzing the Issue, Communicating

SKILLS HANDBOOK 4.A., 4.B.

Do you know that the price and cost of producing electrical energy changes throughout the day? The most expensive time for electrical energy production is generally weekdays from 9 a.m. to 5 p.m. During these "peak times," fossil fuel plants are run in addition to nuclear and hydro-electric generating stations to supply the extra energy. This means that you can save money and reduce pollution if you use electrical energy outside of peak times.

Devices known as Smart Meters encourage consumers to reduce energy use during peak hours by allowing them and the company providing electricity to track when and how much energy they use.

Use the Internet to help you find answers to the following questions.

A. What does "time-of-use-pricing" mean? T/I

B. What times of day are considered "off-peak," "mid-peak," and "on-peak"?

C. What household appliances can most easily be used during off-peak hours? T/I

D. What appliances are most necessary during on-peak hours? T/I

E. How can a family save money on their electricity bill by taking advantage of the Smart Meter? T/I

GO TO NELSON SCIENCE

11.5 Wrap Up

- Electrical devices vary greatly in the amount of electrical energy they consume.
- We can reduce energy consumption if we understand how electrical devices use energy at home and outside the home.
- Electric meters keep track of how much electrical energy is used in units of kilowatt hours.
- Some devices are more efficient at using electrical energy than others.

CHECK YOUR LEARNING

1. Give two examples of "hidden" ways that devices use electricity. K/U
2. Identify three ways that you and your friends use electrical energy at home and in school. In your opinion, which use is most important to running homes and your school? Why? C A
3. What is the main advantage of a compact fluorescent light bulb? K/U
4. A more energy-efficient television costs $150 more than a less-efficient model. What are some reasons you might choose to buy each television? A
5. A clothes dryer uses 4000 W. You use the dryer for 4 h a week. Electricity costs $0.20/kW•h. How much money could you save over 4 weeks if you hung laundry to dry outdoors instead of using the dryer? T/I

Storing Energy in Salt

While renewable energy processes help the environment, there can be disadvantages to them as well. For instance, solar panels generate electricity only when the Sun shines. What happens when the Sun disappears behind a layer of clouds for days at a time? How would we generate electricity at night?

To make solar energy more practical, we need ways to store it. What if we could store the Sun's energy as thermal energy, and use that energy to produce steam rather than changing it immediately to electrical energy?

It may sound strange, but after trying air, water, and mineral oil, some researchers have been looking at storing solar energy as thermal energy in salt. Mirrors focus the Sun's energy on towers where salt is stored in liquid form (**Figure 1**). Molten salt can absorb and store large amounts of thermal energy.

The stored thermal energy can then be used as needed to change water to steam. The steam then turns a turbine. This provides energy for a generator, and there you have it—electricity! **Figure 2** shows this process.

Not Your Average Table Salt

The salt these researchers are using is not common table salt. They use a combination of sodium and potassium nitrate. This mixture melts at a comparatively low temperature (220 °C) and can be kept liquid at 290 °C at normal air pressure in an insulated storage tank. These are ideal temperatures for use in steam turbines. The mixture is non-toxic and non-flammable. This system is designed to store enough thermal energy to run the turbine for a week.

Figure 1 Mirrors reflect the solar radiation, concentrating it up to 600 times on a receptor located on the upper part of the tower. This thermal energy is transferred to a fluid.

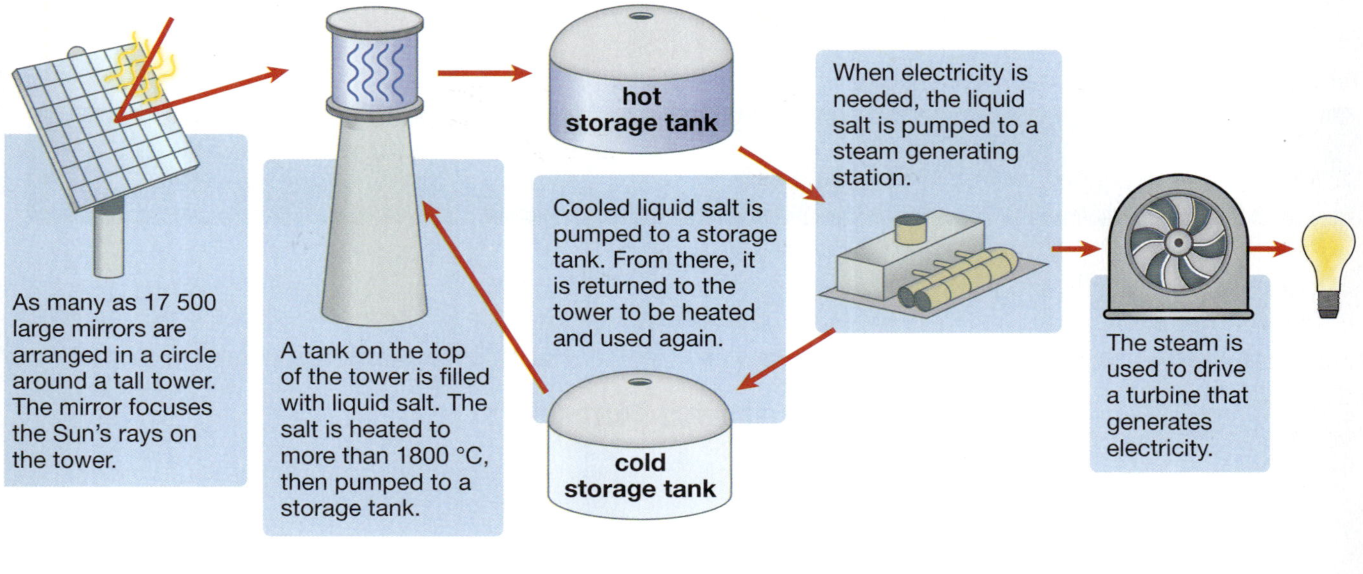

Figure 2 This diagram shows how solar energy can be stored in salt and then used to generate electricity.

11.6 PERFORM AN ACTIVITY

Evaluating Electrical Energy Use and Its Cost

SKILLS MENU

- ▶ Questioning
- ▶ Hypothesizing
- ▶ Predicting
- ▶ Planning
- ▶ Controlling Variables
- ▶ Performing
- ▶ Observing
- ▶ Analyzing
- ▶ Evaluating
- ▶ Communicating

Have you ever considered how much electrical energy you use? In this activity, you will find out how much electrical energy some of the electrical devices in your home use each week. You will then use what you learned about calculating the cost of electricity to figure out the cost of that electrical usage.

SKILLS HANDBOOK 6.A., 6.C.

Purpose

To determine the amount of electrical energy you use in one week when you operate certain electrical devices

Procedure

1. Choose five electrical devices, such as a clothes washer, clothes dryer, television, or computer, that are used every week in your home.

2. List your devices in a table like **Table 1**. Use the table to keep track of the number of hours each device is in use each day.

Table 1 Electrical Energy Used in One Week

Date	Device	Wattage of device (kW)	Time used (h)	Kilowatt hours of energy (kW•h)	Operating cost

Equipment and Materials

graph paper

3. Identify the wattage of each device. This information can be found on the device.
4. Convert the wattage to units of kilowatts (divide watts by 1000). Record these values in your table.
5. Calculate the kilowatt hours of electrical energy used each day by each device. Use this equation:

$$\text{energy used (kW•h)} = \text{wattage (kW)} \times \text{time (h)}$$

6. Tally the amount of electrical energy used each day. Plot your data on a bar graph showing the total amounts for each day.
7. After seven days, calculate the total energy used by each appliance for the week. Then calculate the total amount of energy used for all of the appliances.
8. Suppose the cost of electricity is \$0.10/kW•h. Calculate the cost of electrical energy for each device for one week. Use the following equations:

$$\text{energy used (kW•h)} = \text{wattage (kW)} \times \text{time (h)}$$

$$\text{cost to operate (\$)} = \text{energy used (kW•h)} \times \text{cost of electricity (\$/kW•h)}$$

Analyze and Evaluate

(a) Which device used the most energy? Which used the least energy? T/I

(b) Analyze your graph to determine which day you used the most energy. What do you think is the reason for the higher energy usage? T/I

Apply and Extend

(c) Multiply your total electrical usage for the week by 52 to estimate how much energy these devices would use in one year. T/I

(d) Calculate the yearly cost of using these appliances. T/I

(e) How can you change the way you use these devices to conserve energy? List at least one way to reduce the cost of using each device. A

UNIT TASK

Bookmark

You can apply what you learned in this activity to the Unit Task described on page 536.

11.7 Energy Conservation

When we use less electrical energy in our daily lives, we reduce our personal energy consumption. This is an important part of energy conservation. Conserving energy now will reduce environmental damage and help ensure a supply of energy in the long term.

DIG DEEPER

Writing Tip

Use Evidence

To support the main idea, use facts, examples, or reasons. For example, in this section the author has used facts about greenhouse gas emissions to show that we must try to use electrical energy wisely instead of generating more.

Energy Use and the Environment

These days, our lives seem filled with more and more electrical devices. Using more electrical devices causes a growing demand for electrical energy. Generating more electrical energy is not a preferred option because it is costly and can harm the environment. For example, every kilowatt hour of energy that comes from the burning of coal adds about 1 kg of the greenhouse gas carbon dioxide to the atmosphere. The high levels of greenhouse gases in the atmosphere contribute to global climate change. If, however, you save 1000 kW•h of coal-generated electrical energy, you prevent 1 tonne of carbon dioxide from entering the air. A Canadian family can use more than 20000 kW•h of electrical energy in one year!

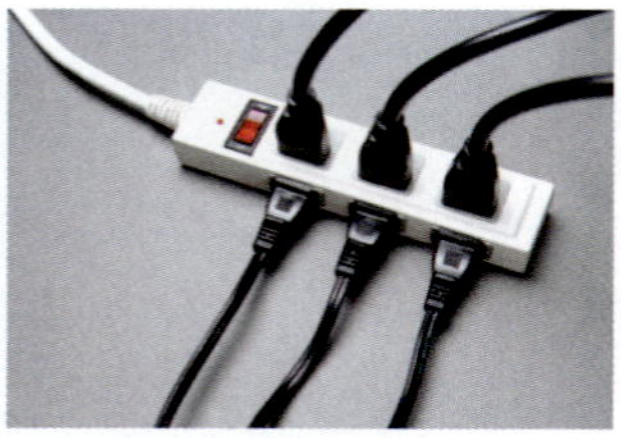

Figure 1 Using a power bar allows you to cut off power to multiple devices at the same time.

Using Energy Wisely

- **Unplug your chargers.** Cellphones, MP3 players, laptops, power tools, digital cameras, and portable gaming systems should be removed from the charger as soon as they are fully charged, and the charger itself should then be unplugged.
- **Use power bars.** Plug all TVs, DVD players, gaming systems, and stereos into power bars (**Figure 1**). Turn the power bars off when you are not using these devices. In standby mode, each of these devices can use as much electrical energy as a 60-W light bulb that is left on all the time.
- **Put it to sleep.** Look for the "sleep," "energy saver," or "hibernate" settings on your computer, and use them. When you have finished for the day, shut your computer off.
- **Turn off lights and devices.** You waste electricity when you leave on lights and other devices when no one is in the room.

- **Use appliances only when necessary and use them appropriately.** You can use a regular can opener rather than an electric one. You can chop vegetables rather than using a food processor. Do not leave a hair dryer running or the refrigerator door open.
- **Raise the temperature in summer, lower it in winter.** In summer, raise the temperature of your air conditioning by one degree. Even one degree can make a difference of 3 % on an electricity bill. Similarly, lowering the temperature in winter helps save energy.
- **Take shorter showers. Do only full loads of laundry, and use only cold water.** It takes a great deal of energy to heat water.
- **Change your light bulbs.** Switch to the most efficient light bulbs available.

DIG DEEPER

Web Link

For more information on how to conserve energy,

GO TO NELSON SCIENCE

Appliances and Their Efficiency

Another way that people can conserve electrical energy is to choose energy-efficient appliances. When you purchase any appliance, consider how much it will cost to use. A new energy-efficient refrigerator may cost more than a less efficient model, but will save money in energy costs over time. Also, by reducing your demand for energy, you will be helping to reduce pollution associated with energy production. To help you choose energy-efficient appliances, look for energy-efficient appliances labelled with an Energy Star symbol (**Figure 2**).

Figure 2 The Energy Star symbol is used to identify products that are energy efficient. They use less energy to accomplish the same task.

EnerGuide

The Energy Star rating system complements the EnerGuide rating system developed by the Office of Energy Efficiency of Natural Resources Canada. EnerGuide rates the energy use and efficiency of

- household appliances
- heating and cooling equipment
- ventilating equipment
- new houses
- personal vehicles

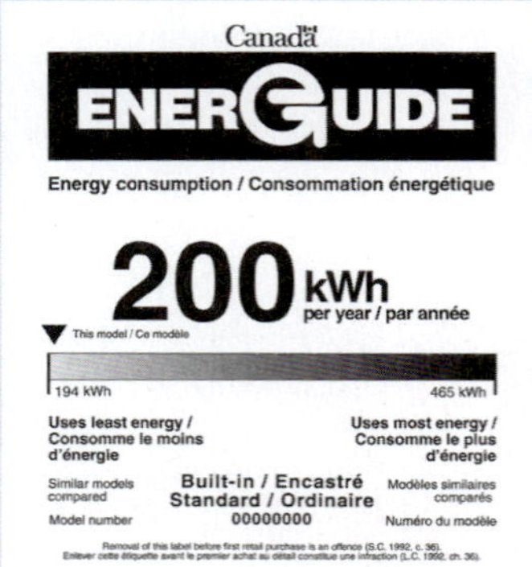

Figure 3 The EnerGuide label provides a great deal of information about an appliance. This information is determined by standardized test procedures.

The goal is to have people use the EnerGuide label (**Figure 3**) to make informed buying decisions by comparing the efficiency of different appliances, equipment, vehicles, and houses.

New homes and vehicles come with their own type of EnerGuide labels that help potential buyers choose the most energy-efficient model to suit their needs.

Homeowners who want to improve the energy efficiency of their homes can apply for grants to pay for an ecoENERGY Retrofit. This retrofit may include

- replacing inefficient heating and air conditioning systems
- replacing or improving insulation
- replacing windows, doors, shower heads, toilets, and light bulbs with Energy Star qualified versions

DIG DEEPER

Writing Tip

State Key Points

When writing persuasive text, use the body of the text to explain the key points of your argument. State your point clearly, and then support it with evidence. Use connecting words to show the relationship between ideas.

Producing Your Own Electrical Energy

Another way we can reduce our use of fossil fuels is by producing our own electrical energy. By using renewable energy sources such as solar and wind energy, home and business owners can generate some of the electrical energy they need. In areas where wind blows steadily and consistently, small-scale wind turbines can be used to generate electricity for a home or small business (**Figure 4**). Solar panels can also be installed on homes and offices.

Figure 4 A small wind turbine can generate electricity for a home or business.

More and more people and communities are producing their own electrical energy using renewable resources. Ontario's Green Energy Act (GEA) was passed to help them get funding.

These methods reduce the amount of energy people need to draw from their electrical energy distribution grid. In some cases, they may even be able to connect their solar or wind energy systems to the electrical energy distribution grid. Sometimes a personal generating system produces more electricity than one home needs. In this case, the extra energy can be sold to the local energy company. This reduces a household's energy costs even further.

DIG DEEPER

Web Link

For more information about the Green Energy Act,

GO TO NELSON SCIENCE

CITIZEN ACTION

Lights Out!

Grocery stores and malls use a lot of electrical energy to stay brightly lit. Next time you walk into one, ask yourself if you would be able to see the products as well if one-quarter, or even half, of the lights were turned off. Write a letter to a store manager, explaining the importance of conserving energy, and suggesting ways the store might save energy by using less light.

GO TO NELSON SCIENCE

11.7 Wrap Up

- Electrical energy should be conserved and used wisely.
- Taking simple steps, such as unplugging chargers or using energy-efficient lighting, can help conserve energy.
- When purchasing an electrical device, you should consider the price, the cost of operating it, and its environmental impact.
- People can generate their own electrical energy.

CHECK YOUR LEARNING

1. Why should we find ways to reduce our electrical energy consumption? A
2. When purchasing an electrical appliance, what are the two financial costs you need to consider? K/U
3. Of all the ways suggested for reducing energy use at home, which ones are new to you? Which ones do you think would be easiest to do? A
4. What are the financial benefits of producing your own electrical energy? What are the environmental benefits of producing your own electrical energy? A
5. (a) What household products and equipment does EnerGuide rate? K/U
   (b) Why should consumers buy products with Energy Star symbols? A

11.8 PERFORM AN ACTIVITY

Evaluating Electrical Energy Use at Home

SKILLS MENU
- Questioning
- Hypothesizing
- Predicting
- Planning
- Controlling Variables
- Performing
- Observing
- Analyzing
- Evaluating
- Communicating

Over the course of a year, the amount of electricity we use can increase or decrease, depending on the month. Our consumption usually follows a pattern. In this activity, use the information provided in **Table 1** and **Table 2** to graph the energy consumption of a family and a school. Then, use your graphs to help interpret the data.

Table 1 Home Electrical Energy Use by Month

Month	Electrical energy use (kW•h)
January	410
February	456
March	399
April	350
May	679
June	868
July	1119
August	1178
September	880
October	299
November	350
December	385

SKILLS HANDBOOK 6.A., 6.B.

Purpose

To graph and interpret electrical energy consumption over time for a home and a school

Procedure

Part A: Home Electrical Energy Use

1. Examine **Table 1**. The table shows one family's monthly electrical energy use in kilowatt hours for one year.
2. Use the information in **Table 1** to graph the family's monthly energy use. You can draw a bar graph or a line graph.

Equipment and Materials

graph paper

Part B: School Electrical Energy Use

1. Examine **Table 2**. The table shows one school's electrical energy use in kilowatt hours for one year.
2. Use the information in **Table 2** to graph the school's electrical energy use. You can draw a bar graph or a line graph.

Table 2 School Electrical Energy Use by Month

Month	Electrical energy use (kW•h)
January	220 000
February	198 000
March	145 000
April	174 000
May	135 000
June	148 000
July	68 000
August	71 000
September	140 000
October	190 000
November	230 000
December	215 000

Analyze and Evaluate

(a) In Part A, when was the family's electrical energy use highest? When was its electrical energy use lowest? What do you think accounts for this difference? T/I

(b) In Part B, when was the school's electrical energy use highest? When was its electrical energy use lowest? What do you think accounts for this difference? T/I

(c) How did the electrical energy use for the family and the school compare? List the major uses of electricity in a school, then list the major uses of electricity in a home. Compare your lists. Think about what the major uses of electricity are in both locations for lighting, air conditioning, heating, refrigeration, computers, and so on. T/I

Apply and Extend

(d) Most newly built homes use dual systems to provide heating. They combine gas heating with an electric heat pump. The heat pump does not heat the air. Also known as air-to-air heat exchangers, they reduce the amount of energy lost when cold air is brought into the house. This system uses much less energy than gas-only systems or electric-only systems. If this trend continues, how will it affect the energy use in winter months? T/I A

(e) The average Ontario home uses about 1000 kW•h of electricity per month. Find out what your household electrical energy consumption was for one or more months of the year. (You might ask to see your family's monthly electricity bills.) How does your household's electrical use compare to the average Ontario home? T/I A

11.9 PERFORM AN ACTIVITY

Reducing Energy Use

SKILLS MENU

- Questioning
- Hypothesizing
- Predicting
- Planning
- Controlling Variables
- Performing
- Observing
- Analyzing
- Evaluating
- Communicating

Conserving energy means making smart choices when purchasing devices that use electricity. We rely on electrical appliances in the home to cook, clean, entertain, and store food safely. Responsible energy use may mean choosing to replace older, less efficient appliances with ones that use less electricity.

DIG DEEPER

Career Link

Energy auditors help homeowners find ways to decrease their energy consumption. For more information about a career in energy auditing,

GO TO NELSON SCIENCE

SKILLS HANDBOOK 3.B.7., 3.B.8.

Purpose

To identify the best appliance to buy to reduce the amount of energy used

Procedure

1. Create a table in your notebook similar to **Table 1**.
2. Computers, televisions, clothes washers, and clothes dryers are among the electrical appliances many people use each day. List the electrical appliances you use in your home.

Table 1 Energy Use Comparison

Type of appliance	Energy use of existing appliance (kW•h)	Cost per year ($)	Energy use of new appliance (kW•h)	Cost per year ($)

Equipment and Materials

calculator (optional)

3. Choose two of the major appliances in your home. Older ones are preferable. Look for the wattage. The label with this information is usually on the inside of the door or on the back or bottom of the appliance.
4. Estimate how much time (in hours) each appliance is in use per week. Then multiply by 52 to estimate how much time each appliance is used per year. You can use the information in the appliances' EnerGuide label if it is available.
5. If electricity costs $0.10 per kilowatt hour, calculate the approximate cost to operate each appliance per year.
6. You could save energy by replacing an older appliance with a newer, more energy-efficient one. Use the Internet or visit a local appliance store. For example, if one of your choices is an older, top-loading washer, find out how much electricity a new energy-efficient front-loading washer uses. If your computer has a CRT monitor, find out how much electricity an LCD monitor uses. Record the make and model of the appliances you examine.
7. Use the EnerGuide estimates or calculate the cost of using each new appliance for one year.
8. Write calculations and the data you gather in your table. Compare the energy use of the appliances you have with the energy use of newer or more efficient appliances.

DIG DEEPER

Learning Tip

CRT stands for cathode ray tube and LCD stands for liquid crystal display.

Analyze and Evaluate

(a) Study the information in your table. Based on your data, which make and model of a new appliance would make the biggest difference in your home's energy cost per year? T/I

(b) Write a proposal in which you recommend this change. Appliances are expensive. Make sure you present your recommendations based on facts and figures. T/I C

Apply and Extend

(c) Your refrigerator stopped working, and your family is shopping for a new one. One refrigerator costs $460 and is estimated to use 468 kW•h per year. A similar, Energy Star® refrigerator costs $546 but uses 365 kW•h per year. What is the yearly cost of using each one? How many years would you need to use the Energy Star® refrigerator to make up the difference in cost? T/I A

(d) List some environmental concerns that should be considered when you replace old appliances with new ones. A

Chapter 11 KEY CONCEPTS SUMMARY

Electrical energy is generated using a variety of technologies.

- A generator is a device that rotates a wire near a magnet to produce electricity.
- There are many different ways of supplying the motion needed to produce electricity.

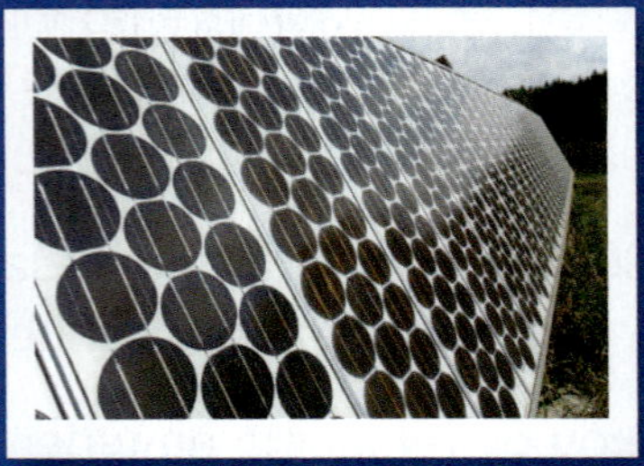

Electrical energy is generated using both renewable and non-renewable sources.

- Renewable sources include falling water, wind, solar, biomass, and geothermal.
- Non-renewable sources include fossil fuels, which are burned to produce steam, and uranium, used for nuclear power.

Electrical energy production has an effect on the environment and society.

- Using fuel causes pollution and contributes to global change.
- Even renewable sources of energy can change ecosystems and have negative effects on animal habitats.

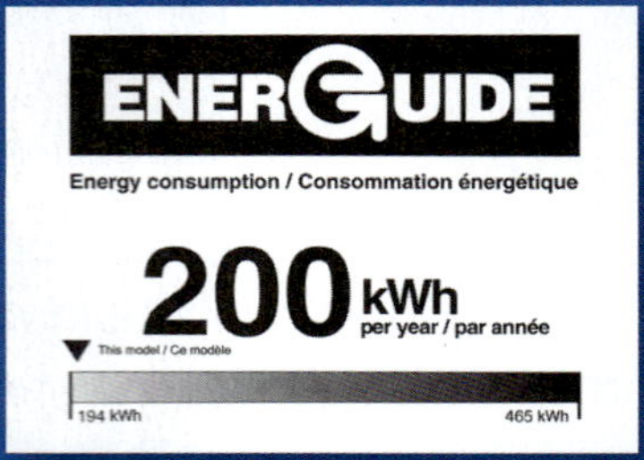

Electrical energy should be conserved and used wisely.

- Efficient devices use less energy to do the same tasks.
- Replacing non-renewable sources of electricity with renewable sources can help reduce the use of fossil fuels.

WHAT DO YOU
THINK NOW?

You thought about the following statements at the beginning of the chapter. You may have encountered these ideas in school, at home, or in the world around you. Consider them again and decide whether you agree or disagree with each statement.

1 Using electricity creates pollution.
Agree/disagree?

4 When some electrical devices are turned off, they continue to use energy.
Agree/disagree?

2 It is easy to produce more electricity when needed.
Agree/disagree?

5 Rechargeable batteries are better for the environment than disposable batteries.
Agree/disagree?

3 We will run out of electricity someday.
Agree/disagree?

6 All known methods of generating electricity have both advantages and disadvantages.
Agree/disagree?

How have your answers changed since then?
What new understanding do you have?

BIG Ideas

- Electricity is a form of energy produced from a variety of non-renewable and renewable sources.
- The production and consumption of electrical energy has social, economic, and environmental implications.

DIG DEEPER

Vocabulary

electrical energy p. 420
electrical energy distribution grid p. 422
generator p. 422
kilowatt p. 423
kilowatt hour p. 423
non-renewable energy source p. 424
thermal generation p. 424
renewable energy source p. 428
photovoltaic cell p. 429
electrical energy consumption p. 434
efficiency p. 436

CHAPTER 11

SELF-QUIZ

The following icons indicate the Achievement Chart category addressed by each question.

K/U Knowledge/Understanding
T/I Thinking/Investigation
C Communication
A Application

For each question, select the best answer from the four alternatives.

1. Which of the following has chemical energy? (11.1) K/U
   (a) music
   (b) wind
   (c) lightning
   (d) gasoline

2. Which action involves transforming chemical energy into thermal energy? (11.1) K/U
   (a) using a toaster
   (b) having a campfire
   (c) sailing a boat
   (d) beating a drum

3. What three things do you need in order to generate electricity using a generator? (11.1) K/U
   (a) wind, steam, and the Sun
   (b) wire coil, a turbine, and wind
   (c) motion, wire coil, and a magnet
   (d) light energy, thermal energy, and sound energy

4. What are kilowatt hours used to describe? (11.1) K/U
   (a) energy costs
   (b) type of energy
   (c) energy use
   (d) energy sources

5. Which of the following is a fossil fuel? (11.2) K/U
   (a) coal
   (b) carbon dioxide
   (c) steam
   (d) uranium

6. Which of the following statements describes a renewable energy source? (11.3) K/U
   (a) It always increases greenhouse gases.
   (b) It is always formed from the remains of living things.
   (c) It will eventually be used up.
   (d) It can be replaced by natural processes.

Indicate whether each of the statements is TRUE or FALSE. If you think the statement is false, rewrite it to make it true.

7. Using fossil fuels to produce electrical energy contributes to global climate change. (11.2) K/U

8. A cellphone charger stops using electricity as soon as the cellphone is unplugged from it. (11.5) K/U

9. All brands of refrigerators use the same amount of energy to keep food cold. (11.7) K/U

10. People can use solar panels and small wind turbines to generate electricity for their homes. (11.7) K/U

Copy each of the following statements into your notebook. Fill in the blanks with a word or phrase that correctly completes the sentence.

11. ________ is one of the substances used to produce nuclear energy. (11.2) K/U

12. Thermal energy may be used to produce ________, which then turns a turbine. (11.2) K/U

13. ________ power plants produce radioactive wastes. (11.2) K/U

14. A photovoltaic cell converts ________ energy into electrical energy. (11.3) K/U

Match each term on the left with the most appropriate description on the right.

15. (a) electrical energy consumption
    (b) thermal generation
    (c) fossil fuels
    (d) greenhouse gases

    (i) formed from plants, animals, and micro-organisms that lived millions of years ago
    (ii) how much electrical energy is used
    (iii) generating electricity using steam
    (iv) substances that trap the energy radiated by Earth's surface (11.2, 11.5, 11.7) K/U

Write a short answer to each of these questions.

16. How much would it cost to run a video game player 4 h per day for a week? The player is rated at 1 500 W. Electricity costs \$0.10/kW•h. (11.5) T/I

17. A CD player, rated at 100 W, cost \$75 new. It was used for a total of 9 000 h before it was replaced. How much did it cost to operate the CD player during its full lifetime of use? Electricity costs \$0.10/kW•h. (11.5) T/I

18. The school district is going to buy 100 new computers for each of 8 high schools in a small city. They have to decide between laptops and desktop models. The desktop has a wattage of 175 W, and the laptop has a wattage of 45 W. Would this difference be enough to make the school district choose a laptop even if it were more expensive? Explain your reasoning. (11.5) T/I

19. A family wants to reduce the energy they use from the grid. They ask you for advice. What do you suggest? (11.7) A

20. How can you use a power bar to help reduce your energy use? (11.7) K/U

21. A family has decided to stop using their air conditioner. (11.8) C A
    (a) How will this affect the pattern of electrical use over the course of one year? How would a graph of this family's electrical energy use and a graph of a family that uses air conditioning be different? Sketch what you think each family's graphs would look like.
    (b) What are some other actions the family could consider taking to conserve electrical energy? Explain why these actions would help.

22. Compare fossil fuels and biomass. How are they similar? How are they different? (11.2, 11.3) K/U

23. Suppose you live in a large apartment complex. It is totally dependent on the local grid. All of the generating stations on this grid burn fossil fuels to produce electricity. How could the complex switch from fossil fuels to using as many different types of renewable sources of electrical energy as possible? (11.2) K/U

24. A friend says that using solar panels to generate electricity at your school would have no disadvantages at all. Is she correct? Explain your answer. (11.3) A

Self-Quiz

To do an online self-quiz,

GO TO NELSON SCIENCE

CHAPTER 11 REVIEW

The following icons indicate the Achievement Chart category addressed by each question.
K/U Knowledge/Understanding
T/I Thinking/Investigation
C Communication
A Application

What Do You Remember?

1. Explain how a turbine and a generator work together to generate electricity. (11.1) K/U
2. Identify two main advantages common to all renewable energy sources used in generating electrical energy. (11.3) K/U
3. Identify how you could generate electrical energy for each of the following locations or devices. Use renewable sources whenever possible. Justify your answer. (11.2, 11.3) A
   (a) two cities near a river
   (b) your home
   (c) a cattle ranch
   (d) an electric car
   (e) a flashlight that is used infrequently
4. What is one disadvantage of using moving water to produce electrical energy? (11.3) K/U

What Do You Understand?

5. A farmer wants to install a renewable energy system to help generate electricity for his farm. What are his options? What advantages and disadvantages does each option have? (11.3) A
6. Why do you think generating stations that use fossil fuels to produce electrical energy are more common than other kinds of generating stations? (11.2) A
7. Compare one non-renewable energy source with one renewable energy source. Consider pollution, cost, and how the electrical energy is produced. (11.2, 11.3) A
8. You have just been elected mayor of Futuretown. The suburban town has 100 000 residents and a small factory. There is a wide, slow-moving river nearby. Your residents need a source of electrical energy. In two or three paragraphs, describe what sources you would propose using and why. (11.3) C A
9. EnerGuide labels (**Figure 1**) are on two similar refrigerators that you are looking at. How can the information on the labels help you decide which device to purchase? (11.7) T/I

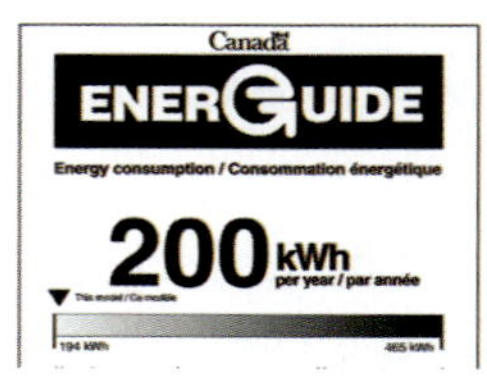

Figure 1

Solve a Problem

10. (a) Jack has a large aquarium with a heater that runs 12 h a day. The heater is rated at 150 W. Electricity costs $0.10 per kilowatt hour. How much would it cost to heat the aquarium for a year? (11.5)
    (b) In 2015, the cost of electricity increases to $0.15 per kilowatt hour. How much more will it cost Jack to heat his aquarium for a year? (11.5) T/I
11. You have decided to replace one of your older household appliances with a new, energy-efficient model. (11.7) A
    (a) How will you choose which one to replace?
    (b) What resources will help you make the decision?

12. (a) Calculate the annual cost of operating the electrical appliances listed in **Table 1** below. The cost of electricity is $0.10/kW·h.
(b) Look at **Table 1** on page 446. Do the data look similar to your household? How are they the same? How are they different? Are the total hours used per year reasonable? Explain your answers.
(c) A homeowner wants to reduce the amount of money she spends on electricity. If the data in the table represent this homeowner's electrical consumption, what steps can she take to reduce her energy use? (11.5, 11.7, 11.8, 11.9) T/I

Table 1 Electrical Appliance Data

Electrical appliance	Wattage (W)	Time per year (h)
stove	12 000	300
microwave	1 000	120
TV	300	700
computer	250	700
washer	1 300	60
dryer	5 600	70

Create and Evaluate

13. Is it better to generate more electrical energy from renewable energy sources or conserve electrical energy? Discuss your opinion in a paragraph with supporting information. (11.2, 11.3, 11.7) C A

14. Develop a poster or radio announcement to promote energy conservation in your community. Explain the steps people can take to reduce the amount of energy they use daily. (11.7) K/U

Reflect on Your Learning

15. Describe an idea about the generation of electrical energy that was new to you. Will you change anything in your everyday life because of this idea? Explain your answer. C

Web Connections

16. Ontario is banning the incandescent light bulb because it is very inefficient. Find out how this ban will affect average electrical energy usage in the average home. How will this ban affect the environment? Write a paragraph about what you find. T/I C A

17. Sir Adam Beck was an important Canadian. Two hydro-electric generating stations are named after him. Research to find out how Adam Beck's work affected the well-being of the people of Ontario. Write a short report about your findings. T/I C

18. Mercury is an element that is poisonous to organisms. Using compact fluorescent light (CFL) bulbs reduces the burning of coal, which is a major source of mercury pollution. CFL bulbs contain a very small amount of mercury. Research the overall impact of using CFL bulbs on mercury pollution. Present a proposal on the proper disposal of these bulbs. T/I C A

19. Ontario's Green Energy Act (GEA) is one way Ontario is encouraging people to produce their own electrical energy. T/I C A
(a) What is the Green Energy Act?
(b) How will the Green Energy Act encourage people to use renewable energy sources?

GO TO NELSON SCIENCE

CHAPTER 12

Static and Current Electricity

KEY QUESTION

What is electricity?

KEY CONCEPTS

- Electricity describes how electrons move from one place to another.
- Static electricity is the result of an imbalance of charges on the surface of an object.
- There are many useful applications of static electricity.
- Objects can be charged by contact or by induction.
- Current electricity is generated by moving electrons in a closed path.
- Some materials allow electrons to move through them easily (conductors), while others do not let electrons move through them easily (insulators).

ENGAGE IN
SCIENCE

Electrostatic Discharges

The dramatic bolt of lightning you see in the sky is one of the most spectacular displays of electrostatic discharge (ESD). ESD is the rapid transfer of charge between two objects.

Most people know that lightning is dangerous. A single lightning strike can have more than 5 billion joules of energy! This is enough energy to turn sand into glass instantly. How do scientists study lightning when it is so dangerous?

Scientists at the University of Toronto use the CN Tower to study lightning strikes. They want to know when and why it strikes and to understand how lightning causes damage to buildings and electronics. Lightning hits the CN Tower about 75 times a year. Other scientists use machines like van de Graaff generators to create their own lightning. The sparks from a van de Graaff generator are another form of electrostatic discharge.

Sparks fly off a van de Graaff generator.

Not all forms of ESD are as dramatic as lightning strikes. The small spark you feel when you touch a doorknob after walking across a carpet is ESD. It causes the crackling noise you hear when you pull clothes out of the dryer. A tiny ESD can cause your MP3 player or laptop to stop working. This is why technicians who work on electrical devices wear special clothing to prevent ESD.

WHAT DO YOU THINK?

Many of the ideas you will explore in this chapter are ideas that you have already encountered. You may have encountered these ideas in school, at home, or in the world around you. Some of the statements are not true. Consider each statement and decide whether you agree or disagree with it.

1 You use static electricity to recharge your cellphone.
Agree/disagree?

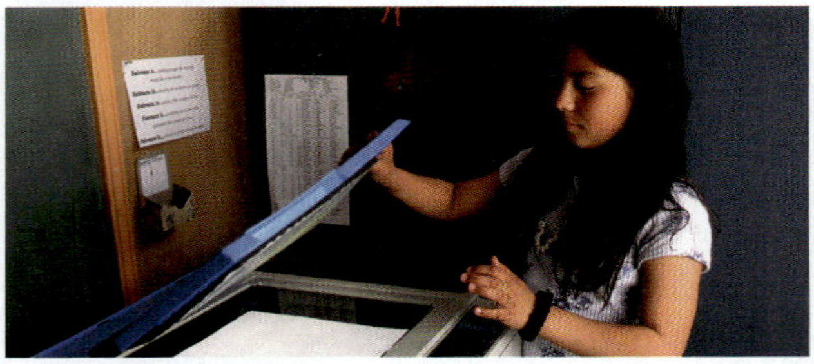

2 Both photocopiers and lightning involve static electricity.
Agree/disagree?

3 Electricity can travel over your skin without harming you.
Agree/disagree?

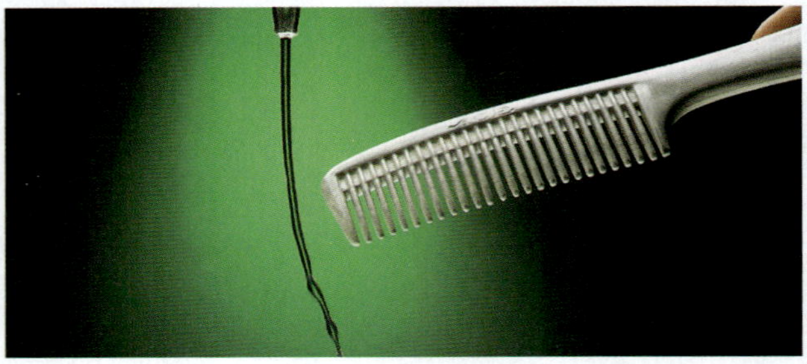

4 You can use static electricity to move objects without touching them.
Agree/disagree?

5 Current electricity moves through a wire like water flows through a pipe.
Agree/disagree?

6 Static electricity is a useful form of electricity.
Agree/disagree?

FOCUS ON

READING

Evaluating

When you evaluate a text, you make a judgment about it. You express whether you agree or disagree with the ideas and information. Is the author biased? Is there information that is missing? Does it fit with your opinion or what you know?

DIG DEEPER

Reading Tip

As you work through this chapter, look for tips like this. They will help you develop literacy strategies.

The Grid

Electric current travels to your house through an amazing system called the electrical energy distribution grid. **Figure 6** shows how the electric energy is produced at the generating plant. Fossil fuels (coal, natural gas, or oil) are burned, or nuclear energy is used to heat water into steam. Steam, water, or diesel engines spin fan-like turbines. The spinning turbines spin coils of wires between magnets inside the generators. This produces an **alternating current** that flows first in one direction through the wires, then reverses, and goes in the other direction. This is the current used in our homes and offices. This current reverses itself 60 times each second.

▸ **alternating current (AC):** an electric current that repeatedly reverses direction

Evaluating *in Action*

Here is how one student asked and answered questions to evaluate the selection about the electrical grid on page 487.

Questions	Answers
What do I already know about the topic?	I have seen transmission towers and hydro poles and wires.
Is the information accurate and does it support the main idea?	The text does not explain why the current reverses.
Is the text biased?	The author calls the grid "amazing." The text does not explain the dangers of nuclear energy.
Opinion/Judgment: The text helps me understand how energy is produced, but it does not help me understand why current reverses direction.	

12.1 Electric Charges

Have you ever walked across a carpet in stocking feet and reached for the doorknob only to be "zapped" by an unexpected shock? Why does your hair mysteriously stand on end after you pull off a wool hat on a winter day? These events are actually caused by the same thing.

Charges

To understand what is happening, we need to look at the tiny particles that make up the atoms of the rug, the hat, the socks, and us. You read in Chapter 6 that these small particles are called protons, electrons, and neutrons.

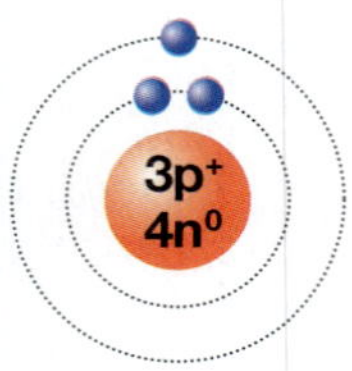

Figure 1 A lithium atom. Electrons (blue) can move from atom to atom.

Atomic Structure and Electric Charge

Figure 1 is an atomic model of lithium. Recall that neutrons and protons are located in the nucleus, or centre, of the atom and cannot move from atom to atom. Electrons move around the space surrounding the nucleus and can move from atom to atom. Each element has a different number of protons, neutrons, and electrons. For example, lithium has three protons, four neutrons, and three electrons.

Protons repel (push away) each other. Electrons also repel each other. Protons and electrons attract (pull together) each other. This force of attraction or repulsion is known as **electric force**. Objects that exert an electric force are said to have an **electric charge**. Protons have a positive electric charge. Electrons have a negative electric charge. Neutrons do not have an electric charge (**Table 1**).

When there is an equal number of protons and electrons in an atom, the number of positive and negative charges is equal. The charges balance each other. The atom is neither positive nor negative—it is **neutral**.

- **electric force:** force of attraction or repulsion between charged objects
- **electric charge:** positive or negative charge on objects that exert an electric force
- **neutral:** having no charge or an equal number of positive and negative charges

Table 1 Properties of the Particles of an Atom

Particle	Electric charge	Location	Particle symbol
proton	positive	nucleus	p^+
neutron	no charge	nucleus	n^0
electron	negative	outside nucleus	e^-

Charged and Neutral Objects

Objects are made up of billions and billions of atoms. Whether an object has a charge depends upon the charge of its atoms. The outer electrons of some atoms can move from one atom to another. When atoms gain electrons, they become negatively charged. When atoms lose electrons, they become positively charged. When atoms become charged, the object becomes charged.

As Figure 2(a) shows, a **neutral object** has an equal number of protons and electrons. An object that has more electrons than protons is **negatively charged** (Figure 2(b)). An object that has fewer electrons than protons is **positively charged** (Figure 2(c)).

- **neutral object:** an object that has an equal number of protons and electrons
- **negatively charged object:** an object that has more electrons than protons
- **positively charged object:** an object that has fewer electrons than protons

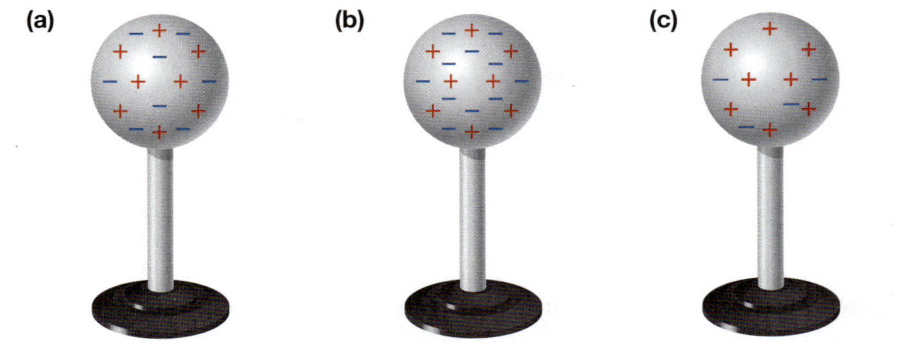

Figure 2 (a) A neutral object (b) Negatively charged objects have more electrons than protons. (c) Positively charged objects have fewer electrons than protons. (Note: Each charge symbol represents a very large number of protons (+) or electrons (–).)

DIG DEEPER

Did You Know?

Charge and Recycling

Recycling centres use the fact that different materials can have different charges to make separating them faster and easier. For example, plastic can be separated from paper because each can be given a different charge.

TRY THIS TESTING CHARGED OBJECTS

SKILLS HANDBOOK 3.B.3., 3.B.6., 3.B.9.

SKILLS: Hypothesizing, Observing, Communicating

In this activity, you will examine the interactions of objects that have been given an electric charge.

Equipment and Materials: acetate strips (2); ring clamp; retort stand; vinyl strips (2); thread; tape; paper towels

1. Use a piece of thread and tape to suspend and balance an acetate strip from a ring clamp as shown in Figure 3.
2. Rub the hanging acetate strip at both ends with the paper towel. Assume that the acetate strip has become positively charged.

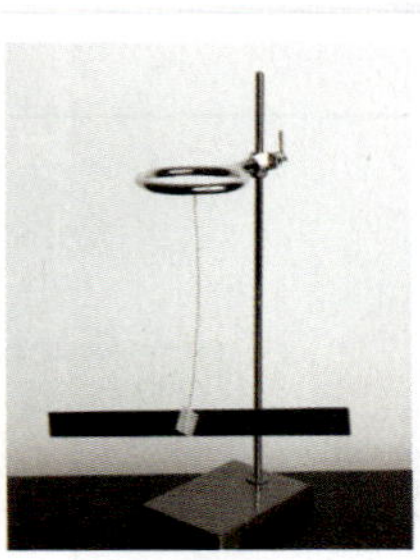

Figure 3

3. Rub the second acetate strip with a paper towel and bring the second strip near the hanging acetate strip. Do not let them touch. Record your observations.
4. Rub a vinyl strip with a paper towel. Assume that the vinyl strip has become negatively charged. Bring the vinyl strip near the hanging acetate strip. Record your observations.
5. Repeat Steps 1 through 4 with a hanging vinyl strip. Record your observations.

A. Based on your observations, how do objects with similar charges behave? T/I

B. How do objects with different charges behave? T/I

C. Make a hypothesis to explain why this occurs. T/I

Law of Electric Charges

Charged objects exert forces on each other. In the 1700s, scientists tested charged objects with many different materials. They observed that this force may be attractive or repulsive. They generalized their observations in the **Law of Electric Charges**, which states the following:

- Objects that have opposite charges attract each other. Negative attracts positive and positive attracts negative (**Figure 4(a)**).
- Objects that have the same charges repel each other. Negative repels negative and positive repels positive (**Figure 4(b)**).

▶ **Law of Electric Charges:** objects that have opposite charges attract each other; objects that have like charges repel each other

DIG DEEPER

Learning Tip

Opposites Attract

To remember the Law of Electric Charges, remember the phrase "opposites attract." Opposite charges attract and like charges repel.

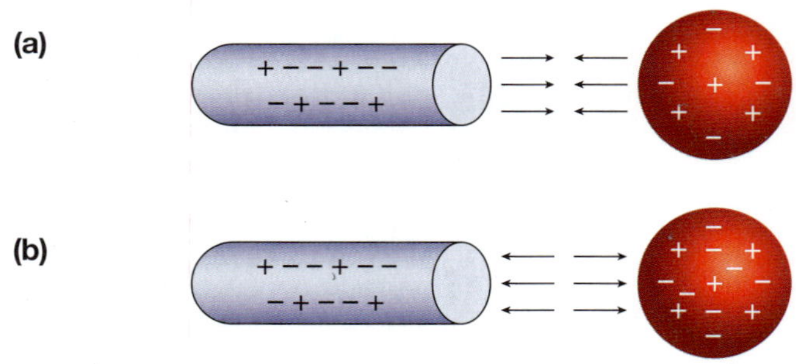

Figure 4 (a) Count the charges in each object. The rod is negatively charged and the ball is positively charged. They attract each other. (b) Both the rod and the ball are negatively charged. They repel each other.

The strength of the force between charged objects is related to the distance between objects and the amount of charge on each object. The electric force decreases as distance between the objects increases (**Figure 5(a)**). The electric force increases as electric charge increases (**Figure 5(b)**).

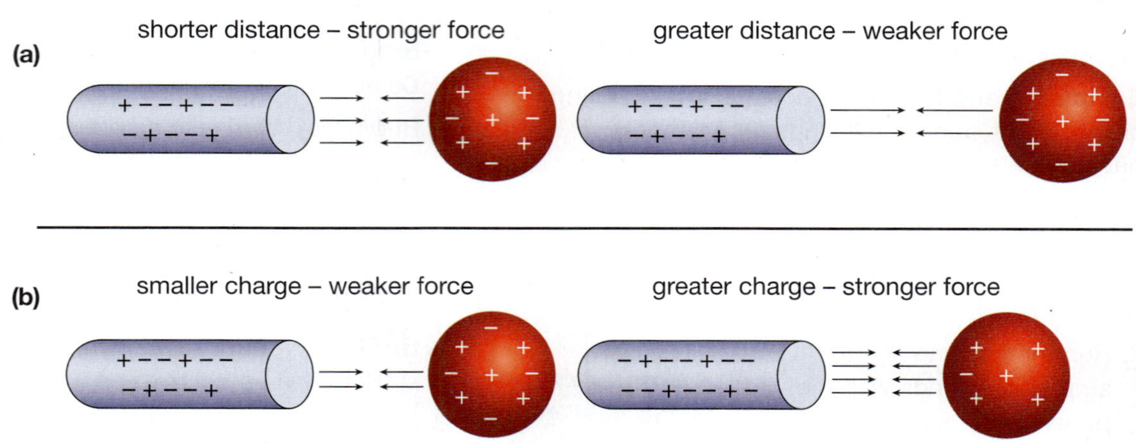

Figure 5 (a) For a given amount of charge, the amount of force decreases as distance increases. (b) For a given distance, the amount of force increases as the amount of charge increases.

TRY THIS MOVING OBJECTS BY CHARGING

SKILLS HANDBOOK 3.B.

SKILLS: Predicting, Performing, Observing, Communicating

In this activity, you will investigate the attractive and repulsive forces between charged objects.

Equipment and Materials: balloon; wool cloth; foam pellets (or puffed-rice cereal); packets of salt and pepper; blank white paper

1. Blow up the balloon and rub it with the wool cloth.
2. Place a handful of the foam pellets on a white piece of paper on your tabletop.
3. Start from a distance away, then slowly bring the balloon close to the foam pellets. Observe what happens as the balloon gets closer to the foam pellets. Record your observations.
4. Clean off the piece of paper. Pour the packet of salt and an equal amount of pepper onto the piece of paper. Mix the salt and pepper well with your finger.
5. Recharge the balloon by rubbing it with the wool cloth again. Slowly bring the charged part of the balloon over the salt and pepper mixture. Record your observations.

A. How did the distance of the balloon from the salt, pepper, and foam pellets affect your results? T/I

B. Why do you think you use lightweight materials? T/I

C. How could you use the properties of the balloon to separate the mixture? T/I

12.1 Wrap Up

- Charged objects exert an electric force on each other. Objects with like charges repel each other. Objects with opposite charges attract each other.
- Protons have a positive charge. Electrons have a negative charge. Neutrons have no charge.
- Electrons can move from one atom to another. When electrons move, they transfer electric charges.
- Protons do not move from one atom to another.
- Neutral objects have an equal number of protons and electrons.
- When a neutral object gains electrons, it becomes negatively charged. When a neutral object loses electrons, it becomes positively charged.
- Electric force increases with an increase in electric charge and decreases with increasing distance between the objects.

CHECK YOUR LEARNING

1. Explain why atoms are electrically neutral. K/U
2. (a) Which particles of an atom move most easily from one material to another?
   (b) How does the movement of charge affect the overall charge on objects? K/U
3. How does understanding atomic structure help explain how charges can be transferred between objects? Draw a diagram to help explain your answer. K/U C
4. How is a neutral object different from a negatively charged object? K/U
5. You rub a neutral object with a piece of wool. Some of its electrons are transferred to the wool. Is the object now neutral, negatively charged, or positively charged? How do you know? T/I
6. Object A is negatively charged. It is attracted by object B. Object C is repelled by object B. What charge does object C have? How do you know? T/I

12.2

Static Electricity and Discharges

What do lightning and a photocopier have in common? The surprising answer is static electricity.

▶ **static electricity:** an imbalance of electric charge at rest on the surface of an object

Static Electricity

Static electricity is an imbalance of electric charge on the surface of an object. "Static" means not moving. This form of electricity is called static because the charges stay in one place until they move suddenly from one object to another. The device the girl is touching in **Figure 1** is called a van de Graaff generator. It generates static electric charge. The student's hair in the picture shows what can happen when static electricity builds up on someone touching the generator. Each strand of hair has the same static charge. This makes the strands repel one another and spread out as far from one another as possible.

DIG DEEPER

Reading Tip

Challenging Beliefs

You may read explanations that challenge a belief you hold about something you have experienced. Evaluate what you read carefully. You may learn something that will change what you think about static electricity.

Figure 1 When you touch a van de Graaff generator as it is charging, a static charge builds up on your hair. This causes the strands to repel each other and stand on end.

▶ **electric discharge:** the rapid transfer of electrons from one object to another

Electric Discharge

An **electric discharge** (also known as electrostatic discharge or ESD) is the sudden movement of static charge between two objects. When you rub your socks across a carpet, electrons from the carpet may move to you. When you then touch a metal doorknob, the electrons can move from you to the doorknob. You feel a shock and you may even see the small spark of the discharge.

Small electric discharges are not harmful to you, but they can be strong enough to damage delicate computer parts such as hard drives or USB memory sticks. In fact, the microchips and circuitry in smaller consumer electronics can be damaged by much smaller electric discharges.

Lightning

A lightning bolt is an example of a natural electric discharge. It is caused by the difference in charge between a cloud and the ground, between two clouds, or between parts of a cloud. What causes this imbalance in charge?

The process is not fully understood, but scientists do have a theory. Swirling winds in a storm cloud cause water and ice particles to come in contact and become charged. Positive charges move to the top of the cloud where ice and snow crystals form. Negatively charged water droplets move to the bottom of the cloud (**Figure 2**).

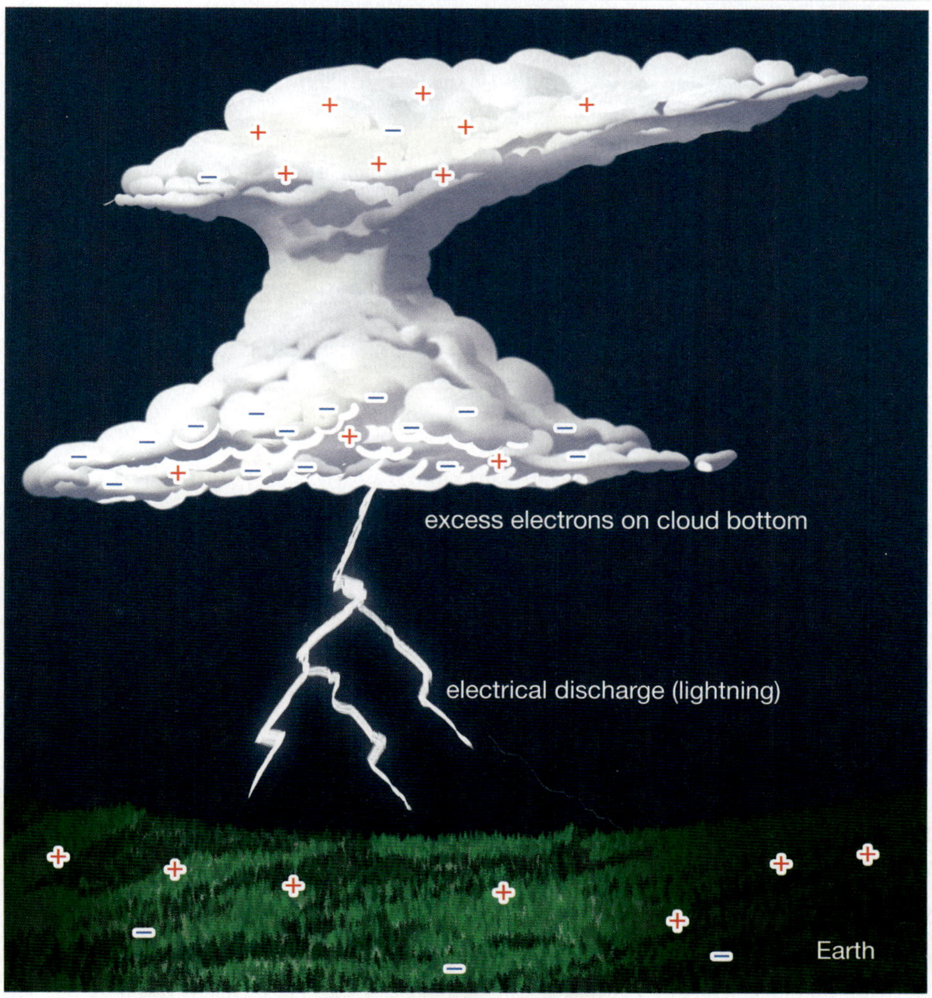

Figure 2 The imbalance of charge between a cloud and the ground results in a brilliant flash of lightning.

DIG DEEPER

Web Link

To learn more about how to prevent electrostatic discharge (ESD) and to investigate the spectacular natural phenomena of lightning,

GO TO NELSON SCIENCE

DIG DEEPER

Did You Know?

Anti-static Packaging

The packaging used to ship computers, video games and other electronic devices is specially designed to protect the devices from harmful electrostatic charges. You may have seen these special non-conductive bags made of plastic, but, in an effort to go green, more and more manufacturers are switching from oil-based plastics to paper-based packaging.

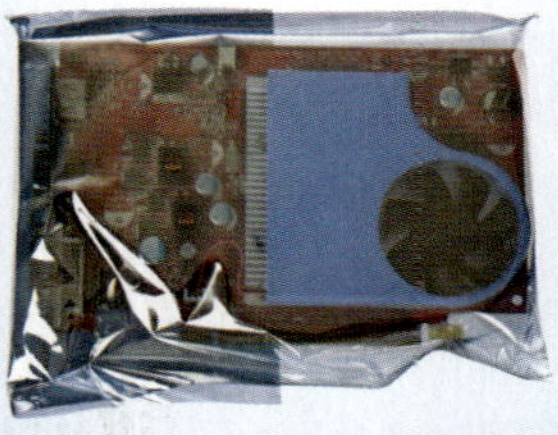

DIG DEEPER

Did You Know?

Amber

The word *electricity* comes from the Greek word (*ēlektron*) for amber, a type of fossilized tree sap. Ancient Greeks observed that if you rubbed materials such as wool or fur against amber, it would attract small objects.

The ground below the cloud develops a positive charge. When the difference in charge between the cloud and Earth becomes great enough, a path forms between the cloud and the ground. When the two parts meet, a surge of electrons travels along the path. This produces a giant flash called lightning (**Figure 3**). The lightning heats the surrounding air, making it expand very rapidly. This causes the sudden, loud sound that we hear as thunder.

Figure 3 A cloud-to-ground lightning strike

Figure 4 Vinyl decals use static electricity to cling to surfaces.

Uses of Static Electricity

Most of the time we think of static electricity as an annoyance. It makes our hair look messy and can make our clothes stick together. There are, however, many practical uses for static electricity.

Static-cling Film

Advertisers use static-cling vinyl film for brightly coloured removable signs, decals, and stickers like the one shown in **Figure 4**. The windows of many homes, office buildings, and cars are covered with filters that screen out sunlight. These screens use static electricity to cling to the window.

Photocopiers

Photocopiers convert an image captured by a scanner into a laser beam. This beam is projected onto a drum or belt, creating a negatively charged pattern. The positively charged ink, called toner, is attracted to the negative charges in the image. As negatively charged paper moves past the drum, the toner particles are transferred to the paper, recreating the image (**Figure 5**).

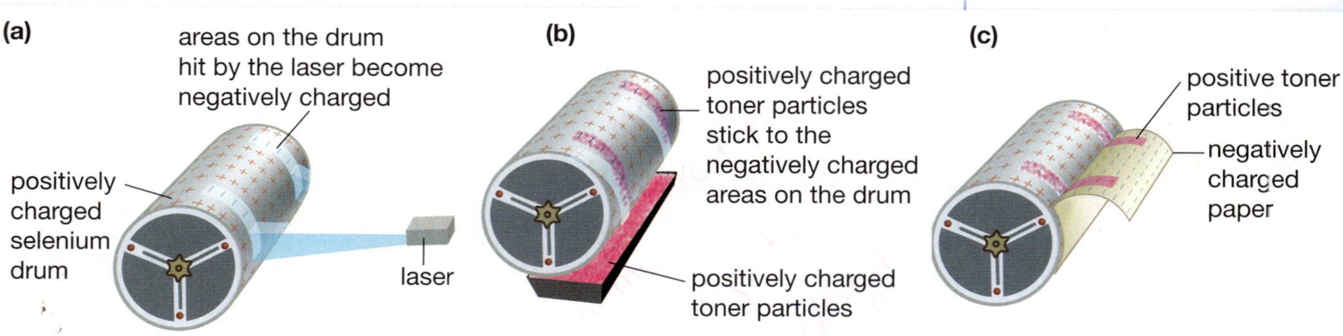

Figure 5 (a) The laser "draws" a negatively charged image on the drum. (b) The drum rolls across the positively charged toner, which sticks to the negatively charged laser "drawing." (c) The drum rolls across paper with a higher negative charge, and the toner particles "stick" to the paper.

12.2 Wrap Up

- Static electricity is an imbalance of charge at rest on the surface of an object.
- An electric discharge is a sudden movement of electrons between two objects with different charges.
- Lightning is a natural discharge of electrical energy.
- Static electricity has many real-world applications.
- Static electric discharges can damage electronic devices.

CHECK YOUR LEARNING

1. What causes the unexpected spark when you walk across a rug and touch a metal doorknob? K/U
2. Why do computer technicians have to make sure they do not carry any static electric charge? A
3. Why does someone's hair stand on end when they are touching a charging van de Graaff generator? A
4. (a) What is the theory as to how static charge builds up in a thundercloud? You may wish to draw a diagram to illustrate your explanation.
   (b) Explain what causes the explosive release of energy in a lightning bolt. K/U C
5. How do photocopiers make use of static electric forces? K/U
6. Write an advertisement about the advantages of static cling signs and decals. C A

12.3 Charging by Contact

Hat hair and clingy clothes from the dryer are two everyday examples of static electricity. The static electricity is the result of two neutral objects made of different materials that have become charged after being rubbed together. How does this happen?

Charging by Friction

When neutral objects of different materials are brought together, electrons are transferred from one object to the other. In the process, each object becomes charged. One object is more likely to attract extra electrons and become negatively charged. The other object gives up electrons and becomes positively charged. Rubbing an object or surface against another object creates friction, which speeds up the transfer of electrons. The attraction between objects charged by friction shows that they have opposite charges. The process of charging neutral objects by rubbing is known as **charging by friction**.

▶ **charging by friction:** the charging of two neutral objects made of different materials by rubbing them together

For example, when you comb your hair, both the comb and your hair start out neutral (**Figure 1(a)**). When your hair and the comb are rubbed together, the atoms in the plastic comb have a slightly greater ability to attract electrons than the atoms in your hair. Therefore, some electrons move from your hair to the comb. Your hair is now attracted to the comb (**Figure 1(b)**).

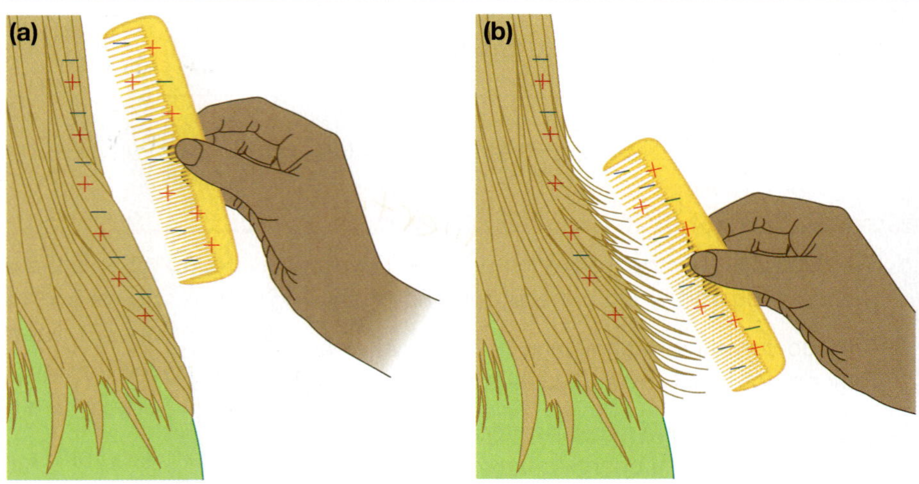

Figure 1 (a) The hair and the comb start out electrically neutral. (b) After combing, the comb is negatively charged and the hair is positively charged.

DIG DEEPER

Reading Tip

Evaluating

To evaluate the main idea, think about examples from your life. Have you experienced charging by friction when combing your hair? Apply the explanations you read to your own experience. This will help you determine whether the main idea is reasonable.

Effects of Humidity

You may have noticed that your hair does not stick to your hat on wet, humid days. In humid air, water molecules have more opportunity to bump into nearby objects. During these "bumps," water molecules are able to transfer electrons from negatively charged objects to positively charged objects. As a result, charged objects tend to lose their charge quickly in humid weather. In dry weather, there are fewer water molecules in the air, so charged objects are less likely to lose charge by bumping into water molecules. Charged objects stay charged in dry weather, making your hair more prone to static electricity.

TRY THIS CHARGING BY CONTACT

SKILLS HANDBOOK 3.B.5., 3.B.6.

SKILLS: Performing, Observing, Communicating

You do not need to rub neutral objects together for them to be charged. In this activity, you will explore charging objects by simple contact.

Equipment and Materials: roll of clear adhesive tape

1. Pull two 8 cm to 10 cm pieces of tape from the roll.
2. Hold one piece of tape in each hand and bring the two shiny (non-sticky) sides close together without letting them touch (**Figure 2**). Record your observations.

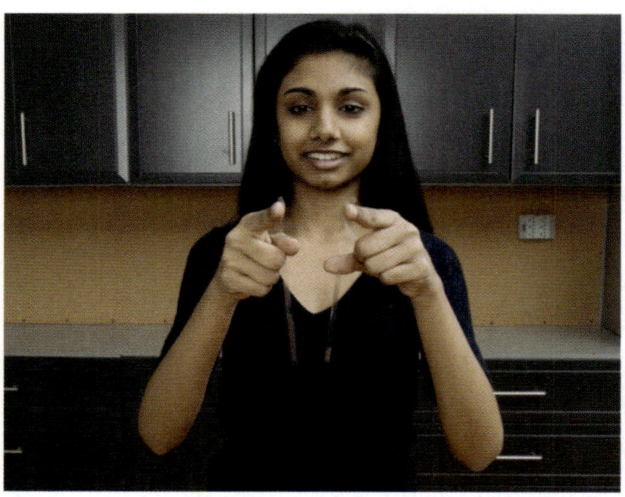

Figure 2

3. Breathe onto both sides of each piece of tape several times (over its entire length). Bring the two shiny sides of tape close together again without letting them touch. Record your observations.
4. Using the same two pieces of tape, allow each piece of tape to stick to the top of a clean desk. Do not rub the tape. Quickly pull the pieces of tape off the desk.
5. Bring the shiny, non-sticky side of one of the pieces of tape close to the edge of the desk without letting it touch the desk. Record your observations.
6. Quickly bring the shiny, non-sticky sides of both pieces of tape close to each other without letting them touch. Record your observations.

A. What do your observations in Step 2 tell you about the electric charge on the pieces of tape when they were first pulled off the roll? Explain. T/I

B. What do your observations in Step 3 tell you about the charge on the pieces of tape? Explain. Why is there a difference between Steps 2 and 3? T/I

C. How did touching the desk surface affect the charge on the tape? Explain. T/I

D. Based on your observations, what do expect to happen if you let the two pieces of tape touch after Step 6? Explain. T/I

Charging by Conduction

A neutral object can also become charged by direct contact with another object that is already charged. **Charging by conduction** occurs when two objects with different amounts of electric charge come in contact and electrons move from one object to the other.

▶ **charging by conduction:** charging an object by contact with a charged object

Conduction Between Charged and Neutral Objects

Assume you have a negatively charged metal rod and an uncharged or neutral metal sphere (**Figure 3**). The negatively charged rod has more electrons than it does protons. These extra electrons repel each other and try to spread away from each other. When you allow the rod to touch the neutral sphere, some of the electrons from the negatively charged rod move onto the metal sphere. These electrons give the sphere a negative charge. The rod loses some of its negative charge so the two objects end up with the same overall charge.

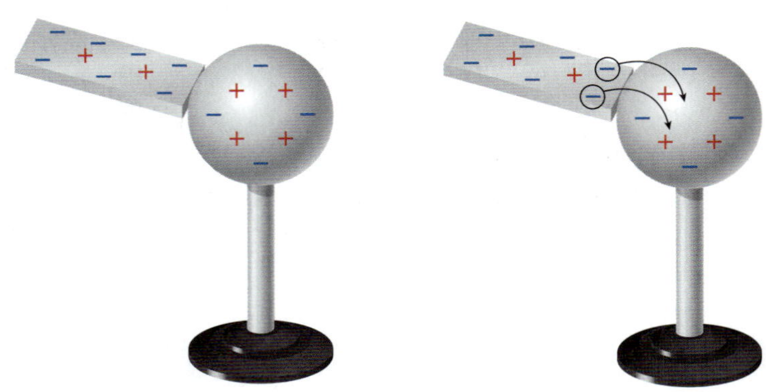

Figure 3 The negatively charged rod transfers electrons to the sphere by direct contact. The sphere gains a negative charge by conduction.

The opposite occurs if you touch a positively charged rod to a neutral sphere. Electrons will be more attracted to the positively charged rod. They move from the neutral sphere to the positively charged rod until there is a balanced distribution of charge between the two objects. Having lost electrons, the sphere now has a positive charge. Gaining electrons raises the amount of negative charge on the rod. The rod is still positively charged, but the amount of positive charge is lower. If the rod gains enough electrons, it can become neutral. The charges are evenly distributed across the two objects, so they have the same charge.

Conduction Between Two Charged Objects

Charging by direct contact does not always involve a charged object and a neutral object. If there is a difference in charge between two charged objects that are the same size, electrons will move from the object with a larger negative charge (the less positive one) to the object with the smaller negative charge (the more positive one) (**Figure 4**).

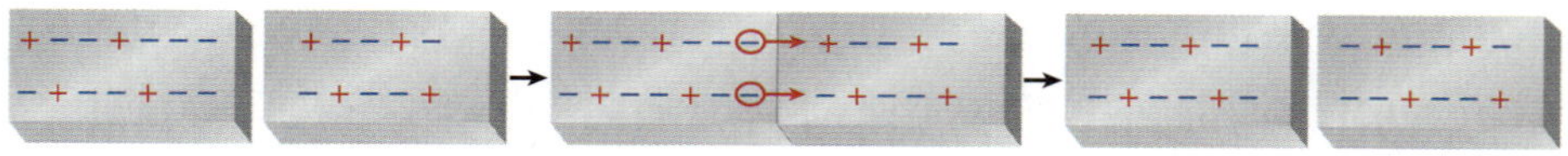

Figure 4 Electrons will move from the rod with the greater negative charge to the one with the lesser charge until the charges are evenly distributed. Both rods still have a negative charge, but now they both have the same amount of charge.

12.3 Wrap Up

- When two different neutral materials rub against each other, electrons are transferred between them. This is called charging by friction.
- Charging by friction results in charged objects that have opposite charges.
- When a neutral object is charged by contact with a charged object, the neutral object becomes charged with the same type of charge as the object that touched it.
- When there is a difference in charge between two objects that are the same size, contact causes electrons to be transferred between the objects until the charge is balanced.

CHECK YOUR LEARNING

1. Suppose your cat rubs up against a balloon. K/U T/I
   (a) What is this process called?
   (b) What can you tell about the charges on the cat's fur and the balloon after they have come in contact?
2. (a) What will happen if you bring a negatively charged metal pie plate in contact with a neutral glass rod?
   (b) What will be the charge on the glass rod after charging? T/I
3. Explain why it is not a good idea to try to put up static-cling decals on a very humid day. K/U
4. Two positively charged rods are the same size. One has more positive charge than the other. Describe what will happen when the rods touch. T/I
5. **Figure 5** shows a neutral metal sphere about to be touched by a positively charged rod. In your notebook, write a word or phrase to complete each sentence. K/U
   (a) When the rod touches the sphere, electrons will move from the ________ to the ________.
   (b) The sphere will acquire a(n) ________ charge.

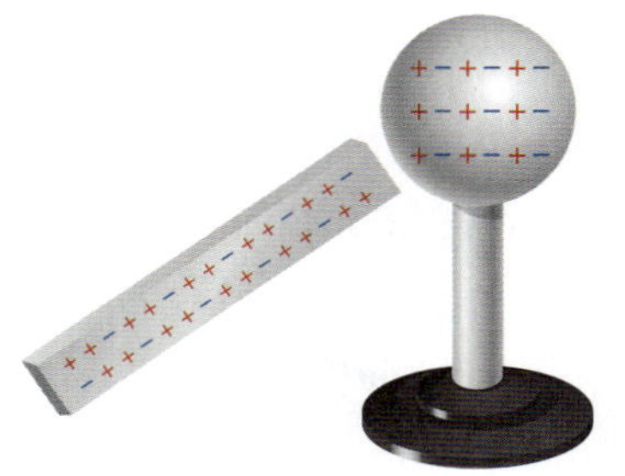

Figure 5

12.4 CONDUCT AN INVESTIGATION

Charging Objects by Contact

SKILLS MENU

- Questioning
- Hypothesizing
- Predicting
- Planning
- Controlling Variables
- Performing
- Observing
- Analyzing
- Evaluating
- Communicating

Rubbing an ebonite (hard rubber) rod with wool cloth gives the ebonite a negative charge. Rubbing Lucite (a type of plastic) with nylon cloth will give the Lucite a positive charge. A pith ball electroscope can be used to detect a charge imbalance.

In this investigation, you will transfer a charge from a charged object to a neutral object by contact. Then you will use the Law of Electric Charges to investigate the type of charge the neutral object gets.

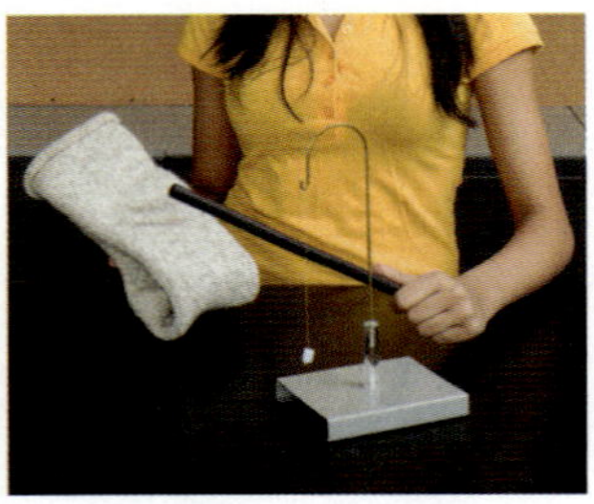

Figure 1 Ebonite rod and wool cloth

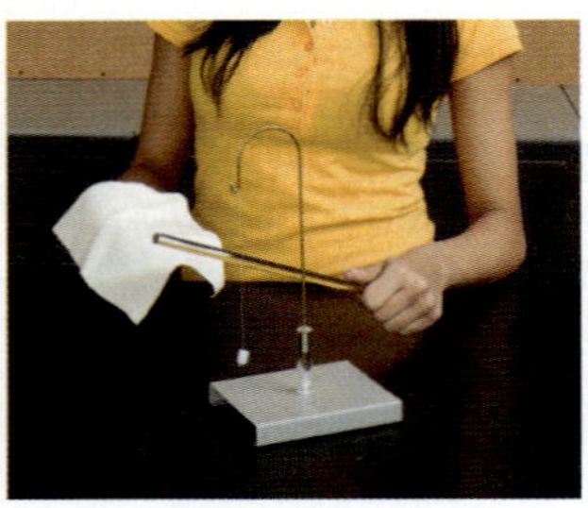

Figure 2 Lucite rod and nylon cloth

SKILLS HANDBOOK 3.B.3., 3.B.5.–3.B.8.

Testable Question

Can you use the Law of Electric Charges to determine what charge a neutral object has after it has been touched by a charged object?

Hypothesis/Prediction

Create a hypothesis based on the Testable Question. Your hypothesis should include a prediction and the reasons for your prediction.

Experimental Design

You will observe what happens when you bring a charged object near a neutral pith ball (**Figure 1**). Then you will charge the pith ball by touching it with a negatively charged object. You will predict and then observe what happens when you bring differently charged objects near the negatively charged ball. You will discharge the ball and repeat the experiment using a pith ball with a positively charged object (**Figure 2**).

Equipment and Materials

pith ball electroscope

ebonite rod

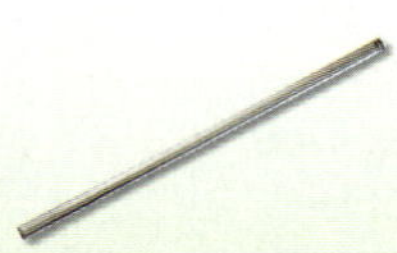

Lucite rod

wool cloth

nylon cloth

Procedure

1. Read through the Procedure and create a table in which to record your observations.
2. Set up a pith ball electroscope. Suspend the pith ball from a ring stand with thread. Remove any charge from the ball by touching it with your fingertips. After each series of tests, you must discharge the pith ball and rod by touching them with your fingertips.
3. Bring the uncharged ebonite rod close to the pith ball to test it. Record your observations and the pith ball charge (neutral (+ –), negative (–), or positive (+)) in the table.
4. Charge the ebonite rod by rubbing it with the wool cloth. Bring the rod close to, but not touching, the pith ball. Record your observations and the charge of the pith ball.
5. Touch the charged ebonite rod to the pith ball. Record your observations and conclusions about the pith ball charge.
6. Charge the ebonite rod again and bring it near the pith ball. Record your observations and conclusions.
7. Charge the Lucite rod by rubbing it with the nylon cloth. Bring the rod close to, but not touching, the pith ball. Record your observations and the charge in the table.
8. Remove all remaining charge from the pith ball. Make a prediction about what will happen when you charge it using the Lucite rod.
9. Repeat Steps 3 to 6 with the Lucite rod.
10. Repeat Step 7 with the ebonite rod. Record all your observations and conclusions about charge in your table. Remove any charge from the pith ball with your hand.
11. Repeat Steps 3 to 7 for any additional materials that your teacher provided. Record your observations.

CAUTION

Allergies

Warn your teacher if you are allergic to any of the materials in this activity so alternatives can be provided. Do not handle materials to which you are allergic.

Analyze and Evaluate

(a) Explain whether your observations support your hypothesis. Remember to include your evidence. T/I

(b) What type of charge did the pith ball have after you touched it with the ebonite rod? What charge did the pith ball have after you touched it with the Lucite rod? How do you know? T/I

(c) How did you use the Law of Electric Charges to determine the charge of the pith ball? T/I

12.5 Charging by Induction

Have you ever been able to stick a balloon onto a wall after rubbing it on your sweater? How is this possible? You know that the balloon gains a static charge because of the rubbing. You also know that opposite charges attract each other. However, the wall is neutral. How does the balloon stick to it?

- **induced charge separation:** a shift in the position of electrons in a neutral object that occurs when a charged object is brought near it
- **charging by induction:** charging a neutral object by bringing another charged object close to, but not touching, the neutral object

Induced Charge Separation

When you rub a balloon against your sweater, the balloon gains a negative charge due to contact. The wall has an equal number of positive and negative charges. It is neutral. Bringing the negatively charged balloon near to the wall causes the electrons in the balloon to repel the electrons in the wall. This induces some electrons in the wall to move away from the balloon. This is known as an **induced charge separation**. **Figure 1** shows how induced charge separation allows a negatively charged balloon to stick to a neutral wall.

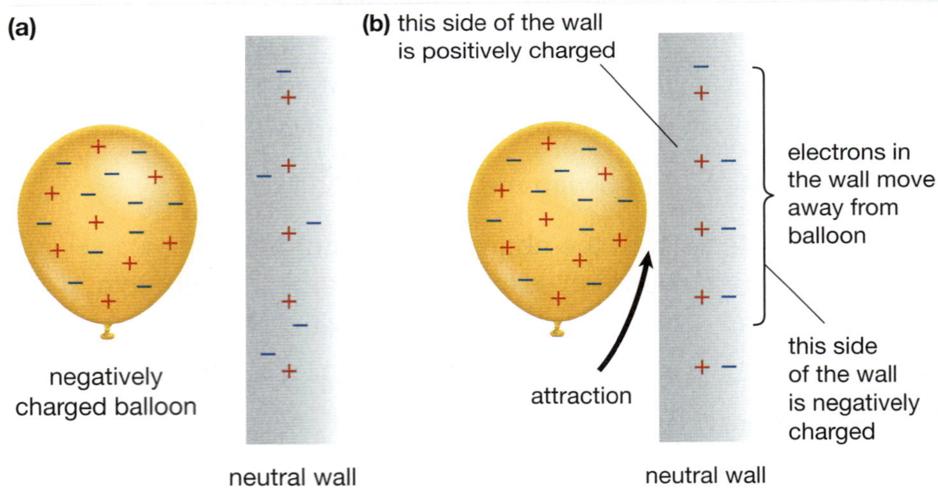

Figure 1 (a) When a negatively charged balloon is brought near a neutral wall, (b) the part of the wall near the balloon is temporarily charged by induction. The positive charges on the surface of the wall attract the balloon.

Induced charge separation leaves a positive charge on the surface of the wall closest to the balloon. This temporary charge in the wall is an example of **charging by induction**. A portion of the neutral object (the wall) was charged by bringing another charged object (the balloon) close to it.

DIG DEEPER

Web Link

Discover more about the charged balloon and how it sticks to the wall after being rubbed on your sweater in the "Balloons and Static Electricity" simulation. To learn more about charging by induction,

GO TO NELSON SCIENCE

The negatively charged balloon is attracted to the positively charged section of the wall. This is why the balloon can stick to the wall.

Both the balloon and the wall are poor conductors. Very few electrons move from the balloon to the wall. The wall still has the same number of positive and negative charges, so it is still neutral overall. This induced charge is only temporary. The electrons in the wall will move back to their original position once the charged balloon is moved away.

Charging More Than One Object by Induction

You can use a positively charged object to induce a charge in a neutral object. You can even charge two objects at the same time and make the charge permanent. When you rub a balloon with a polyester cloth, the balloon loses electrons and gains a positive charge by contact. If you bring this positively charged balloon close to two neutral metal spheres (**Figure 2(a)**), it produces an induced charge separation.

- The electrons on sphere A are attracted by the positive charge on the balloon and move to the left. This creates a positive charge on the right side of sphere A, which causes some of the electrons to move from sphere B onto sphere A (**Figure 2(b)**).
- Keeping the balloon in place and separating spheres A and B prevents the electrons from moving back onto sphere B (**Figure 2(c)**).
- After the balloon is removed, the electrons spread out evenly on spheres A and B. Sphere A is negatively charged and sphere B is positively charged (**Figure 2(d)**).

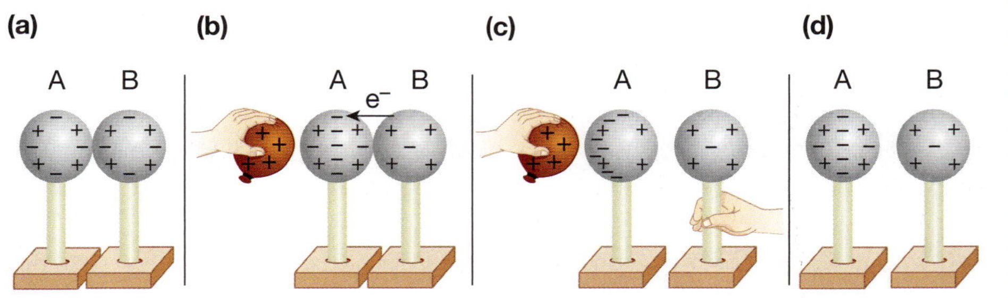

Figure 2 (a) Two identical neutral spheres are in contact. (b) A charged balloon induces the movement of electrons. (c) Separating the spheres prevents further electron transfer. (d) Both spheres have been charged by induction.

DIG DEEPER

Did You Know?

Static and Dust

When most electronic devices such as televisions and computer monitors are turned on, a slight charge develops on the outside of the device. This charge can induce an opposite charge on the near side of any passing dust particles. This results in dust being attracted to the device. A buildup of dust can lead to overheating or damage to delicate circuits.

Grounding

You can return any charged object to neutral by adding or removing electrons. The electrons can be transferred between the charged object and a large neutral object such as Earth (the ground). Large objects like Earth can gain or lose electrons and remain neutral because the charges are spread over a huge area. The process of removing charges from objects by contact with a large, neutral object is called **grounding**. You can use the symbol shown in **Figure 3** to show that an object is grounded.

▶ **grounding:** connecting an object to a large body, like Earth, that can remove an electric charge from the object

When a negatively charged object is grounded, electrons travel from the object into the ground until the object is neutral. When a positively charged object is grounded, electrons from the ground travel up to the positively charged object until the object is neutral.

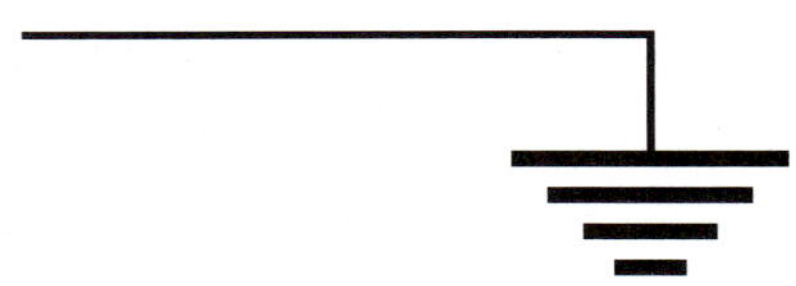

Figure 3 This symbol indicates an object is grounded.

Discharge and Grounding

Figure 4 shows what happens when a person carrying a negative charge on his hand comes in contact with a neutral faucet that is grounded. There is an electric discharge when his hand gets close enough to the faucet. The excess electrons are rapidly transferred from his hand to the faucet. He receives a shock and may even see a spark. The faucet is grounded, so the electrons do not stay in the faucet. They travel from the faucet to the ground until both the hand and the faucet are neutral again.

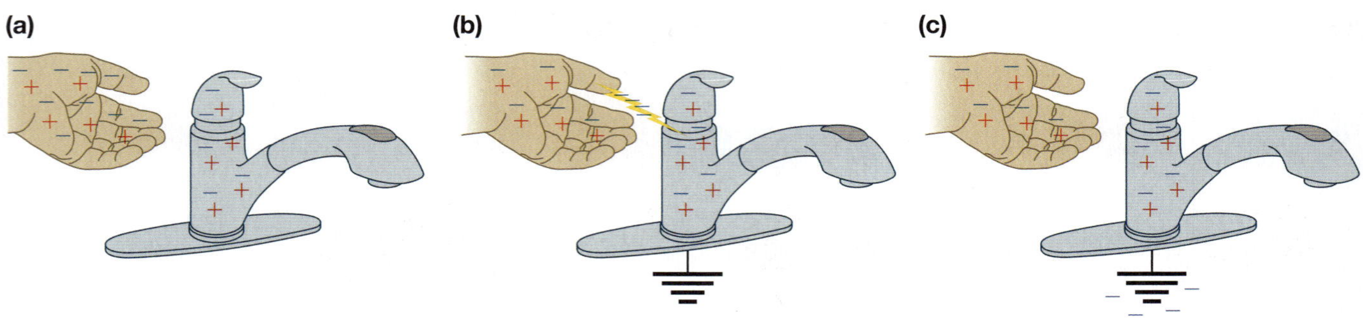

Figure 4 (a) The person's hand is negatively charged. The faucet is neutral. (b) The excess electrons move from the hand through the faucet to the ground. (c) Both the hand and the faucet are neutral.

Technicians working on electronic devices always make sure they are grounded to avoid the buildup of electrostatic charge that can damage equipment. Objects can be grounded without being attached to Earth's surface. Ships at sea and planes in the air can be grounded. A connection to any neutral object that can store enough excess charge can serve as a "ground." In the case of ships, the ocean acts as a ground. The body of the airplane acts as a ground for the electronic devices on the plane.

DIG DEEPER

Reading Tip

Checking for Bias

When you read, check for clues of bias. Are both advantages and disadvantages described? If not, does the text show a bias toward one side of the issue?

Permanent Charges and the Law of Electric Charges

You can apply the Law of Electric Charges to understand how you can permanently charge an object by induction by momentarily grounding it. In **Figure 5(a)**, as the distance between a negatively charged rod and the side of a neutral pith ball decreases, the force with which the electrons in the rod repel the electrons in the pith bath increases. This repulsion induces a temporary positive charge on the side of the ball closest to the rod because the repelled electrons there move to the other side of the ball. The side of the pith ball farthest from the rod now has a temporary negative charge. Following the Law of Electric Charges, the negative rod and the positive side of the pith ball are attracted to each other because they have opposite charges.

If you ground the negatively charged side of the pith ball by touching it with your hand or with a wire connected to a ground (**Figure 5(b)**), some of the electrons are transferred from the pith ball. They pass through your hand or wire to the ground. The pith ball is left with a positive charge. The charge is made permanent if you remove the ground before the charged rod is moved away (**Figure 5(c)**). Charging by induction causes a neutral object to gain a charge opposite to that of the object which induced the charge.

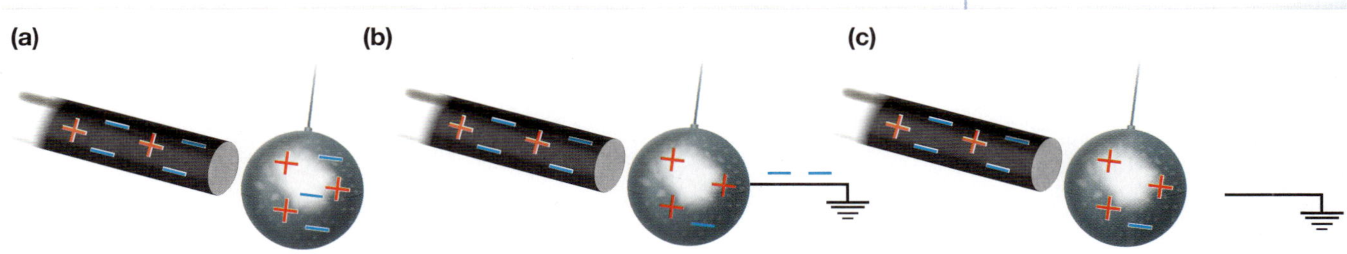

Figure 5 (a) A negatively charged rod induces a positive charge on one side of the pith ball. (b) Electrons that moved to the other side of the ball can travel to the ground. (c) Removing the ground before the charged rod is moved away leaves the ball permanently charged.

TRY THIS STORING STATIC ELECTRICITY

SKILLS HANDBOOK 3.B.5., 3.B.6.

SKILLS: Observing, Communicating

You can apply charging by induction to build a device that stores static charge for use in other experiments. In this activity, you will build a device called an *electrophorus*, or charge carrier. Then you will investigate what happens when it is discharged.

Equipment and Materials: clear plastic tumbler or Styrofoam cup; glue or tape; aluminum pie plate; Styrofoam dinner plate; piece of wool cloth; paper clip; modelling clay

1. Place the tumbler or cup upside down on the aluminum pie plate. Glue or tape it in place. It will be used as a handle.
2. Rub the Styrofoam plate with the wool cloth for at least 1 min to charge it. Place the Styrofoam plate on the table. Make sure the side of the plate you rubbed is facing up.
3. Place the aluminum pie plate on top of the Styrofoam plate. Briefly touch the top of the pie plate with your finger as shown in **Figure 6**. Record your observations.

Figure 6 A hand-made electrophorus

4. Insert the end of the paper clip in some modelling clay so that it is sticking up. Carefully lift up the pie plate by the handle only. Do not touch the pie plate itself.
5. Bring the rim of the pie plate slowly toward the paper clip until they are almost touching as shown in **Figure 7**. Record your observations. (It may be easier to see if you can dim the room lights.)

Figure 7

6. Repeat Steps 2 and 3. Observe and record what happens when you bring the edge of the pie plate near other objects in your classroom.

A. What happened when you brought your finger near the pie plate after you had put it on the Styrofoam plate? T/I

B. What happened as the pie plate and the paper clip were brought closer together? T/I

C. What happened when you brought the edge of the pie plate close to other objects? T/I

D. Why should you carefully hold the pie plate only by the handle after you touched it the first time? T/I

12.5 Wrap Up

- Charging by induction refers to charging a neutral object by bringing it close to, but not touching, a charged object.
- Charging by induction can be either temporary or permanent.
- Charging by induction helps explain why neutral objects are always attracted to charged objects.
- Charging permanently by induction always produces an object with the opposite charge to the original charged object.

CHECK YOUR LEARNING

1. Which of the following are examples of charging by induction? Explain how you know. K/U
   (a) static cling on clothes
   (b) bringing a charged rod near a neutral electroscope
   (c) attracting pieces of foam with a charged comb
   (d) hair standing on end after you have taken off a hat
2. **Figure 8** shows a Styrofoam block that has a negative charge after being rubbed with fur. A neutral aluminum pot lid is brought close to the charged plate. T/I

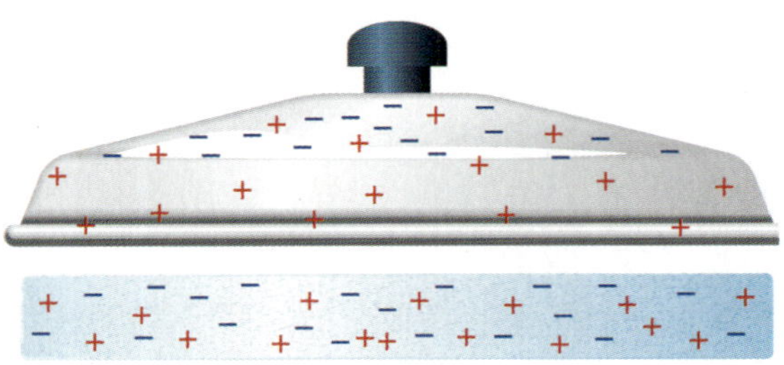

Figure 8

   (a) What is the charge on the side of the aluminum lid closest to the plate? How do you know?
   (b) Describe what happens if you touch the aluminum lid with your finger. What charge will it have?
   (c) What charge will the aluminum lid have after you have touched it the first time then moved it away from the foam plate?
   (d) Which object was charged by contact?
   (e) Which object was charged by induction?
3. You notice the hair on your arm begins to rise when a negatively charged balloon is nearby. Using the concept of induction and the Law of Electric Charges, explain why you think this happens. T/I
4. Someone who has a positive charge on their hands touches a faucet. The faucet was neutral before it was touched. The faucet is grounded. Draw a picture to show the transfer of charge between the person, the faucet, and the ground. T/I C
5. Suppose you approach a neutral pith ball with a rod of unknown charge. You ground the pith ball, then remove the rod. You observe that the ball repels a negatively charged strip of plastic. What was the charge of the rod that induced a charge in the pith ball? Explain. T/I
6. Use the drawing shown in **Figure 9** to complete the following questions. T/I

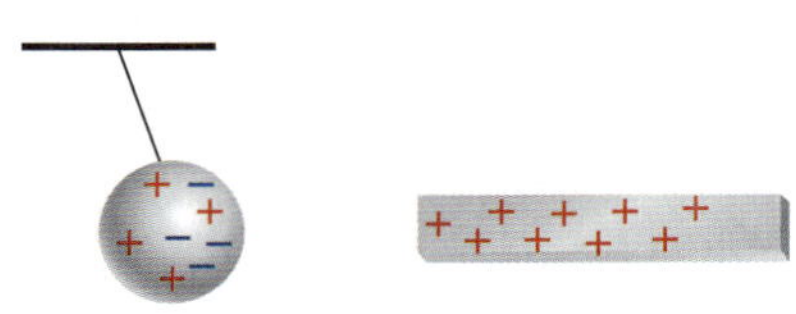

Figure 9

   (a) Explain why the neutral pith ball is attracted to a positively charged rod.
   (b) Describe what would happen if you were to touch the ball while the rod was near.

12.6 PERFORM AN ACTIVITY

Charging Objects by Induction

SKILLS MENU

- Questioning
- Hypothesizing
- Predicting
- Planning
- Controlling Variables
- Performing
- Observing
- Analyzing
- Evaluating
- Communicating

You have already investigated charging objects by contact. In this activity, you will investigate charging different materials by induction.

SKILLS HANDBOOK 3.B.5.–3.B.7.

Purpose

To investigate the effects of charging by induction

Procedure

1. Create a table like **Table 1** in your notebook to record your observations.

Table 1 Electrostatic Testing

Electric charge tester	Effect on paper circles	Effects on thin stream of water	Effect on metal leaf electroscope (before touching)	Effect on metal leaf electroscope (after touching)
ebonite rod charged with wool cloth				
plastic rod charged with polyester cloth				

2. Bring the ebonite rod close to, but not touching, the paper circles. Then bring the rod close to a stream of water from the sink. Observe carefully and record your observations.

Equipment and Materials

metal leaf electroscope

ebonite rod

wool cloth

polyester cloth

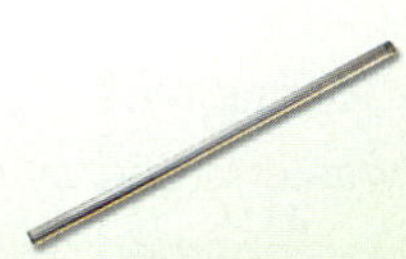
clear plastic rod

10 paper circles (from a hole punch)

3. Charge the ebonite rod by rubbing it with the wool cloth. Bring the rod close to, but not touching, the paper circles. Observe carefully and record your observations.
4. Bring the rod close to, but not touching, a stream of water. Vary the distance and record your observations.
5. Bring the rod close to, but not touching, the knob of the electroscope. Observe what happens to the leaves. Note what happens when the rod is taken away.
6. Bring the rod near, but not touching, the knob again. Briefly touch the electroscope with your finger. Observe what happens to the leaves. Note what happens when the rod is taken away.
7. Bring the plastic rod close to, but not touching, the paper circles. Then bring the rod close to the stream of water. Observe carefully and record your observations.
8. Charge the plastic rod by rubbing it with the polyester cloth. Bring the rod near, but not touching, the knob of the charged electroscope. What happens to the leaves? Note what happens when the rod is taken away. Discharge the electroscope by touching the knob with your finger before continuing.
9. Charge the plastic rod again by rubbing it with the polyester cloth. Repeat Steps 3 to 6 with the charged plastic rod.
10. Repeat Step 8 using the ebonite rod.

CAUTION

Allergies

Warn your teacher if you are allergic to any of the materials in this activity so alternatives can be provided. Do not handle materials to which you are allergic.

Analyze and Evaluate

(a) Compare how the rods affected the paper circles, stream of water, and the leaves of the electroscope. T/I

(b) In which step(s) was the electroscope charged by induction? T/I

(c) What conclusions can you draw about what happens when charged objects are placed near uncharged ones? T/I

(d) Based on your observations, what did you notice as the distance between the charged and uncharged objects decreased? T/I

(e) Describe a test that you could perform to determine whether an object is charged or uncharged. T/I

(f) Could you tell whether a charged electroscope was positively or negatively charged by looking at it? Explain. T/I

(g) Suppose you know the charge on one rod. How could you use an induced charge on an electroscope to figure out what charge a different rod has? T/I

12.7 PERFORM AN ACTIVITY

Testing Materials for Electron Flow

SKILLS MENU

- Questioning
- Hypothesizing
- Predicting
- Planning
- Controlling Variables
- Performing
- Observing
- Analyzing
- Evaluating
- Communicating

Some materials allow electrons to flow through them easily. Other materials do not. When a metal leaf electroscope is charged, its leaves carry the same charge and repel each other. If a material that resists electron flow touches the charged leaf electroscope, the electroscope will keep its charge. The leaves will remain raised.

If a material that allows electrons to flow comes into contact with the knob on the charged electroscope, electrons flow either into or out of the material depending on the charge on the electroscope. After it discharges, the electroscope becomes neutral, and its leaves fall.

In this activity, you will test materials to classify them as allowing electron flow or not.

SKILLS HANDBOOK 3.B.5.–3.B.7.

Purpose

To classify materials as allowing electron flow or not allowing it

Procedure

1. Work with a partner or small group. Obtain the objects you want to test and place them in your work area.

Equipment and Materials

metal leaf electroscope

ebonite rod

wool cloth or piece of fur

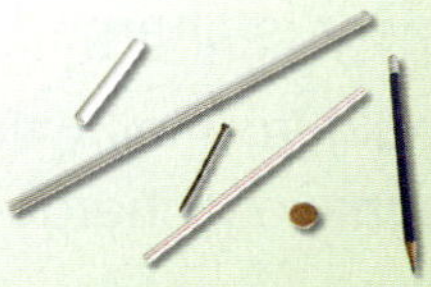
testing materials

2. Construct a table similar to **Table 1** in your notebook to record your observations.

Table 1 Test Results

Object	Position of electroscope leaves (raised/lowered)	Electroscope stays charged (yes/no)	Material allows electron flow (yes/no)
copper penny			
pencil lead			

3. Charge the ebonite rod by rubbing it with the wool cloth (or fur).
4. Touch the ebonite rod to the knob on top of the electroscope to charge it. The leaves should be in the raised position. Do not touch the top of the electroscope with your hands during the testing process.
5. Repeat Steps 3 and 4 if the leaves of the electroscope return to the neutral position during the tests.
6. Test each of the objects in turn by touching them to the knob on top of the electroscope. Observe what happens. Record your results in the table.

CAUTION

Allergies

Warn your teacher if you are allergic to any of the materials in this activity so alternatives can be provided. Do not handle materials to which you are allergic.

Analyze and Evaluate

(a) Which of the items you tested allowed electrons to flow, and which did not? T/I

(b) Can you classify the objects that allow electron flow in another way? How? What about the objects that did not allow electron flow? T/I

Apply and Extend

(c) Choose other objects to test. Use your observations to predict whether the object will allow electrons to flow or not. Then test your prediction. T/I

UNIT TASK

Bookmark

You can apply what you have learned about electron flow in different materials to the Unit Task described on page 536.

12.8 Moving Charges

Technicians who work with delicate electronic parts are required to wear shoes with rubber soles instead of leather soles. They do this to reduce the chance of any buildup of static charge as they walk around. This works because rubber resists the movement of electrons more than leather does. In this section, you will learn more about how different materials allow electrons to flow.

Conductors

In Activity 12.7, you tested different materials to see if they allowed electrons to flow. A **conductor** of electricity is a material that allows electrons to move through it easily. Metals such as the copper wire shown in **Figure 1** are considered good conductors. Metals hold their outer electrons loosely, so in general metals are good conductors of electricity. Other common metal conductors of electricity are aluminum, silver, and gold. As you read in Chapter 6, being a good conductor is one of the properties used to classify metals.

▶ **conductor:** a material that transmits thermal energy or electrical energy easily

Figure 1 Copper wire is used in many electrical applications because it is a good conductor of electricity, does not easily corrode, and is inexpensive compared to gold or silver.

You have probably seen or read warning labels cautioning you not to use electric appliances, such as the hair dryer, near water (**Figure 2**). Pure water is generally considered a non-conductive substance. However, tap water usually contains dissolved minerals and ions. These substances make water a good conductor of electricity. So, touching an appliance near a bath or shower could put you in danger of serious injuries resulting from conducting electric charges through your skin.

Figure 2 Never use electric appliances such as this hair dryer near water because of the danger of electric shock.

Insulators

An **insulator** is a material, such as the ceramic electric fixture in **Figure 3**, which prevents the flow of electrons. Insulators are materials that have more tightly bound electrons than those in conductors. Examples of common insulators are plastic, glass, rubber, air, and wood. Notice that these are all non-metals that you read about in Chapter 6.

In Section 12.5 you picked up a plastic or Styrofoam cup attached to an aluminum pot lid that sat on a Styrofoam plate. In this activity, the aluminum pot lid was the conductor. You were able to pick up the pie plate without grounding it because plastic and Styrofoam are good insulators.

Insulators are used to prevent conductors from transferring charges into their surroundings. This is why the outside of most electronic devices such as cellphones, CD players, and video game players are made of plastics. Plastics help to protect the delicate parts inside the device from electrostatic discharge.

Insulators are also used to protect us from the dangerous effects of electricity flowing through conductors. For this reason, a plastic or rubber coating is placed around electric wires used to conduct electricity. **Table 1** lists examples of common electrical conductors and insulators.

▶ **insulator:** a material that resists or blocks the flow of electrons through it

Figure 3 Glass and ceramic insulators protect people and structures from the electricity flowing through power lines.

Table 1 Conductors and Insulators

Conductors	Insulators
silver	glass
copper	rubber
gold	ceramics
aluminum	quartz
iron/steel	paper
graphite (carbon)	wood
human body	plastic
nickel	wax
salt water	pure water
platinum	dry air

Conductivity

Separating objects into conductors and insulators is not entirely accurate. Instead, a range of conductivity might be a more accurate way to classify materials. Electrical **conductivity** is a material's relative ability to conduct electricity and allow electrons to flow through it. Different materials fall along a range that starts with those that are very conductive and going to those that are least conductive (best insulators). **Figure 4** shows a range of items with different conductivity.

▶ **conductivity:** a measure of a material's ability to conduct electricity

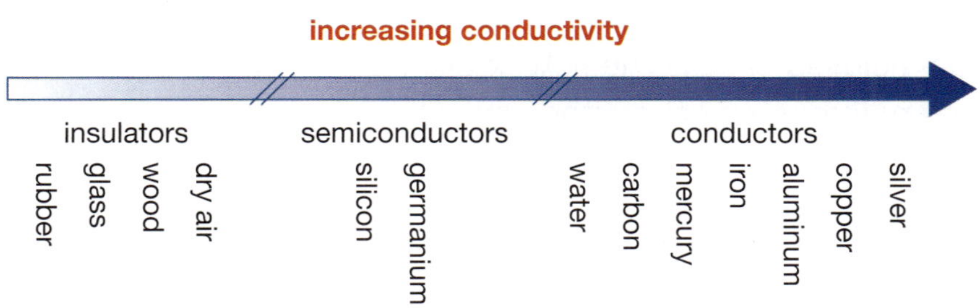

Figure 4 Insulators and conductors actually fall along a range based on increasing levels of conductivity.

Current Electricity

A power failure can make you realize just how important electricity is in your life. All the lights go out, and the television, computer, and other electronic devices and appliances do not work. Most of these devices do not use static electricity. They are dependent on electricity that flows from an energy source through conductors to the devices and back again. The movement of electrons through a conductor in this sort of closed path is called an **electric current**. The path the current takes must be closed. This means that the current must be able to return to its source or it does not flow.

▶ **electric current:** the flow of electricity through a conductor in a closed path

Electric current might be compared to water flowing at high pressure through pipes. A pipe filled with water is like a conducting material, such as copper wire, that carries current electricity.

DIG DEEPER

Did You Know?

Batteries Not Needed

New types of flashlights and radios can be run without batteries. These appliances are designed so that shaking or cranking them generates an electrical current that powers them. It also charges batteries so the appliance can continue to run after you stop cranking or shaking.

Batteries

The most common source of electric current to power small electrical devices such as flashlights is a dry cell battery. A battery is shown in cross-section in **Figure 5**.

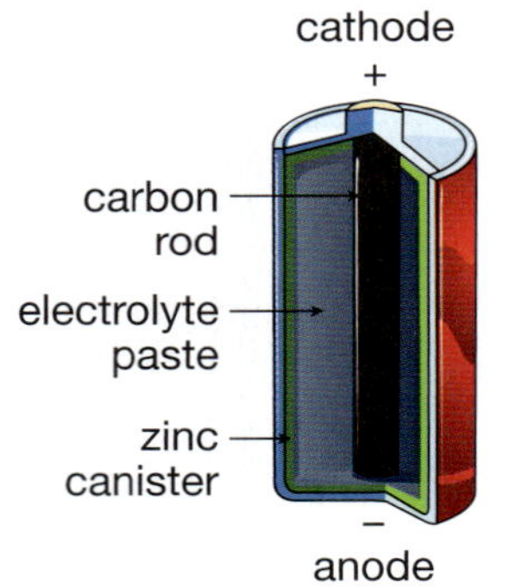

Figure 5 A dry cell battery converts the chemical energy generated by reactions between chemicals inside it into electrical energy.

Chemical reactions in the electrolyte of a battery give electrons energy as they move internally from the positive terminal (cathode) to the negative terminal (anode). These electrons are now attracted back to the positive terminal and can only get there by moving through a closed circuit. The flow of electrons in one direction from the battery through a conductor (wires) to a device and then back to the other end of the battery is known as **direct current (DC)**.

▶ **direct current (DC):** an electric current in which the flow of electrons travels in one direction only

The wires and the device resist the flow of electrons through them. To continue the water pipe analogy, resistance is like having a partial clog that obstructs the flow of water through the pipe.

The Grid

Electric current travels to your house through an amazing system called the electrical energy distribution grid. **Figure 6** shows how the electric energy is produced at the generating plant. Fossil fuels (coal, natural gas, or oil) are burned, or nuclear energy is used to heat water into steam. Steam, water, or diesel engines spin fan-like turbines. The spinning turbines spin coils of wires between magnets inside the generators. This produces an **alternating current** that flows first in one direction through the wires, then reverses, and goes in the other direction. This is the current used in our homes and offices. This current reverses itself 60 times each second.

▶ **alternating current (AC):** an electric current that repeatedly reverses direction

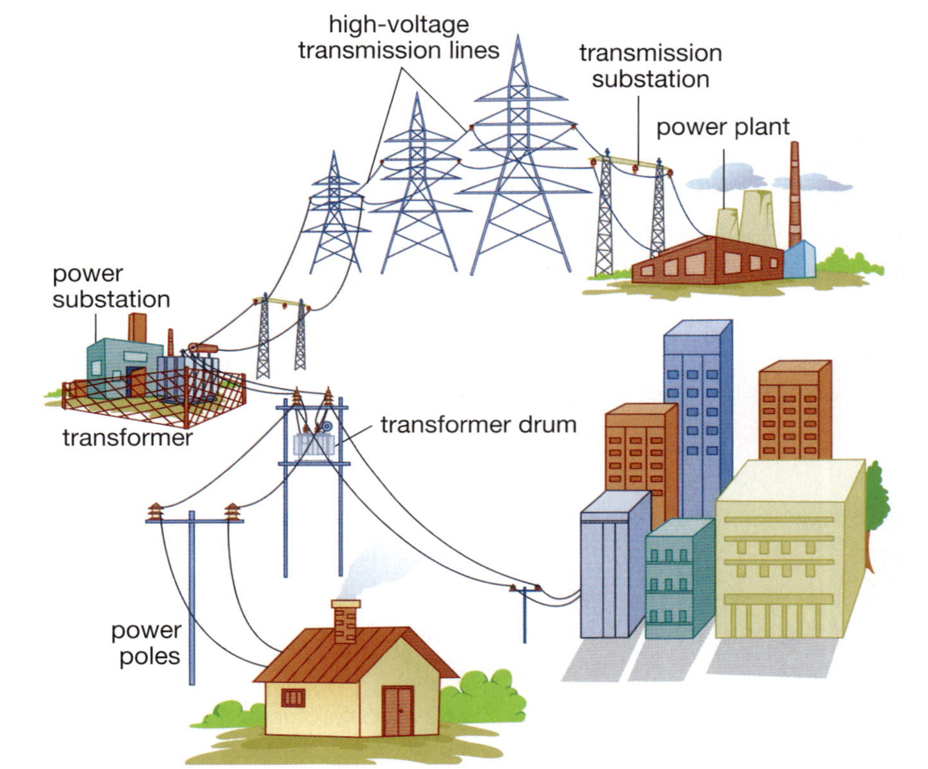

Figure 6 The grid consists of a generating plant with its electric generator, the substation, transmission lines, transformers at the substation, and the poles leading electrical wires into your home.

The alternating current produced by the generator must pass through devices called transformers so that it can be transmitted over long distances. At a substation, other transformers reverse the process so that the current can travel safely over transmission lines closer to the homes and businesses that will use the electricity. It is transformed one more time by pole transformers or transformer boxes (in areas where the lines run underground) so that it is safe for use in homes, schools, and offices.

DIG DEEPER

Reading Tip

Evaluating Diagrams

Examine diagrams and captions carefully to improve your understanding of what you read. Make connections to what you already know. For example, you may have observed the wires that deliver electrical energy to your home.

DIG DEEPER

Career Link

Lineworkers install and maintain the transmission lines that carry electric current. Analysts monitor power usage and determine if more electrical generating plants are needed. For more information about working in the electrical energy production industry,

GO TO NELSON SCIENCE

CITIZEN ACTION

Replacing Old Wiring

Many homes and public buildings still have old wiring that can be a fire hazard. Several types of old wires are found in older buildings. Some of the most dangerous are wrapped with insulators such as cloth or rubber, which break down over time. To prevent fires in older homes and buildings, modern building codes require that electricians install and replace wires with new ones wrapped in long-lasting insulating materials.

Contact an electrician or an energy company. Find out how to look for signs of old wiring and what should be done to replace it. Decide on a method to inform your friends and neighbours of what you learned and what they should do to protect their homes.

GO TO NELSON SCIENCE

UNIT TASK

Bookmark

You can apply what you have learned about current electricity and the electrical energy distribution grid to the Unit Task described on page 536.

12.8 Wrap Up

- Electrical conductors are materials that allow electrons to flow through them easily.
- Insulators are materials that resist or block the flow of electrons.
- Conductivity is a measure of a material's ability to conduct electricity.
- Electric current is the flow of electrons through a conductor.
- Direct current is current that flows through a conductor in one direction only.
- Alternating current is current that repeatedly reverses direction.
- Electric current travels over transmission lines from a generating plant to your home by the electrical energy distribution grid.

CHECK YOUR LEARNING

1. Is each of the following items a conductor or an insulator? Explain your reasoning. K/U
   (a) iron nail
   (b) glass jar
   (c) dry air
   (d) gold ring
   (e) carbon rod
   (f) wooden spoon
2. Why is it dangerous to use appliances near water, even though pure water is considered non-conductive? K/U
3. Why can dry-cell batteries be thought of as energy converters? K/U
4. Currents of water and electricity flow from a source to where they will be used. What happens next in the current of electricity that makes the flow different from that of water? T/I
5. Compare static electricity and current electricity. How are they alike? How are they different? T/I
6. Which source of electrical energy do you prefer for your personal electronic devices: batteries or a power cord? Why? C A

TECH CONNECT OSSLT

Superconductors

Imagine a train "floating" above its tracks and travelling at speeds of nearly 500 km/h. That technology is already here, but how does it work?

The answer lies in what happens to certain materials when they are cooled to very low temperatures. At room temperature, even the best regular conductors such as silver still lose some of the electrical energy that passes through them as thermal energy. However, if you make certain materials very, very cold (–150 °C), no electrical energy is lost as it passes through them. They become *superconductors*.

Superconductors can be used to make magnets powerful enough to allow even heavy objects to levitate or float above them (**Figure 1**). Germany and China have already built maglev (magnetic levitation) train systems (**Figure 2**).

Figure 1 A magnet floats, or levitates, above a superconductor cooled by liquid nitrogen to extremely low temperatures.

Superconductors are also having a tremendous impact in medicine. They are used to generate powerful magnetic fields. These fields are used to make magnetic resonance images (MRI) of the inside of the human body (**Figure 3**).

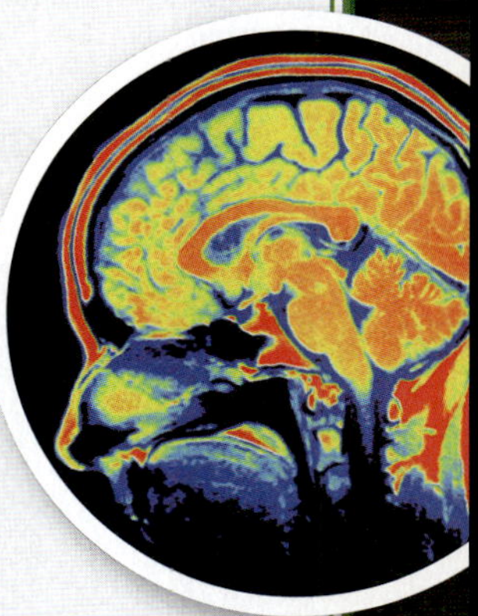

Figure 3 This MRI image inside the human brain was made possible by superconductivity.

Scientists believe that in the near future superconductors will also play a major role in efficient energy production, storage, and transmission. They will also be used to build faster, more powerful computers.

To achieve these goals, Canadian researchers, such as Michael Thwart at Simon Fraser University in British Columbia, are working to develop a room-temperature superconductor. This would make superconductors easier to handle and avoid the need for cooling systems.

Figure 2 This Maglev Train in Shanghai is a new form of high-speed transport. It uses electromagnets made with superconductors to levitate and to push itself forward at great speed without friction to slow it down.

Chapter 12 KEY CONCEPTS SUMMARY

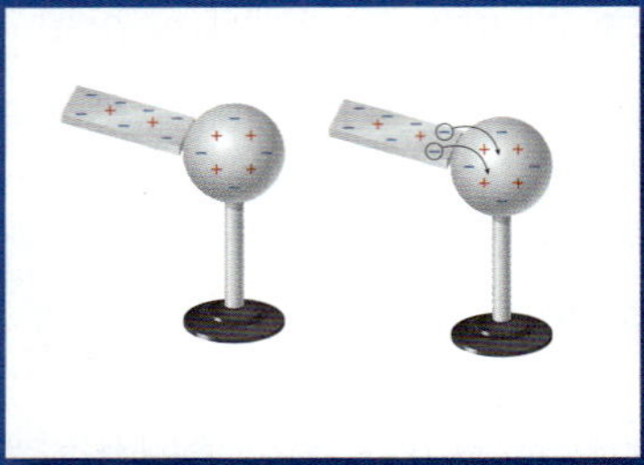

Electricity describes how electrons move from one place to another.

- Electrons move from atom to atom.
- The law of electric charges states that opposite charges attract each other, while like charges repel each other.
- Charged objects attract neutral objects.

Static electricity is the result of an imbalance of charges on the surface of an object.

- An electric discharge is the rapid transfer of electrons from one object to another.
- Examples of static electricity include static cling and lightning.

There are many useful applications of static electricity.

- Advertisers use static cling vinyl film for brightly coloured removable signs.
- Windows can be covered with filters that depend on static cling.
- Photocopiers use static electricity to make copies.

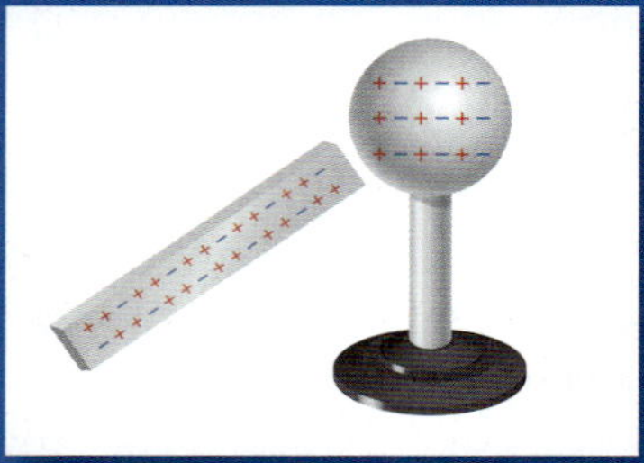

Objects can be charged by contact or by induction.

- Neutral objects can be charged by friction.
- Bringing a charged object near a neutral object can induce a temporary, opposite charge.
- An object is grounded if electrons can flow to or from it to a large, neutral object such as Earth.

Current electricity is generated by moving electrons in a closed path.

- Electrons move from an energy source through conductors to a device and then back to the source.
- Batteries produce direct current.
- Electrical energy generating stations generate alternating current.

Some materials allow electrons to move through them easily (conductors), while others do not let electrons move through them easily (insulators).

- Conductors are used to transmit electrical energy.
- Insulators protect us from current electricity.
- You can use an electroscope to test conductivity.

WHAT DO YOU

THINK NOW?

You thought about the following statements at the beginning of the chapter. You may have encountered these ideas in school, at home, or in the world around you. Consider them again and decide whether you agree or disagree with each statement.

1 You use static electricity to recharge your cellphone.
Agree/disagree?

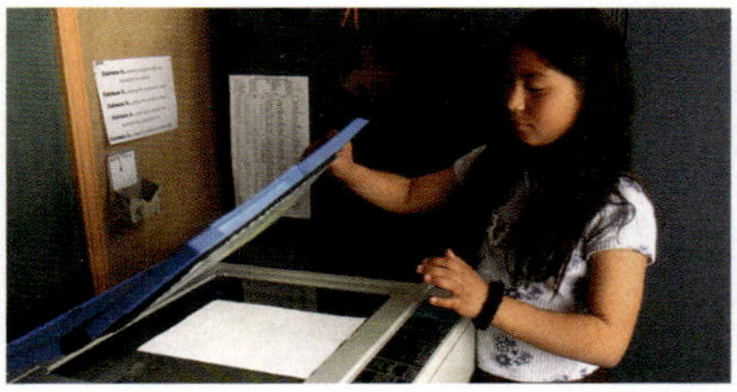

2 Both photocopiers and lightning involve static electricity.
Agree/disagree?

3 Electricity can travel over your skin without harming you.
Agree/disagree?

4 You can use electricity to move objects without touching them.
Agree/disagree?

5 Current electricity moves through wires like water flows through a pipe.
Agree/disagree?

6 Static electricity is a useful form of electricity.
Agree/disagree?

How have your answers changed since then?
What new understanding do you have?

BIG Idea

- Static and current electricity have distinct properties that determine how they are used.

DIG DEEPER

Vocabulary

electric force p. 460
electric charge p. 460
neutral p. 460
neutral object p. 461
negatively charged object p. 461
positively charged object p. 462
Law of Electric Charges p. 462
static electricity p. 464
electric discharges p. 464
charging by friction p. 468
charging by conduction p. 470
induced charge separation p. 474
charging by induction p. 474
grounding p. 476
conductor p. 484
insulator p. 485
conductivity p. 485
electric current p. 486
direct current (DC) p. 486
alternating current (AC) p. 487

CHAPTER 12

SELF-QUIZ

The following icons indicate the Achievement Chart category addressed by each question.

K/U Knowledge/Understanding
T/I Thinking/Investigation
C Communication
A Application

For each question, select the best answer from the four alternatives.

1. Which of the following materials would make the best insulator? (12.8) K/U
   (a) pure water
   (b) wooden peg
   (c) iron nail
   (d) glass rod

2. Which of the following is an example of charging by induction? (12.5) K/U
   (a) static cling on clothes
   (b) attracting paper with a charged rod
   (c) rubbing a balloon on a sweater
   (d) charging an electroscope by touching it

3. A positively charged rod is brought close to a neutral sphere. If you touch the sphere with your finger before pulling the rod away, what type of charge will the sphere have? (12.5) T/I
   (a) temporarily positive
   (b) permanently positive
   (c) temporarily negative
   (d) permanently negative

4. Neutral objects can become charged by
   (a) gaining or losing neutrons
   (b) gaining or losing protons
   (c) gaining or losing electrons
   (d) gaining or losing protons, neutrons, or electrons (12.1) K/U

5. Which factor best explains why lightning strikes the ground? (12.2) K/U
   (a) metals conduct the current electricity to the ground
   (b) friction between the cloud and the ground
   (c) electrostatic discharge between cloud and Earth
   (d) charging by friction and direct contact

Indicate whether each of the statements is TRUE or FALSE. If you think the statement is false, rewrite it to make it true.

6. Tap water is a good insulator of electricity because it contains dissolved salts and charged particles. (12.8) K/U

7. In charging by direct contact, electrons are transferred between the objects. (12.3) K/U

8. Computers are shipped in special anti-static packaging to protect against electric current. (12.2) K/U

9. Static electricity refers to the movement of electrons in a closed path. (12.2, 12.8) K/U

10. Metals make good conductors because they have electrons that can move from atom to atom. (12.8) K/U

11. You must touch an object to permanently give it a charge. (12.5) K/U

12. Neutral objects are attracted to negatively charged objects but not positively charged objects. (12.1, 12.5) K/U

Copy each of the following statements into your notebook. Fill in the blanks with a word or phrase that correctly completes the sentence.

13. Charging by _______ causes electrons in a neutral object to move. (12.5) K/U

14. Objects with like charges _______ each other. (12.1) K/U

15. A(n) _______ is a material that prevents the flow of current. (12.8) K/U

16. A rapid release of static energy is a(n) _______. (12.2) K/U

17. Permanent charging by induction results in two charged objects that have _______ charges. (12.5) K/U

Match each term on the left with the most appropriate description on the right.

18. (a)	neutral object	(i)	material that transmits electrical energy easily
(b)	conductivity	(ii)	attraction or repulsion between charged objects
(c)	alternating current	(iii)	has an equal number of protons and electrons
(d)	electric force	(iv)	measure of ability to allow electric current to pass through it
(e)	induction	(v)	electric current that repeatedly reverses direction
(f)	electrical conductor	(vi)	brings about charge separation without touching (12.1, 12.5, 12.8) K/U

Write a short answer to each of these questions.

19. Explain what happens at the atomic level when a neutral object becomes positively charged. Draw a picture to help explain your answer. (12.1) K/U C

20. An oil tanker car has arrived at its destination. Before it starts to empty the fuel with a pump, a metal wire is connected between the tanker and the ground. Suggest a reason why this would be done. (12.5) K/U

21. Use a diagram to explain what causes the electric discharge called lightning. (12.2) K/U C

22. (a) What causes clothes to cling when they are tumbled together in a dryer?
    (b) What causes the crackling sounds you sometimes hear when you pull them out? (12.2) K/U

23. What causes your hairs to repel each other when you take off a wool hat? (12.2) K/U

24. **Figure 1** shows an experiment with a pith ball electroscope. A rod is negatively charged by rubbing with a wool cloth. How will the pith balls react when you touch one of the balls with the charged rod? Redraw the picture to show what happens to the pith balls. (12.3, 12.5) T/I C

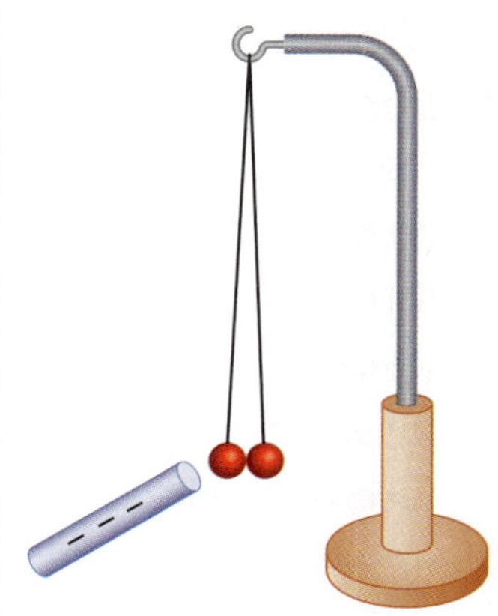

Figure 1

25. Explain why a coating of rubber or plastic is placed around the wires in an extension cord and why the wire is made of metal. How does this make use of the properties of conductors and insulators? (12.8) A

26. (a) How can you tell whether a metal leaf electroscope (**Figure 2**) is charged?
    (b) How could you use this tool to tell whether a material is a conductor or an insulator? (12.6, 12.7, 12.8) T/I

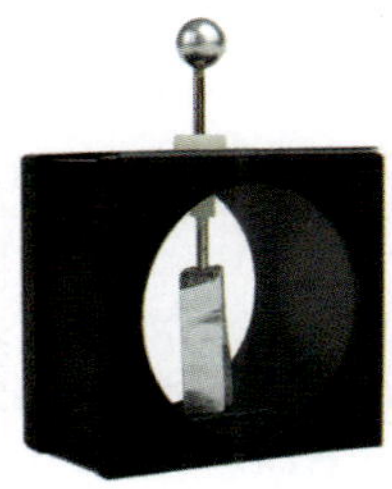

Figure 2

Self-Quiz

To do an online self-quiz,

GO TO NELSON SCIENCE

The following icons indicate the Achievement Chart category addressed by each question.

K/U Knowledge/Understanding | T/I Thinking/Investigation | C Communication | A Application

What Do You Remember?

1. Define the following terms in your own words. (12.1, 12.2, 12.5, 12.8) K/U
   (a) electrically neutral
   (b) charging by induction
   (c) electrical insulator
   (d) electrical conductivity
   (e) electric discharge
   (f) electric current
2. Describe the electrical energy distribution grid. Explain how electric energy gets from the generating plant to your home. (12.8) K/U C
3. Give two examples of how modern technology makes use of static electricity. (12.2) A
4. Restate the Law of Electric Charges. Explain how electric force is affected by distance. (12.1) K/U C
5. Give three examples of charging by contact. (12.3) K/U
6. Draw diagrams that show appropriate charges on the objects when
   (a) a positively charged rod touches a neutral pith ball
   (b) a negatively charged balloon approaches a wall
   (c) a leaf electroscope is charged by a negatively charged ebonite rod (12.3, 12.5, 12.6) K/U C
7. Explain why it is more difficult to charge objects on a humid summer day than when the air is dry. (12.3) K/U
8. Which holds onto its electrons more strongly—an insulator or a conductor? How can you tell? (12.8) K/U

What Do You Understand?

9. (a) Explain the difference between alternating and direct current.
   (b) Give examples from your own experience of where each form of electricity is used. (12.8) K/U A
10. Explain why static electricity can build up on clothes in a dryer. (12.1, 12.2, 12.3) K/U
11. Static means "not moving." Explain why "static" is an appropriate word to describe this type of electricity. (12.2) K/U
12. Use a diagram to show why electrons move easily between materials but protons do not move. (12.1) K/U C
13. Compare and contrast static and current electricity. How are they alike and how do they differ? (12.2, 12.8) K/U C
14. Explain why neutral objects such as bits of tissue paper would be attracted to a charged object. (12.1, 12.5) K/U
15. Explain the difference between how a metal leaf electroscope and a pith ball electroscope detect charge imbalances. (12.4, 12.6, 12.8) K/U
16. During a lightning storm, the ground below negatively charged clouds becomes positively charged. Use your understanding of induction to explain how this might happen. (12.2, 12.3, 12.5) K/U
17. Why is it important to use an anti-static device (**Figure 1**) when working on electronic devices? (12.2) K/U

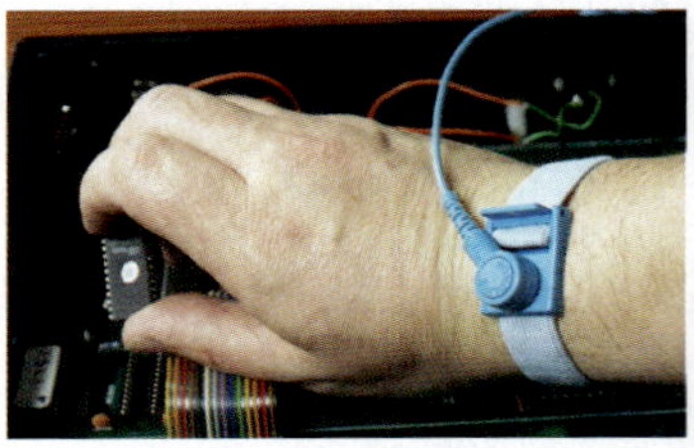

Figure 1 Antistatic device

Solve a Problem

18. A student rubs a plastic plate with a cloth and then brings the plate close to a stream of water. (12.1, 12.2, 12.5) T/I
    (a) What causes the stream of water to bend toward the plate?
    (b) When he places the plate on a metal rack, he sees a spark. What causes the spark?

19. Lightning can be very dangerous if it strikes a building. It can damage electronic devices inside. Lightning rods are metal rods attached to the top of buildings such as the CN Tower. There is a wire attached that goes into the ground. Explain how a device like this uses the properties of conductors to protect electrical devices from fire. (12.2, 12.5, 12.8) A

20. Examine the experimental set-up shown in **Figure 2**. (12.5, 12.8) K/U
    (a) Identify the conductors and the insulators in this experiment.
    (b) Use the drawing to explain why charging by induction results in two objects with opposite charges.

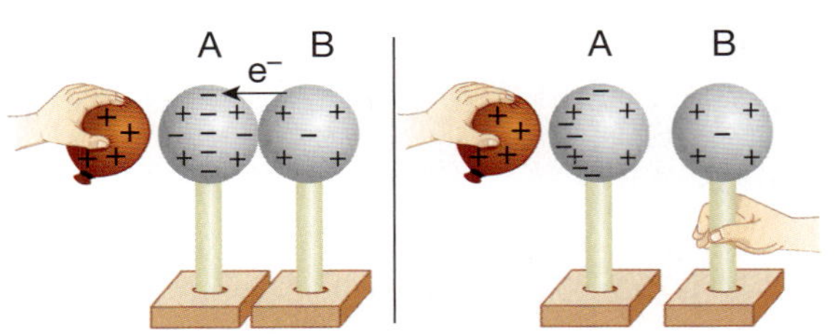

Figure 2

21. A student decided to experiment by bringing a negatively charged balloon near some floating bubbles. With a little practice, the student was able to make the floating bubbles rise and fall repeatedly almost as if a string were attached. The balloon never touched the bubbles. Explain what was happening. (12.1, 12.2, 12.5) T/I

Create and Evaluate

22. Design a procedure to determine whether an unknown material is charged negatively or positively. You may use any of the equipment used in this chapter. (12.1, 12.2, 12.4, 12.6, 12.8) T/I C

23. A variety of products are used to remove dust from furniture in the home. Many of these products make use of static charges. Evaluate the effectiveness of several of these products using your own criteria. Decide which one is the most environmentally friendly. (12.2) T/I A C

Reflect on Your Learning

24. What information presented about static or current electricity did you find most interesting? Explain why. C

25. Has any of the information presented in this chapter caused your attitudes toward the use of electricity or its hazards to change? Explain your answer. C

Web Connections

26. Maglev trains make use of superconducting magnets for high-speed travel. This new form of transportation could revolutionize public transportation in the future. Find out more about how electricity in the tracks and cars makes maglev trains work. How are attractive and repulsive forces involved? Write a summary or make a poster to share what you learn. T/I C A

27. Electrostatic discharges occur in space. Satellites in Earth orbits are especially affected by this. Find out how NASA deals with this problem on Moon and Mars missions such as the *Sojourner* and *Pathfinder* missions. Prepare a presentation in the form of your choosing to report what you have learned. T/I C

GO TO NELSON SCIENCE

CHAPTER 13

Circuits

KEY QUESTION

How do we control current electricity in circuits?

KEY CONCEPTS

- A circuit contains a source of electrical energy, a load, conducting wires, and usually a switch.
- Circuits can be connected in series or in parallel.
- Current is measured using an ammeter connected in series to the circuit.
- Electric potential difference (voltage) is measured using a voltmeter connected in parallel to the circuit.
- Resistance is the opposition to the flow of electrons.
- When you increase the voltage in a circuit, the current also increases.

Bursting Batteries

Computer battery bursts into flames! Laptop explodes! Major battery recall!

Headlines such as these appeared around the world.

Computer manufacturers had to recall thousands of the batteries used to run laptop computers. The batteries were recalled because a small number of them overheated to the point that they caused burns, started fires, or even exploded!

If the batteries are potentially dangerous, why were they used in laptops in the first place?

People want their laptops to be lightweight with big, bright screens and powerful processors. They want them to be rechargeable and work a long time between charges. This means the batteries must be lightweight and provide a lot of energy for a long time.

Lithium-ion batteries fit this description and have advantages over other kinds of rechargeable batteries:

- They have high energy density. This means they are very light but deliver a lot of charge.
- They hold their charge longer.
- They can be recharged many more times.

Unfortunately, building the batteries creates tiny pieces of metal that float in the flammable, lithium-containing liquid inside the batteries. This liquid can become very hot when the computer is used or recharged.

- If there is a spark, the flammable liquid can ignite, causing a fire.
- If the temperature in the battery rises rapidly, the battery can explode due to the increased pressure.
- If the temperature in the battery rises slowly, the battery can melt, and the liquid inside can leak out.

Manufacturers can reduce the amount of lithium in the fluid, but they cannot eliminate it altogether. Instead, they have created another type of battery that uses a non-flammable gel instead of the liquid.

This new type of battery, called a lithium-ion polymer battery, is now commonly used in laptops, cellphones, and portable music players. Larger versions of the battery can power electric cars.

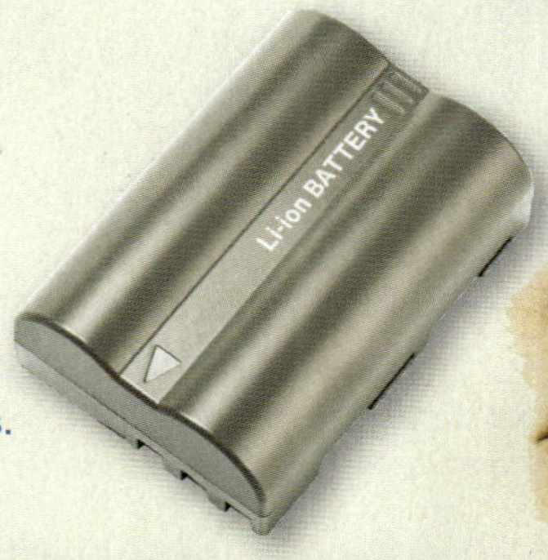

The lithium-ion polymer battery (top) has largely replaced the lithium-ion battery (bottom) in electronic devices.

WHAT DO YOU THINK?

Many of the ideas you will explore in this chapter are ideas that you have already encountered. You may have encountered these ideas in school, at home, or in the world around you. Some of the statements are not true. Consider each statement and decide whether you agree or disagree with it.

1 Electrons travel in one direction through the wires in a building.
Agree/disagree?

2 It is possible to control the flow of electricity in your home.
Agree/disagree?

3 The electric current in a circuit increases when more devices are connected to the circuit.
Agree/disagree?

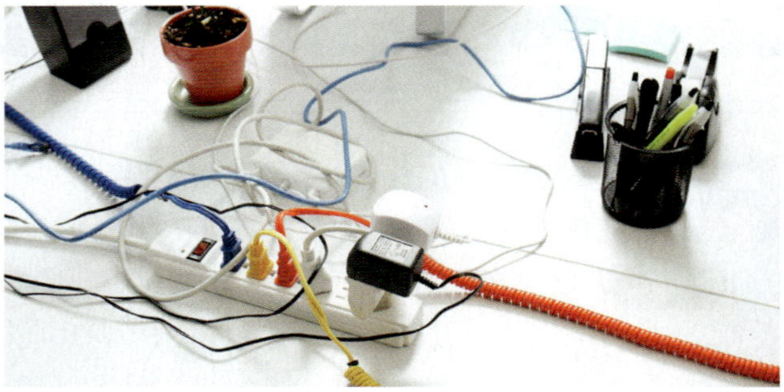

4 Longer wires resist the flow of electrons more than shorter wires.
Agree/disagree?

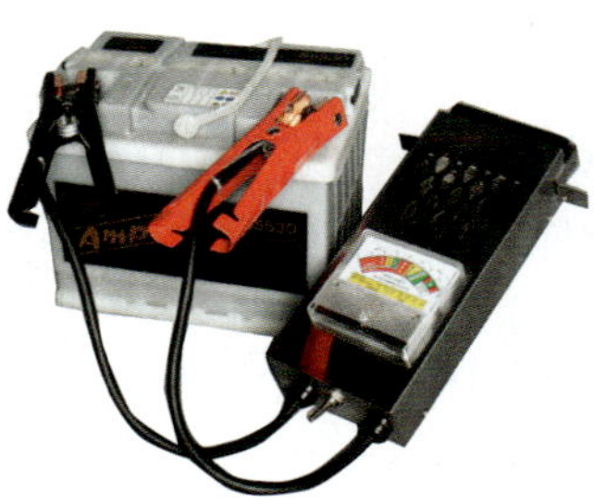

5 Electrons flow from the negative terminal of a battery, through the circuit, to the positive terminal.
Agree/disagree?

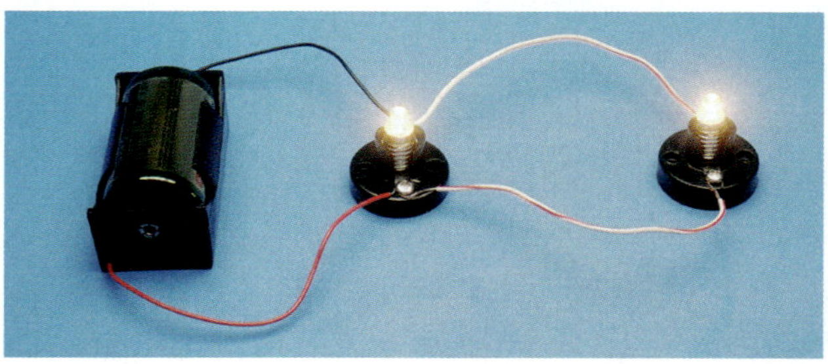

6 There is only one way to connect the components of a circuit so that charges can flow.
Agree/disagree?

FOCUS ON

WRITING

Writing a Critical Analysis

When you write a critical analysis of a text, you ask questions the text has not answered. Find evidence from other sources. The following is a critical analysis of *Engage in Science: Bursting Batteries* on page 497. Use the strategies beside the text to improve your critical writing skills.

> DIG DEEPER
> **Writing Tip**
> As you work through this chapter, look for tips like this. They will help you develop literacy strategies.

Better Batteries

Say why you are doing the critical analysis in the first paragraph. Ask the question that you will answer.

I read the Engage in Science about lithium-ion laptop batteries that caught on fire or exploded when overheated. Lithium-ion polymer (LiPo) batteries are safer because they use a gel that does not burn. Are LiPos the best choice for laptops?

Describe the results of your initial research.

The first website I researched showed a video of a punctured LiPo bursting into flames. Two other websites warned that overcharging LiPos can cause them to get very hot. I decided to find out if there were any safer alternatives to LiPos.

Describe the information you found by digging deeper. If your question is still unanswered, say what you are going to do next.

I discovered a new type of rechargeable battery that uses silver-zinc (AgZn). This type of battery has been used by the military and NASA. The newest version will not burn easily. Test results show the new AgZn batteries charge faster and run longer than LiPos. Overcharging or puncturing the battery cell did not cause fires.

Give additional information to answer the question.

New AgZn rechargeable batteries are also greener than LiPos. Materials used for LiPos cannot be reused and new lithium must be mined to create new LiPos. In contrast, 95 % of the materials in AgZn rechargeable batteries can be recycled and reused. Only the high cost of silver prevents us from using AgZn.

Answer the question in the concluding paragraph.

After I read the Engage in Science, I wondered if LiPos were the best rechargeable batteries for laptops. Tests show that AgZn rechargeable batteries are safer, more efficient, and better for the environment.

13.1

What Is an Electric Circuit?

To turn on a lamp, you flip a switch and the light turns on. What is happening inside the lamp?

Electric Circuits

As soon as a device is turned on, electrons begin to flow through it. An **electric circuit** is a continuous, or closed, path in which electrons can flow. The path is only complete when the circuit is closed. Turning on a device closes the circuit.

- **electric circuit:** a continuous path along which electrons flow that includes an energy source, a load, conductors, and usually a switch
- **load:** an electrical device that converts electrical energy into another form of energy
- **switch:** a control device that allows you to safely open or close a circuit
- **energy source:** a device that provides electrical energy in a circuit, such as a battery or an electrical outlet

Parts of a Circuit

Every electric circuit requires an energy source, a load, and conductors. A **load** is an electrical device such as a light bulb that converts electrical energy into another form of energy. Often, a switch is also added to the circuit. A **switch** is a control device that allows you to safely open or close a circuit. A flashlight is an example of a simple circuit (**Figure 1**).

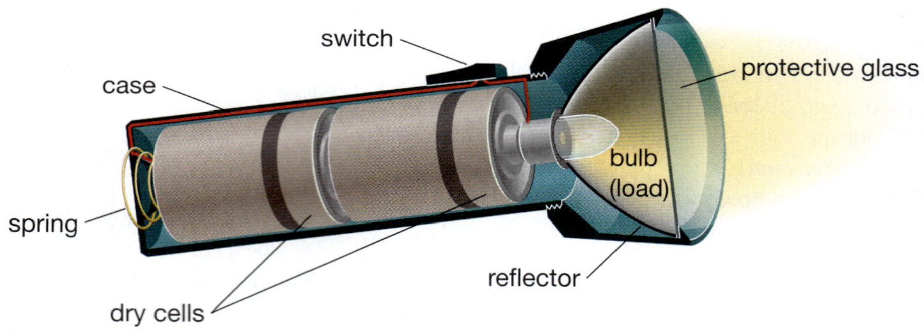

Figure 1 Diagram of a simple circuit in a flashlight

Energy Sources

An **energy source** provides the electrical energy in a circuit. For example, a simple circuit may use a dry cell battery as the energy source. As you read in Section 12.8, electrons flow from a battery in a direct current. The electrons flow in one direction through a circuit and back to the battery.

Other devices such as televisions require more energy for longer periods of time. These devices use electricity from a wall outlet rather than from a battery. The electrical energy that comes from a wall outlet is usually produced at an electrical generating station far from where the energy is used. Electrons flow back and forth between the generating station and the devices plugged into outlets in an alternating current.

Conductors

Look at the flashlight circuit in **Figure 1**. Notice that a piece of metal is attached to each end of the batteries. The metal acts as the conductor, allowing electrons to flow easily through the circuit. When the flashlight is switched on, the metal ends of the batteries make contact with a wire or piece of metal attached to the bulb. This contact closes the circuit, allowing electrons to flow from the batteries through the circuit to the bulb.

In many circuits, the conductors are wires. For example, in your home, conducting wires in the walls connect the energy source (the distant generating station) to each outlet. When you plug a lamp into an outlet, an electric circuit is created with the energy source.

Load

Fans, ovens, motors, light bulbs, and doorbells are all loads. Each type of load converts electrical energy into a different form of energy. **Table 1** shows the main energy conversions in each of these loads.

Table 1 Energy Conversions in Some Electrical Loads

Type of load	Electrical energy is converted into ...
fan	mechanical energy
oven	thermal energy
motor	mechanical energy
light bulb	light energy
doorbell	sound energy

Switch

When you turn on a flashlight, the switch closes the circuit that connects the battery to the light bulb. This causes the bulb to glow. When you turn off the flashlight, the switch opens the circuit. Now, electrons from the battery cannot flow through the bulb. A circuit breaker is a special type of switch that you will read about in Section 13.8. Circuit breakers are found in most electrical panels in homes.

DIG DEEPER

Writing Tip

Writing a Critical Analysis

When writing a critical analysis related to circuits, you might give context to your analysis by describing the parts of a circuit.

Circuit Symbols and Circuit Diagrams

▶ **circuit diagram:** a way of drawing an electric circuit using standard symbols

A **circuit diagram** uses symbols to describe each part of the circuit. **Table 2** shows some common circuit diagram symbols. You will find a more complete list of these symbols in the Skills Handbook.

Table 2 Common Circuit Diagram Symbols

	Part of circuit and symbol		
Sources of electrical energy	electric cell	three-cell battery	variable DC power supply
Electrical conductor	connecting wire		
Control device	open switch		closed switch
Loads	lamp		electric motor

Series Circuits

▶ **series circuit:** a circuit in which two or more loads are connected one after another so that electrons follow only one path

In a **series circuit**, two or more loads are connected so that the electrons can follow only one path. **Figure 2(a)** shows a series circuit with an energy source, a conductor, two loads (lamps), and a switch. Notice that the battery, switch, and lamps are connected one after the other. **Figure 2(b)** is a circuit diagram of this circuit. Electrons flow from the negative terminal of the battery, through the circuit, toward the positive terminal.

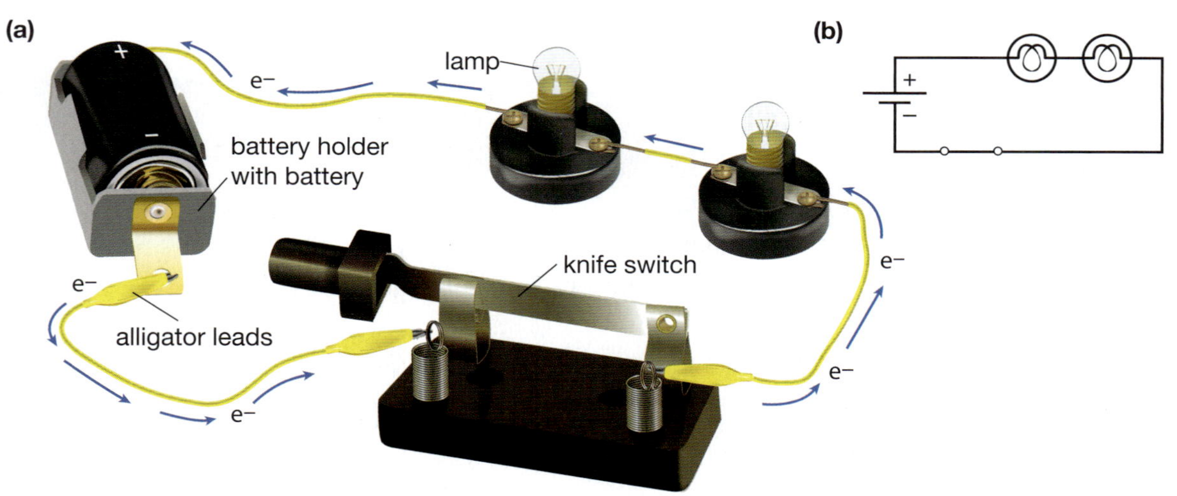

Figure 2 (a) A series circuit with a battery, conducting wires, two lamps, and a switch (b) Circuit diagram of the same series circuit

Parallel Circuits

In a **parallel circuit** (Figure 3(a)), two or more loads are arranged so there are two or more paths for electrons to follow. Figure 3(b) shows the circuit diagram for the same circuit.

▶ **parallel circuit:** a circuit in which two or more loads are connected so that electrons can follow more than one path

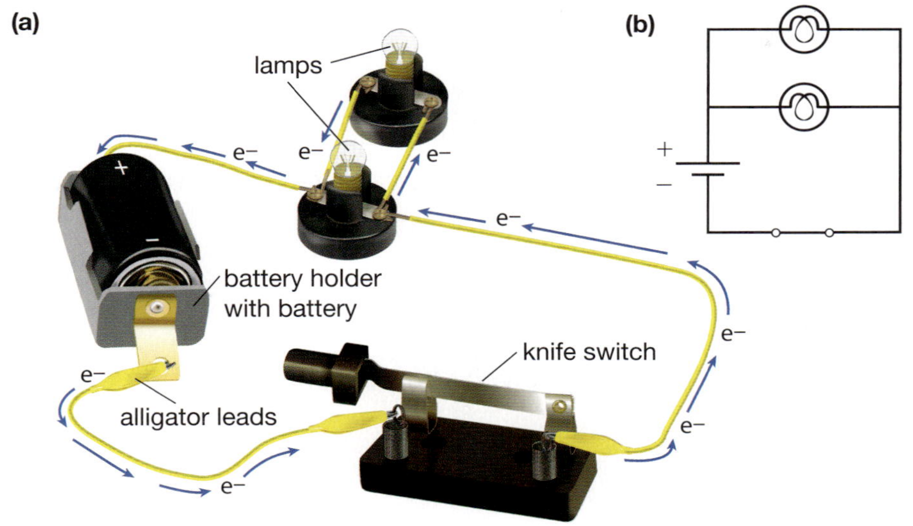

Figure 3 (a) Parallel circuit with a battery, conducting wires, two lamps connected in parallel, and a switch (b) Circuit diagram of the parallel circuit

DIG DEEPER

Learning Tip

Is It Series or Parallel?

Trace the path electrons would follow, without touching the circuit, starting from the energy source. If you make it back to the energy source without coming to a branch in the path, it is a series circuit. If there are one or more branches, it is a parallel circuit.

13.1 Wrap Up

- An electric circuit is a continuous path along which electrons flow. A simple circuit is made up of an energy source, a load, conductors, and usually a switch.
- A switch controls the flow of electrons through a circuit.
- A load is a device that converts electrical energy into another form of energy.
- In a series circuit, electrons follow only one path.
- In a parallel circuit, electrons can follow more than one path.

CHECK YOUR LEARNING

1. Which part of a flashlight is the load? K/U
2. Draw the circuit diagram symbols for
   (a) switch
   (b) battery
   (c) lamp K/U C
3. Use circuit diagram symbols to draw a series circuit containing a battery, conductors, an open switch, and three lamps. K/U C
4. What kind of circuit is shown in Figure 4? K/U

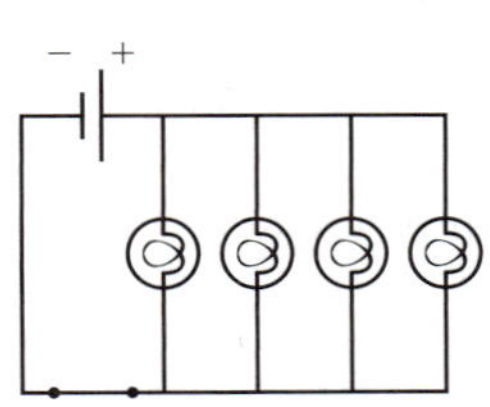

Figure 4

5. Give an example of a load that converts electrical energy into mechanical energy. K/U

13.2 PERFORM AN ACTIVITY

Drawing and Building Circuits

SKILLS MENU
- Questioning
- Hypothesizing
- Predicting
- Planning
- Controlling Variables
- Performing
- Observing
- Analyzing
- Evaluating
- Communicating

In this activity, you will draw circuit diagrams to illustrate given circuits. You will also build your own circuits using electrical components.

SKILLS HANDBOOK 2.D., 3.B.4.–3.B.7.

Purpose

To draw circuit diagrams of given circuits, and to build circuits given circuit diagrams

Procedure

Part A: Drawing a Circuit Diagram

1. Draw a circuit diagram of each circuit provided by your teacher.
2. Copy **Table 1** into your notebook. Fill in the columns to describe each circuit.

Table 1 Observations

Circuit	Power source	Number of switches	Number and type of loads	Type of circuit (series/parallel)
Part A				

Equipment and Materials

white paper

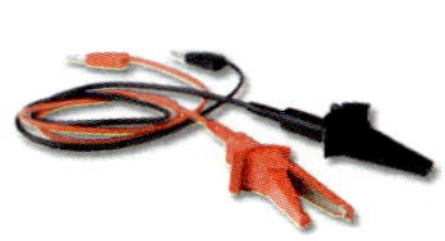
alligator leads

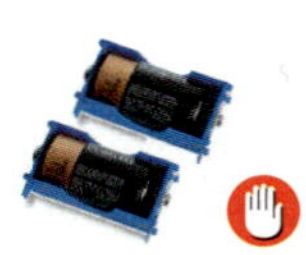
2 battery holders with batteries or power supply

2 knife switches

4 lamps

Part B: Building Circuits

3. Evaluate the circuits in **Figure 1** and **Figure 2**.
4. Use the electrical components provided to build each circuit.
5. After getting your teacher's approval, close the switch to complete the circuit. Observe what happens.
6. Complete **Table 1** for these circuits.

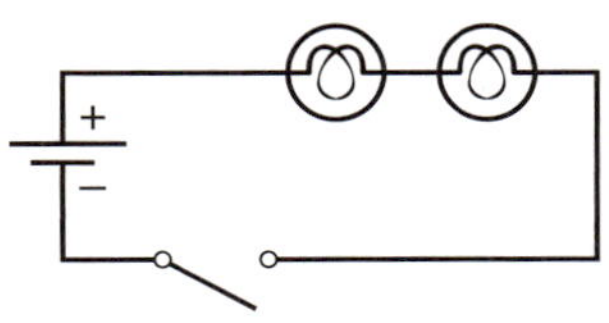

Figure 1 Circuit 1

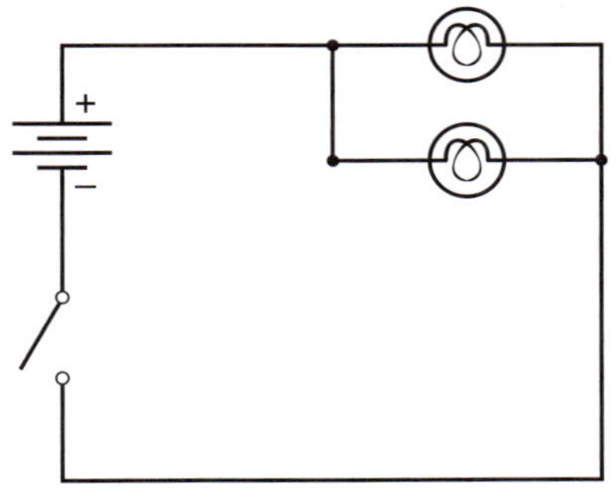

Figure 2 Circuit 2

CAUTION

Electricity

Avoid shocks. Check your circuits with your teacher before turning them on. Never operate electrical equipment with wet hands.

Analyze and Evaluate

(a) How many paths are there for electrons to flow through each circuit you drew for **Part A**? Write a description for each circuit. T/I C

(b) How many paths are there for electrons to flow through each circuit you built for **Part B**? Write a description for each circuit. T/I C

Apply and Extend

(c) Based on what you know about circuits, if a lamp at home or at school was not turning on, what might be the reason? A

(d) Why do you think operating instructions for electrical devices sometimes include circuit diagrams? A

UNIT TASK

Bookmark

You can apply what you learned in this activity about circuits to the Unit Task described on page 536.

13.3 PERFORM AN ACTIVITY

Designing Circuits

In this activity, you will design and build circuits based on descriptions. You will explore how the brightness of the lamps in each circuit varies as you add or remove loads.

SKILLS MENU

- Questioning
- Hypothesizing
- Predicting
- Planning
- Controlling Variables
- Performing
- Observing
- Analyzing
- Evaluating
- Communicating

SKILLS HANDBOOK 2.D., 3.B.5.–3.B.8.

Purpose

To compare the brightness of each lamp in series and parallel circuits

Procedure

1. Gather the equipment for all parts of this activity. Your teacher may assign groups to build and share circuits. Copy **Table 1** into your notebook to record observations. If you are using a variable DC energy source, your teacher will explain how to use it.

Part A: Simple Circuit: One Bulb, One Switch

2. Follow **Figure 1** to connect the parts of the circuit. This is Circuit 1.
3. If you are using a DC power supply, set the voltage as directed by your teacher. Have your teacher approve your circuit. Close the switch to complete the circuit. Observe the brightness of the lamp.
4. Turn off the power supply. Then open the switch. Do not disassemble until the activity is complete.

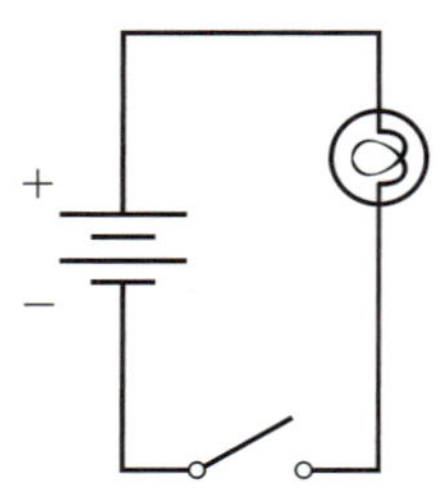

Figure 1 Circuit 1

CAUTION
Electricity

Always disconnect the energy source before connecting or disconnecting parts to a circuit. Once you have connected a circuit, have your teacher approve it before you turn it on. If your circuit does not work, ask your teacher for assistance.

Table 1

Circuit	Bulb 1 brightness	Bulb 2 on/off	Brightness
1		—	—
2		on	
2		off	
3		on	
3		off	

Equipment and Materials

white paper

alligator leads

2 battery holders with batteries or power supply

2 knife switches

4 lamps

Part B: Series Circuit: Two Bulbs, One Switch

5. Build the series circuit (Circuit 2) shown in **Figure 2**.
6. If you are using a power supply, set the same voltage as in **Part A** as directed by your teacher. Have your teacher approve your circuit. Close the switch to complete the circuit.
7. Turn Circuit 1 back on.
8. Compare the brightness of each lamp in Circuit 2 to the brightness of the lamp in Circuit 1. Record your observation in **Table 1**.
9. Unscrew the bulb in one lamp. Observe what happens to the other lamp. Record your observation.
10. Turn off the power supply. Open the switches for both circuits.

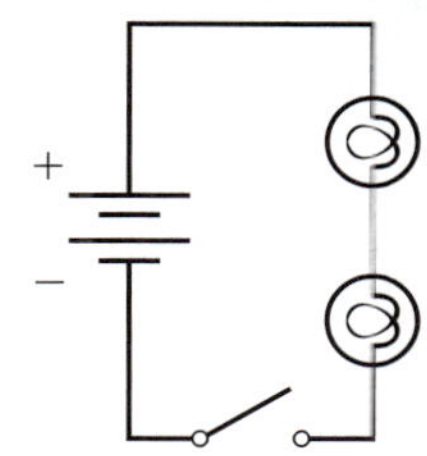

Figure 2 Circuit 2

Part C: Parallel Circuit: Two Bulbs, One Switch

11. Build the parallel circuit (Circuit 3) shown in **Figure 3**.
12. If you are using a power supply, set the same voltage as in **Part A** as directed by your teacher. Have your teacher approve your circuit. Close the switch to complete the circuit.
13. Turn Circuit 1 back on.
14. Compare the brightness of each lamp in Circuit 3 to the brightness of the lamp in Circuit 1. Record your observation.
15. Unscrew the bulb in one lamp in Circuit 3. Observe what happens to the other lamp. Record your observations.
16. Turn off the power supply. Open the switches for both circuits.

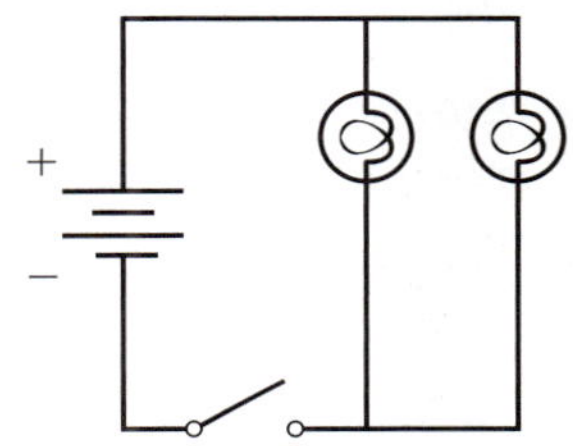

Figure 3 Circuit 3

Analyze and Evaluate

(a) How does the brightness of the lamp compare in each of the circuits? T/I

(b) What happened when you unscrewed a bulb in Circuits 2 and 3? What happened when you opened the switch in each circuit? T/I

(c) Based on the brightness you observed in the lamps, which lamp(s) do you think used more energy? T/I

Apply and Extend

(d) Which type of circuit, series or parallel, would you use to wire the lights in your home? Why? A

(e) Why do you think most homes have a switch wired in series with all of the circuits in the house? A

UNIT TASK

Bookmark

You can apply what you learned about circuits in this activity to the circuits you design as part of the Unit Task described on page 536.

13.4

Measurements in Electric Circuits

In everyday life, you cannot often use the brightness of a lamp to check if a circuit is working. Instead, three circuit variables can be measured. The variables are electric current, potential difference (or voltage), and resistance.

- **electric current (*I*):** the rate of flow of electrons past a certain point in a circuit
- **ampere (A):** the unit of electric current
- **ammeter:** a device used to measure current

Measuring Energy

You read in Chapter 12 that electric current is the flow of electrons through a circuit. The rate at which electrons flow past a certain point in the circuit is also called **electric current**. The symbol for current is *I*.

Imagine that you are watching a bicycle race. If you counted the number of cyclists passing you within a certain period of time, that measurement would be similar to an electric current (**Figure 1**). The difference between an electric current in a circuit and the rate of cyclists going by you is that the number of cyclists is extremely small, while huge numbers of electrons flow in a circuit.

Figure 1 The number of cyclists moving past you within a period of time is similar to the rate of electron flow.

Measuring Electric Current

Electric current is measured in units called **amperes (A)**. The French scientist André-Marie Ampère was the first person to figure out a way to measure the electric current in a circuit. He built a device called an **ammeter** (**Figure 2**). To measure current, an ammeter must be connected in series with a load in a circuit (**Figure 3**). That way, all the electrons that flow through the load also pass through the ammeter.

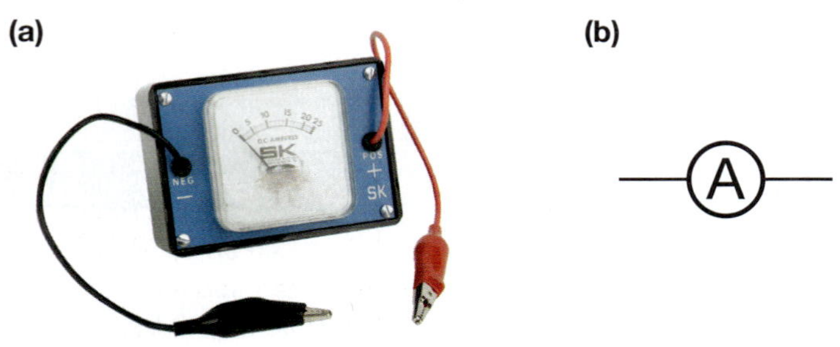

Figure 2 (a) When this type of ammeter is connected to a circuit, the current is measured by reading the needle position on the dial. (b) The circuit diagram symbol for an ammeter

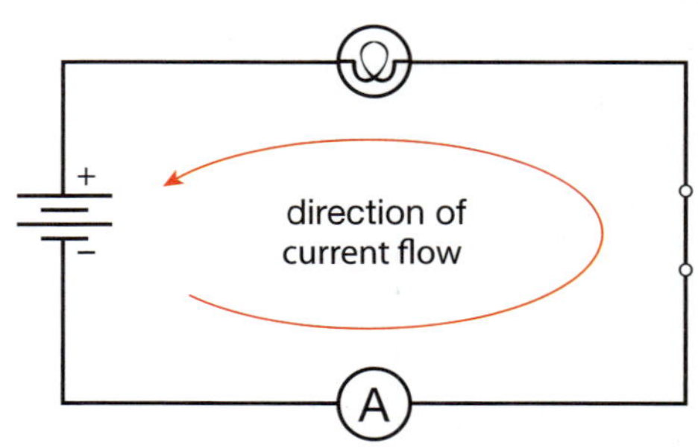

Figure 3 To measure the current through the lamp, an ammeter must be connected in series with the lamp.

Potential Difference or Voltage

A ball at the top of a hill has potential energy because of the difference in height between two points: the top of the hill and the bottom of the hill. As the ball rolls downhill, the potential energy is changed to kinetic energy, the energy of motion.

Within a circuit, electrons flow through the wire from the negative terminal of the battery to the positive terminal with a certain amount of energy (**Figure 4**). This energy, called electric potential energy, is used by loads connected to the battery such as lamps, motors, or other electrical devices. The **potential difference** in an electric circuit is the difference in electric potential energy per unit of charge measured between two points in a circuit. Another name for potential difference is **voltage**. The symbol for potential difference is *V*.

To increase the amount of electric potential energy in a circuit, batteries are connected in series. This is why some flashlights use more than one battery.

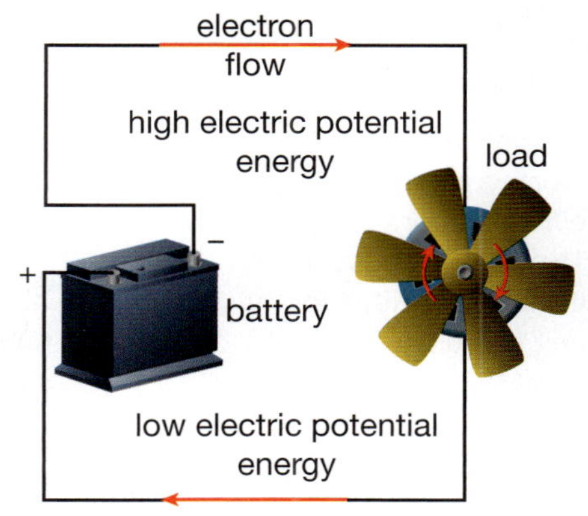

Figure 4 Electrons flow through a circuit with a certain amount of energy that can be used by loads connected to the circuit.

- **potential difference** or **voltage (*V*):** the difference in electric potential energy per unit of charge measured between two points
- **volt (V):** the unit of potential difference
- **voltmeter:** a device used to measure potential difference

Measuring Potential Difference

The unit for potential difference is called the **volt (V)**, in honour of Italian scientist Alessandro Volta. Volta built the first electric battery. The device used to measure potential difference is called a **voltmeter** (**Figure 5**). A voltmeter must always be connected in parallel with either an energy source or a load in a circuit (**Figure 6**).

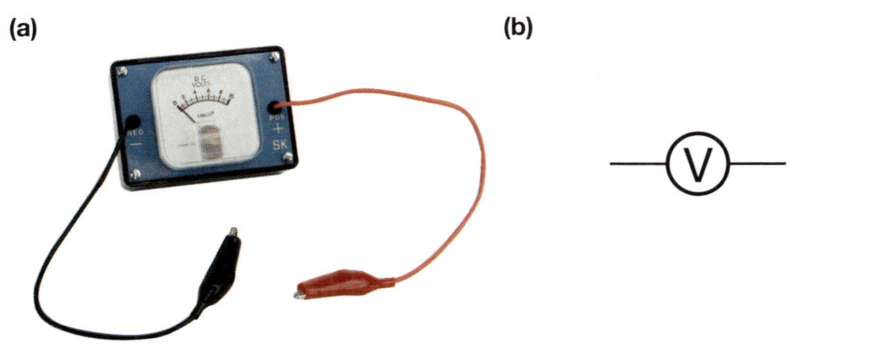

Figure 5 (a) When this type of voltmeter is connected to a circuit, the voltage is measured by reading the needle position on the dial. (b) The circuit diagram symbol for a voltmeter

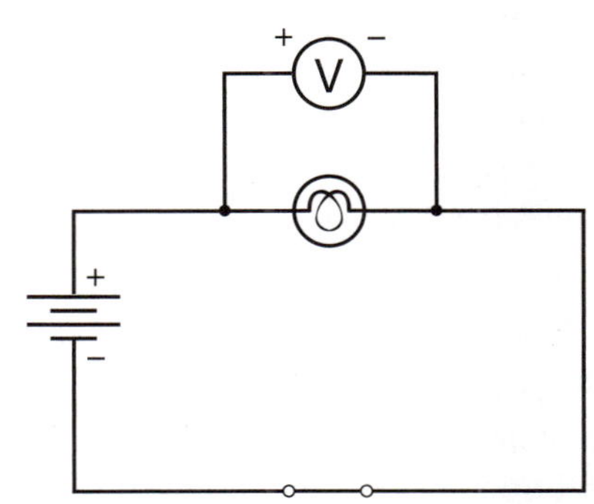

Figure 6 To measure the potential difference across the lamp, a voltmeter must be connected in parallel with the lamp.

DIG DEEPER

Learning Tip

Volts and Voltage

In science, an italic *V* represents potential difference, or voltage. This is to avoid confusing it with the symbol for the unit for volts (V).

- **electrical resistance (*R*):** the ability of a material to oppose the flow of electrons through it
- **ohm (Ω):** the unit of electrical resistance
- **ohmmeter:** a device used to measure electrical resistance

Electrical Resistance

The third variable associated with circuits is electrical resistance. **Electrical resistance** is the opposition to the flow of electrons through a material. The symbol for electrical resistance is *R*.

For example, when you run along the beach, you can move very easily. If you run into the water, you have to apply a lot of effort to keep running. This means that the water has a greater resistance than air. Just as running through different materials affects your ability to run, the electrical resistance of a material affects how easily electrons can flow through the material.

Measuring Electrical Resistance

The unit for electrical resistance is called the **ohm (Ω)**, in honour of German scientist Georg Ohm. Ohm discovered a way to determine the electrical resistance of a material by using current and voltage measurements. You will learn more about this in Section 13.7. The device used to measure electrical resistance is called an **ohmmeter** (**Figure 7**).

(a) (b)

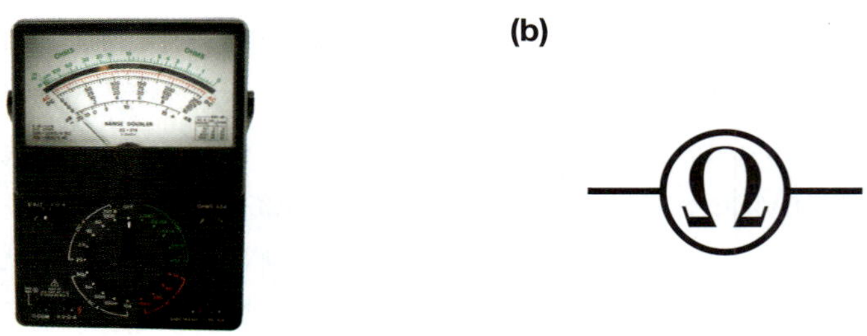

Figure 7 (a) Ohmmeter (b) The circuit diagram symbol for an ohmmeter

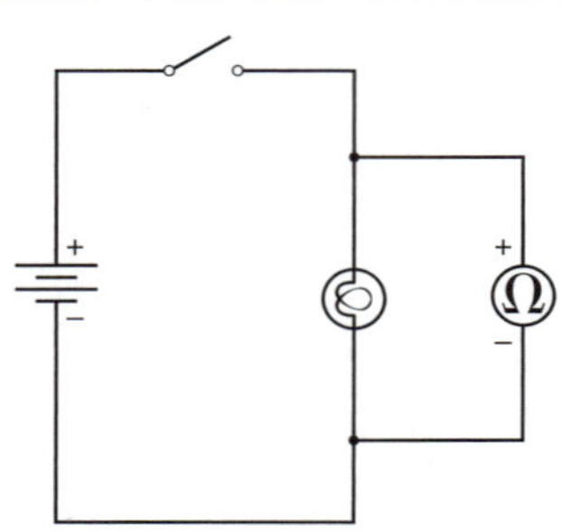

Figure 8 Notice that the switch of the circuit is left open when using an ohmmeter.

An ohmmeter must always be connected in parallel with a load in a circuit (**Figure 8**). The ohmmeter must be connected to each end of the load to measure its electrical resistance. The load cannot be connected to anything else that forms a closed circuit, or else the resistance of the entire circuit will be measured. The circuit or switch does not need to be closed to measure the resistance across a load. An ohmmeter has its own energy source, which provides a flow of electrons through the load.

Factors That Affect Electrical Resistance

All materials have some electrical resistance. The electrical resistance of a material depends on four main factors: type of material, length, thickness, and temperature.

Type of Material

Some materials have much less internal electrical resistance than others. For example, when electrons flow through a wire, they repel the electrons on the atoms that make up the wire. This makes the atoms making up the wire vibrate. In the process, some of the electric potential energy stored in the electrons gets converted into thermal energy, and the wire becomes warm. If the wire is made of copper, a flow of electrons has less chance of making the atoms in the wire vibrate than if the wire were made of carbon. Copper has less internal resistance than carbon.

Length

The longer a wire is, the more electrical resistance the wire has. Electrons flowing through a long wire have more material to go through than if the wire were shorter. Electrons have a greater chance of making the atoms of a longer wire vibrate (**Figure 9**).

Thickness

A thick wire has less electrical resistance than a thin wire. A thicker wire is like a multi-lane highway with lots of space. A thicker wire allows more electrons to flow per second, just like a multi-lane highway allows more cars to drive on it at one time.

Temperature

When the temperature of a material increases, its electrical resistance usually increases. When a wire gets warmer, the atoms that make up the wire gain energy and vibrate faster. The increased vibration increases the chance that the electrons in the current will collide with the moving atoms of the material.

Figure 9 The circuits on a computer chip are extremely small. The wires that connect each part of a circuit are as short as possible to reduce electrical resistance. This makes the chip operate faster and more efficiently.

DIG DEEPER

Writing Tip

Supporting the Main Idea

A clearly expressed main idea is a powerful communication tool. Main ideas need supporting details to help readers understand what they mean. When writing a critical analysis, use examples, or even diagrams, to help explain the main idea.

Materials Used in Circuit Components

Differences in electrical resistance make some materials better suited than others for certain components in a circuit.

Conductors

The best materials to use as conductors are those that have low electrical resistance. Copper and silver are two examples of conductors with low electrical resistance. Silver is much more expensive than copper, so copper is used more often as a conductor in circuits.

Insulators

Materials that have a very high electrical resistance are best used as insulators. Glass, rubber, and most plastics are examples of insulators. These materials are used to cover the conducting wires in all electrical wires and extension cords (**Figure 10**). Insulators protect you from an electric shock when you handle the wires.

Loads

Loads are usually made of materials that have high electrical resistance. However, the electrical resistance must be less than that of the insulators used in the circuit, otherwise no electrons will flow through the load. When electrons flow through the wires of a toaster, the electric potential energy stored in the electrons gets converted into thermal energy and light energy (**Figure 11**). The thermal energy is what toasts bread.

Figure 10 Rubber and plastic are often used to wrap conductors.

Figure 11 The glowing wires of a toaster indicate that their electrical resistance is high.

Circuit Measurements in Everyday Life

There are many situations in everyday life where it is necessary to measure the current, voltage, and resistance in a circuit. One example is a circuit that will not switch on when it is connected to an energy source. An electrician measures the current, voltage, and resistance at different points in the circuit to find the source of the problem. For this task, electricians use a multimeter. A **multimeter** is a device that allows you to measure current, voltage, and resistance by changing its settings. Multimeters may be small, inexpensive hand held devices for home use (**Figure 12**).

▶ **multimeter:** a device used to measure different circuit quantities

Figure 12 You can use multimeters to take current, voltage, and resistance measurements.

13.4 Wrap Up

- Electric current is a measure of the rate of electrons flowing past a certain point in a circuit. Current is measured in amperes using an ammeter.
- Potential difference, or voltage, is the difference in electric potential energy per unit of charge measured between two points. Voltage is measured in volts using a voltmeter.
- Electrical resistance is the opposition to the flow of electrons through a material. Resistance is measured in ohms using an ohmmeter.
- All materials have some electrical resistance. Materials with the least resistance are used as conductors. Materials with the highest resistance are used as insulators.

DIG DEEPER

Career Link

Electricians are trained to install and repair electrical circuits. To learn more about becoming an electrician,

GO TO NELSON SCIENCE

CHECK YOUR LEARNING

1. Hakim connects a voltmeter in parallel to the load on a circuit. Compare the voltage of the electrons before and after they pass through the load. K/U T/I
2. The label on the bottom of a toaster says it is a 120 V device that draws 12 A at 10 Ω. Identify the potential difference, current, and resistance. K/U
3. You want to buy a small electric space heater. You consider two different heaters. One has a longer, thinner cord and the other has a shorter, thicker cord. Which one should you buy? Why? A
4. Explain why your computer needs a cooling fan. A
5. (a) Draw a circuit diagram that shows a battery connected in series with a switch and three lamps connected in series.
   (b) Show how the meters should be connected to test the third lamp for
   (i) resistance
   (ii) current
   (iii) potential difference T/I C

13.5 PERFORM AN ACTIVITY

Measuring Quantities in Series Circuits

SKILLS MENU

- Questioning
- Hypothesizing
- Predicting
- Planning
- Controlling Variables
- Performing
- Observing
- Analyzing
- Evaluating
- Communicating

In this activity, you will use an appropriate meter to measure current, potential difference, and electrical resistance in a series circuit. You will then determine if your circuit works.

SKILLS HANDBOOK 2.D., 2.E.

Purpose

To measure electric currents, potential differences, and electrical resistance in circuits using appropriate meters

Procedure

1. Read through the procedure, and create a table in which to record your observations.
2. Obtain the required materials from your teacher. Make sure the energy source is turned off. Connect the circuit shown in **Figure 1**.
3. If using a variable DC energy source, make sure it is at its lowest setting as instructed by your teacher. After your teacher has approved your circuit, close the switch.
4. If using a variable DC energy source, slowly increase the voltage until the lamps turn on. Once they are lit, leave the energy source at this setting.

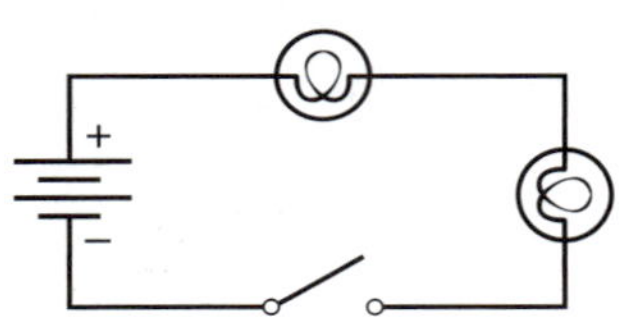

Figure 1

Equipment and Materials

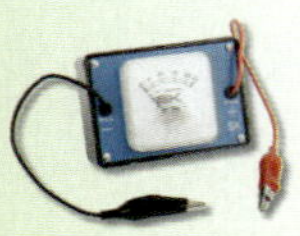
ammeter

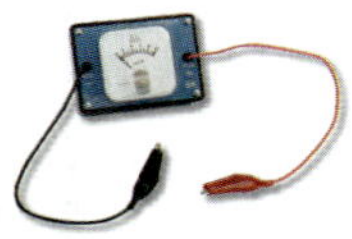
voltmeter

ohmmeter

multimeter

battery holders with batteries or variable DC energy source

2 lamps

knife switch

6 alligator leads or connecting wires

5. Open the switch. If using a DC energy source, turn off the energy source. Connect the ammeter in series at A1 in **Figure 2**. Close the switch and turn on the energy source. Record the measurement in your table. Turn off the energy source.
6. Turn off the energy source and leave the switch open. Repeat Step 5, connecting the ammeter at A2, then at A3.
7. Use the voltmeter to measure the potential difference across the energy source. Connect the voltmeter as shown at V1 in **Figure 3**. Turn the energy source on and close the switch. Record the measurement. Turn off the energy source.
8. Repeat Step 7, connecting the voltmeter at V2, then at V3.
9. Turn the energy source on. Remove one of the bulbs from its socket. Note any difference in the brightness of the other lamp. Use an ammeter to measure the current at A1, A2, and A3. Then, use a voltmeter to measure the potential difference at V1, V2, and V3. Remember to open the switch and, if necessary, turn off the energy source each time you are connecting a meter. Record the measurements.
10. Open the switch, then turn off the energy source. Replace the first bulb in its socket. Use an ohmmeter to measure the total resistance of the circuit (**Figure 4**). Record the measurement. Disconnect your circuit.

CAUTION

Electricity

Always disconnect the energy source before connecting or disconnecting any meters or parts to a circuit. To unplug the source, do not pull the cord. Pull the plug itself. Ask your teacher to check your circuit before you turn the energy source on. If the circuit or the meter does not work, ask your teacher for assistance.

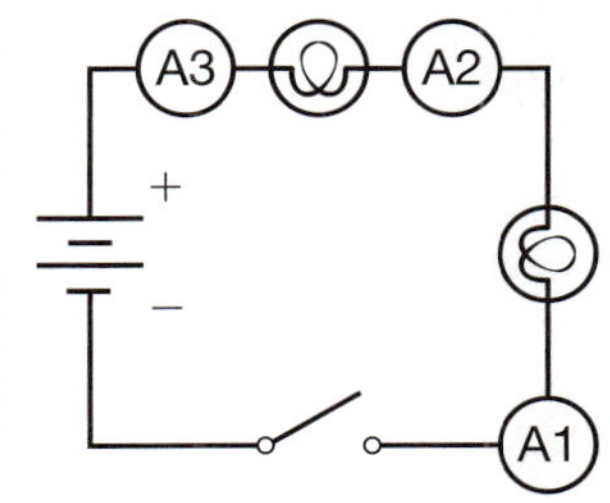

Figure 2 Connection points for the ammeter

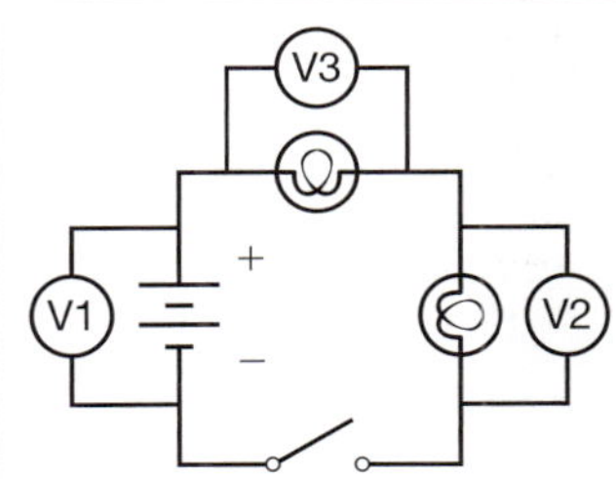

Figure 3 Connection points for the voltmeter

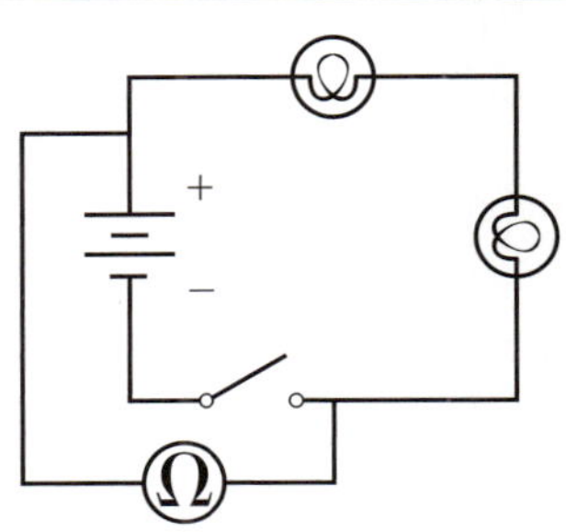

Figure 4 Connection points for the ohmmeter

Analyze and Evaluate

(a) Compare how you had to connect the different kinds of meters. What was the same? What was different? T/I

(b) What conclusion can you make about the electric current in a series circuit? T/I

(c) How does unscrewing the first bulb affect the brightness of the second lamp in the circuit? T/I

Apply and Extend

(d) Suppose you had another type of device connected to your circuit that was not functioning properly. What would you do to determine what might be wrong? T/I

13.6 PERFORM AN ACTIVITY

Measuring Quantities in Parallel Circuits

SKILLS MENU

- Questioning
- Hypothesizing
- Predicting
- Planning
- Controlling Variables
- Performing
- Observing
- Analyzing
- Evaluating
- Communicating

In this activity, you will use an appropriate meter to measure current, potential difference, and electrical resistance in a parallel circuit. You will then use meters to determine if your circuit works.

SKILLS HANDBOOK 2.D., 2.E.

Purpose

To measure electric current, potential difference, and electrical resistance in circuits using appropriate meters

Procedure

1. Read through the Procedure, and create a table in which to record your observations. Obtain the required materials from your teacher.
2. Keep the switch open. If using a DC energy source, make sure the energy source is turned off. Connect the circuit shown in **Figure 1**.

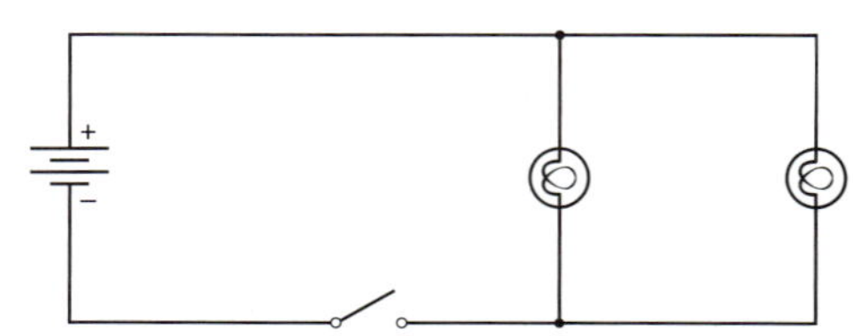

Figure 1

Equipment and Materials

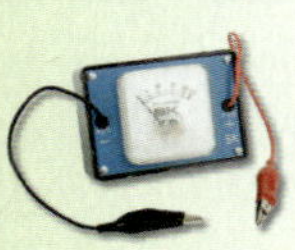
ammeter

voltmeter

ohmmeter

multimeter

battery holders with batteries or variable DC energy source

2 lamps

knife switch

6 alligator leads or connecting wires

3. Make sure the energy source is at its lowest setting as instructed by your teacher. After your teacher has approved your circuit, close the switch.
4. If using a DC energy source, slowly increase the voltage until the lamps turn on. Once they are lit, leave the energy source at this setting.
5. Use an ammeter to measure the electric current at position A1 (**Figure 2**). Remember to turn off the energy source each time you are connecting a meter. Record the measurement. Repeat this step for A2, A3, and A4.
6. Use a voltmeter to measure the voltage at V1, V2, and V3 (**Figure 3**). Remember to turn off the energy source each time you are connecting a meter. Record the measurement.
7. Remove one of the bulbs from its socket. Note any difference in the brightness of the other lamp. Use an ammeter to measure the current at A1, A2, and A3. Then use a voltmeter to measure the potential difference at V2 and V3. Remember to turn off the energy source each time you are connecting a meter. Record the measurements.
8. Open the switch then turn off the power supply. Replace the first bulb in its socket. Use an ohmmeter to measure the total resistance of the circuit (**Figure 4**). Record the measurement.
9. Disconnect your circuit.

CAUTION

Electricity

Always disconnect the energy source before connecting or disconnecting a circuit. Ask your teacher to check your circuit before you turn the energy source on.

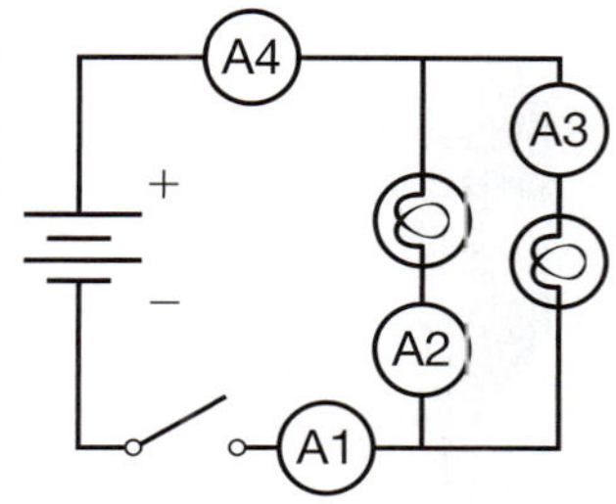

Figure 2 Connection points for the ammeter

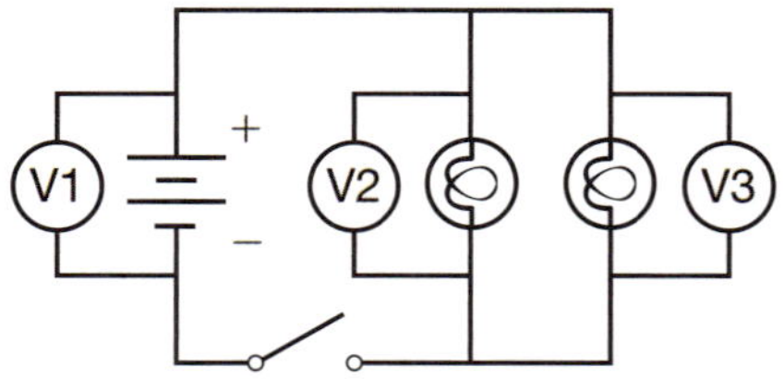

Figure 3 Connection points for the voltmeter

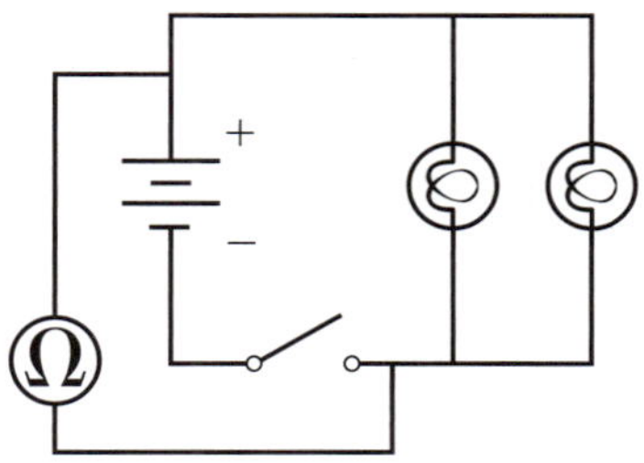

Figure 4

Analyze and Evaluate

(a) Compare how you had to connect the different kinds of meters. What was the same? What was different? T/I

(b) What conclusion can you make about the electric current in a parallel circuit? T/I

(c) How does unscrewing the first bulb affect the brightness of the second lamp in the circuit? T/I

(d) Refer back to your observations in Section 13.5. Compare the resistance in the series circuit with the parallel circuit. What did you observe? T/I

UNIT TASK

Bookmark

You can apply what you learned in this activity about parallel circuits to the Unit Task described on page 536.

13.7

Factors Affecting Series Circuits

In Sections 13.5 and 13.6, you explored what happens to the potential difference (voltage) and current when loads (resistance) are removed from a series circuit and a parallel circuit. These activities are very similar to those performed by Georg Ohm.

Figure 1 Ohm built this device, called a thermocouple, to study the relationship between current, potential difference, and resistance.

Ohm's Law

In 1827, Georg Ohm discovered the relationship between current, potential difference, and resistance. He described this relationship mathematically. Ohm conducted many experiments during which he kept the potential difference (voltage) across a conductor constant, but varied the length of the conductor (**Figure 1**). He then measured the current through the conductor.

Ohm discovered that when the voltage is constant, as the length of the conductor increases, the current through the conductor decreases. Recall from Section 13.4 that the longer a wire is, the more electrical resistance it has. The current through the wire will decrease if the wire is longer.

The relationship that Ohm discovered became known as **Ohm's law**. Ohm's law states that, for a given resistance, as the electric potential difference across a load increases, so does the current.

▶ **Ohm's law:** the relationship between resistance, current, and potential difference; for a given potential difference, as the resistance increases, the current decreases

Increasing Resistance

The unit for resistance is the ohm (Ω). An ohm is a measure of the opposition to the flow of electrons through a material.

You can compare the flow of electrons through a circuit to people running through an obstacle course (**Figure 2**). The more obstacles there are in their way, the slower the runners will be as they go through the course. Similarly, the wider the course is, the more runners will be able to go through it at the same time.

Figure 2 Runners try to make it past the obstacles as quickly as possible.

Just as in the obstacle course, the amount of resistance in a circuit affects the amount of current that can flow through it. If the resistance is increased by making a conductor longer or thinner, the amount of current that can flow through the conductor in a given amount of time will decrease. Similarly, if the conductor is replaced by another conductor with a higher internal resistance, the amount of current that can flow through the conductor in a given amount of time will decrease.

The amount of current that flows through a series circuit is also affected by the number of loads in the circuit. Think of each load as an obstacle. The more obstacles, or loads, in a circuit, the greater the resistance and the lower the current will be.

The following factors may increase the total resistance in a circuit and thus decrease the current:

- connecting more loads in series
- connecting loads in series rather than in parallel
- making the conductor longer
- making the conductor thinner
- using a material with a higher resistance

DIG DEEPER

Did You Know?

Park and Plug In

The provincial government in Ontario wants to encourage the development of all electric cars. Besides offering rebates to people who purchase electric cars, the government is working with developers to make charging stations available in all new housing and office buildings. The rechargeable batteries in some electric cars can either be fully charged in 1 h or replaced with another charged battery in less than 5 min.

Decreasing Resistance

Think about the obstacle course again. If the course is changed, it could affect how quickly runners can complete the course or how many can complete the course at the same time. The fewer the obstacles on the course, the faster the runners will be able to go. If the course becomes narrower, fewer runners will be able to complete the course at the same time.

Electric circuits work in a similar way. The less the opposition to electron flow through a material, the greater the current will be. The following factors may help lower resistance in a circuit and thus increase the current:

- connecting fewer loads in series
- connecting loads in parallel rather than in series
- making the conductor shorter
- making the conductor thicker
- using a material with a lower resistance

Using Materials with Low Resistance

Engineers use knowledge of Ohm's law and the resistance of different materials when designing the electronic systems in satellites (**Figure 4(a)**) or planetary rovers (**Figure 4(b)**). For example, the solar cells used on satellites or rovers can generate only a certain amount of electrical energy, so these systems need to reduce resistance as much as possible because it wastes energy. This is why they use metals such as silver, platinum, and rhodium or special alloys that have very low internal resistance for conductors.

Figure 4 (a) The worker is testing a satellite's circuits before launch. (b) Satellites and rovers need to function using as little energy as possible. To make this possible, their conductors are materials with very low resistance.

DIG DEEPER

Writing Tip

Conducting Research

When preparing to write a critical analysis, conduct research to improve your understanding of the issue. It is important to be aware of both supporting evidence and opposing views before completing the analysis.

Changing Electric Potential Difference

Changing the resistance in a circuit changes the amount of current in the circuit. What if the resistance in a circuit stays the same? Another way to change the amount of current between two points in a circuit is to change the potential difference between the two points.

Increasing the potential difference increases the electric current. Likewise, lowering the potential difference between two points on a circuit lowers the amount of current, as long as resistance stays the same.

Batteries

You may have noticed that batteries include a voltage rating (**Figure 5**). This rating is related to the potential difference between the two terminals of the battery. The greater the potential difference across the terminals, the greater the current that will flow when the battery is connected to a circuit. If you try to use a battery with too little potential difference in an electric device, it will either not work well or not work at all. If you switch to a battery with too much voltage, circuits in the device could overheat because of the amount of current flowing through them, causing damage or even a fire.

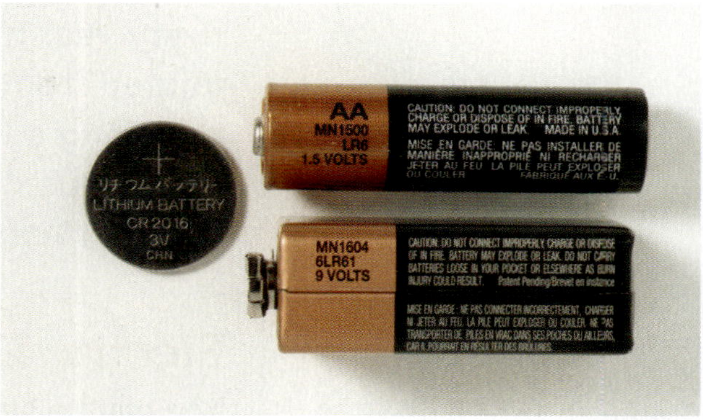

Figure 5 Check the voltage rating of batteries before using them in electronic devices.

13.7 Wrap Up

- Ohm's law states that as the potential difference (voltage) across a load increases, so does the current.
- As the potential difference across a load decreases, so does the current.
- As the resistance increases, the current decreases for a given potential difference.
- As the resistance decreases, the current increases for a given potential difference.

CHECK YOUR LEARNING

1. What needs to decrease in order to increase the current in an electric circuit? How can this be accomplished? T/I
2. Georg installs a dimmer switch in his living room. He connects the switch in series with the lights there. What happens to the resistance of the dimmer as the dial is turned, dimming the lights more and more? T/I
3. The wiring in an electric motor on a lunar rover is made of copper. What can the engineer do to lower the voltage needed to run the motor? T/I
4. Explain to another student why it is important to always use batteries with the proper voltages in electric devices. K/U C
5. Elaine connects an ammeter to a lamp, which is connected to a battery. She records the current. She connects another battery in series to the circuit and records the ammeter reading again. She repeats this step two more times. T/I
   (a) Explain which variable Elaine is not changing, which one she is changing, and which one she is investigating.
   (b) What do you expect Elaine's readings will show?
6. Mary measures the current in a long, thin extension cord. She unplugs the cord and switches it for a shorter, thicker one. What will Mary observe if she measures the current through the new extension cord? T/I

13.8 Safety Issues with Circuits

Every year hundreds of people in Canada are killed or injured from contact with electricity. Did you know that even changing a light bulb without unplugging the lamp can be hazardous? The more you learn about electricity, the better you can keep yourself and others safe.

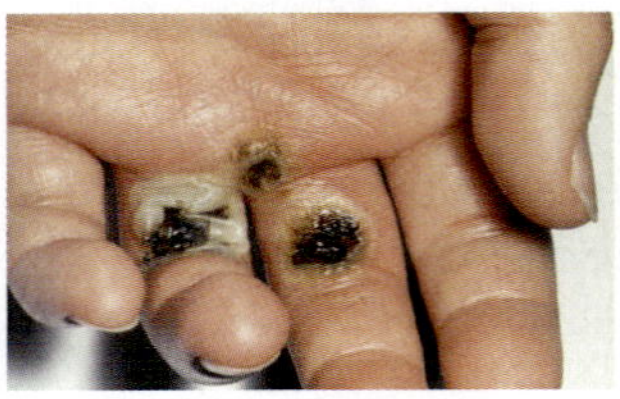

Figure 1 An electrical burn

Figure 2 Frayed electrical cords on appliances are very dangerous.

Electric Shocks

You can be severely injured or even killed by electric shocks. Your body uses electrical signals to communicate with your muscles, including your heart. A current passing through your body would interfere with these signals. Your muscles would contract and jerk violently. Your body tissues could even burn (**Figure 1**).

Electric shocks can be caused by poorly insulated wires (**Figure 2**). An electric shock could also occur when you change the electrical resistance of your body. If you wet your skin with tap water, electrons can flow over a larger area and more current can flow through your body if you come in contact with a current. Once past the skin, current flows easily because we are filled with chemicals that conduct electricity.

Figure 3 An overloaded circuit can overheat and start a fire.

Overloaded Circuits

One of the most common dangers with circuits in homes involves overloading a circuit. A circuit can become overloaded when you plug too many loads into the same circuit (**Figure 3**).

Series Connections in Household Wiring

The electrical panel in most homes has a group of circuit breakers. A circuit breaker is a special type of switch. Each circuit breaker is connected in series with the switches and outlets in a certain room or rooms in a home. When the circuit breaker is open, no current flows through that circuit.

Parallel Connections in Household Wiring

All the circuit breakers in an electrical panel are connected in parallel with each other. This arrangement allows a circuit breaker to be opened without affecting the other circuits in the house.

Recall from Section 13.6 that the potential difference across each parallel load is the same as the potential difference across the energy source. Each load on a household circuit is connected in parallel. This means each load has the same drop in potential difference as the energy source. Each circuit is wired this way so that the electric devices connected to the circuit will function properly.

As you plug in more and more loads in parallel, the overall resistance of the circuit decreases and the current increases because there are more and more paths that electrons can follow. If there is too much current, the main conducting wires in the circuit can overheat and start a fire.

Figure 4 Before you reset a circuit breaker, examine the circuit to find out what caused the breaker to trip, and correct the problem.

Safety Devices in Circuits

To prevent overheating and fires, safety devices such as circuit breakers are installed.

Circuit Breakers and Fuses

What makes circuit breakers different from other switches is that they will automatically open, or "trip," when too much current passes through them (**Figure 4**). This disconnects the circuit and stops the current. Before resetting a circuit breaker, make sure you examine the circuit to determine what caused the breaker to trip. Use caution and ask an adult before touching any part of a circuit or a circuit breaker.

Older homes usually have fuses instead of circuit breakers in electrical panels. A fuse contains a thin piece of metal that is designed to melt when too much current passes through it. Once the metal melts, the circuit is broken and the current stops. Household fuses have different current ratings depending on the circuits that they are intended to control (**Figure 5**).

Figure 5 When a fuse must be replaced, never use one with a higher current rating.

GFCI Outlets

By Canadian law, outlets near sources of water (bathrooms, kitchens, outdoors) contain a special safety device called a ground fault circuit interrupter (GFCI) (**Figure 6**). These safety devices detect changes in current, such as those that can happen when an electrical device is in contact with water. When a change in current is detected, the circuit is immediately disconnected.

Figure 6 Most bathrooms in modern homes have GFCI outlets. A GFCI outlet can be reset by pressing the red button in the middle.

Figure 7 The wider prong is a safety feature which prevents you from plugging in the device the wrong way.

Figure 8 Most ovens are designed to safely handle up to 40 A of current.

AFCI Circuit Breakers

If the conductor is exposed because of damaged or worn insulation on wires and extension cords, current through the conductor can arc and jump onto another material. This arcing releases a great deal of thermal energy that can start a fire. A special type of circuit breaker called an arc fault circuit interrupter (AFCI) is designed to detect arcing and disconnect the circuit.

Plug and Outlet Design

The shape, size, and position of the prongs on plugs are designed to safely handle the currents and voltages that are needed to operate an electrical device. As a safety feature, most plugs only fit one way in an outlet. For example, one prong is wider, so you can only plug in the device one way in the appropriate outlet (**Figure 7**). The plug on an oven is very thick and large because it handles a large amount of current (**Figure 8**). This type of plug also is designed to be plugged in only one way. Electrical outlet covers are another safety feature. They prevent children and pets from accidentally coming into contact with electricity.

13.8 Wrap Up

- Poor or worn insulation can lead to electric shocks.
- Electric shocks can interfere with signals within our bodies and can burn body tissues.
- Overloaded circuits can cause too much current to flow through a circuit, which can cause wires to overheat and start fires.
- Circuit breakers are special switches that "trip" or open a circuit when too much current is detected.
- Circuit breakers, fuses, GFCI outlets, arc fault circuit interrupters, and the size and shape of plugs can help prevent fires and shocks.

CHECK YOUR LEARNING

1. Explain why overloaded circuits on home circuits are so hazardous. K/U
2. Explain why a fuse or circuit breaker must be connected in series with the rest of the circuit to be effective. K/U
3. What could happen if a blown fuse were replaced with one that blew at a much higher current than the original? K/U
4. The circuit breaker for one of the circuits in Mary Ellen's kitchen trips every time she tries to run the toaster and the microwave at the same time. What can Mary Ellen do to prevent this? K/U
5. Explain why circuits near pools need both ground fault circuit interrupters (GFCI) and arc fault circuit interrupters (AFCI). A

TECH CONNECT OSSLT

Touchscreens

One of the most fascinating innovations in how we use computers in recent years has been the introduction of touchscreens. How we use cellphones, personal digital assistants (PDAs), portable video games, and other electronic devices has changed dramatically (**Figures 1–3**). Rather than interacting with these devices indirectly, we simply use our fingers to point.

There are several different ways these screens work. Some involve the user becoming part of the circuit. In others, the user affects the circuit but does not become a part of it. The two most common touchscreens are resistive touchscreens and capacitive touchscreens.

Resistive Touchscreens

Resistive touchscreens are pressure sensitive. They are made of a glass panel covered with two thin metallic layers. One metallic layer is conductive and the other is resistive. An electric current passes through both layers. When you touch the screen, the layers come in contact with each other and vary the current.

Capacitive Touchscreens

Capacitive systems are not pressure sensitive, so they are better for cellphones and PDAs. They are also brighter and clearer. An electric charge is placed on a glass panel. When you touch the screen, a small amount of charge is transferred to you.

A multi-touch screen works by dividing its display into hundreds of small screens. This way the computer can detect more than one touch at a time. You can change the size of a picture on the screen by simply pinching it between your fingers.

Figure 1 Internet-connected cellphone

Figure 2 Bank ATM touchscreen display

Figure 3 Student using Smartboard

DIG DEEPER
Web Link
To learn more about touchscreens,
GO TO NELSON SCIENCE

13.9 Thermal Energy in Circuits

In every circuit, it is very important to control the amount of thermal energy produced by the circuit components. Some devices are actually designed to produce thermal energy. For example, devices such as hair dryers and heat lamps are designed to get very, very hot. These devices have heating elements. Some heating elements are exposed, as in a toaster, whereas some are sealed, as in a clothes iron.

Figure 1 The heating elements of a toaster are made of nichrome wires.

Exposed Heating Elements

Toasters and hair dryers are two devices with exposed heating elements. If you look inside a toaster, you will see that the elements look like flat ribbons or coils of wire (**Figure 1**). These exposed wires are made from metals that have a relatively high electrical resistance. When electrons flow through these metals, a great deal of the stored electric potential energy of the electrons is converted into thermal energy.

Ceramics

Newer hair dryers and room heaters do not use exposed wires. Instead they use long, thin pieces of a special type of ceramic that has been combined with barium or lead and titanium. These ceramics can be made so that at a specific temperature, their resistance increases so much that current stops flowing. This is a tremendous safety advantage. Heating elements made with this kind of ceramic can only get so hot and no hotter. (**Figure 2**).

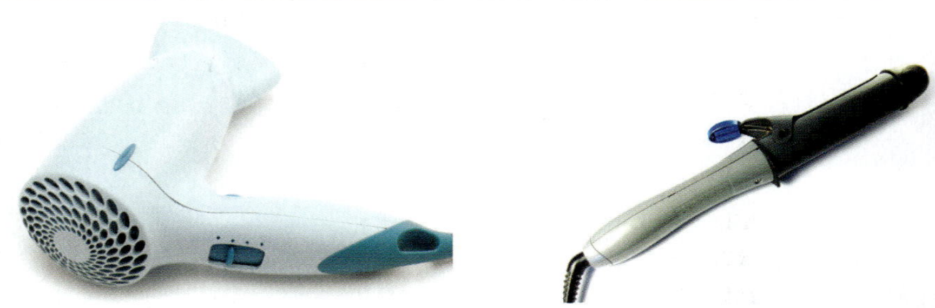

Figure 2 The ceramic heating elements in hair dryers and curling irons can heat up and cool down quickly for safer use.

Sealed Heating Elements

In devices such as clothes irons and electric grills, the heating elements are sealed (**Figure 3**). This prevents the material being heated, such as food, from touching the heating element directly.

Sealed heating elements are made from the same metal combinations as exposed heating elements. However, the wires are sealed with special insulators, such as certain plastics and ceramics (**Figure 4**).

Figure 3 Sealed heating elements are used in many household devices.

Safety with Heating Elements

Whether or not a heating element is sealed, you should never touch the heating element in any device. A heating element that is on, or was recently turned on, can cause severe burns. With exposed heating elements, you could also experience a severe electric shock. Always unplug any heating element and allow it to cool down before touching it.

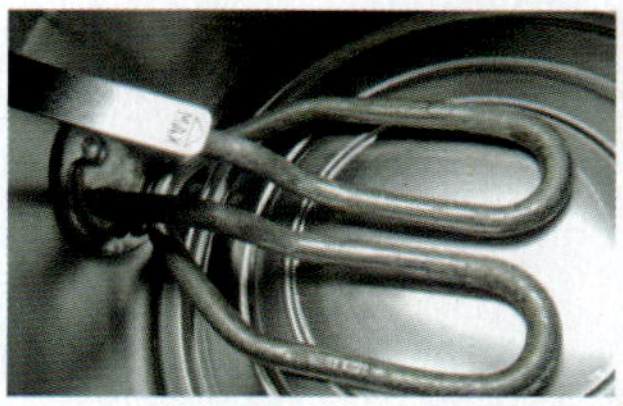

Figure 4 Sealed heating element

13.9 Wrap Up

- Devices such as toasters, clothes irons, and hair dryers have circuit components, called heating elements, that are designed to become very hot.
- Heating elements may be exposed or sealed depending on the device in which the element will be used.
- All heating elements are made from special metal mixtures. The resulting materials have a relatively high electrical resistance.

CHECK YOUR LEARNING

1. What happens as the current flowing through the wire in a heating element increases? K/U
2. Explain why the heating element in a toaster is not covered with an insulator to protect us from shocks. K/U
3. Which type of heating element is used in electric blankets? Why? A
4. What is the advantage of using ceramics for heating elements? K/U
5. Why should you never touch a heating element? K/U
6. Imagine you are designing a waffle maker. Which type of heating element would you use? Explain your reasoning. A
7. Suppose a piece of toast gets stuck in the slot of the toaster. What do you need to do to safely remove it? K/U A

Chapter 13 KEY CONCEPTS SUMMARY

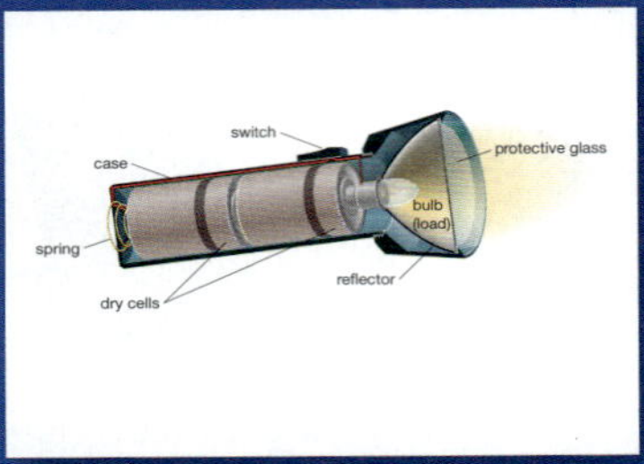

A circuit contains a source of electrical energy, a load, conducting wires, and usually a switch.

- An energy source increases the electric potential energy of electrons.
- The load is a device that converts electrical energy into another form of energy.

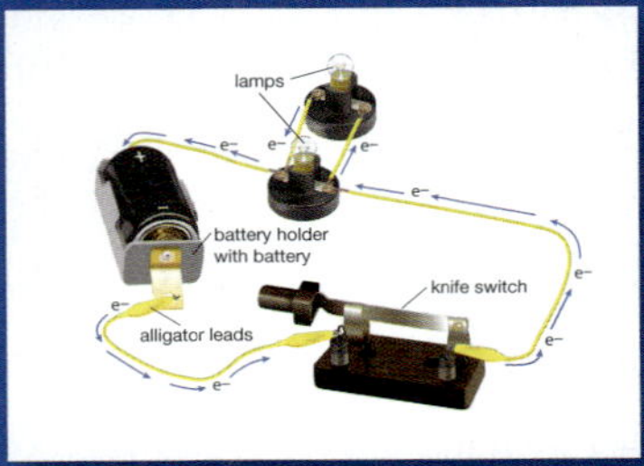

Circuits can be connected in series or in parallel.

- Series circuits have only one path through which electrons can flow.
- Parallel circuits have more than one path through which electrons can flow.

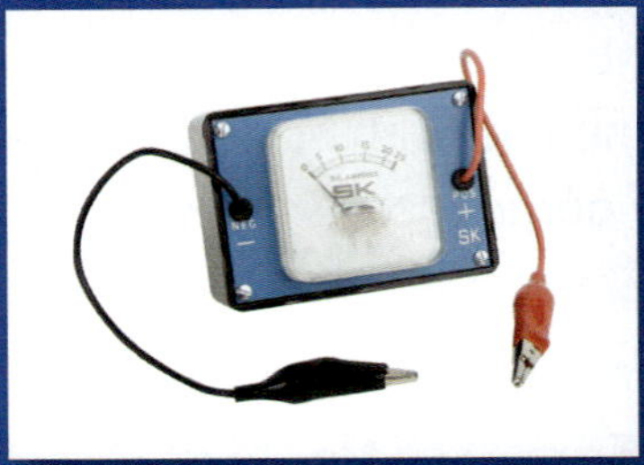

Current is measured using an ammeter connected in series to the circuit.

- Electric current (*I*) is the amount of charge passing a point in a unit of time.
- Current is measured in amperes (A).

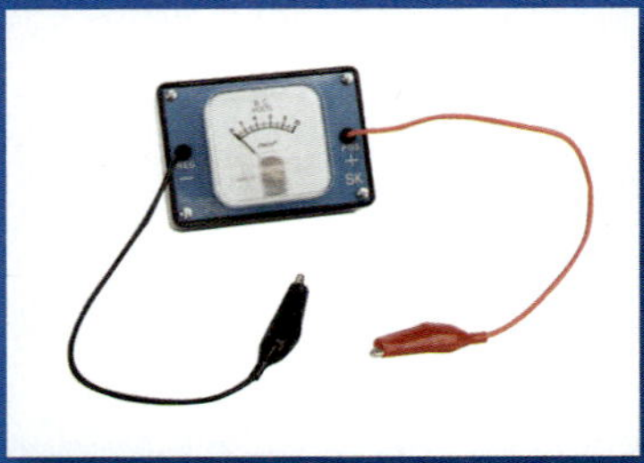

Electric potential difference (voltage) is measured using a voltmeter connected in parallel to the circuit.

- The potential difference between two points in a circuit affects electron flow.

Resistance is the opposition to the flow of electrons.

- Resistance is measured in ohms (Ω) using an ohmmeter.
- The resistance of a conductor is affected by the thickness of the wire, the type of material used, the temperature, and the length.

When you increase the voltage in a circuit, the current also increases.

- Ohm's law describes the relationship between voltage, current, and resistance.
- Increasing the voltage increases the current for a given resistance in a circuit.

WHAT DO YOU THINK NOW?

You thought about the following statements at the beginning of the chapter. You may have encountered these ideas in school, at home, or in the world around you. Consider them again and decide whether you agree or disagree with each statement.

1 Electrons travel in one direction through the wires in a building. **Agree/disagree?**

2 It is possible to control the flow of electricity in your home. **Agree/disagree?**

3 The electric current in a circuit increases when more devices are connected to the circuit. **Agree/disagree?**

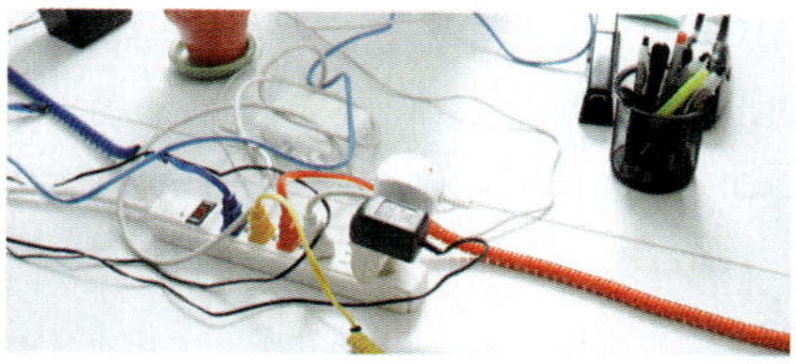

4 Longer wires resist the flow of electrons more than shorter wires. **Agree/disagree?**

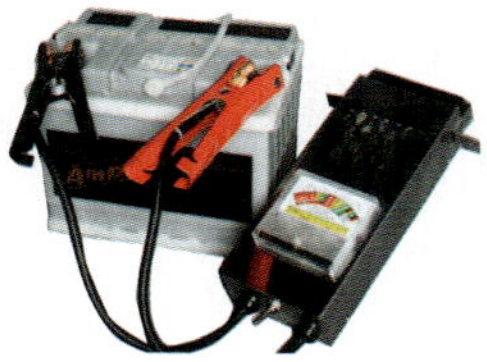

5 Electrons flow from the negative terminal of a battery, through the circuit, to the positive terminal. **Agree/disagree?**

6 There is only one way to connect the components of a circuit so that charges can flow. **Agree/disagree?**

How have your answers changed since then? What new understanding do you have?

BIG Ideas

- Static and current electricity have distinct properties that determine how they are used.

DIG DEEPER

Vocabulary

electric circuit p. 500
load p. 500
switch p. 500
energy source p. 500
circuit diagram p. 502
series circuit p. 502
parallel circuit p. 503
electric current (*I*) p. 508
ampere (A) p. 508
ammeter p. 508
potential difference or voltage (*V*) p. 509
volt (V) p. 509
voltmeter p. 509
electrical resistance (*R*) p. 510
ohm (Ω) p. 510
ohmmeter p. 510
multimeter p. 513
Ohm's law p. 518

CHAPTER 13

SELF-QUIZ

The following indicate the Achievement chart category addressed by each question.

K/U Knowledge/Understanding T/I Thinking/Investigation C Communication A Application

For each question, select the best answer from the four alternatives.

1. Which of the following refers to the path along which electrons can flow? (13.1) K/U
   (a) circuit
   (b) current
   (c) potential difference
   (d) resistance
2. Which device is used to measure the potential difference across a load? (13.4) K/U
   (a) ammeter
   (b) ohmmeter
   (c) voltmeter
   (d) circuit breaker
3. Which of the following is measured in amperes? (13.4) K/U
   (a) current
   (b) potential difference
   (c) resistance
   (d) temperature
4. Which safety device needs to be replaced each time it stops excess current? (13.8) K/U
   (a) arc fault circuit interrupter
   (b) circuit breaker
   (c) knife switch
   (d) fuse
5. Which of the following is a measure of the amount of opposition to the flow of electrons through a material? (13.4) K/U
   (a) current
   (b) potential difference
   (c) resistance
   (d) voltage
6. Which device does not need a current flowing in a circuit because it has its own power supply? (13.4) K/U
   (a) ammeter
   (b) ohmmeter
   (c) voltmeter
   (d) heating element

Indicate whether each of the statements is TRUE or FALSE. If you think the statement is false, rewrite it to make it true.

7. When the thickness of a wire increases, the resistance of the wire decreases. (13.4) K/U
8. An ammeter must be connected in parallel with a circuit to take accurate readings. (13.4) K/U
9. Ceramic heaters are not very safe because their resistance increases rapidly when they become very hot. (13.9) K/U

Copy each of the following statements into your notebook. Fill in the blanks with a word or phrase that correctly completes the sentence.

10. An energy _______ is a device that provides electrical energy in a circuit. (13.1) K/U
11. Electrons only have one path to follow from the source through the loads and back to the source in a(n) _______ circuit. (13.1) K/U
12. Electrons have more than one path to follow from the source through the loads and back to the source in a(n) _______ circuit. (13.1) K/U
13. When the resistance increases, the current _______ according to Ohm's law. (13.7) K/U

Match each term on the left with the most appropriate description on the right.

14. (a) volt — (i) the unit of electric resistance
    (b) ohm — (ii) device that measures several electrical quantities including voltage, current, and resistance
    (c) multimeter — (iii) an electric current in which the flow of charged particles travels in one direction only
    (d) direct current — (iv) the unit of electric potential (13.1, 13.4) K/U

Write a short answer to each of these questions.

15. Classify each of the following circuits as a series circuit or as a parallel circuit. (13.1, 13.2, 13.3) K/U

(a)

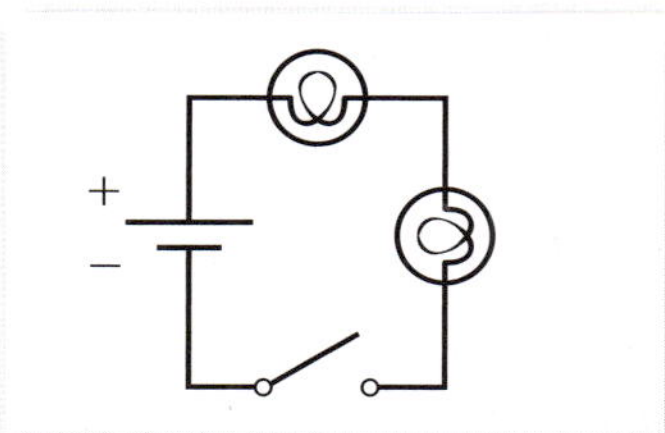

(b)

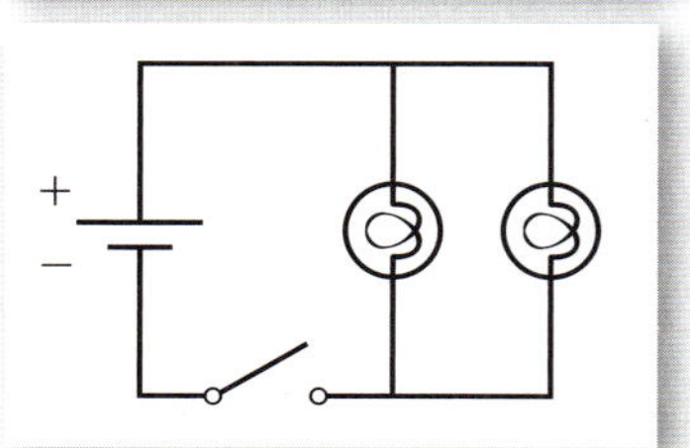

(c)

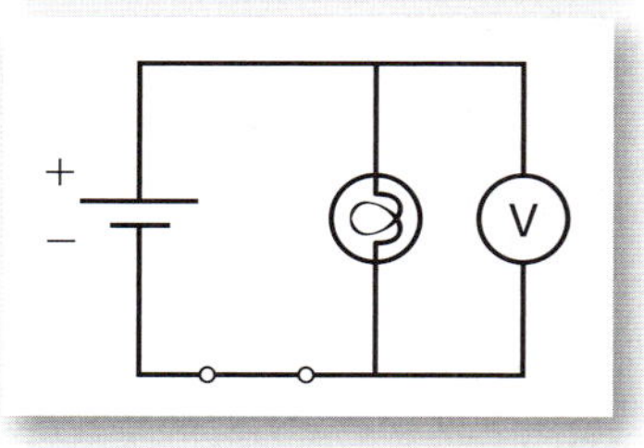

(d)

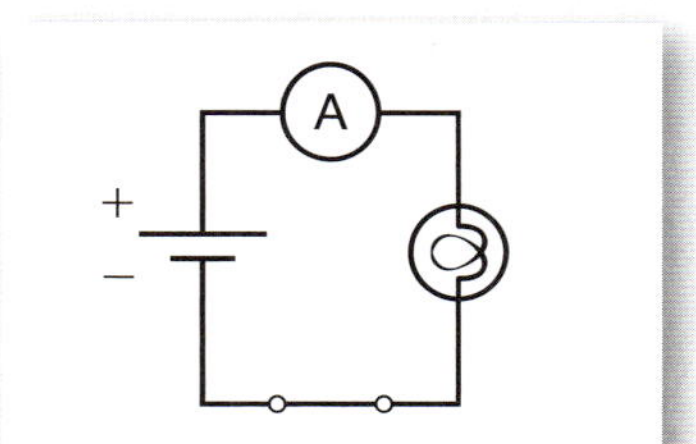

16. You connect a battery, a switch, and two lamps in a parallel circuit. One of the bulbs burns out. What do you observe? Explain why. (13.3) T/I

17. A series circuit has a battery, a switch, two lamps, and an ammeter. You add in series another two lamps to the circuit. What will the ammeter show? Why? (13.1, 13.4) T/I

18. Will the resistance of a copper wire extension cord increase or decrease if its length is unchanged but it is made out of thicker copper wire? Explain. (13.4, 13.9) T/I

19. You take away two loads from a parallel circuit. What will you observe about the voltage on the voltmeter across one of the remaining loads? Why? (13.4, 13.6) T/I

20. A circuit breaker in your house trips, cutting off the current. What should you do before resetting the circuit breaker? (13.8) K/U

21. Name two things that help prevent severe shocks from electrical devices. (13.8) K/U

22. What will an ohmmeter show after you add another load to a parallel circuit? (13.4, 13.6) K/U

23. Why does a voltmeter have to be connected in parallel to the load? (13.4) K/U

24. (a) What happens to the resistance in a series circuit if you add another load to the circuit?
    (b) How does this affect the current? (13.7) K/U

25. Why are the heating elements of a slow cooker sealed? (13.9) A

Self-Quiz

To do an online self-quiz,

GO TO NELSON SCIENCE

CHAPTER 13 REVIEW

The following icons indicate the Achievement Chart category addressed by each question.

K/U Knowledge/Understanding   T/I Thinking/Investigation
C Communication   A Application

What Do You Remember?

1. What happens to the current when the switch is opened in a parallel circuit? (13.1) K/U
2. Which type of circuit, series or parallel, is used most often in buildings? (13.8) K/U
3. What must you do before using an ohmmeter? Explain. (13.4) K/U
4. When the resistance in a circuit decreases, what happens to the current? (13.7) K/U
5. Explain why current is able to pass through the human body. (13.6, 13.7) K/U
6. What unit is used to measure resistance in a circuit? (13.4) K/U
7. If the potential difference increases for a given resistance, what happens to the current? (13.7) K/U
8. What happens inside a fuse when too much charge flows through it? (13.9) K/U
9. The potential difference in a circuit is also known as its ________. (13.4) K/U
10. Which type of meter must be connected in parallel with a circuit with the current flowing in order to take a reading? (13.4) K/U
11. Name three safety devices used in electrical circuits. Tell how each one works. (13.8) K/U
12. Why does the brightness of a bulb stay the same when another lamp is added in parallel? (13.1, 13.6) K/U
13. Should a toaster be made with wire that has high or low electrical resistance? Why? (13.9) K/U
14. Draw a circuit diagram for each description.
    (a) A series circuit showing a battery, a closed switch, an ammeter, and an electric motor. (13.1, 13.2, 13.4) K/U T/I C
    (b) A parallel circuit showing a battery, two lamps, and three switches. Each lamp should have its own switch. The third switch should control both lamps.
    (c) A parallel circuit showing a battery with a switch that will control all loads on the line. The loads are a lamp and an electric motor with a voltmeter connected across the motor.

What Do You Understand?

15. Classify each circuit as a series circuit or as a parallel circuit. Explain how you know in each case. (13.1, 13.2) K/U

(a)

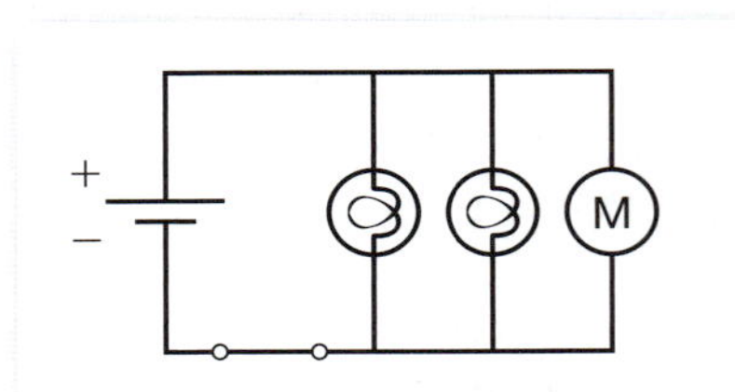

(b)

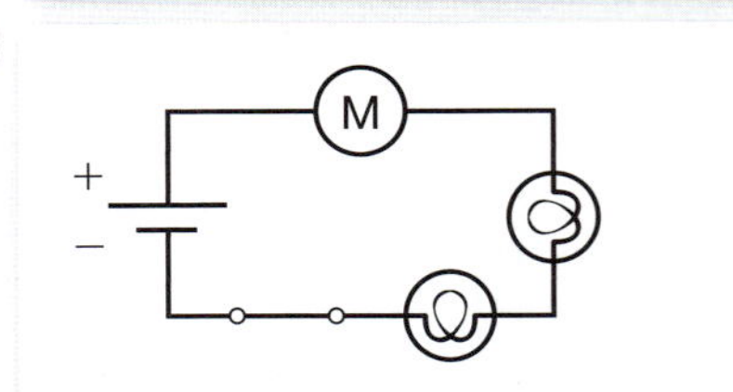

(c)

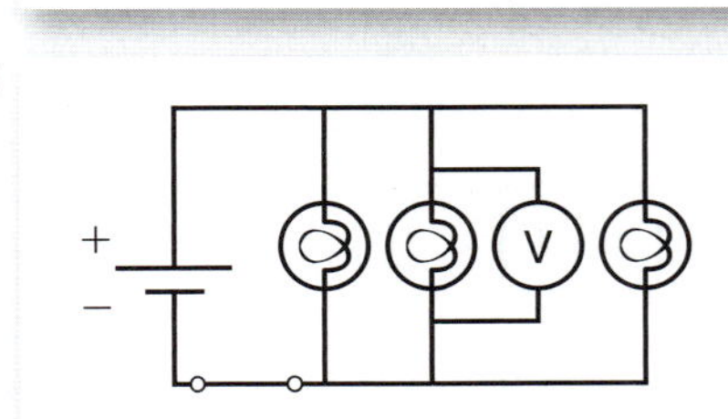

(d)

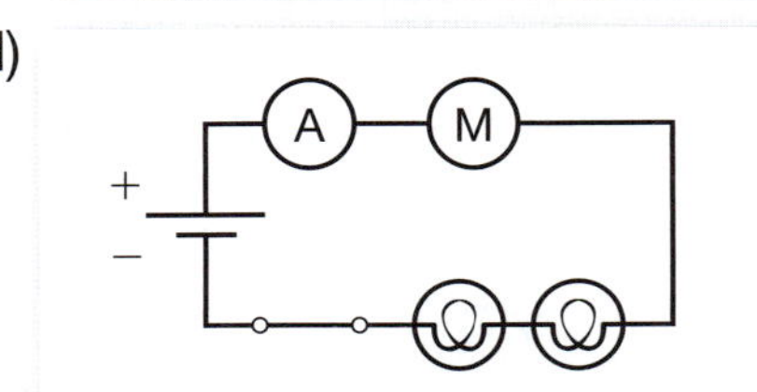

16. Every time you use an iron you notice that the extension cord gets very hot. Why does this happen and what should you do about it? (13.4, 13.8) A

17. The rear-window defroster on a car is made from silver–ceramic material. It takes only a small voltage through the circuit to make it warm. Is the resistance of this material high or low? How do you know? (13.4) K/U

18. When you plug in an old string of decorative lights connected in series, none of the bulbs light. What can you do to find out what is wrong? (13.1) A

19. Tap water has low resistance. Explain why it is dangerous to handle electronic devices with wet hands. (13.8) K/U

20. People often use an electrical device called a hot dipstick in areas with very cold climates. It keeps the oil in a car warm. Would you expect the heating element of the hot dipstick to be exposed or sealed? Explain. (13.9) A

21. Cold weather can slow down a chemical reaction in a battery. What might happen if you try to turn on a flashlight that has been out all winter? (13.7) A

Solve a Problem

22. Your portable video game player has stopped working. Describe how a technician might use a multimeter to determine what is wrong with the device. (13.4, 13.5) T/I C

Create and Evaluate

23. Design a device to heat individual cups of water or hot cocoa. Describe the following parts of the device and explain your decisions: (13.1, 13.8, 13.9) T/I C
    (a) power source
    (b) the kind of wires you would use
    (c) the type of circuit and how the parts would be arranged
    (d) the type of heating element: exposed or sealed
    (e) an indicator light and any safety devices

Reflect on Your Learning

24. The relationships between resistance, current, and potential difference in a circuit can be difficult to understand. Do you still have questions about any of these relationships? What strategies can you use to find the answers to your questions? C

25. Has any of the information presented in this chapter caused you to rethink your approach to using electrical devices in your home? Explain your answer. C

Web Connections

26. Some of the safety devices discussed in this chapter were developed in recent years. Not all homes may have them. Use the Internet to find out more about these safety devices. Then create a safety device survey for your household. T/I C

GO TO NELSON SCIENCE

Electrical Applications

CHAPTER 11

Electrical Energy

KEY CONCEPTS

Electrical energy is generated using a variety of technologies.

Electrical energy is generated using both renewable and non-renewable sources.

Electrical energy production has an effect on the environment and society.

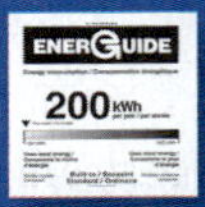

Electrical energy should be conserved and used wisely.

CHAPTER 12

Static and Current Electricity

KEY CONCEPTS

Electricity describes how electrons move from one place to another.

Static electricity is the result of an imbalance of charges on the surface of an object.

There are many useful applications of static electricity.

Objects can become charged by contact or by induction.

Current electricity is generated by moving electrons in a closed path.

Some materials allow electrons to move through them easily (conductors), while others do not let electrons move through them easily (insulators).

CHAPTER 13

Circuits

KEY CONCEPTS

A circuit contains a source of electrical energy, a load, conducting wires, and usually a switch.

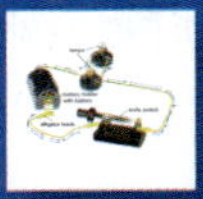
Circuits can be connected in series or in parallel.

Current is measured using an ammeter connected in series to the circuit.

Electric potential difference (voltage) is measured using a voltmeter connected in parallel to the circuit.

Resistance is the opposition to the flow of electrons.

When you increase the voltage in a circuit, the current also increases.

MAKE A SUMMARY

A summary does not have to be just a paragraph or list of questions. You can make summarizing the unit into a game.

1. Divide a stack of index cards into two equal piles.
2. Write a vocabulary word or key concept on each card in one pile. Be sure to include words and concepts from each chapter.
3. Write the definition or explanation of each concept on a card from the other pile. You can also draw a picture illustrating a concept or vocabulary word on a card from the other pile.
4. Put the two piles of cards together and shuffle them.
5. Arrange the cards into a rectangular array of columns and rows.
6. Choose a player to go first.
7. Each player turns over two cards. If they match, the player keeps the cards. If they do not match the cards are turned face down again.
8. Play until all of the cards are gone.
9. To play a faster version of the game, you can make separate sets of cards for each chapter.

Questions

1. List the words or concepts you found difficult to match with their explanation.
2. What can you do to improve your understanding of the difficult vocabulary and concepts?

CAREER LINKS

List the careers mentioned in this unit. Choose two of the careers that interest you, or choose two other careers that relate to Electrical Applications. For each of these careers, research the following information:

- educational requirements (secondary and post-secondary)
- skill/personality/aptitude requirements
- potential employers
- salary
- duties/responsibilities

Assemble the information you have discovered into a brochure or poster. Your brochure or poster should compare and contrast your two chosen careers, and explain how they connect to Electrical Applications.

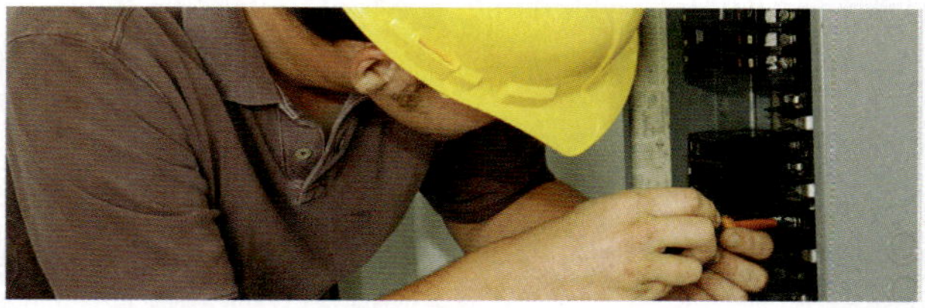

Figure 1 An electrician

GO TO NELSON SCIENCE

Model Home Powered by an Alternative Source of Energy

In this activity, you will design and build a model of the circuitry needed for a home powered by an alternative source of energy. You will research alternate sources of energy. Then you will report your results and conclusions.

SKILLS MENU

- Questioning
- Hypothesizing
- Predicting
- Planning
- Controlling Variables
- Performing
- Observing
- Analyzing
- Evaluating
- Communicating
- Researching
- Identifying Alternatives
- Defending a Decision

CAUTION

Electricity

Always disconnect the energy source before constructing or modifying a circuit or when connecting a meter. To unplug the energy source, do not pull the cord. Pull the plug itself. Do not adjust the power supply unless instructed to do so by your teacher.

SKILLS HANDBOOK 2.D., 2.E.

Purpose

To design and build the circuits needed for a model home, including series and parallel circuits with multiple loads and switches

Procedure

Part A: Build a Model Circuit

1. Draw a circuit diagram for the circuits needed in a model home with the following specifications:
    - In the front, there are two side-by-side rooms. One is a living room. The other is a kitchen with a dining area.
    - There are ceiling lights in both of these rooms. Both lights are controlled by a switch just inside the front door.
    - In the back, there are another two side-by-side bedrooms. There are ceiling lights in both of these rooms. Each light is controlled by its own switch
    - There is also a main switch for the whole home.

Equipment and Materials

white paper

4 alligator leads

battery holders with batteries or variable DC energy source

4 knife switches

4 lamps

2. Show your diagram to your teacher for approval.
3. Once you have an approved plan, obtain the required materials from your teacher. Construct the circuit you designed.
4. If you are using batteries, you will need to connect at least two in series. If you are not using batteries, turn off the power supply.
5. Make sure the power supply is at its lowest setting, or as instructed by your teacher. Close the switch.
6. Slowly increase the power until the lamps turn on. Once the lamps are lit, leave the power supply at this setting.

Part B: Research Alternative Sources of Energy

7. Research three types of electrical energy generation that use renewable resources. Decide which one is the best one to use to supply electrical energy for your house.
8. Gather information to answer questions such as the following:
   - For what do you need electricity? Make a list of the appliances you will have in the house. Estimate the total amount of electrical energy you will need for a year.
   - What are the costs to install and operate each system?
   - What is the environmental impact of each system?
   - What are the available renewable resources in your area?

GO TO NELSON SCIENCE

Analyze and Evaluate

(a) Describe your circuit. Explain the decisions you made while designing and building your circuit.

(b) How does your circuit compare to circuits used in real homes?

(c) Which system did you choose to generate electricity? Give reasons for your choice.

Apply and Extend

(d) Choose one of the following to present your conclusions:

- Write a report or create a PowerPoint presentation describing your circuit and how you overcame the challenges involved in building it.
- Make an oral presentation and/or write a brochure marketing your design to the class.

CHECKLIST

Assessment

Your completed Performance Task will be assessed according to the following criteria:

KNOWLEDGE/ UNDERSTANDING

- Understand how to draw circuit diagrams to give descriptions of them.
- Understand how to connect loads in series and in parallel.
- Build a working electrical circuit.

THINKING/ INVESTIGATION

- Research and gather information on alternative sources of energy.
- Test alternative energy sources in a circuit.

COMMUNICATION

- Express ideas using an appropriate format in oral, written, and or visual form.

APPLICATION

- Apply knowledge to an unfamiliar context.
- Make connections between models and real-world applications of circuits.

DIG DEEPER

Web Link

To find the typical energy use of common household appliances and information about renewable sources of energy,

GO TO NELSON SCIENCE

UNIT E

SELF-QUIZ

The following icons indicate the Achievement Chart category addressed by each question. K/U Knowledge/Understanding T/I Thinking/Investigation C Communication A Application

For each question, select the best answer from the four alternatives.

1. A battery converts chemical energy into which of the following? (11.1, 12.8) K/U
   (a) sound energy
   (b) mechanical energy
   (c) electrical energy
   (d) light energy

2. Which of the following units is used to measure how much electrical energy a household uses? (11.1) K/U
   (a) amperes
   (b) volts
   (c) watts
   (d) kilowatt hours

3. Object A has a negative charge. Object B is attracted to object A and repelled by object C. What is the charge on object C? (12.1) K/U
   (a) neutral
   (b) negative
   (c) positive
   (d) any of the above

4. Which material is a good conductor? (12.8) K/U
   (a) rubber
   (b) plastic
   (c) copper
   (d) glass

5. Which device measures the potential difference across a load? (13.4) K/U
   (a) ammeter
   (b) ohmmeter
   (c) switch
   (d) voltmeter

6. Which of the following describes charging a neutral object by bringing a charged object close to it, but not touching it? (12.2, 12.3, 12.5, 12.8) K/U
   (a) induction
   (b) friction
   (c) discharge
   (d) current

Copy each of the following statements into your notebook. Fill in the blanks with a word or phrase that correctly completes the sentence.

7. A measure of the opposition to the movement of electrons through a material is _______. (13.4) K/U

8. An electric _______ is a very rapid transfer of electrons from one object to another. (12.2) K/U

9. Energy resources such as coal, oil, and natural gas are known as _______ fuels. (11.2) K/U

10. A device used to detect static charge is called a(n) _______. (12.4, 12.6) K/U

Indicate whether each of the statements is TRUE or FALSE. If you think the statement is false, rewrite it to make it true.

11. When you add loads to a series circuit, the total resistance in the circuit remains the same. (13.4, 13.7) K/U

12. Solar radiation is a non-renewable source of electrical energy. (11.2, 11.3) K/U

13. When the thickness of a wire is decreased, the resistance in the wire increases. (13.4) K/U

14. Electrostatic discharges are always completely harmless. (12.2) K/U

15. The wires in the heating element of a hair dryer have relatively high resistance. (13.9) K/U

Match each term on the left with the most appropriate description on the right.

16. (a)	conductor	(i)	an electric current that repeatedly reverses direction
(b)	static electric charge	(ii)	a circuit in which two or more loads are connected one after another so that electrons follow only one path
(c)	alternating current	(iii)	a device that converts light energy from a light source directly into electrical energy
(d)	series circuit	(iv)	charges on objects that are at rest, rather than flowing as a current
(e)	parallel circuit	(v)	a material that transmits thermal energy or electrical energy easily
(f)	solar cell	(vi)	a circuit in which two or more loads are connected so that electrons can follow more than one path (11.3, 12.2, 12.8, 13.1) K/U

Write a short answer to each of these questions.

17. You connect a battery, a switch, and two lamps in a series circuit. You close the switch. Both lamps light up, then one of the bulbs burns out. Explain what might have happened. (13.1, 13.2, 13.3) T/I

18. A parallel circuit has a battery, a switch, two lamps, and a voltmeter across each lamp. You add another lamp in parallel to the circuit. What will the voltmeters show? Explain. (13.4) T/I

19. Make a list of at least five things your school could do to conserve energy. (11.6) (11.7) A

20. Explain why your hair stands up after you pull off a woollen cap. (12.1, 12.2, 12.3) K/U

21. Why are the heating elements in a dishwasher sealed rather than exposed? (13.9) K/U

22. **Figure 1** shows three spheres hanging from strings. In your notebook, redraw the diagram to show how the spheres would hang if the centre sphere were positively charged and the spheres on the left and right were negatively charged. Use the symbols "+" and "–" to mark the charge of each sphere. (12.1, 12.2, 12.5) T/I C

Figure 1

23. Some electric stoves have exposed heating elements, while others have sealed heating elements. Which type would you choose? Explain your reasoning. (13.8, 13.9) T/I

REVIEW

The following icons indicate the Achievement Chart category addressed by each question.

K/U Knowledge/Understanding
T/I Thinking/Investigation
C Communication
A Application

What Do You Remember?

For each question, select the best answer from the four alternatives.

1. Which form of energy is electrical energy changed into by a kettle? (11.1) K/U
   (a) chemical energy
   (b) mechanical energy
   (c) thermal energy
   (d) light energy

2. Which energy source is non-renewable? (11.2) K/U
   (a) coal
   (b) solar
   (c) wind
   (d) geothermal

3. Which of the following units is used to measure current? (11.4, 11.5, 13.4) K/U
   (a) amperes
   (b) volts
   (c) watts
   (d) joules

4. If electrons move to an object, that object will have what kind of charge? (12.1) K/U
   (a) neutral
   (b) negative
   (c) positive
   (d) both negative and positive

5. Which material is a good insulator? (12.8) K/U
   (a) aluminum
   (b) nichrome
   (c) copper
   (d) glass

6. Which device is used to measure electrostatic charge? (12.4, 12.6) K/U
   (a) ammeter
   (b) ohmmeter
   (c) electroscope
   (d) voltmeter

7. A balloon is rubbed on a carpet. It becomes negatively charged. What sort of charge will the balloon induce in a wall when it is brought close to the wall? (12.5) K/U
   (a) neutral
   (b) negative
   (c) positive
   (d) both negative and positive

8. Which of the following energy sources produces the least air pollution when used to generate electricity? (11.2, 11.3) K/U
   (a) falling water (c) oil
   (b) natural gas (d) coal

9. Loads connected in parallel
   (a) increase the potential difference across each load
   (b) decrease the potential difference across each load
   (c) do not affect the potential difference across each load
   (d) all of the above (13.6) K/U

10. What will happen to the potential difference across each load if you add a battery to a series circuit? (13.4) K/U
   (a) increase
   (b) decrease
   (c) remain unchanged
   (d) increase in one part of the circuit and decrease in another part

11. Rubbing two different objects together produces objects with what kind of charge? (12.3) K/U
(a) both positive
(b) both negative
(c) both neutral objects
(d) one positive, one negative

12. Which device has a sealed heating element? (13.9) K/U
(a) toaster
(b) hair dryer
(c) portable room heater
(d) clothes iron

13. If one lamp bulb on a series circuit burns out, what will happen to other lamps on the same circuit? (13.1) K/U
(a) They will all continue to glow at the same brightness.
(b) They will all continue to glow and will get brighter.
(c) Some will get brighter and some will go out.
(d) They will all go out.

14. Why does a circuit breaker trip? (13.8) K/U
(a) too much current in a circuit
(b) too little current in a circuit
(c) current switches from AC to DC
(d) no current in a circuit

15. What happens when oppositely charged objects are brought near to each other? (12.1) K/U
(a) They repel each other.
(b) They attract each other.
(c) They do not affect each other in any way.
(d) They both become neutral.

16. These are required on any outlet that is located near water. (13.8) K/U
(a) fuse
(b) circuit breaker
(c) ground fault circuit interrupter
(d) arc fault circuit interrupter

17. Why do homeowners switch from incandescent bulbs to compact fluorescent light bulbs (CFLs)? (11.5) K/U
(a) They are less expensive.
(b) They produce more pleasing light.
(c) They contain no harmful chemicals.
(d) They use less electrical energy to produce the same brightness.

18. Which device uses static electricity to function? (12.2) K/U
(a) refrigerator
(b) photocopier
(c) television
(d) MP3 player

19. What should you do if the fuses on a circuit keep burning out? (13.8) K/U
(a) Use a fuse with a higher current rating.
(b) Replace the wiring in the circuit.
(c) Reduce the number of devices that use the circuit.
(d) Replace the entire fuse box.

20. Which of these fuels is a renewable resource? (11.3) K/U
(a) coal
(b) biomass
(c) uranium
(d) natural gas

21. What is the energy source for photovoltaic cells? (11.3) K/U
(a) biomass
(b) geothermal fields
(c) sunlight
(d) nuclear fission

22. Electrostatic charge can be built up when what particles are transferred from one object to another? (12.1, 12.3) K/U
(a) atoms
(b) protons
(c) neutrons
(d) electrons

REVIEW

The following icons indicate the Achievement Chart category addressed by each question.

K/U Knowledge/Understanding
C Communication
T/I Thinking/Investigation
A Application

Write a short answer to each of these questions.

23. What might have increased if the current increases for a given resistance? (13.7) K/U

24. How many watts are in a kilowatt? (11.1) K/U

25. What type of meter must be connected in series with the circuit with the current flowing in order to take a reading? (13.4) K/U

26. How can you make an induced charge on an object permanent? (12.5) K/U

27. What has to be able to flow through a material for an electric current to be generated? (12.8) K/U

28. An object that has an equal number of positive and negative charges can be described this way. (12.1) K/U

What Do You Understand?

29. Draw a diagram to show three charged objects. Two are attracted to each other. The third object attracts both of them. (12.1) K/U C

30. Draw a diagram to show how an object can be positively charged by induction permanently. (12.5, 12.6) K/U C

31. Draw a diagram to show how the distribution of an object's charge can be changed by induction even though the object still remains neutral. (12.5) K/U C

32. Draw a diagram to show a series circuit with two batteries, an open switch, an ammeter, a lamp, and an electric motor. (13.1, 13.2, 13.4) K/U C

33. Draw a diagram to show a battery, two lamps, an electric motor, and four switches. One switch should control the entire circuit. Both of the lamps and the motor should each have its own switch. (13.1, 13.2) T/I C

34. Draw a diagram to show a parallel circuit with a battery connected to two electric motors. One motor has a lamp connected to it in series. The other motor is connected correctly to a voltmeter. Each motor has its own switch. (13.1, 13.2) T/I C

35. Classify the circuit in **Figure 1** as a series circuit or as a parallel circuit. Explain how you know. (13.1, 13.4) T/I C

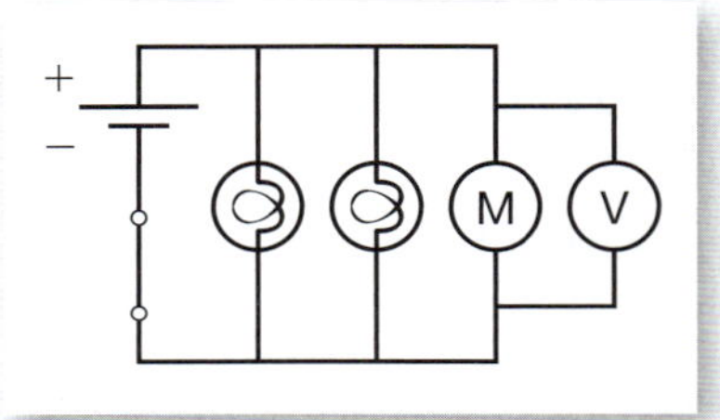

Figure 1

36. Explain why some governments want to ban incandescent light bulbs. (11.5) A

37. When you plug in an iron while watching the television, a circuit breaker trips. What should you do so you can finish ironing? (13.8) K/U

38. Explain why rooms containing multiple computers are often on a separate air conditioning system. (11.5) K/U

39. Explain why it is dangerous to use an appliance such as a hair dryer or radio near water, even though pure water is not a good conductor of electricity. (12.8, 13.8) K/U

Solve a Problem

40. All technicians who troubleshoot electronic equipment are required to wear a wrist strap to prevent electrostatic discharge. Explain how you think the strap works. (12.2, 12.8) K/U

41. A decorative light company has received the following complaints. Draft a plan for how the company can deal with these complaints. (13.1, 13.4, 13.7) T/I C
    (a) When a bulb on their lights burns out, the entire string goes out.
    (b) The bulbs get very hot when lit for a long time.
    (c) The wire for the lights get very hot.

42. Object A has a positive charge. Object B is attracted to object A and repelled by object C. Object D is repelled by object C. Which three objects have the same charge? Explain your reasoning. (12.1) T/I

Create and Evaluate

43. Suppose you lived in a riverside community known for its steady winds and sunny days. Create a list of the ways the community could switch to renewable sources of energy. At the end of the list include a few of the problems with each source. (11.2) (11.3) A

44. A school wants to lower its energy bills by doing something about all the electronic equipment that keeps drawing energy after school hours. Create a plan for how it can achieve this goal. (11.5) A

45. Design a line of shoes and clothing for people who work with electronics. Explain how you chose the materials you would use in each item of clothing. (12.2, 12.8, 13.8) C A

Reflect on Your Learning

46. Has any of the information presented in this unit changed how you think about electrical devices in your home? Explain your answer. C

47. Do you support efforts to convert to renewable energy sources in your area? Discuss the issue with someone whose opinion differs from yours. Then describe the similarities and differences in your points of view. A C

Web Connections

48. You have learned about the commercial uses of static electricity in such things as photocopiers. Use the Internet to research industries, that use equipment such as paint sprayers, dusting products, or filtration equipment. Select one industry and find out how they make use of static electricity. Choose a method to present a short summary of your findings. T/I C A

49. In comics and science fiction movies, people harness the electricity in the human body to do amazing things. In reality, scientists and doctors have worked together to develop devices that use the electricity produced in our bodies in amazing ways. Research some of these devices. Choose a method to present a short summary of your findings. T/I C A

50. Pacemakers and defibrillators both use electrical energy to deal with heart rhythm disorders. Research how each of these devices works. Choose a method to present a summary of your findings. T/I C A

GO TO NELSON SCIENCE

SKILLS HANDBOOK

Contents

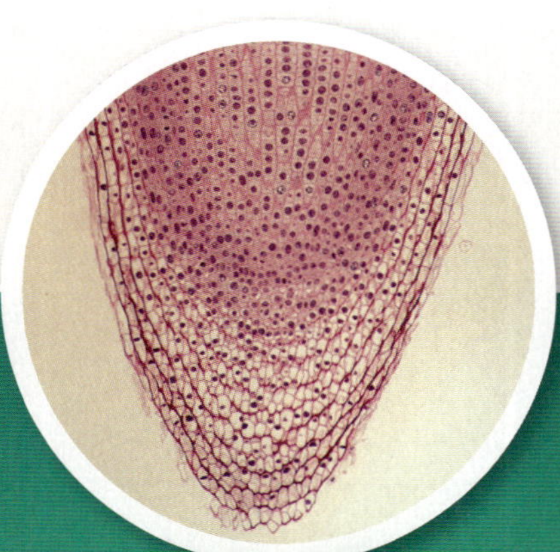

1

Safe Science

1.A. Having a Safe Attitude

Science investigations can be a lot of fun, but certain dangers exist in any laboratory. Before starting an activity, you should know how to reduce the risk of an accident. Always follow proper procedure and handle equipment and materials correctly.

The activities and investigations in this textbook are safe, as long as you follow proper safety precautions and instructions and use common sense. General laboratory safety rules are outlined below (**Table 1**). Your teacher may provide you with additional safety rules for specific tasks.

Table 1 Practise Safe Science in the Classroom

Be science ready	Follow instructions	Act responsibly
• Come prepared with your textbook, notebook, pencil, and anything else you need. • Tell your teacher about any allergies or medical problems. • Keep yourself and your work area tidy and clean. Keep aisles clear. • Wear closed shoes (not sandals). • Do not wear contact lenses while doing investigations. • Read all written instructions carefully before you start an activity or investigation.	• Do not enter a laboratory unless a teacher is present or you have permission to do so. • Listen to your teacher's directions. Read written instructions. Follow them carefully. • Ask your teacher for directions if you are not sure what to do. • Wear eye protection or other safety equipment when instructed by your teacher. • Never change anything, or start an activity or investigation on your own, without your teacher's approval. • Get your teacher's approval before you start an investigation that you have designed yourself.	• Pay attention to your own safety and the safety of others. • Know the location of MSDS (Material Safety Data Sheet) information, exits, and all safety equipment, such as the first-aid kit, fire blanket, fire extinguisher, and eyewash station. • Alert your teacher immediately if you see a safety hazard such as broken glass, a spill, or unsafe behaviour. • Stand while handling equipment and materials. • Avoid sudden or rapid motion in the laboratory, especially near chemicals or sharp instruments. • Never eat, drink, or chew gum in the laboratory. • Do not taste, touch, or smell any substance in the laboratory unless your teacher asks you. • Clean up and put away any equipment when finished. • Wash your hands with soap and water when done.

1.B. Specific Safety Hazards

Follow these instructions to use materials and equipment safely in the science classroom.

1.B.1. Chemicals

Some of the chemicals recommended for use in this course are dangerous if used incorrectly. Be sure to follow these rules to avoid accidents.

- Assume that all chemicals are hazardous.
- Never touch any chemicals with your hands. Always use a scoopula.
- Always read the label before you take any chemical out of its container. Make sure you are taking the correct substance. Replace the lid securely when you have taken what you need.
- Never use the contents of a container that has no label or whose label you cannot read.
- Place test tubes in a rack before pouring liquids into them. When using a test tube, tilt it away from you and others before pouring in a liquid.
- Pour liquid chemicals carefully to avoid splashing (the liquid should slide down the inside of the receiving container or down a stirring rod).
- If your teacher instructs you to smell a chemical, first take a deep breath, then waft the vapours toward your nose.
- Do not return chemicals to stock bottles, and do not pour them down the drain. Dispose of excess chemicals as directed by your teacher.
- If any part of your body comes in contact with a chemical, wash the area immediately and thoroughly with cool water. Rinse affected eyes for at least 15 min. Alert your teacher.

1.B.2. Heat Sources

Heat sources, such as hot plates, light bulbs, and Bunsen burners, can cause painful burns. Use caution when there are hot objects around.

- Secure your Bunsen burner to a retort stand with a clamp before lighting it.
- Always follow your teacher's directions for lighting and adjusting a Bunsen burner.
- Never leave a lighted burner unattended. A clean blue flame is almost invisible and is much hotter than an orange flame.
- Never heat a flammable material over a Bunsen burner. Keep flammable materials away from the flame.
- Never look down the barrel of a burner.
- Liquids should be heated only in heat-resistant glassware. When boiling a liquid, use boiling chips to prevent "bumping." Keep the open end of the container away from yourself and others. Never allow a container to boil dry.
- Always use a test-tube clamp and spurt cap when heating a test tube over a Bunsen burner. Hold the test tube pointing away from you and others. Heat the upper half of the liquid first and then gently move it back and forth through the flame to heat it evenly.
- Always turn off the gas at the valve, not using the gas-adjustment screw of the Bunsen burner.
- If you burn yourself, immediately shut off the gas and apply cool water to the affected area. Inform your teacher.

1.B.3. Glass and Sharp Objects

Handle glass with care.

- Never use glassware that is broken, cracked, or chipped.
- Never touch broken glass with your fingers. Use gloves as well as a broom and a dustpan to remove broken glass from the area.
- Dispose of glass fragments in special containers marked "Broken Glass."
- All cuts should be reported to your teacher immediately.
- Never use a knife when scissors would work better.
- Never carry a scalpel with an exposed blade; always carry it in a dissection tray.
- Make sure your cutting instruments are sharp. A dull blade might slip. If your scalpel blade needs to be changed, ask your teacher to change it for you.
- Always cut away from yourself and others. Cut downward on a tray, cutting board, or paper towel.

1.B.4. Electricity and Light

- Never touch an electrical device, electrical cord, or outlet with wet hands.
- Keep water away from all electrical equipment.
- Do not use the equipment if wires or plugs are damaged or if the ground pin has been removed.
- If using a light source, check that the wires of the light fixture are not frayed and that the bulb socket is in good shape and is well secured to a stand.
- Make sure that you do not trip on electrical cords.
- When unplugging equipment, hold the plug to remove it from the socket. Do not pull on the cord.
- When using variable-power supplies, start at low voltage and increase slowly.
- Do not look directly at any bright source of light. You cannot rely on the sensation of pain to protect your eyes.
- Never point a laser beam (directly or as a reflection) at anybody's eyes.

1.B.5. Living Things

- Treat all living things with care and respect. Never treat an animal in a way that would cause it injury or harm.
- Animals that live in the classroom should be kept in a clean, healthy environment.
- Wear gloves and wash your hands before and after feeding or handling an animal, touching materials from the animal's cage or tank, or handling bacterial cultures.

1.B.6. Observing Space

- Never look directly at the Sun or a solar eclipse. Special filters must be used when looking directly at the Sun.
- When going out at night to observe stars, always go with an adult. Be sure to let others know where you are going and when you plan to return.

1.C. Accidents Can Happen

The following guidelines apply if an injury occurs.

- Inform your teacher immediately.
- If the injury is from contact with a chemical, wash the affected area with a continuous flow of cool water for at least 15 min. Remove contaminated clothing. Consult the Material Safety Data Sheet (MSDS) for the first-aid requirements. If the chemical is splashed into your eyes, have another student assist you in getting to the eyewash station immediately. Rinse with your eyes open for at least 15 min.

- If you have ingested or inhaled a hazardous substance, consult the MSDS for the first-aid requirements for the substance.
- If the injury is from a burn, immediately immerse the affected area in cold water. This will reduce the temperature and prevent further tissue damage.
- In the event of electrical shock, do not touch the affected person or the equipment they were using. Switch off the source of electricity or remove the plug.

1.D. Safety Symbols

The activities and investigations in *Nelson Science Connections 10* are safe, provided precautions are taken. General safety hazards are identified with caution symbols (**Figure 1**).

Figure 1 Potential safety hazards are identified with caution symbols

Make sure that you read the information and associated instructions carefully. You must understand and follow these instructions to perform the activity or investigation safely. Check with your teacher if you are unsure.

1.D.1. WHMIS Symbols

The Workplace Hazardous Materials Information System (WHMIS) provides workers and students with complete and accurate information about hazardous products. Clear and standardized labels must be present on the product's original container or must be added to other containers if the product is transferred. If the material is hazardous, the label will include one or more of the WHMIS symbols (**Figure 2**).

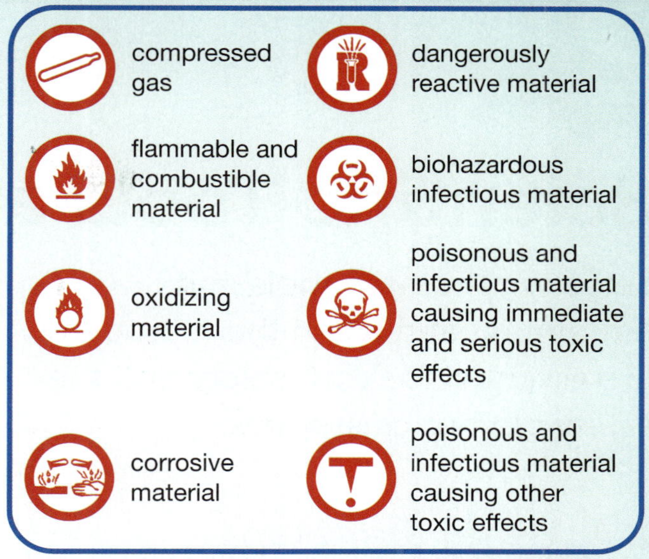

Figure 2 WHMIS symbols identify dangerous materials.

1.D.2. Hazardous Household Products Symbols (HHPS)

The *Canadian Hazardous Products Act* requires manufacturers of consumer products to include a symbol that specifies both the nature and the degree of any hazard. The Hazardous Household Products Symbols (HHPS) were designed to do this. Each symbol is made up of a picture and a frame. The picture tells you the type of danger. The frame tells you whether it is the contents or the container that poses that hazard (**Figure 3**).

Symbol	Danger
	**Explosive** This container can explode if it is heated or punctured.
	**Corrosive** This product will burn skin or eyes on contact, or throat and stomach if swallowed.
	**Flammable** This product, or its fumes, will catch fire easily if exposed to heat, flames, or sparks.
	**Poisonous** Licking, eating, drinking, or sometimes smelling, this product is likely to cause illness or death.

Figure 3 Household Hazardous Products Symbols (HHPS) appear on many products. A triangular frame indicates that the container is potentially dangerous. An octagonal frame indicates that the contents pose a hazard.

2

Scientific Tools and Equipment

Selecting the right tools and equipment and using them properly are essential for your safety and the safety of your classmates.

2.A. Using Star Maps

2.A.1. What Is a Star Map?

A star map is a map of the night sky that shows the positions of stars. Each star map is designed for a specific latitude, so a star map designed for Canada cannot be used in Australia.

The stars we can see in the night sky also depend on the time and date. Different star maps are designed for different seasons or months (**Figure 1**). Planispheres combine a star map with a frame. They can be adjusted to show stars that are visible at a particular time and date.

2.A.2. Stargazing Trips

Consider the following tips when planning a stargazing trip.

- Plan ahead of time by checking the weather forecast and considering safety, transportation, location, clothing, and things to bring.
- Choose a location that is far or screened from bright lights.
- Be prepared to record your observations.
- Before viewing, allow your eyes at least 10 min to adapt to the dark.
- Use a flashlight covered with red cellophane to view your star map and take notes.

2.A.3. Using a Star Map

To use a seasonal star map, follow these steps:

- Hold the map above your head.
- Position the part marked "Facing North" so that it is facing north.
- Compare what you see in the sky to the map.
- Notice any planets or objects besides stars.

2.A.4. Keeping Records

Use a table similar to **Table 1** to record your observations. Keep in mind that the date changes after midnight.

Table 1

Date	Object seen	Description and diagram	Location and angles	Questions to ask

2.A.5. Star Map

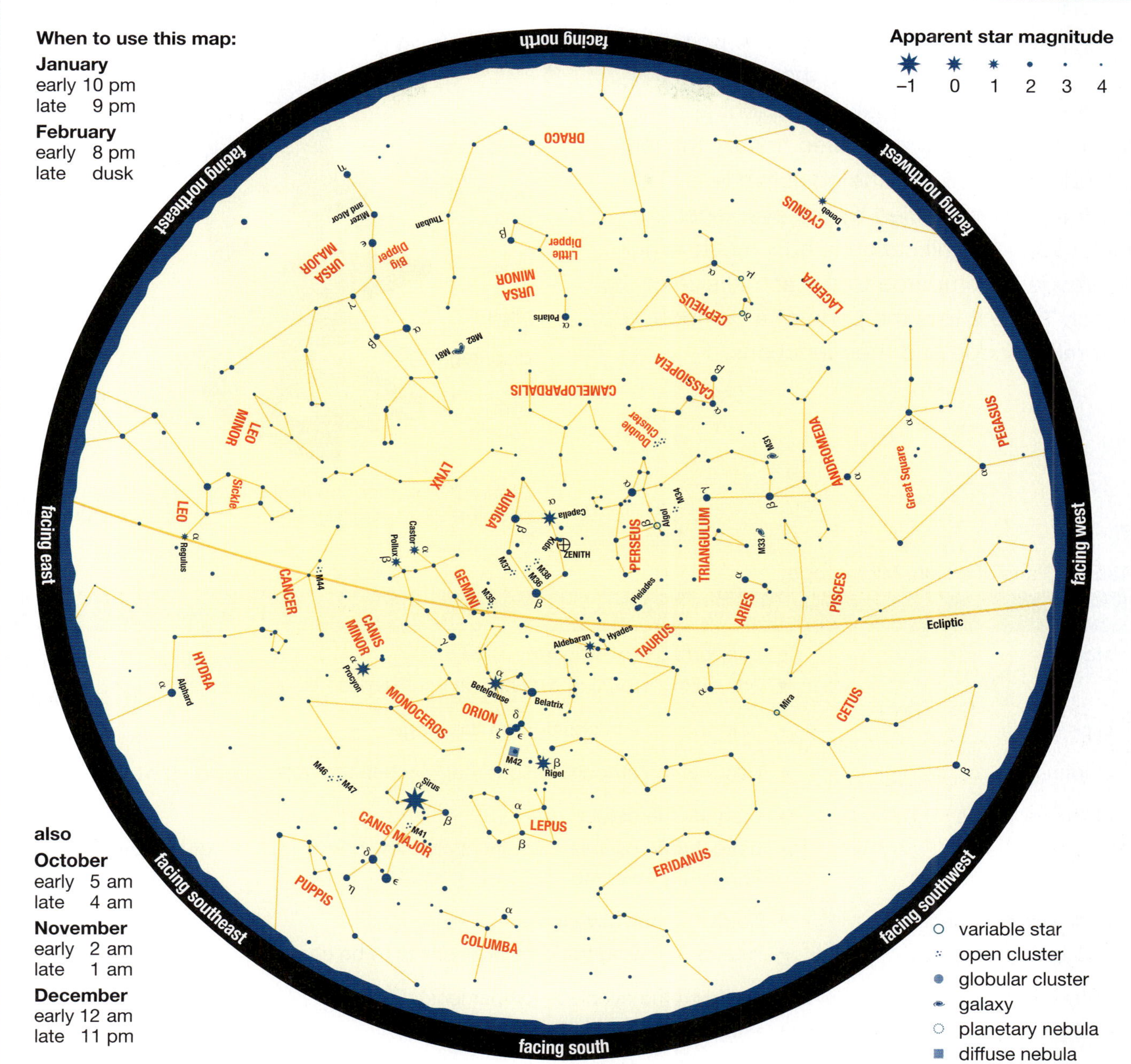

Figure 1 A star map drawn specifically for 45° north of the equator

2.B. Using the Microscope

You will use a microscope to see cells and small objects up close (**Figure 2**). A compound light microscope uses more than one lens and a light source to make an object appear larger. The object is magnified first by the lens just above the object: the objective lens. Then that image is magnified by the eyepiece: the ocular lens. Magnification is how much larger the image is compared to the actual size of the object. **Table 2** lists the parts of a compound light microscope and their functions.

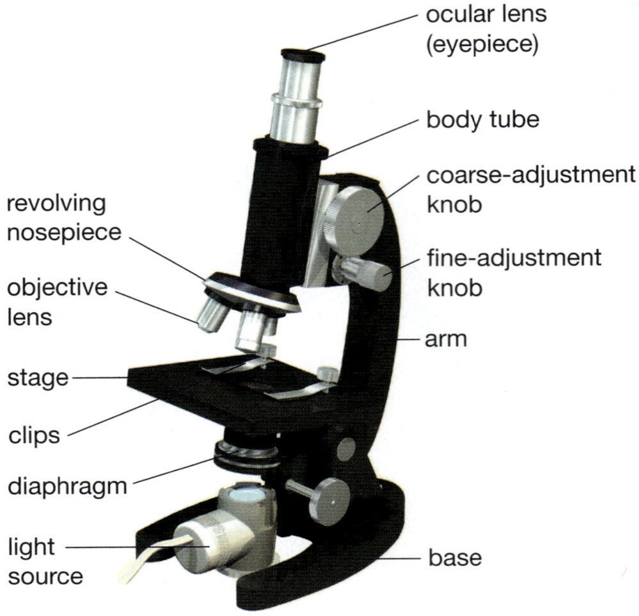

Figure 2 A compound light microscope

Table 2 Parts of a Light Microscope

Part	Function
stage	• supports the microscope slide • has central opening that allows light to pass through the slide
clips	• hold the slide in position on the stage
diaphragm	• controls the amount of light that reaches the object being viewed
objective lenses	• magnify the object • have three possible magnifications: low power (4×), medium power (10×), and high power (40×)
revolving nosepiece	• holds the objective lenses • rotates, allowing the objective lenses to be changed
body tube	• contains the eyepiece (ocular lens) • supports the objective lenses
eyepiece (ocular lens)	• the part you look through to view the object • magnifies the image of the object, usually by 10×
course-adjustment knob	• moves the body tube up or down to get the object into focus • is used with the low-power objective lens only
fine-adjustment knob	• moves the body tube to get the object into sharp focus • is used with medium-power and high-power magnification • is used only after the object has been located, centred, and focused under low-power magnification using the coarse-adjustment knob
light source	• may be an electric light bulb or a mirror that can be angled to direct light through the object being viewed

2.B.1. Microscope Skills

The basic microscope skills are presented as instructions. Practise these skills before you need to use them in the activities in *Nelson Science Connections 9*.

Viewing Objects under the Microscope

1. Position the low-power objective lens over the diaphragm. Either raise the objective lens or lower the microscope stage as far as possible. Put your slide in the centre of the stage. Use the stage clips to hold the slide in position. Turn on the light source (**Figure 3**).

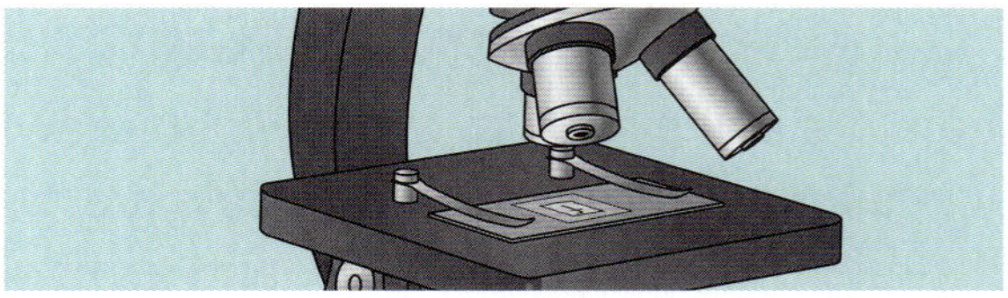

Figure 3

2. View the microscope stage from the side. Then, using the coarse-adjustment knob, lower the low-power objective lens until it is close to the object. (Some microscopes have movable stages, rather than movable lenses.) Do not allow the lens to touch the cover slip (**Figure 4**).

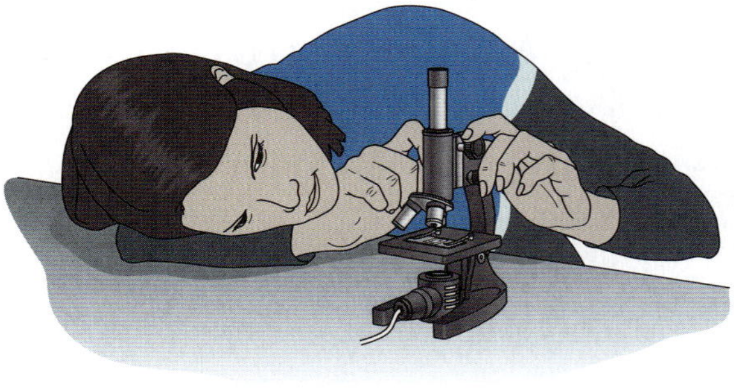

Figure 4

3. Look through the eyepiece. Slowly raise the objective lens using the coarse-adjustment knob until the image is in focus. The area you can see is called the field of view.
4. Draw a circle in your notebook to represent the field of view, and draw what you see. Make sure the object fills the same proportion of area in your drawing as it does in the microscope.
5. While you look through the microscope, slowly move the slide away from you. Note that the object appears to move toward you. Now move the slide to the left. Notice that the object appears to move to the right.
6. Turn the nosepiece to the medium-power objective lens. Use the fine-adjustment knob to focus the object. Notice that the object appears larger. Always use the fine adjustment when using the medium- or high-power objectives; the coarse adjustment may damage the slide or lenses.
7. Place the object directly in the centre of the field of view. Rotate the high-power objective lens into place. Again, use the fine-adjustment knob to focus the image. Notice that you see less of the object than you did under medium-power magnification and the object appears larger.

Putting Away the Microscope

After you have completed an activity using a microscope, follow these steps.

1. Rotate the nosepiece to the low-power objective lens.
2. Raise the lenses (or lower the stage) as far as possible.
3. Remove the slide and the cover slip (if applicable).
4. Clean the slide and the cover slip and return them to their appropriate location.
5. Use two hands to return the microscope to the storage area.

2.C. Testing for Electrical Conductivity

Devices used to test electrical conductivity vary considerably. Ask your teacher for specific operating instructions on your school's equipment (**Figure 5**). Two metal electrodes are inserted into the sample to be tested.

Figure 5 You should use only low-voltage (battery-powered) conductivity testers.

In many cases, if the sample conducts electricity, the conductivity tester gives a positive result (for example, a light bulb turns on). Some conductivity testers give variable results (**Table 3**)

Table 3 Results of Conductivity Testing

Observation	Sample
bright glow	good conductor
faint glow	poor conductor
no glow	insulator

2.D. Drawing and Constructing Circuits

2.D.1. Sources of Electrical Energy

To provide the electrical energy in most of the circuits you create, you will be using dry cells or a special device called a power supply. These are both direct current (DC) sources of electrical energy, where current flows in only one direction. Wall outlets provide a different kind of electrical energy known as an alternating current (AC) source. In an AC circuit, the electric current reverses its direction 60 times a second.

2.D.2. Drawing Circuit Diagrams

There are some rules to follow when drawing a circuit diagram. Connecting wires are shown as straight lines or 90° angles, and symbols are used to represent all the components (**Table 4**).

2.D.3 Safety Considerations

It is important to follow safety procedures, especially when you see a WHMIS safety symbol.

- Always make sure your hands are dry and that you are standing on a dry surface.
- Do not use faulty dry cells, and do not connect different makes to the same circuit. Avoid connecting old, used dry cells to fully charged cells.
- Do not use frayed or damaged connectors.
- Do not operate a circuit without your teacher's approval.

2.D.4. Constructing a Circuit

Always follow the instructions when working with circuits. If you are not sure what to do, ask for help. Check that all of the components work properly.

- Check the connections of any circuit carefully before connecting a dry cell. Incorrect connections can cause shorted circuits or explosions. Ask your teacher if you are unsure.
- When attaching connecting wires to a meter, connect a red wire to the positive (+) terminal and a black wire to the negative (–) terminal of the meter.
- Use approved connectors, such as alligator clips, and be careful not to let them touch one another when they are connected to the circuit.
- Open the switch (turn off the circuit) before changing or adding anything such as a lamp, motor, or meter.
- If the circuit is not working correctly, open the switch and check the wiring and connections. Ask your teacher if you cannot find the problem.

Table 4 Circuit Diagram Symbols

	Part of circuit	Symbol
sources of electrical energy	electric cell	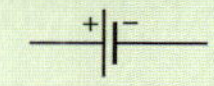
	three-cell battery	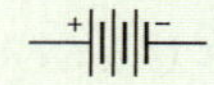
	AC power supply	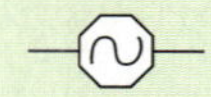
	variable DC power supply	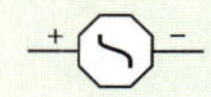
electrical conductor	connecting wire	
control device	switch (open)	 switch (open)
	switch (closed)	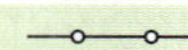 switch (closed)
loads	lamp	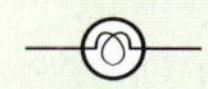
	electric motor	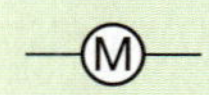
	resistor	
meters	ammeter	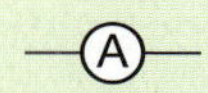
	voltmeter	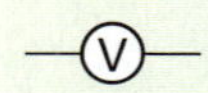
	ohmmeter	

2.E. Using the Voltmeter, Ammeter, and Multimeter

We rely on instruments that can detect and measure electricity. Two common instruments are the voltmeter and the ammeter. These instruments may be digital (providing a digital readout) or analog (showing values by the motion of a needle across a scale). You may also use a digital multimeter, which measures voltage, current, and resistance.

2.E.1 The Voltmeter

A voltmeter measures the voltage difference between two different points in a circuit. A voltmeter can be connected across the terminals of a cell to measure voltage output, or across another component of a circuit to measure the voltage drop across this component. A voltmeter is always connected in parallel with the component you want to investigate.

2.E.2. Reading an Analog Voltmeter

The needle on a voltmeter usually moves from left to right, with zero voltage on the left side of the scale and the maximum voltage on the right. If the voltmeter scale has only one set of numbers, it is relatively easy to measure the voltage. If the voltmeter has several sets of numbers, identify which set of numbers matches the voltage range selected by the switch on the voltmeter.

The negative terminal of the voltmeter must be connected to the more negative part of the circuit. If the leads are attached incorrectly, the needle will not move or give a reading.

2.E.3. The Ammeter

An ammeter measures the amount of electric current that is flowing in a circuit. To measure the electric current, connect the ammeter directly into the circuit. Disconnect a wire from the part of the circuit you want to measure. Connect the ammeter, in series, to complete the circuit. Reading an analog ammeter is similar to reading a voltmeter. The unit of current is the ampere (A) or milliampere (mA).

2.E.4. The Digital Multimeter

A digital multimeter can measure the voltage, current, or resistance in a circuit (**Figure 6**). Using the selector knob, select the electrical value and range that you want to measure. Then connect the wire leads properly to the appropriate place in the circuit.

Remember to connect the meter in parallel for measuring voltage and resistance, or in series for measuring current. The multimeter will provide a digital readout.

Figure 6 A digital multimeter

Scientific Inquiry Skills

3.A. Thinking as a Scientist

Imagine that you are planning to buy a new MP3 player. First, you write a list of questions. Then you check print and Internet sources, visit stores, and talk to your friends to find out which is the best purchase. Solving problems in a step-by-step way is thinking like a scientist.

- Scientists investigate the natural world to test their ideas. For example, biologists raise questions about the impact of climate change on water in the Arctic (**Figure 1**). They also think of hypotheses to answer their questions. They design investigations to test their hypotheses. This process leads scientists to come up with new ideas to be tested and more questions to be answered.

Figure 1 Water samples can be tested for pH, dissolved substances, and other impurities.

- Scientists investigate the natural world to describe it. For example, climatologists study the layers inside ice cores to make inferences about what the climate was like in the past (**Figure 2**).

Figure 2 Scientists can learn what the climate was like long ago by studying the layers inside ice cores.

- Scientists investigate objects to classify them. For example, chemists classify substances according to their properties (**Figure 3**).

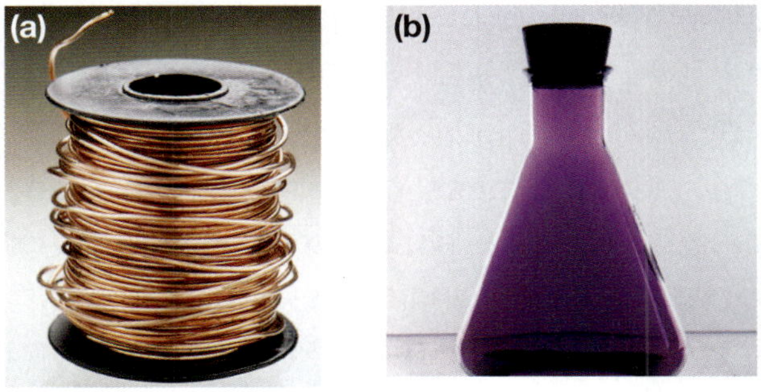

Figure 3 (a) Copper is considered ductile. (b) Iodine is a deep violet colour as a gas.

3.B. Scientific Inquiry

You need to use a variety of skills when you do scientific inquiry and design or carry out an experiment. Refer to this section when you have questions about any of the following skills and processes:

- questioning
- hypothesizing
- predicting
- planning
- controlling variables
- performing
- observing
- analyzing
- evaluating
- communicating

3.B.1. Questioning

Curiosity about the natural world leads to scientific investigations. We might ask, "Why does that happen?" or "What would happen if I ...?"

Before you conduct an investigation, you must have a clear idea of what you want to know. This will help you make up questions that will lead you to the right information. Avoid questions that have a simple "yes" or "no" answer. Think of questions that are testable and lead to an investigation.

Sometimes, an investigation starts with a special type of question called a "cause-and-effect" question. This question asks whether something is causing something else to happen. It might start in one of the following ways: What causes ...? How does ... affect ...?

3.B.2. Predicting and Hypothesizing

A prediction states what is likely to happen as the result of a controlled experiment. Scientists make predictions based on their observations and knowledge. They look for patterns in the data to help them understand what might happen next. A prediction may be written as an "if ... then ..." statement.

A prediction states what you think will happen. Predictions are not guesses. They are suggestions based on prior knowledge and logical reasoning (**Figure 4**).

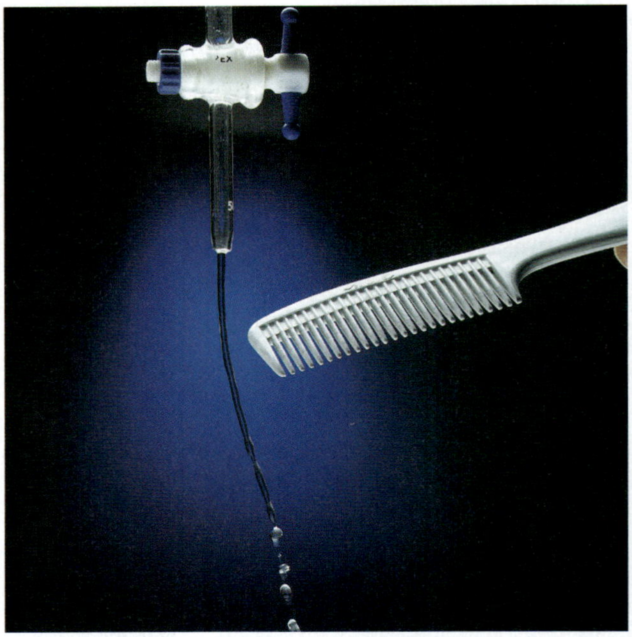

Figure 4 A positively charged plastic comb causes a stream of tap water to bend. What would you predict might happen if a negatively charged object were brought near the stream of water?

For example, we may want to find out what effect salt has on the boiling point of water. For this investigation, your prediction might be, "If the amount of dissolved salt is increased, then the boiling point will also increase."

A prediction can be used to write a hypothesis. A hypothesis is a possible answer about the outcome of a controlled experiment along with an explanation for the outcome. A hypothesis may be written in the form of an "if … then … because …" statement. *If* the cause variable is changed in a particular way, *then* the effect variable will change in a particular way, and this change occurs *because* of certain reasons.

For example, "If the amount of salt is increased, then the boiling point will also increase." If your observations confirm your prediction, then they support your hypothesis.

You can create more than one hypothesis from the same question or prediction. Another hypothesis might be, "If the amount of salt is increased, then the boiling point will not change." Of course, both hypotheses cannot be correct. When you conduct an investigation, sometimes your results show your hypothesis is not correct.

An investigation that does not support your hypothesis is not a bad investigation or a waste of time. It has contributed to your scientific knowledge. You can re-evaluate your hypothesis and design a new investigation.

3.B.3. Controlling Variables

Considering the variables involved is an important step in designing an effective investigation. Variables are any factors that might affect the outcome of an investigation.

There are three kinds of variables in an investigation:

1. The variable that is changed is called the independent variable, or cause variable.
2. The variable that is affected by a change is called the dependent variable, or effect variable. This is the variable you measure to see how it was affected by the independent variable.
3. All the other conditions that remain unchanged and did not affect the outcome of the investigation are called the controlled variables.

For example, for the boiling point investigation,

- the mass of dissolved salt is the independent variable
- the boiling point of water is the dependent variable
- the amount of water in the beaker and the type of salt used are two controlled variables

A controlled experiment involves only one independent variable at a time. This way, you are sure that the results are caused by the one variable that was changed and not by any of the other variables.

3.B.4. Planning

You have been asked to design and carry out the boiling point investigation described in 3.B.2. First, you create an experimental design. To conduct a controlled experiment that tests your hypothesis, you decide to change only one variable—the amount of dissolved salt. This is your independent variable. You dissolve 5.0 g, 10.0 g, 15.0 g, and 20.0 g of salt in four identical beakers containing exactly 100 mL of water. The fifth beaker should contain water but no salt. This is your control. Each solution is heated using the same hot plate until it boils. A thermometer suspended in each solution measures the temperature of the solution when it starts to boil (**Figure 5**). Temperature is the dependent variable.

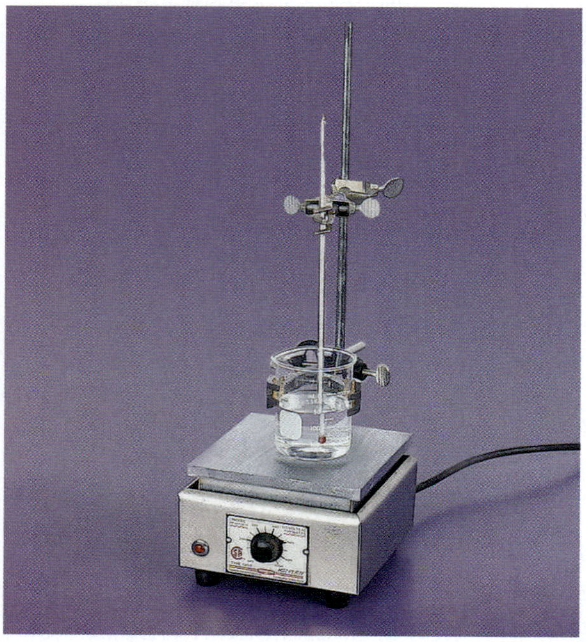

Figure 5 Measuring the temperature at which water containing a known mass of salt boils.

Now consider the equipment and materials you need. You must include any safety equipment, such as an apron or eye protection, in your equipment and materials list. How will you secure the beaker so that it does not accidentally fall off the hot plate as it boils? Will you use tap water or distilled water?

For the boiling point investigation, your equipment and materials list might include:

- eye protection
- apron
- gloves
- hot plate
- stand with beaker clamp
- beaker
- thermometer
- wire gauze
- distilled water
- boiling chip
- salt

Next, you need to write a procedure: a step-by-step description of how you will perform your investigation. The first step in a procedure usually refers to any safety precautions that need to be addressed, and the last step relates to any cleanup that needs to be done.

Your teacher must approve your procedure and list of equipment and materials before you begin. A procedure should be written as a series of numbered steps, with only one instruction per step.

For example, for the boiling point investigation, your procedure might include:

1. Wear safety goggles, a protective apron, and heat-resistant gloves.
2. Set up the equipment as shown in the diagram. (Include a clearly labelled diagram.)
3. Add 100 mL of distilled water to the beaker.
4. Add one boiling chip to the beaker to ensure that the water boils gently and does not bump.

5. Dissolve 5.0 g of sodium chloride in the water.
6. Set the heating control on the hot plate to 50 %.
7. Wait for the mixture to boil.
8. Turn off the heat and allow the beaker to cool.
9. Repeat Steps 3 to 8 using 10.0 g, 15.0 g, and 20.0 g of sodium chloride.
10. Allow the apparatus to cool before dismantling it.
11. Once cool, return all equipment to the storage cabinet.

Your procedure must be clear enough for someone else to follow, and it must explain how you will deal with each of the variables in your investigation.

3.B.5. Performing

An investigation involves a step-by-step procedure, which must be followed carefully. Check with your teacher if you need to make changes in your procedure. All equipment and materials must be used safely, accurately, and with precision. Be sure to take detailed notes and to record all of your observations. Numerical data should be recorded in a table.

3.B.6. Observing

When you observe something, you use your senses to learn about it. You can also use tools, such as a balance or a microscope. Observations can be divided into two categories: quantitative and qualitative.

Quantitative observations consist of anything that is measured, such as temperature, volume, and mass. Numerical data from quantitative observations are usually recorded in data tables or graphs.

Qualitative observations are made using your senses. Colour, smell, clarity, and state of matter are examples of qualitative observations (**Figure 6**). Qualitative observations can be recorded using words, pictures, or labelled diagrams.

Figure 6 Qualitative observations such as a colour change, bubbles, an irritating odour, or a sizzling sound suggest that a chemical change is occurring when certain chemicals are mixed.

As you work on an investigation, be sure to record all of your observations, both qualitative and quantitative. If appropriate, organize your observations and measurements in a data table. (See Data Tables and Graphs, page 583.)

Your final lab report must include all observations and measurements. It is important to remain impartial when recording observations. Record only what you observe. Observations from an experiment may not always be what you expect.

Scientific Drawings

Scientific drawings are used to record your observations. They must be clear, well labelled, and easy to understand. Below are some tips that will help you produce useful scientific drawings.

GETTING STARTED

The following materials and ideas will help you get started:

- Use plain, blank paper. Lines might obscure your drawing or make your labels confusing.
- Look closely at your specimen or equipment, noting details and sizes, before you begin the drawing (**Figure 7**).

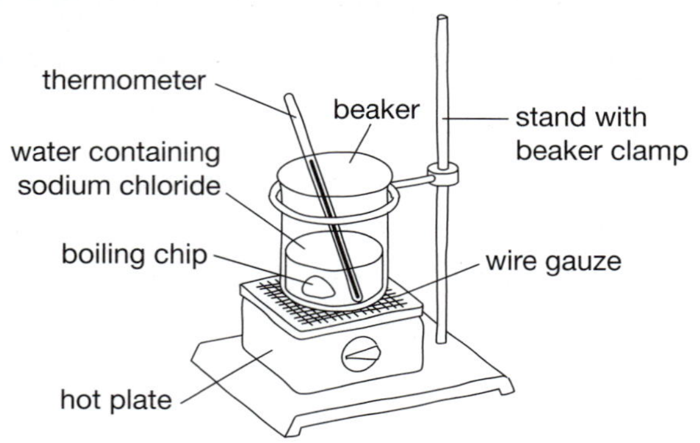

Figure 7 Your drawing of an experimental set-up should show how the equipment was assembled.

- Use a sharp pencil rather than a pen or marker in case you need to erase part of your drawing.
- Your drawing should be large enough to show details. For example, a diagram of a single cell or a unicellular organism would use about one-third to half of a page.
- Label your drawing clearly. Use a ruler to draw the label lines.

Horizon Diagrams

Horizon diagrams are useful for recording observations of the night sky relative to landmarks on the horizon (**Figure 8**). Refer to the section on Using Star Maps on page 550 when planning a stargazing trip.

Follow these steps to create a horizon diagram:

- Draw a line at the bottom of a sheet of paper to represent the horizon.
- Sketch any landmarks, such as trees or buildings, you see along the horizon.
- Sketch all of the objects you see in the night sky relative to the landmarks.
- Record the date and exact time of your observations. Include a detailed description of the location and cloud conditions.

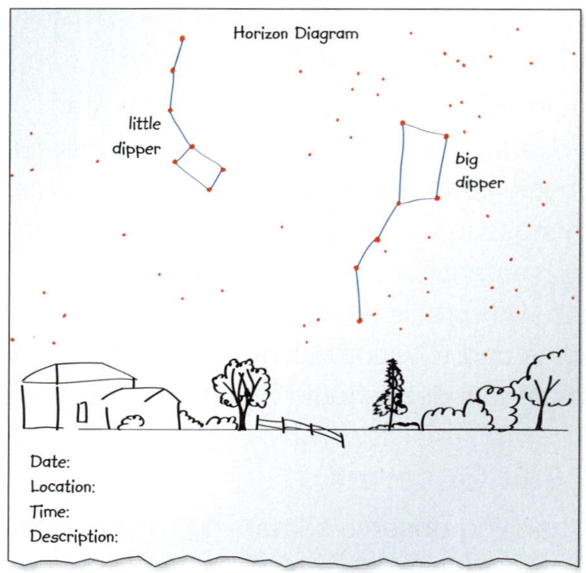

Figure 8 Sample horizon diagram

CHECKLIST FOR SCIENTIFIC DRAWING

- ✓ Use plain, blank (unlined) paper and a sharp, hard pencil.
- ✓ Draw as large as necessary to show details clearly.
- ✓ Do not shade or colour.
- ✓ Draw label lines that are straight and run outside your drawing. Use a ruler.
- ✓ Include labels, a title or caption, and, if appropriate, the magnification of the microscope you are using.

3.B.7. Analyzing

You analyze data from an investigation to make sense of it. You examine and compare the measurements you have made. You look for patterns and relationships that will help you explain your results and give you new information about the question you are investigating (**Figure 9**).

Figure 9 Analyzing the results of your investigation will help you to see patterns in the data.

Often, making tables or graphs of your data will help you see patterns and relationships more easily. Turn to the section Using Mathematics in Science to learn more about creating data tables and graphing your results.

Once you have analyzed your data, you can tell whether your prediction or hypothesis is correct. You can write a conclusion that either supports your hypothesis or not. You might have to come up with a new hypothesis that can be tested in a new investigation.

3.B.8. Evaluating

How useful is the evidence from an investigation? Was it helpful in answering your prediction or hypothesis? The answers to these questions are important because we need quality evidence before we can accurately evaluate our prediction or hypothesis. If the evidence is poor or unreliable, you can still identify areas for improvement for when the investigation is repeated

Below are some things to consider when evaluating the results of an investigation:

- *Plan:* Were there any problems with the way you planned your experiment or your procedure? Did you control for all the variables except the independent variable?
- *Equipment and Materials*: Could better or more accurate equipment have been used? Was something used incorrectly? Did you have difficulty with a piece of equipment?
- *Observations:* Did you record all the observations that you could have? Or did you ignore or overlook some observations that might have been important?
- *Skills:* Did you have the appropriate skills for the investigation? Did you have to use a skill that you were just beginning to learn?

Once you have identified areas in which errors could have been made, you can judge the quality of your evidence.

3.B.9. Communicating

When you plan and carry out your own investigation, it is important to share both your process and your findings. Other people may want to repeat your investigation, or they may want to use or apply your findings in another situation. Your write-up or report should reflect the process of scientific inquiry that you used in your investigation.

Sample Lab Report

Write the title of your investigation at the top of the page.

Raising the Boiling Point of Water

List the question(s) you are trying to answer. This section should be written in sentences.

Testable Question

What effect does the mass of sodium chloride have on the boiling point of water?

Write your hypothesis or prediction.

Hypothesis

If the amount of sodium chloride is increased, then the boiling point will also increase because sodium and chloride ions attract water molecules, preventing them from changing into gas.

Write your experimental design, briefly outlining what you will be doing and identifying your independent, dependent, and controlled variables (if appropriate).

Experimental Design

Various amounts of salt will be added to water to create the same volume of solution. Each solution will be heated and its boiling point measured. The independent variable is the mass of salt added. The dependent variable is the boiling temperature. The controlled variables are: type of salt used, equipment used, and volume of solution heated.

Write the equipment (items that can be re-used) and materials (substances that will be used up) in a list. Give the amount or size, if this is important. Remember to include safety equipment.

Equipment and Materials

- eye protection
- lab apron
- heat-resistant gloves
- 250 mL beaker
- boiling chip
- hot plate
- wire gauze
- retort stand with beaker clamp
- thermometer
- 50 g sodium chloride
- water

Draw a large, labelled diagram, if necessary, to show how equipment was set up.

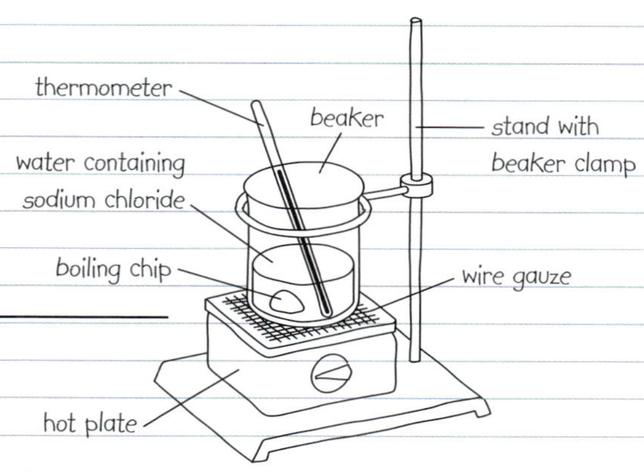

Procedure

1. Safety goggles, a protective apron, and heat resistant gloves were obtained.
2. The equipment was set up as shown in the diagram.
3. 100 mL distilled water was added to the beaker.
4. One boiling chip was added to the beaker to prevent the solution from bumping.
5. 5.0 g sodium chloride was dissolved in the water.
6. The hot plate heater control was set to 50 %.
7. The temperature at which the water boiled was recorded.
8. The heat was turned off and the beaker was allowed to cool.
9. Steps 3–8 were repeated using 10.0 g, 15.0 g, and 20.0 g of sodium chloride.
10. The apparatus was allowed to cool before it was returned.

Observations

Table of Observations

Mass of sodium chloride (g)	Boiling point (°C)
5	100.4
10	100.8
15	101.4
20	101.8

In each case, the water boiled gently. Small bubbles formed throughout the solution as it boiled.

Analysis and Evaluation

The boiling point of the solution increased as more salt was dissolved in it. Water condensing on the thermometer made some of the temperatures difficult to read. This problem could be overcome by using a temperature probe instead of a thermometer. The data clearly support the hypothesis that the boiling point of water increases as the mass of dissolved sodium chloride increases.

Applications and Extensions

Adding table salt (sodium chloride) to boiling water does not increase the boiling point of the water a great deal. Therefore, adding a little salt to boiling water when cooking will not speed up the cooking process.

Describe the procedure using numbered steps. Each step should start on a new line. Write the steps as they occurred, using past tense and passive voice. Make sure that your steps are clear so that someone else could repeat your investigation. Include any safety precautions.

Present your observations in a form that is easily understood. Quantitative observations should be recorded in one or more tables, with units included. Qualitative observations can be recorded in words or drawings.

Analyze your results and evaluate your procedure. If you have created graphs, refer to them here. If necessary, include them on a separate piece of graph paper. Write a conclusion indicating whether or not your results support your hypothesis or prediction. Answer any Analyze and Evaluate questions here.

Describe how the knowledge you gained from your investigation relates to real-life situations. How can this knowledge be used? Answer any Apply and Extend questions here.

4

Research Skills

We are exposed to a great deal of information every day. Some of this information is reliable; some is not. You must learn how to search efficiently and assess the credibility of the information you find.

4.A. General Research Skills

4.A.1. Identify the Information You Need

- Identify your research topic.
- Identify the purpose of your research.
- Identify what you already know about the topic.
- Develop a list of key questions that you need to answer.
- Identify categories based on your key questions.
- Use these categories to identify key search words.

4.A.2. Identify Sources of Information

Where can you look for information about your topic? Places might include television programs, people in your community, print sources (**Figure 1**), and electronic sources (such as CD-ROMs and Internet sites).

Remember, gathering information from a variety of sources will improve the quality of your research.

Figure 1 Your school library and local library are both excellent sources of information.

4.A.3. Evaluate the Sources of Information

Read through your sources of information and decide whether they are useful and reliable. Here are five things to consider:

- *Authority*: Who wrote the information or who sponsors the website? What are their qualifications?
- *Accuracy*: Are there any obvious errors or inconsistencies in the information? Does the information agree with other reliable sources?
- *Currency*: Is the information up to date? Has recent scientific information been included?
- *Suitability*: Does the information make sense to someone with your experience or of your age? Do you understand it?
- *Bias*: Are facts reported fairly? Are there reasons why your sources might express some bias? Are facts deliberately left out?

4.A.4. Record and Organize the Information

After you have gathered your information, you can start organizing your research. Identify categories or headings for note taking. Use point-form notes to record information in your own words under each heading in your notebook.

Do not copy information directly from your sources. If you quote a source, you must use quotation marks, record the title, author, publisher, page number, and date for each of your sources. For websites, record the URL (website address). All of these details are necessary to help you keep track of all the information (**Figure 2**). You will also need this information to create a bibliography.

To help you organize the information further, you may want to use pictures, graphic organizers, and diagrams. (See the Literacy section on page 590.)

Figure 2 Keep track of your sources so you do not forget where you found your information.

4.A.5. Make a Conclusion

Look at your original research question. What did you learn from the information you gathered? Can you state a conclusion based on that information? What further information do you need? Where would you look for that information?

You should now have an informed opinion on the research topic that you did not have before you started. If you do not have an informed opinion, what additional information do you need to reach such an opinion?

4.A.6. Evaluate Your Research

Now that your research is complete, think of ways to improve the research process for next time. How valuable were the sources of information you selected?

4.A.7. Communicate Your Conclusions

Choose a format for communication that suits your audience, your purpose, and the information you gathered (**Figure 3**). Consider using labelled diagrams, graphs, or charts, if appropriate.

Figure 3 Share your information in a way that makes it easy for your audience to understand.

4.B. Using the Internet

The Internet is a vast and constantly growing network of information. You can use search engines to help you find what you need, but keep in mind that not everything you find will be useful, reliable, or true.

4.B.1. Search Results

Once you have entered your search word or phrase, a list of web page "matches" will appear. Be sure to make your keywords specific or you will get too many matches to go through. Most search engines provide online help and search tips.

Every web page has a URL (universal resource locator), which provides clues about the organization hosting the web page (**Table 1**). For example, a URL that includes "ec.gc.ca" indicates that the content is hosted by Environment Canada—a reliable source.

Table 1 Common URL Codes and Organizations

Code	Organization
ca	Canada
com or co	commercial
edu or ac	educational
org	non-profit
net	networking provider
mil	military
gov	government
int	international organization

4.B.2. Evaluating Internet Resources

Anyone can post information on the Internet without verifying its accuracy. You must decide whether the information comes from dependable and legitimate sources. Use the following questions to help determine the quality of an Internet source. The more questions you answered "yes" to, the more likely it is that the source is of high quality.

- Is it clear who is sponsoring the page, and is it a reputable organization? Is the site permanent?
- Is contact information about the sponsoring organization available, such as a phone number or address?
- Is it clear who wrote the information? Are the author's qualifications given?
- Are the sources for factual information given so that they can be checked?
- Are there dates to indicate when the page was written, placed online, or last revised?
- Is the page presented as a public service? Does it present balanced points of view?

4.B.3. Using School Library Resources

Many schools have access to online encyclopedias with science sections in them. Find out if your school has a website where you can access these resources. You may need a password.

4.B.4. Using the Nelson Website

When you see the Nelson Science Web Link icon in your textbook, you can go to the Nelson website and find links to useful sources of information.

DIG DEEPER

Web Link

GO TO NELSON SCIENCE

4.C. Exploring an Issue Critically

An issue is a situation in which several points of view need to be considered in order to make a decision. When a decision affects many people or the environment, it is important to look at all sides of the issue. Think about all the possible solutions and try to understand all the different sides—not just your own point of view. Consider the pros and cons of each possible solution. Put yourself in the place of several of the stakeholders to try to understand their positions. Exploring an issue critically also means researching and investigating your ideas and sharing them with others. **Figure 4** shows all the steps in the process.

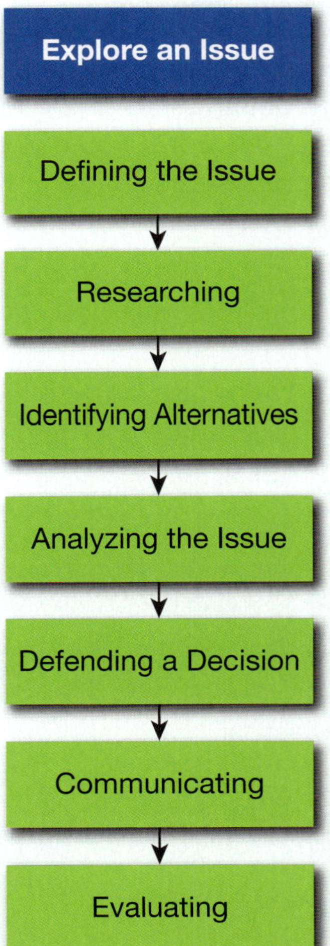

Figure 4 You may perform some or all of these steps as you explore an issue critically.

4.C.1. Defining the Issue

An issue has more than one solution, and there are different points of view about which solution is the best. Rephrase the issue as a question: "What should ...?"

The issue can also include information about the *role* a person takes when thinking about an issue. For instance, you may think about the issue from someone else's point of view—you may take the role of a landowner or a government worker.

The issue can also include a description of who your audience will be—will it be other students, government officials, or your parents? Be sure to take into account your role and the audience when defining your issue.

4.C.2. Researching

To reach the best decision, you have to have a good understanding of the issue. You must choose the most appropriate solution. To do this, you need to gather all the facts about the different points of view: come up with good questions and a plan for your research. Your research may include talking to people, reading about the topic in newspapers and magazines, and doing Internet research.

Make sure your information is reliable, accurate, and current. Avoid biased information that leans to only one side of the issue. It is important to gather information that represents all aspects of an issue. Think about how valuable the sources are and whether you could you find better information elsewhere.

4.C.3. Identifying Alternatives

Consider all possible solutions to the issue. For example, the town council has to decide how best to use the vacant land next to the school. Students would like to have a nature park. Another group suggests that the land be used for building a seniors' home. The school board would like to use the land to build a track for sporting events.

After defining the issue and researching, you can now make a list of possible solutions:

- Turn the land into a nature park for the community and the school.
- Use the land as a playing field and track for the community and the school.
- Create a combination of a nature park and a playing field.
- Use the land to build a seniors' home with a nature park.

4.C.4. Analyzing the Issue

How will you figure out which possibility is the best? Should the solution be the one that has the most community support, or should it be the one that best protects the environment? Decide which criteria you will use to make a decision.

4.C.5. Defending a Decision

At this stage, everyone gets a chance to share ideas and information. Then the group needs to evaluate all the possibilities and decide on one solution.

Pros and Cons

When there are several options, look at the pros and cons of each solution. Try to decide how important each of the pros and cons is. Remember that your opinions should be based on facts.

1. Create a table similar to **Table 2**. List pros and cons.
2. Rate each pro and con on a scale from 1 to 5, where 1 is the least important and 5 is the most important.
3. After rating each pro and con, add up the results to obtain totals. If the total pros are higher than the cons, you can go ahead with the proposed solution.

Table 2 The Pros and Cons of Using Land to Build a Seniors' Home with Nature Park

Pros		Cons	
Possible result	Benefit of result (rate 1 to 5)	Possible result	Cost of result (rate 1 to 5)
seniors' home provides necessary housing	5	land cannot be used for sports	2
park preserves some habitat for plants and animals	4	expensive to maintain	4
park increases value of seniors' home	3	nature park will be very small	3
total pros value	12	total cons value	9

4.C.6. Communicating

Your teacher may tell you how to communicate your decision. For example, your class might hold a formal debate. Or, you might be free to choose your own method of communication. You could choose one of the following methods of communicating your decision:

- Write a report.
- Give an oral presentation (**Figure 5**).
- Make a poster.
- Prepare a slide show.
- Create a video.
- Organize a town hall or panel discussion.
- Create a blog or webcast.
- Write a newspaper article.

Figure 5 Giving an oral presentation with visuals is an effective way to communicate information about an issue.

Choose the presentation that is best suited for your audience. For example, if your audience is small, it might be easiest to present your decision in person. If your audience consists of many people, an oral presentation with visuals is a good way to present your decision to many people at one time. Make sure that visuals are large and clear enough for your audience to see. A poster or a blog allows people to read your recommendation on their own, but you must find a way to let others know where to find this information.

Whatever means you use, however, you should

- state your position clearly, considering your audience
- support it with objective data and with a persuasive argument
- be prepared to defend your position against opposition

4.C.7. Evaluating

The final step of the decision-making process includes evaluating the decision itself and the process used to reach the decision. After you have made a decision, carefully examine the thinking that led to this decision. Some questions to guide your evaluation include the following:

- What was my initial perspective on the issue? How has my perspective changed since I first began to explore the issue?
- How did I gather information about the issue? What criteria did I use to evaluate the information? How satisfied am I with the quality of my information?
- What information did I consider to be the most important when making my decision?
- How did I make my decision? What process did I use? What steps did I follow?
- To what extent were my arguments factually accurate and persuasively made?
- In what ways does my decision resolve the issue?
- What are the likely short-term and long-term effects of my decision?
- How might my decision affect the various stakeholders?
- To what extent am I satisfied with my decision?
- If I had to make this decision again, what would I do differently?

5

Using Mathematics in Science

Communication is an important part of science. To avoid confusion, certain rules must be followed when using or reporting measurements.

5.A. SI Units

The scientific communities of many countries, including Canada, have agreed on a system of measurement called SI (Systeme international d'unites). This system has seven main SI units, called base units (**Table 1**).

Table 1 The Seven SI Base Units

Quantity name	Unit name	Unit symbol
length	metre	m
mass	kilogram	kg
time	second	s
electric current	ampere	A
temperature	kelvin	K
amount of substance	mole	mol
light intensity	candela	cd

All other physical quantities can be expressed as a combination of these seven SI base units. The unit for speed is metres (distance) per second (time), or m/s. These "mixed" units are called derived units. Some derived units have special names and symbols.

Table 2 lists some other common units with special names and symbols. Note that the symbols representing the quantities are italicized, whereas the unit symbols are not.

Table 2 Common Quantities and Units

Quantity name	Quantity symbol	Unit name	Unit symbol
distance	*d*	metre	m
area	*A*	square metre	m^2
volume	*V*	cubic metre	m^3
		litre	L
speed	*v*	metre per second	m/s
concentration	*c*	gram per litre	g/L
temperature	*t*	degree Celsius	°C
pressure	*p*	pascal	Pa
energy	*E*	joule	J
work	*W*	joule	J
power	*P*	watt	W

5.A.1. Converting Units

An important feature of SI is the use of prefixes. These prefixes help to express and compare small or large quantities (**Table 3**). The most common prefixes, such as milli- and kilo-, change the size in multiples of 1000. For example, 1 km = 1000 m. Both values express the same distance—only the unit of measurement has changed.

Table 3 Common SI Prefixes

Prefix	Symbol	Factor by which unit is multiplied	Example
giga	G	1 000 000 000	1 000 000 000 m = 1 Gm
mega	M	1 000 000	1 000 000 m = 1 Mm
kilo	k	1 000	1 000 m = 1 km
hecto	h	100	100 m = 1 hm
deca	da	10	10 m = 1 dam
		1	
deci	d	0.1	0.1 m = 1 dm
centi	c	0.01	0.01 m = 1 cm
milli	m	0.001	0.001 m = 1 mm
micro	μ	0.000 001	0.000 001 m = 1 μm
nano	n	0.000 000 001	0.000 000 001 m = 1 nm

SAMPLE PROBLEM 1 Using Conversion Factors

A block of cheese at a grocery store has a mass of 1256 g. Its price is $15.00/kg. What is the price of the block of cheese?

First, convert the mass from grams to kilograms.

There are two possible conversion factors between g and kg, as shown above.

Always choose the conversion factor that cancels the original unit. In this case, the original unit is g, so the correct conversion factor is $\frac{1\text{ kg}}{1000\text{ g}}$.

$$1256\ \cancel{\text{g}} = \frac{1\text{ kg}}{1000\ \cancel{\text{g}}} = 1.256\text{ kg}$$

The original units, g, cancel (divide to give 1), leaving kg as the new unit.

Now that you have determined the mass in kg, multiply the price per kg by the mass in kg.

$$\$15.00/\text{kg} \times 1.256\text{ kg} = \$18.84$$

The price of the block of cheese is $18.84.

Practice: Conversions

Make the following conversions. Refer to **Table 3**, if necessary.

(a) Write 3.5 s in ms. (b) Change 5.2 A to mA. (c) Convert 7.5 μg to ng.

5.B. Measuring Volume and Mass

5.B.1. Tips for Measuring Volume

- Use a container that is big enough to hold twice as much liquid as you need. You want a lot of space so that you can get an accurate reading.
- Use a graduated cylinder to get the most accurate measurement.
- To measure liquid in a graduated cylinder, beaker, or measuring cup, make sure that your eyes are at the same level as the top of the liquid. You will see that the surface of the liquid curves downward in the middle. This downward curve is called the meniscus. You need to measure the volume from the bottom of the meniscus (**Figure 1**).
- To measure the volume of an irregular object, place the object in water and see how much the water rises. To calculate the volume, subtract the original water level before the object was placed in the water from the water level with the object inside. Use the following formula:

  volume of solid =
  (volume of water + solid) – volume of water

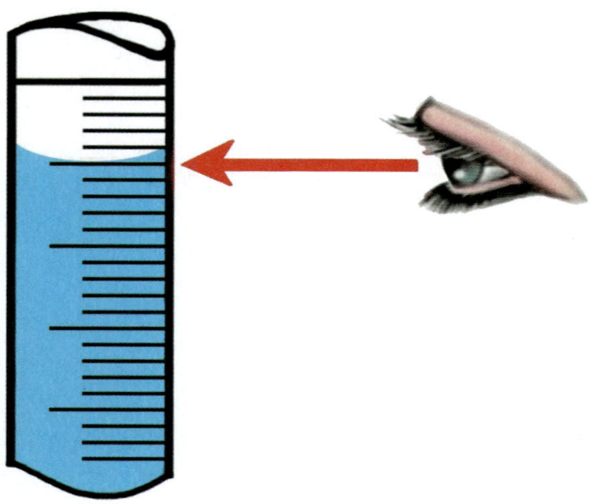

Figure 1 Read the volume from the bottom of the meniscus.

5.B.2. Tips for Measuring Mass

- Mass is usually measured in grams or kilograms.
- To use a triple-beam balance, place the object you are measuring on the pan. Adjust the masses on each beam, starting with the largest, until the pointer on the right side is level with the zero mark. Then add the values of each beam to find the mass.
- To use an electronic balance, place the object you are measuring on the platform and read the digital display. If you want to measure the mass of a substance without the mass of its container, place the empty container on the platform and re-zero the balance. This resets the scale to zero. You can then measure the mass of the substance added to the container.
- To measure the mass of a liquid, first measure the mass of a suitable container. Then measure the mass of the liquid and the container. Subtract the mass of the container from the mass of the liquid and the container.
- To measure the mass of a powder or crystals, first determine the mass of a sheet of paper (or weighing paper). Then place the sample on the paper and measure the mass of both. Subtract the mass of the paper from the mass of the sample and the paper.

5.C. Solving Numerical Problems Using the GRASS Method

In science, you may have to solve problems involving quantities (numbers), units, and mathematical equations. A good way to solve these problems is to use the GRASS method. **GRASS** stands for Given, Required, Analysis, Solution, and Statement.

Given: Read the problem carefully and list all of the values that are given. Remember to include units.

Required: Read the problem again to find the value that the question requires.

Analysis: Look for the relationship between the given values and the required value. Is there a mathematical equation you could use to find the required value?

If so, write the equation down. You can also draw a diagram to help you.

Solution: Use the equation from the Analysis step to solve the problem. Enter the given values into the equation to find the required value. Remember to include units and to round off your answer. (See Sections 5.D. and 5.E. on significant digits and scientific notation for help.)

Statement: Write a sentence to describe your answer to the question from the Required step. Sometimes two of these steps can be done together

SAMPLE PROBLEM 2 Determine the Density of Cooking Oil

An empty container has a mass of 50 g. When 75 mL of oil are placed in it, the total mass is 120 g. Calculate the total density of the oil.

Given: mass of container = 50 g
mass of container + oil = 120 g
volume of oil = 75 mL

Required: density of oil

Analysis: $\text{density} = \frac{\text{mass}}{\text{volume}}$

Solution: mass of oil = (mass of oil + container) − (mass of container)

$$\begin{aligned} &= 120\text{ g} - 50\text{ g} \\ &= 70\text{ g} \end{aligned}$$

$$\begin{aligned} \text{density} &= \frac{\text{mass}}{\text{volume}} \\ &= \frac{70\text{ g}}{75\text{ mL}} \\ &= \frac{0.93\text{ g}}{\text{mL}} \end{aligned}$$

Statement: The density of the oil is 0.9 g/mL

5.D. Scientific Notation

Scientists often work with very large or very small numbers. These numbers can be difficult to work with. For example, the speed of light is about 300 000 000 m/s. There are many zeros to keep track of if you have to multiply or divide this number. Sometimes it is possible to simplify a number by changing the SI prefix. For example, 237 000 000 mm can be converted to 237 km, and 0.000 895 kg can be expressed as 895 mg.

Scientific notation is a type of mathematical abbreviation that omits large numbers of zeros. In scientific notation, the number is written in the form $a \times 10^n$. The letter *a* is the first number of a value. It is a value that is at least 1 and less than 10. The number 10 in the formula never changes, and *n* is the exponent, or the number of times *a* will be multiplied by 10. For example, $10 \times 10 \times 10 \times 10$ is the same as 10^4. For negative exponents, we must also take into account 10^0, which is equal to 1, so 10^{-4} is the same as $1 \div 10 \div 10 \div 10 \div 10$. The 10 and the exponent are read as "10 to the power of *n*."

To write a large number in scientific notation, follow these steps:

1. To find the exponent, *n*, count the number of places you have to move the decimal point to the left to make a number between 1 and 10. For example, to write the speed of light (300 000 000 m/s) in scientific notation, you have to move the decimal point eight places to the left. In this case, $n = 8$.
2. After moving the decimal point, you also have the value for *a*. Now, using the speed of light example, we can remove the extra zeros after the decimal point so that $a = 3.0$.
3. Combine the value for *a* with the base, 10, and the exponent, *n*. For the above example, the speed of light is 3.0×10^8 m/s.

Scientific notation can also be used for very small numbers (less than 1). To find *n*, count the number of places you move the decimal point to the right. You will end up with a number between 1 and 10 for *a*. For very small numbers, *n* is always negative.

For example, a millionth of a second, 0.000 001 s, can be written in scientific notation as 1×10^{-6} s.

Table 4 shows several examples of large and small numbers in scientific notation.

To multiply numbers in scientific notation, multiply all of the *a* values together and add the *n* values. Remember that the final *a* value must be less than 10.

For example,

$$\begin{aligned}(3 \times 10^3)(5 \times 10^4) &= 15 \times 10^{3+4} \\ &= 15 \times 10^7 \\ &= 1.5 \times 10^8\end{aligned}$$

When dividing numbers in scientific notation, divide the *a* values and subtract the *n* values.

For example,

$$\begin{aligned}\frac{8 \times 10^6}{2 \times 10^4} &= 4 \times 10^{6-4} \\ &= 4 \times 10^2\end{aligned}$$

Table 4 Numbers Expressed in Scientific Notation

Large or small number	Common decimal notation	Scientific notation
124.5 million km	124 500 000 km	1.245×10^8 km
154 thousand nm	154 000 nm	1.54×10^5 nm
753 trillionths of a kg	0.000 000 000 753 kg	7.53×10^{-10} kg
315 billionths of a m	0.000 000 315 m	3.15×10^{-7} m

5.E. Uncertainty in Measurement

Numbers used in science are either exact values or measurements. Exact values do not change depending on how they are measured. These include SI values, such as 1 km = 1000 m, and others, such as 1 h = 60 min. Exact values also include counted values, such as 5 beakers or 10 cells. In other words, 1 km is exactly 1000 m, not 999.9 m or 1000.2 m. Similarly, 5 beakers cannot be 4.9 or 5.1 beakers; 5 beakers are exactly 5 beakers.

Measurements, however, are not always the same. The difference depends on the measuring instrument used and the skill of the measurer.

5.E.1. Significant Digits

The number of significant digits in a measurement indicates how certain we are about that measurement. In a measured or calculated value, significant digits are the digits that are certain, plus one estimated (uncertain) digit.

Significant digits include all the digits that are correctly reported from a measurement. For example, 10 different people reading the water volume in a graduated cylinder may all agree that the volume is at least 50 mL. However, they may not agree on the last digit. In other words, the "5" is certain, but there is some uncertainty in the last digit. The people might report the volume as being 56 mL, 57 mL, or 58 mL. The only way to know for sure is to measure with a more precise measuring device. Therefore, we say that a measurement such as 57 mL has two significant digits: one that is certain (5) and the other that is uncertain (7). The last digit in a measurement is always the uncertain digit. For example, 115.6 g contains three certain digits (115) and one uncertain digit (6).

Table 5 provides the guidelines for determining the number of significant digits, along with examples to illustrate each guideline.

Table 5 Guidelines for Determining Significant Digits

Guideline	Example	
	Number	Number of significant digits
Count from left to right, beginning with the first non-zero digit.	345 457.35	3 5
Zeros at the beginning of a number are never significant.	0.235 0.003	3 1
All non-zero digits in a number are significant.	1.1223 76.2	4 2
Zeros between digits are significant.	107.05 0.02094	5 4
Zeros at the end of a number with a decimal point are significant.	10.0 303.0	3 4
Zeros at the end of a number without a decimal point are ambiguous.	5 400 200 000	at least 2 at least 1
All digits in the coefficient of a number written in scientific notation are significant.	5.4×10^3 5.40×10^3 5.400×10^3	2 3 4

Rounding

Use these rules when rounding answers to the correct number of significant digits:

1. When the first digit removed is less than 5, the last digit kept (the one we want as the last significant digit) should not be changed.

   Example:

   3.141 326 rounded to four significant digits is 3.141.

2. When the first digit removed is greater than 5, or when it is 5 followed by at least one digit other than zero, the last digit kept is increased by one unit.

   Examples:

   2.221 372 rounded to five significant digits is 2.2214. 4.168 501 rounded to four significant digits is 4.169.

3. When the first digit removed is 5 followed by only zeros, the last digit kept is increased by 1 if it is odd but not changed if it is even.

   Note that when this rule is followed, the last digit in the final number is always even.

   Examples:

   2.35 rounded to two digits is 2.4
   2.45 rounded to two digits is 2.4
   6.75 rounded to two digits is 6.8

4. When adding or subtracting measured quantities, look for the measurement with the fewest digits after the decimal point. The answer cannot have more digits after the decimal point than this measurement has. In other words, the answer cannot be more precise than the least precise value.

   Example:

   $$\begin{array}{r} 12.52 \text{ g} \\ +\ 349.0 \text{ g} \\ +\ \ \ 8.24 \text{ g} \\ \hline 369.76 \text{ g} \end{array}$$

   Because 349.0 g has the fewest digits after the decimal point, the answer must be rounded to 369.8 g.

   Example:

   $$\begin{array}{r} 157.85 \text{ mL} \\ -\ 32.4 \text{ mL} \\ \hline 125.45 \text{ mL} \end{array}$$

   Because 32.4 mL has the fewest decimal places, the answer must be rounded to 125.4 mL. Note that Rule 3 also applies.

5. When multiplying or dividing, the answer must contain no more significant digits than the value with the fewest significant digits.

   Example:

   $$m = \frac{1.15 \text{ g}}{\cancel{\text{cm}^3}} \times 16 \ \cancel{\text{cm}^3} = 18 \text{ g}$$

   $$\Delta t = 1.25 \ \cancel{\text{h}} \times \frac{60 \text{ min}}{1 \ \cancel{\text{h}}} = 75.0 \text{ min}$$

   In other words, the answer cannot be more certain than the least certain value.

   Note, in the second example, 1.25 h is a measurement and, as a result, contains uncertainty. However, 60 min/h is an exact quantity. Since it has no uncertainty, the number of significant digits in the final answer is based only on the measured value of 1.25 h.

   Rule 5 also applies when multiplying or dividing measurements expressed in scientific notation. For example,

   Example:

   $$\begin{aligned}(3.5 \times 10^3 \text{ km})(7.4 \times 10^2 \text{ km}) &= 25.9 \times 10^5 \text{ km}^2 \\ &= 2.59 \times 10^6 \text{ km}^2 \\ &= 2.6 \times 10^6 \text{ km}^2\end{aligned}$$

5.E.2. Measurement Errors

There are two types of error that can occur when measurements are taken: random and systematic. Random error results when the last significant digit for a measurement needs to be estimated. The size of the random error is determined by the precision of the measuring instrument. For example, when measuring length, it is necessary to estimate between the marks on the ruler. If these marks are 1 cm apart, the random error is greater than if the marks were 1 mm apart (**Figure 2**).

Systematic error is caused by a problem with the way things are measured. This can happen in a number of ways, such as if the equipment is not set up correctly. For example, if a balance is not reset to zero, all the measurements taken with the balance will have a systematic error.

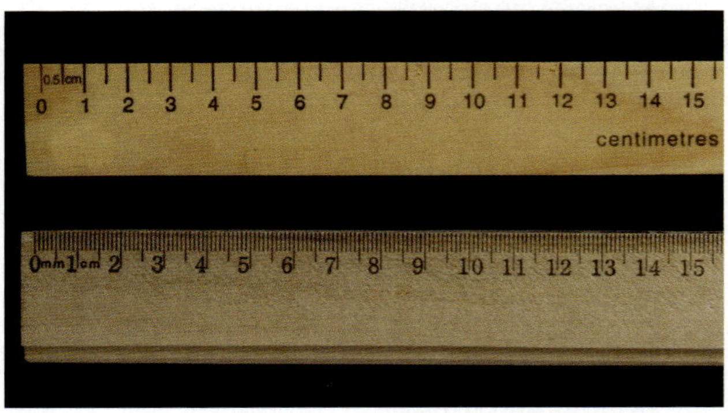

Figure 2 A ruler showing both centimetres and millimetres is more precise than one showing only centimetres.

Precision and Accuracy

The precision of measurements depends on the markings (gradations) of the measuring device. Precision is the place value of the last measurable digit. For example, a measurement of 12.74 cm is more precise than a measurement of 127.4 cm because 12.74 was measured to hundredths of a centimetre, whereas 127.4 was measured to tenths of a centimetre.

A precise measurement is not necessarily accurate. Accuracy refers to how close a value is to its accepted value. If our answers are accurate, we are likely to be correct. Being precise and accurate at the same time means that we are very close to the true answer. **Figure 3** uses the results of a horseshoe game to explain precision and accuracy.

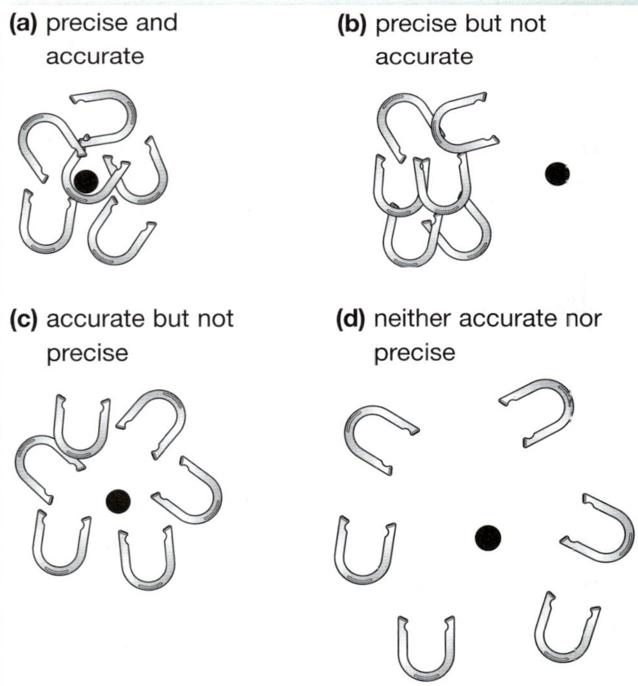

Figure 3 The patterns of horseshoes illustrate the comparison between accuracy and precision.

5.F. Using the Calculator

A calculator makes mathematical calculations easier, faster, and probably more accurate. However, like any other instrument, you need to learn how to use it. These guidelines apply to a basic scientific calculator. If your calculator is different, such as a graphing calculator, some of the instructions and operations may use different keys or sequences, so always check the manual.

General Points

- Most calculators follow the usual mathematical rules for order of operations: multiplication/division before addition/subtraction. For example, if you are calculating *y* using *y* = m*x* + *b*, you can enter the values of m times *x* plus *b* in one sequence. The calculator will "know" that m and *x* must be multiplied first before *b* is added.
- Calculators do not keep track of significant digits. For example, 12.0 is the same as 12 for a calculator.
- Some calculator keys such as [$\frac{1}{x}$], [+/−], [x^2], (and its second function, [$\sqrt{x}$]) apply the operation only to the value in the display regardless of other operations in progress. This means you can quickly change the sign of the number, convert to the reciprocal inverse, square the number, or determine the square root while inputting calculations.
- Do not clear all numbers from the calculator until you are completely finished with a question: the result of one calculation can be reused to start the next one.
- All scientific calculators have at least one memory location where you can store a number (M+ and STO are common keys) and recall it later (usually with MR or RCL). Use it to avoid re-entering many digits.

Multiplication and Division

- Division is the inverse of multiplication. This means that dividing by a number is the equivalent to multiplying by the inverse of the same number. For example, $\frac{12\text{ km}}{0.75\text{ h}}$ is the same as $12\text{ km} \times \frac{1}{0.75\text{ h}}$ and equals $16\ \frac{\text{km}}{\text{h}}$.

This is particularly useful when you want to divide by a number that is currently in the display of your calculator. For example, you have just finished converting 45 min to 0.75 h in your calculator and now you want to calculate the speed.

Display: 0.75

Press: [$\frac{1}{x}$] [×] [1] [2] [=]

New display: 16

- Brackets, ( ), are useful to force the calculator to perform the operation(s) inside the brackets first, before continuing with the calculation. The calculation of the slope of a line is a good example.

$$\begin{aligned}\text{slope} &= \frac{\Delta d}{\Delta t}\\ &= \frac{(15.2 - 4.1)\text{m}}{(6.5 - 3.6)\text{ s}}\\ &= 3.8\ \frac{\text{m}}{\text{s}}\end{aligned}$$

- If you do not use the brackets on your calculator, you will have to calculate the numerator and denominator separately and then divide.

Scientific Notation

On many calculators, scientific notation is entered using a special key, labelled EXP or EE. This key represents "×10" from the scientific notation equation, $a \times 10^n$. You need to enter only the exponent. For example, to enter

7.5×10^4 press [7] [.] [5] [EXP] [4]

3.6×10^{-3} press [3] [.] [6] [EXP] [+/−] [3]

Calculating a Mean

There are many ways of analyzing experimental evidence. One of the most common and important is calculating a mean (often called an average in everyday language). The mean of a set of values is the sum of all reasonable values divided by the total number of values.

Suppose you measure the root growth of five seedlings. Your root measurements after three days are 1.7 mm, 1.6 mm, 1.8 mm, 0.4 mm, and 1.6 mm. What is the mean root growth? Inspection of the measurements shows that the 0.4 mm measurement clearly does not fit with the rest. Perhaps this seedling was infected with a fungus, or some other problem occurred. You should not include this result in your mean but leave it in your evidence table so everyone can see the decision that you made. Using only reasonable values,

mean root growth

$$= \frac{1.7\ \text{mm} + 1.6\ \text{mm} + 1.8\ \text{mm} + 1.6\ \text{mm}}{4}$$

$$= 1.7\ \text{mm}$$

Means are important in all areas of science because multiple measurements or trials are widely used to increase the reliability of the results.

Equations

Algebra is a set of rules and procedures for working with mathematical equations. In general, your equations will contain one unknown quantity. Whatever mathematical operation is performed to one side of an equation must be performed to the other side. To solve for an unknown value, you need to isolate it on one side of the equal sign. To accomplish this, you should follow three rules:

1. The same quantity can be added or subtracted from both sides of the equation without changing the equality.

   The following example illustrates this rule:

   $$100\ \text{m} = 100\ \text{m}$$
   $$100\ \text{m} - 5\ \text{m} = 100\ \text{m} - 5\ \text{m}$$
   $$95\ \text{m} = 95\ \text{m}$$

   The example shows the application of this rule using quantities with numbers and units. The rule works equally well with quantity symbols.

   The example below shows how to isolate *x*:

   $$x + b = y$$
   $$x + b - b = y - b$$
   $$x = y - b$$

2. The same quantity can be multiplied or divided on both sides of the equation without changing the equality. The following examples illustrate this rule. The example below shows the use of this rule with known quantities.

$$120\ \text{m} = 120\ \text{m}$$

$$\frac{120\ \text{m}}{8.0\ \text{s}} = \frac{120\ \text{m}}{8.0\ \text{s}}$$

$$\frac{15\ \text{m}}{\text{s}} = \frac{15\ \text{m}}{\text{s}}$$

Use the same rule with an equation containing quantity symbols to isolate one of the quantities. To solve for *d* in the example below, multiply both sides by *t*. Notice that *t* divided by *t* equals 1. Multiplying or dividing any quantity by 1 does not change the quantity; therefore, $d \times 1 = d$.

$$v = \frac{d}{t}$$

$$v \times t = \frac{d}{\not{t}} \times \not{t}$$

$$vt = d \text{ or } d = vt$$

3. The same power (such as square or square root) can be applied to both sides of the equation without changing the equality.

The following example illustrates this rule using known quantities:

$$25\ \text{s}^2 = 25\ \text{s}^2$$

$$\sqrt{25\ \text{s}^2} = \sqrt{25\ \text{s}^2}$$

$$5.0\ \text{s} = 5.0\ \text{s}$$

Here is another example using quantity symbols:

$$b^2 = A$$

$$\sqrt{b^2} = \sqrt{A}$$

$$b = \sqrt{A}$$

If several of the rules listed above are required to isolate an unknown quantity, then you should apply Rule 1 first, whenever possible, and then apply Rule 2. In general, Rule 3 should be used last.

Data Tables and Graphs

Data tables are an effective means of recording both qualitative and quantitative observations. Making a data table should be one of your first steps as you prepare to conduct an investigation. Doing so will make recording data very easy, especially when there is an independent variable (the cause) and dependent variables (the effects), as shown in **Table 1**.

Table 1 A Running White-Tailed Deer

Time (s)	Distance (m)
0	0
1.0	13
2.0	25
3.0	40
4.0	51
5.0	66
6.0	78

Follow these guidelines to make a data table:

- Use a ruler to make your table.
- Write a title that clearly describes what is being measured.
- Include the units of measurement for each variable, when appropriate.
- List the values of the independent variable in the left column.
- List the values of the dependent variable(s) in the column(s) to the right.

6.A. Graphing Data

A graph shows groups of data in a single diagram. Graphing data often makes it easier to see patterns. If there is a pattern, there may be a relationship between the data. Three of the most useful kinds are bar graphs, circle graphs, and point-and-line graphs. Each kind has its own special uses. You need to decide which type of graph is best for the data you have collected.

Bar Graphs

A bar graph helps you compare data when one variable is in numbers (for example, amount of rainfall) and the other variable is not (for example, month of the year). **Figure 1** shows a bar graph with the distribution of rainfall over a 12-month period in Ottawa. Each bar stands for a different month. This graph clearly shows that rainfall is higher in the summer months and lower in the winter months.

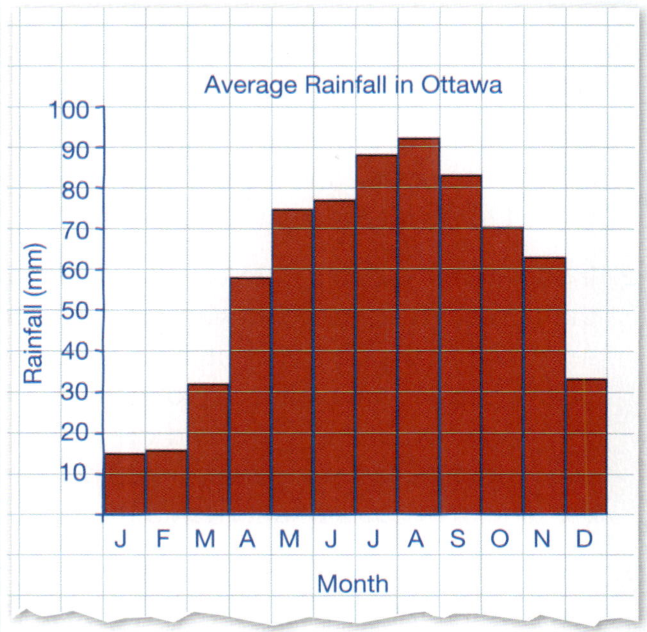

Figure 1 A bar graph

Circle Graphs

Circle graphs (pie charts) are useful for showing how the whole of something is divided into parts. The circle graph in **Figure 2** shows that transportation is the largest source of greenhouse gas emissions by individuals.

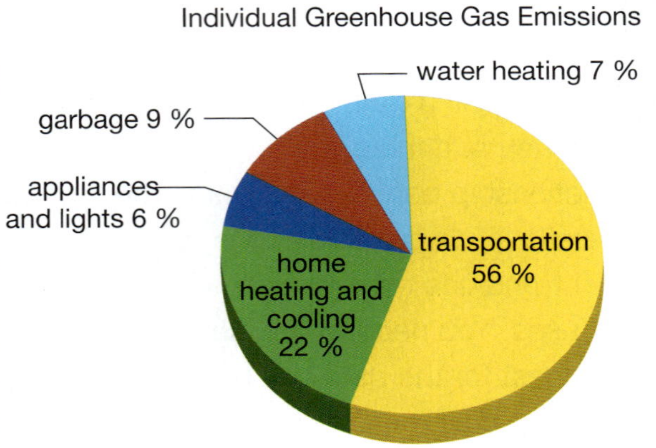

Figure 2 A circle graph

Point-and-Line Graphs

When both variables are quantitative, use a point-and-line graph. This format shows all the data points and a line of best fit indicating any relationship between the variables. For example, we can use the data in **Table 1** on the previous page to construct the line graph in **Figure 3** following the rules below.

MAKING POINT-AND-LINE GRAPHS

1. Use graph paper. Draw a horizontal line close to the bottom of the paper for the *x*-axis. Then draw a vertical line close to the left edge for the *y*-axis.
2. Independent variable data usually go along the *x*-axis. The dependent variable data go along the *y*-axis. The exception is when one variable is time: always plot time on the *x*-axis. Label the *x*-axis and the *y*-axis with the variables being measured. Include the units of measurement beside the label.
3. Give your graph a title: a short, clear description of the data shown on the graph.
4. Find the highest and the lowest values for both variables (the range). Graphs often include a little extra room on each axis.
5. Choose a scale for each axis. The scale shows how the values on each axis increase or decrease. The scale will depend on space and the range of values for each axis. Numbers on a scale usually increase evenly.
6. Use the collected data to plot each point. In **Figure 3**, the first point is 0 on the *x*-axis and 0 on the *y*-axis.

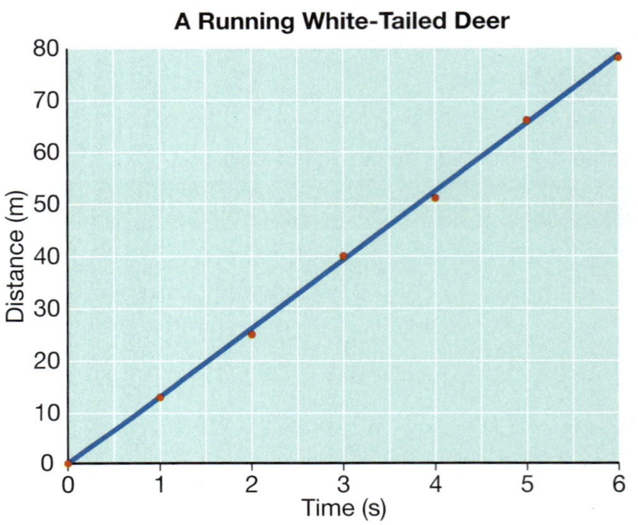

Figure 3 A point-and-line graph of the data from Table 1

7. After all the points are plotted, imagine a line going through the points. Draw the line of best fit—a straight or smoothly curving line through the points so that there is approximately the same number of points on each side of the line. The line's purpose is to show the overall pattern of the data.
8. If you are plotting more than one set of data on the same graph, use different colours or symbols for each. Provide a legend.

6.B. Using Computers for Graphing

You can use spreadsheet or graphing programs on your computer to make bar, circle, and point-and-line graphs. In addition, these programs can calculate the line of best fit. The following instructions explain how to enter values for best-fit line graphs on a spreadsheet.

Steps for Creating a Graph in a Spreadsheet Program

1. Start the spreadsheet program. Enter the data with the values for *x* (the first column in your data table). The column should start in cell A2. Now enter the values for *y* in the next column, starting at B2.
2. Highlight all of the data in the two columns. From the toolbar, select the button for creating graphs. The cursor may change to a symbol, which allows you to "click and drag" to choose the size of your graph.
3. You will have a number of choices now, so you can choose what kind of graph you want to create. A scatter graph is most appropriate for our data.
   - Click in the "Title" box and type the title of the graph.
   - Click in "Value (X)" box and type the label and units for the *x*-axis.
   - Click in "Value (Y)" box and type the label and units for the *y*-axis.
   - Click the "Finish" button.
4. To add a line of best fit, point your cursor to one of the highlighted data points and right click once. Choose the "Trendline" option. Make sure that the "Linear" box is highlighted under "Type."
5. Save your spreadsheet before you close the program.

6.C. Interpreting Graphs

When data from an investigation are shown on a graph, patterns and relationships are easier to see. You can more easily tell if the data support your hypothesis. Looking at the data in the graph may also lead you to a new hypothesis. Or you can learn more from other people's investigations.

What to Look for When Reading Graphs

Here are some questions to help you interpret a graph:

- What variables are represented?
- What is the dependent variable? What is the independent variable?
- Are the variables quantitative or qualitative?
- If the data are quantitative, what are the units of measurement?
- What do the highest and lowest values represent on the graph?
- What is the range between the highest and the lowest values on each axis?
- Are the axes continuous? Do they start at zero?
- What patterns are there between the variables?
- If there is a line of best fit, what might the line tell you?

Using Graphs for Predicting

If a graph shows a regular pattern, you can use it to make predictions. For example, you could use the graph in **Figure 3** on page 584 to predict the distance travelled by the deer in 8.0 s. To do this, you extrapolate the graph (extend it beyond the measured points), assuming that the observed trend would continue. You should be careful when predicting values outside of your measured range. If you are far from the known values, your prediction will not be reliable.

7

Study Skills

7.A. Working Together

Teamwork is just as important in science as it is on the playing field or in the gym. Scientific investigations are usually carried out by teams of people working together. Ideas are shared, experiments are designed, data are analyzed, and results are evaluated and shared with other investigators. Group work is necessary and is usually more productive than working alone.

7.A.1. General Guidelines for Effective Teamwork

- Keep an open mind. Everyone's ideas deserve consideration.
- Divide the task among all the group members. Choose a role best suited to your particular strengths.
- Work together and take turns. Encourage, listen, clarify, help, and trust one another.
- Remember that the success of the team is everyone's responsibility. Every member needs to be able to demonstrate what the team has learned and support the team's final decision.

7.A.2. Exchanging Information Orally

You will be involved in a number of different activities during your science course. These activities help you express your ideas and learn about new ones. Here are three of the more successful discussion formats that your group may use to share ideas:

- In a think-pair-share activity, you and your partner are given a problem. Each of you develops a response (usually within a time limit). Then you share your ideas with each other to resolve the problem. You may also be asked to share your results with a larger group or the class.
- In a jigsaw activity, you are an active member in two teams: your home team and your expert team. Each member of the home team chooses a particular area of research. Each home team member then meets with an expert team in which everyone is working on the same area of research. In your expert team, you may work together to come up with answers to questions in your area of research. Once you have accomplished your task as a team, you return to your home team, and it is your responsibility to teach what you have learned to the members of your home team. Each person on the home team will do the same thing.
- A round table activity can be used to give your group an opportunity to review what they know. Your group is given a pen, paper, and a question or questions. Pass the pen and paper around and take turns writing one line of the solution. You can pass on your turn if you wish. Keep working until the solution is complete. Check to ensure that everyone understands the solution. Finally, working as a team, review the steps to the solution.

7.A.3. Investigations and Activities

This kind of work is best done in small groups.

- Each group member agrees to take part doing a job.
- Take turns doing the work during similar and repeated activities.
- Safety must come first. Be aware of where other group members are and what they are doing.
- Take responsibility for your own learning by making notes and contributing to any discussions about the activities. Make your own observations and compare them with the observations of other group members (**Figure 1**).

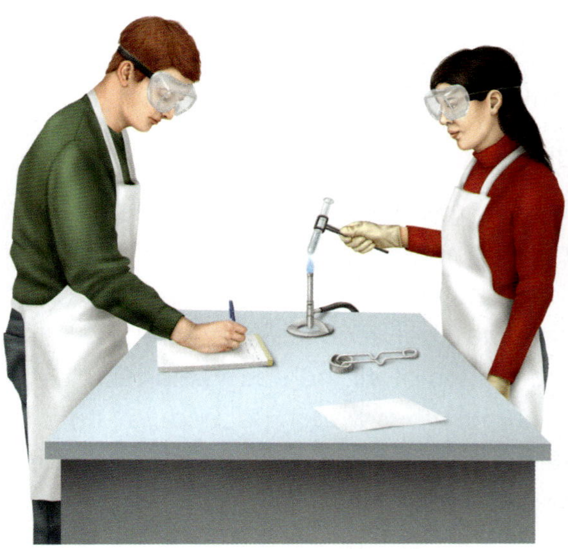

Figure 1 When you are working with one student or several, you can share and compare your results.

7.A.4. Explore an Issue

Follow these guidelines when you are conducting research with a group:

- Divide the topic into several parts and assign one part to each group member.
- Keep records of the sources used by each group member.
- Decide on a format for exchanging information (such as photocopies of notes, oral discussion, electronic files).
- Each group member should contribute to the decision-making process. Make decisions by compromise and consensus.
- Communicating your position should also be a group effort.

7.A.5. Evaluating Teamwork

After you have completed a task with your group, evaluate your team's efforts. Ask yourself the following questions:

- What were the strengths of your teamwork?
- What were the weaknesses of your teamwork?
- What opportunities were provided by working with your group?
- What challenges did you encounter as a member of a group?

7.B. Setting Goals and Monitoring Progress

Think back to your last school year. What classes did you do well in? Why do you think you were successful? In which classes did you have difficulty? What could you do differently this year to improve your performance? Things that you want to accomplish today, this week, and this year are all called goals. Learning to set goals and to make a plan to achieve them takes skill, patience, and practice.

7.B.1. Setting Goals

Assessing Your Strengths and Weaknesses

The process of setting goals starts with honest reflection. Do you do better on projects than tests? Do you do better or worse under pressure? Do you pay attention in class and have good study habits?

Realistic Goals That You Can Measure

Do not set yourself up to fail by setting goals that you cannot possibly achieve. Saying, "I will have the best mark in the class at the end of the semester" may not be realistic. You will find it easier to reach your goals if you can tell whether you are getting closer to them. A goal to increase your test marks by 10 % this semester is easy to measure. When you are thinking of setting goals, remember the acronym SMART: Specific, Measurable, Attainable, Realistic, and Time-limited.

Share Your Goals

People whom you respect can often help you set and clarify goals. Sharing your goals with a trusted friend or adult will often provide needed support to help you reach your goals.

Planning to Meet Your Goals

Once you have made a list of realistic goals, create a plan to achieve them. A successful plan usually consists of two parts: an action plan and target dates.

THE ACTION PLAN

First, make a list of the actions or behaviours that might help you reach your goals (**Figure 2**).

If you have made an honest assessment of your strengths and weaknesses, then you know what you have to do to improve. To improve your test marks, you could work with others to prepare for tests. You could also use a weekly planner or reorganize your study area at home.

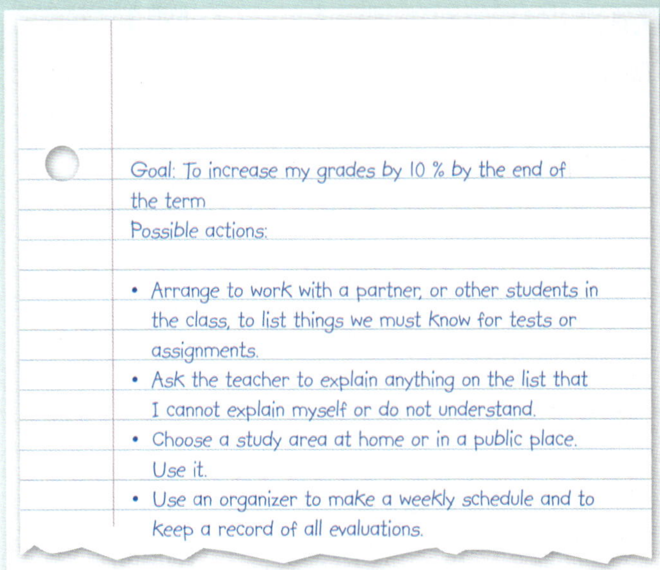

Figure 2 One student's goal and action plan to improve test results.

Identify what is preventing you from achieving your goal. Think of ways to overcome these obstacles. Ask friends for tips.

SETTING TARGET DATES

Suppose you want to improve test results by 10 % by the end of the semester. How much time do you have? How many tests are scheduled between now and then? Write down the dates of all the upcoming tests. This becomes your working schedule and short-term goal. **Figure 3** shows an example of a working schedule.

Once you complete your schedule in your planner, transfer it to a calendar in your study area. Refer to either your planner or your calendar every day.

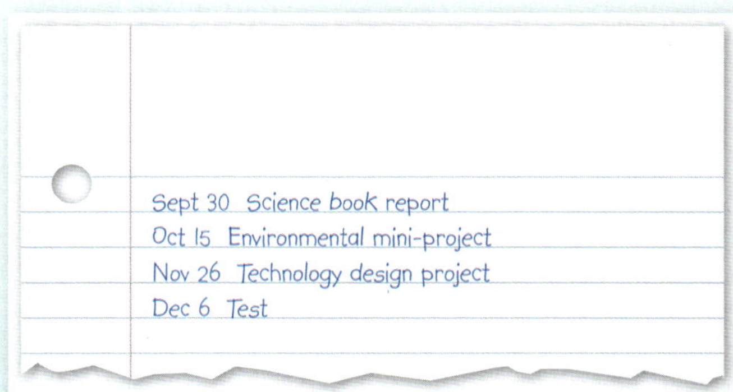

Figure 3 Test dates

7.B.2. Monitoring Progress

Remember to measure your progress along the way. Monitor your achievements after each test. Did the results meet your short-term target to improve test results by 10 % by the end of the semester? If you are not on track to meet your goal, then you may need to change your plan. Perhaps you need to study by yourself or with one friend rather than with a group of friends. It is always possible to change your plan or even adjust your goal. The most important thing is to keep moving forward and to remain committed to improvement.

7.C. Good Study Habits

Studying takes many forms. Developing good study skills can help you study and learn more successfully. Below are some tips to help you achieve better study habits.

7.C.1. Your Study Space

- *Organize your work area*. The place where you study should be tidy and organized. Place all papers, books, magazines, and pictures in appropriate areas of your study area. This will make it easier for you to focus on your school work.
- *Maintain a quiet work area*. Where possible, make sure that your work area is free from distractions—telephone, music, television, and family members. If there are too many distractions at home, find an appropriate space at the school or the public library. Any quiet space, free from interruptions, can be a productive work area.
- *Make sure that you are comfortable in your work area*. If possible, personalize your work area. Make sure that the light is right for your needs. You should have a comfortable chair and a table at the right height.
- *Be prepared—bring everything you need*. Have all the necessary materials you will need at your fingertips. Keep pens, pencils, notebooks, or textbooks in one place near your computer or on your desk. Continually getting up to find something you need decreases your ability to stay on task and be productive.

7.C.2. Study Habits

- *Take notes*. Take notes during class. Outside class, review the appropriate section of the textbook. Read or view additional material on the topic from other sources, such as newspapers, magazines, the Internet, and television. Ask a friend to share notes with you.
- *Use graphic organizers*. You can use a variety of graphic organizers to help you summarize a concept or unit (page 591). They also help you connect different concepts more easily.
- *Schedule your study time*. Use a daily planner and take it with you to class. Write any homework assignments, tests, or projects in it. Use it to create a daily "to do" list. This will help you complete work to hand in and avoid last-minute panic.
- *Take study breaks*. It is important to schedule breaks into your study time. For example, you could take a study break after completing one or two items on your "to do" list. Taking breaks allows you to relax and to recharge your brain so that you can keep focused when you return to your assignments.

8 Literacy

8.A. Reading Strategies

Reading a science book is different from reading a novel because you are reading for information. Here are some strategies to help you read for information.

8.A.1. Before Reading

Skim the section you are going to read. Look at the illustrations, headings, and subheadings.

- *Preview*. What is this section about? How is it organized?
- *Make connections*. What do I already know about the topic? How is it connected to other topics I already know about?
- *Predict*. What information will I find in this section? Which parts provide the most information?
- *Set a purpose*. What questions do I have about the topic?

8.A.2. During Reading

Pause and think as you read. Spend time on the photographs, illustrations, tables, and graphs, as well as on the words.

- *Check your understanding*. What are the main ideas in this section? How would I state them in my own words? What questions do I still have? Should I reread?
- *Determine the meanings of key science terms*. Can I figure out the meanings of terms from context clues in the words or illustrations? Do I understand the definitions of terms in bold type? Which terms should I look up in the glossary?
- *Make inferences*. What conclusions can I make from what I am reading?
- *Visualize*. What pictures can I make to help me understand and remember what I am reading?
- *Make connections*. How is the information in this section like information I already know?
- *Interpret visuals and graphics*. What additional information can I get from the photographs, illustrations, tables, or graphs?

8.A.3. After Reading

This textbook provides summaries and questions at the ends of sections. These will help you check your understanding and make connections to what you have just read.

At the end of each chapter are a summary of Key Concepts and a Vocabulary list, followed by a Chapter Self-Quiz and Chapter Review.

- *Locate needed information*. Where can I find the information I need to answer the questions? Under what heading might I find the information? What details do I need to include in my answers?
- *Synthesize*. How can I organize the information? What graphic organizer could I use?
- *React*. What are my opinions about this information? Do other students agree with my reactions? Why or why not?
- *Evaluate information*. What do I know now that I did not know before? Have any of my ideas changed because of what I have read? What questions do I still have?

8.B. Graphic Organizers

Diagrams that are used to organize and display ideas visually are called graphic organizers. Graphic organizers are especially useful in science and technology studies when you are trying to connect together different concepts, ideas, and data. Different graphic organizers have different purposes. They can be used to

- show processes
- organize ideas and thinking
- compare and contrast
- show properties or characteristics
- review words and terms
- collaborate and share ideas

To Show Processes

Graphic organizers can show the stages in a process (**Figure 1**).

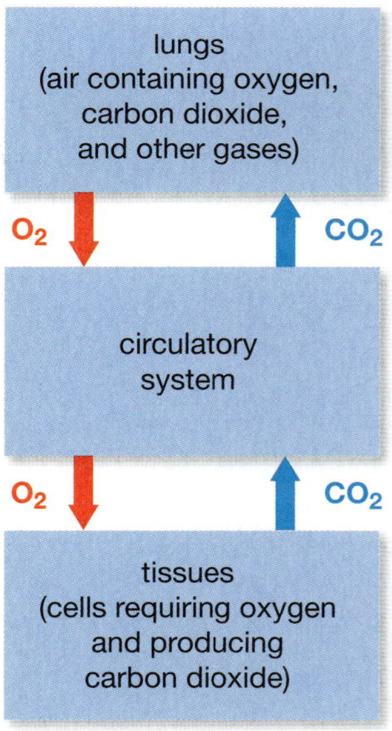

Figure 1 This organizer shows that oxygen and carbon dioxide are transported around the body.

To Organize Ideas and Thinking

A concept map is a diagram showing the relationships among ideas (**Figure 2**). Words or pictures are connected by arrows. You can use a concept map to brainstorm, to map your thinking, or to summarize what you have learned.

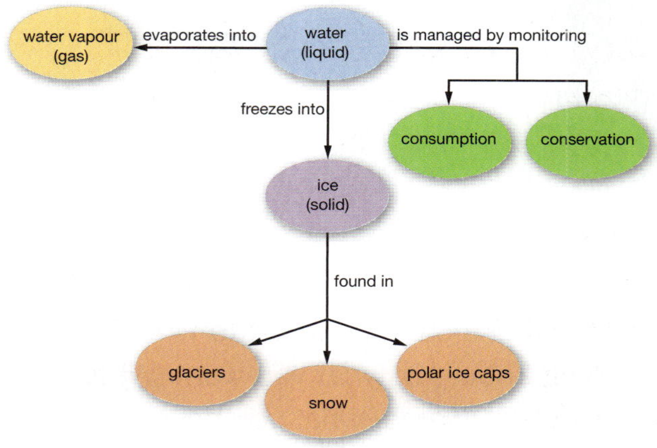

Figure 2 Concept maps help show the relationships among ideas.

Mind maps are similar to concept maps, but they do not have explanations for the connections between ideas. You can use a tree diagram to show concepts that can be broken down into smaller categories (**Figure 3**).

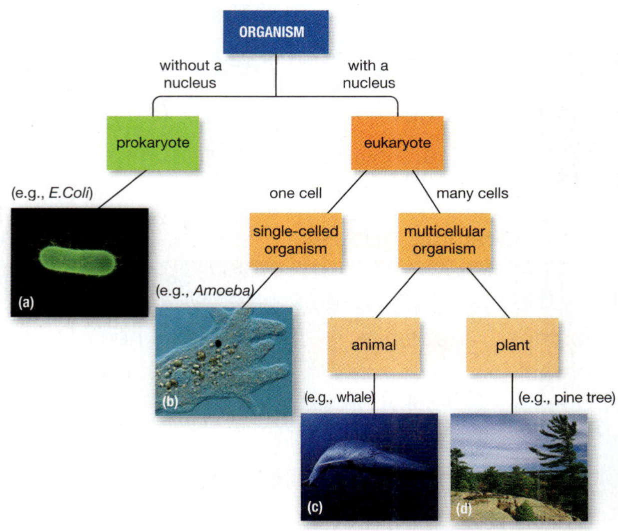

Figure 3 Tree diagrams are very useful for classification.

You can use a fishbone diagram to organize the important ideas under the major concepts of a topic you are studying (**Figure 4**).

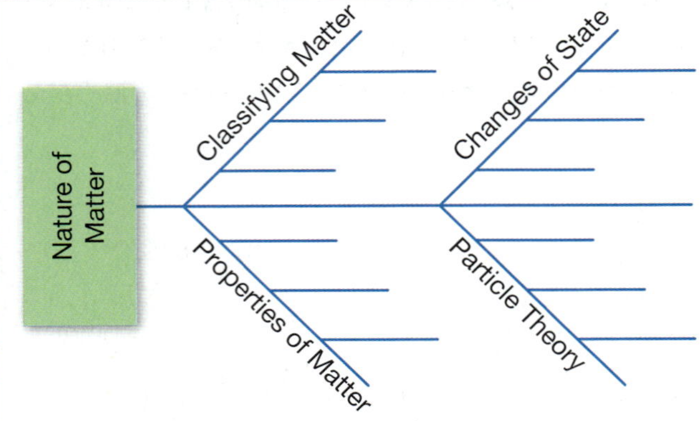

Figure 4 A fishbone diagram

You can use a K-W-L chart to write down what you know (K), what you want (W) to find out, and, afterwards, what you have learned (L) (**Figure 5**).

What do we **Know**?	What do we **Want** to find out?	What did we **Learn**?
Carbon dioxide is a greenhouse gas.	Are there any other important greenhouse gases? If so, where do they come from?	Methane, water, and nitrous oxide are other common greenhouse gases. Methane is released from decaying organic matter and from the digestive tracts of grazing animals. Nitrous oxide is released in automobile emissions.
The greenhouse effect traps solar energy in the atmosphere.	If the energy of the Sun can get through the atmosphere and warm Earth's surface, how is the energy trapped by greenhouse gases?	The atmosphere is transparent to light, allowing light rays from the Sun to strike Earth's surface. This energy is absorbed and then released as infrared waves. Because the atmosphere is not transparent to infrared waves, energy is trapped in the atmosphere and warms Earth.

Figure 5 A K-W-L chart

To Compare and Contrast

You can use a comparison matrix (a type of table) to compare related concepts (**Table 1**).

Table 1 Subatomic Particles

	Protons	Neutrons	Electrons
electrical charge	positive	neutral	negative
symbol	p^+	n°	e^-
location	nucleus	nucleus	orbit around the nucleus

You can use a Venn diagram to show similarities and differences (**Figure 6**).

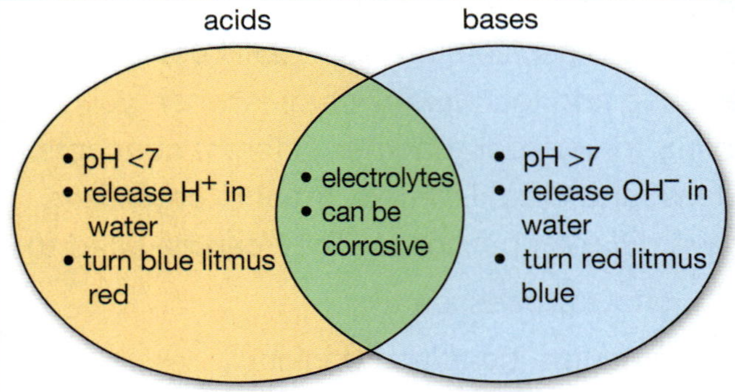

Figure 6 A Venn diagram

You can use a compare-and-contrast chart to show similarities and differences between two substances, actions, ideas, and so on (**Figure 7**).

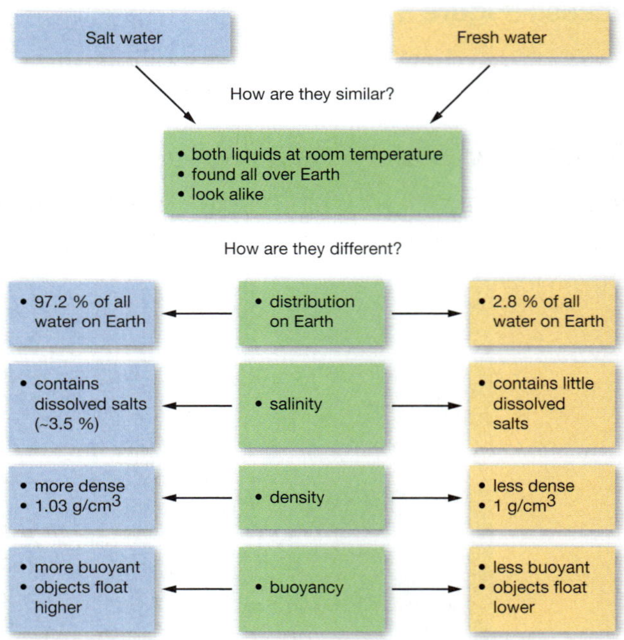

Figure 7 A compare-and-contrast chart

To Show Properties or Characteristics

You can use a bubble map to show properties or characteristics (**Figure 8**).

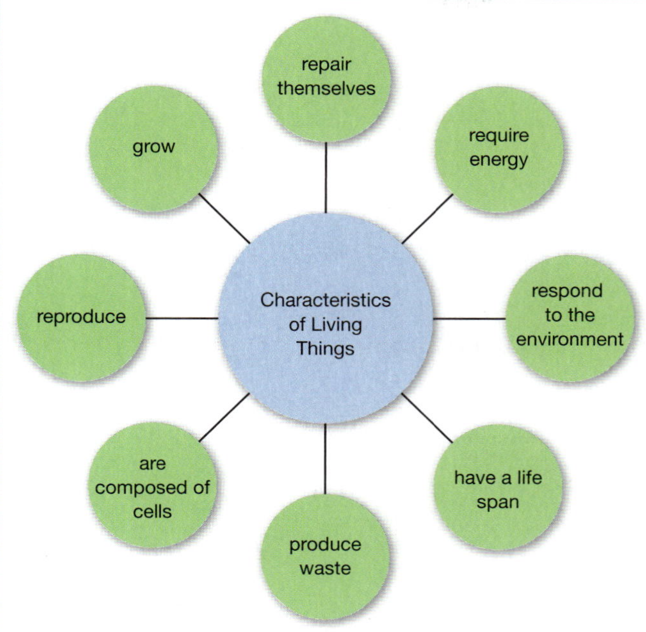

Figure 8 A bubble map

To Review Words and Terms

You can use a word wall to list, in no particular order, the keywords and concepts for a topic (**Figure 9**).

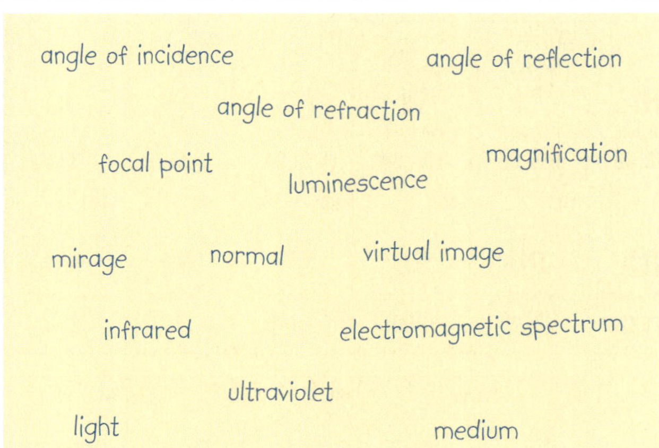

Figure 9 A word wall

To Collaborate and Share Ideas

A placemat organizer gives students in a small group a space to write down what they know about a certain topic. Then group members discuss their answers and write in the middle section what they have in common (**Figure 10**).

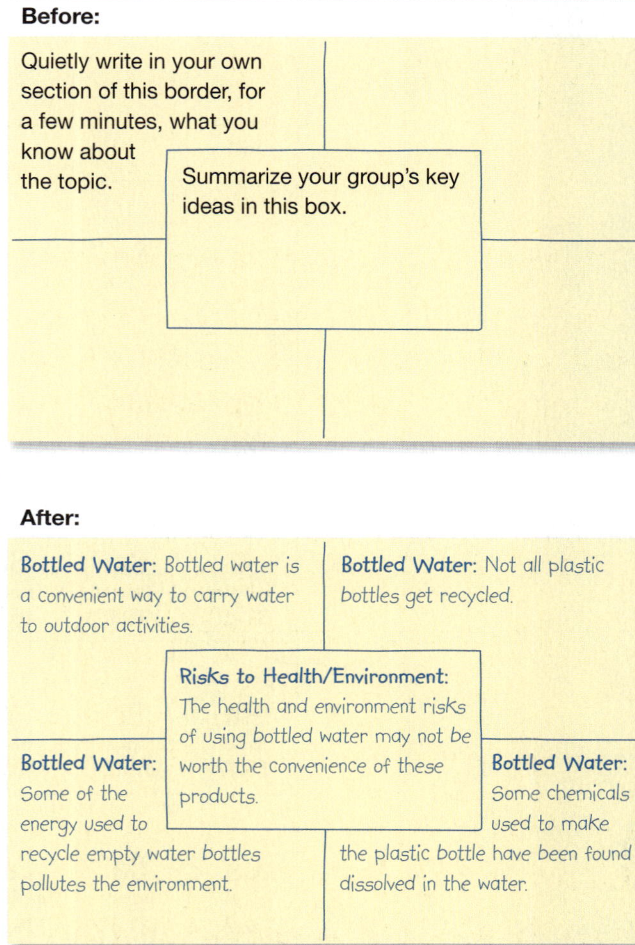

Figure 10 A placement organizer

9

Periodic Table

Nelson Science Connections

Key

- atomic number → 26
- 3+ ← most common ion charge
- 2+ ← other ion charge
- symbol of element → Fe (solids in black, gases in red, liquids in blue)
- iron ← name of element
- 55.85 ↑ atomic mass (u)—based on C-12

Period	1	2	3	4	5	6	7	8	9
1	1 (1+, 1–) **H** hydrogen 1.01								
2	3 (1+) **Li** lithium 6.94	4 (2+) **Be** beryllium 9.01							
3	11 (1+) **Na** sodium 22.99	12 (2+) **Mg** magnesium 24.31							
4	19 (1+) **K** potassium 39.10	20 (2+) **Ca** calcium 40.08	21 (3+) **Sc** scandium 44.96	22 (4+, 3+) **Ti** titanium 47.87	23 (5+, 4+) **V** vanadium 50.94	24 (3+, 2+) **Cr** chromium 52.00	25 (2+, 4+) **Mn** manganese 54.94	26 (3+, 2+) **Fe** iron 55.85	27 (2+, 3+) **Co** cobalt 58.93
5	37 (1+) **Rb** rubidium 85.47	38 (2+) **Sr** strontium 87.62	39 (3+) **Y** yttrium 88.91	40 (4+) **Zr** zirconium 91.22	41 (5+, 3+) **Nb** niobium 92.91	42 (6+) **Mo** molybdenum 95.94	43 (7+) **Tc** technetium (98)	44 (3+, 4+) **Ru** ruthenium 101.07	45 (3+) **Rh** rhodium 102.91
6	55 (1+) **Cs** cesium 132.91	56 (2+) **Ba** barium 137.33	57 (3+, 2+) **La** lanthanum 138.91	72 (4+) **Hf** hafnium 178.49	73 (5+) **Ta** tantalum 180.95	74 (6+) **W** tungsten 183.84	75 (7+) **Re** rhenium 186.21	76 (4+) **Os** osmium 190.23	77 (4+) **Ir** iridium 192.22
7	87 (1+) **Fr** francium (223)	88 (2+) **Ra** radium (226)	89 (3+, 2+) **Ac** actinium (227)	104 **Rf** rutherfordium (261)	105 **Db** dubnium (262)	106 **Sg** seaborgium (266)	107 **Bh** bohrium (264)	108 **Hs** hassium (277)	109 **Mt** meitnerium (268)
	Alkali metals	Alkaline earth metals							

Period					
6	58 (3+) **Ce** cerium 140.12	59 (3+) **Pr** praseodymium 140.91	60 (3+) **Nd** neodymium 144.24	61 (3+) **Pm** promethium (145)	62 (3+, 2+) **Sm** samarium 150.36
7	90 (4+) **Th** thorium 232.04	91 (5+, 4+) **Pa** protactinium 231.04	92 (6+, 4+) **U** uranium 238.03	93 (5+) **Np** neptunium (237)	94 (4+, 6+) **Pu** plutonium (244)

- Metals
- Metalloids
- Nonmetals
- Hydrogen

Periodic Table of the Elements

Measured values are subject to change as experimental techniques improve. Atomic molar mass values in this table are based on IUPAC Web site values (2005).

10	11	12	13	14	15	16	17	18	Period
								2 — He helium 4.00	1
			5 — B boron 10.81	6 — C carbon 12.01	7 3− N nitrogen 14.01	8 2− O oxygen 16.00	9 1− F fluorine 19.00	10 — Ne neon 20.18	2
			13 3+ Al aluminium 26.98	14 — Si silicon 28.09	15 3− P phosphorus 30.97	16 2− S sulfur 32.07	17 1− Cl chlorine 35.45	18 — Ar argon 39.95	3
28 2+ 3+ Ni nickel 58.69	29 2+ 1+ Cu copper 63.55	30 2+ Zn zinc 65.41	31 3+ Ga gallium 69.72	32 4+ Ge germanium 72.64	33 3− As arsenic 74.92	34 2− Se selenium 78.96	35 1− Br bromine 79.90	36 — Kr krypton 83.80	4
46 2+ 3+ Pd palladium 106.42	47 1+ Ag silver 107.87	48 2+ Cd cadmium 112.41	49 3+ In indium 114.82	50 4+ 2+ Sn tin 118.71	51 3+ 5+ Sb antimony 121.76	52 2− Te tellurium 127.60	53 1− I iodine 126.90	54 — Xe xenon 131.29	5
78 4+ 2+ Pt platinum 195.08	79 3+ 1+ Au gold 196.97	80 2+ 1+ Hg mercury 200.59	81 1+ 3+ Tl thallium 204.38	82 2+ 4+ Pb lead 207.2	83 3+ 5+ Bi bismuth 208.98	84 2+ 4+ Po polonium (209)	85 1− At astatine (210)	86 — Rn radon (222)	6
110 Ds darmstadtium (281)	111 Rg roentgenium (272)	112 Uub ununbium (285)	113 Uut ununtrium (284)	114 Uuq ununquadium (289)	115 Uup ununpentium (288)	116 Uuh ununhexium (291)	117 Uus ununseptium	118 Uuo ununoctium (294)	7
							Halogens	Noble gases	

63 3+ 2+ Eu europium 151.96	64 3+ Gd gadolinium 157.25	65 3+ Tb terbium 158.93	66 3+ Dy dysprosium 162.50	67 3+ Ho holmium 164.93	68 3+ Er erbium 167.26	69 3+ Tm thulium 168.93	70 3+ 2+ Yb ytterbium 173.04	71 2+ Lu lutetium 174.97	6
95 3+ 4+ Am americium (243)	96 3+ Cm curium (247)	97 3+ 4+ Bk berkelium (247)	98 3+ Cf californium (251)	99 3+ Es einsteinium (252)	100 3+ Fm fermium (257)	101 2+ 3+ Md mendelevium (258)	102 2+ 3+ No nobelium (259)	103 3+ Lr lawrencium (262)	7

Numerical and Short Answers

This section includes numerical and short answers to questions in Check Your Learning, Chapter Self-Quiz, Chapter Review, Unit Self-Quiz, and Unit Review.

Unit B

Section 2.1, p. 29

1. air; moderates temperature; protects from the Sun

Section 2.2, p. 33

1. abiotic; abiotic; abiotic; biotic; biotic; abiotic, biotic; biotic; abiotic; abiotic
2. only biotic; same species only

Section 2.3, p. 35

1. plant new trees; set limits
3. (a) and (d)

Section 2.5, p. 41

1. 0.023 %; 70 % absorbed, rest reflected
3. sugar, oxygen; for energy
4. (a) photosynthesis
   (b) eating plants
6. (a) eat plants and animals
   (b) dogs, worms, cows, eagles, frogs

Section 2.7, p. 51

3. (a) burning fossil fuels
   (b) more carbon dioxide

Section 2.8, p. 59

1. (a) narrow beak
4. it may be harmless
6. (a) commensalism
   (b) mutualism
   (c) parasitism
   (d) competition
   (e) prey and predator

Section 2.9, p. 64

1. freshwater and salt water ecosystems
3. pollutants enter water cycle

Chapter 2 Self-Quiz, pp. 70–71

1. (c)
2. (d)
3. (a)
4. (c)
5. (d)
6. (c)
7. F
8. F
9. T
10. F
11. F
12. T
13. F
14. F
15. T
16. photosynthesis
17. cellular respiration
18. community
19. population
20. competition
21. limiting
22. (a) (ii)
    (b) (iii)
    (c) (iv)
    (d) (i)
23. consumers; omnivores
24. (a) biting insects and raccoons
    (b) sunlight and rain
    (c) sunblock; bug repellent; raincoat
26. (a) precipitation
    (b) soil fertility
    (c) trees and squirrels
    (d) growing season
27. (a) all consumers but grass
    (b) more snakes
    (c) hawks and snakes decrease, grasshoppers increase
29. (a) greater food supply
    (b) fewer prey species

Chapter 2 Review, pp. 72–73

1. (b)
2. (c)
3. (c)
4. (b)
5. (b)
6. (c)
7. (a)
8. (d)
9. burning fossil fuels
11. energy for living things
13. (a) (i) predation
    (ii) parasitism
15. gulls; mussels
16. all populations increase
17. (a) gulls eat more lobster
    (b) fewer lobsters; more small fish
18. no shared food source
20. (a) boreal forest
    (b) tundra

Section 3.1, p. 81

1. recreation; oxygen; jobs
2. litter, start fires, pollution
4. (a) recreation
   (b) maple syrup
   (c) forestry and ecotourism

Section 3.2, p. 87

1. biotic and abiotic in balance
2. natural disasters; human activities
6. seed release; more sunlight

Section 3.3, p. 89

1. variety of life in an ecosystem
2. near the equator
4. flowers will not reproduce

Section 3.5, p. 95

1. human activity
3. more people live there
7. many unique species

8. deforestation, strip mining, chemical pollution
9. (a) Committee on the Status of Endangered Wildlife in Canada
   (b) monitors status of species
   (c) on the basis of risk

Section 3.6, p. 100

1. less food, water, shelter
2. building roads and buildings; clearing farmland
3. commercial fishing; draining wetlands; increased sediment runoff
6. reduce fragmentation
7. destroys habitats

Section 3.7, p. 103

1. no; most not adapted
2. control crops or other species
5. stop spread of the insect

Section 3.8, p. 107

1. more people, synthetic materials
2. (a) sulfur dioxide, nitrogen oxides
   (b) burning fossil fuels
3. (a) species' numbers decline
   (b) poor plant growth
4. (a) asthma, bronchitis; property damage
   (b) drive less; use less energy

Section 3.10, p. 113

2. plant trees, do not clear-cut
6. managed sustainably

Chapter 3 Self-Quiz, pp. 118–119

1. (b)
2. (b)
3. (c)
4. F
5. F
6. F
7. succession
8. equilibrium
9. acid precipitation
10. biodiversity hotspot
11. (a) (ii)
    (b) (iii)
    (c) (i)
    (d) (iv)
20. get rid of debris; let seeds grow
22. many unique species
25. fossil fuel widely used

Chapter 3 Review, pp. 120–121

1. (a) biodiversity
   (b) succession
   (c) acid precipitation
   (e) sustainable
   (f) stewardship
2. (a) (iii)
   (b) (iv)
   (c) (ii)
   (d) (i)
   (e) (v)
8. (a) cultural
   (b) product
   (c) other benefit
   (d) product
   (e) cultural
   (f) other benefit
   (g) product
9. (a) (ii)
   (b) (i)
   (c) (iii)

Section 4.1, p. 129

1. engineered ecosystem to produce food

Section 4.2, p. 134

1. minerals, humus, air, water
3. provides nutrients
9. compacted soil

Section 4.4, p. 141

1. (a) food; few predators
   (b) pests are competitors
2. (a) herbicides; insecticides; fungicides
   (b) plants; insects; fungi

Section 4.5, p. 143

2. contamination
4. dissolve in fats/oils
5. control or ban use

Chapter 4 Self-Quiz, pp. 154–155

1. (d)
2. (b)
3. (c)
4. (a)
5. (a)
6. (c)
7. (a)
8. T
9. T
10. F
11. T
12. F
13. urban
14. agroecosystem
15. monoculture
16. humus
17. pesticides
18. (a) (iii)
    (b) (ii)
    (c) (iv)
    (d) (i)
19. lack biodiversity
21. crop yields; easy harvesting
24. heavy rains
28. engineered
31. pests; lower costs
33. reduces fossil fuel use

Chapter 4 Review, pp. 156–157

1. (c)
4. some dissolve in fats
6. wild rice, blueberries; corn, cattle
8. organic matter; nutrients
9. less persistent
10. nitrogen, phosphorus, potassium
17. they are producers, consumers, or decomposers

Unit B Self-Quiz, pp. 162–163

1. (b)
2. (c)
3. (d)
4. (a)
5. (b)
6. (b)
7. (c)
8. T
9. F
10. F

11. T
12. T
13. T
14. F
15. F
16. T
17. abiotic
18. photosynthesis
19. extinct
20. humus
21. limiting
22. consumers
23. (a) (iv)
    (b) (ii)
    (c) (i)
    (d) (iii)
25. space for nutrients, oxygen
32. need oxygen for cellular respiration
33. availability of nutrients
34. eat consumers and producers

Unit B Review, pp. 164–167

1. (d)
2. (b)
3. (d)
4. (a)
5. (a)
6. producer
7. (d)
8. temperature; nutrients; light
9. live together; both benefit
11. engineered ecosystem; cities
14. outcompete native species
16. matter; carbon cycle, water cycle
18. farm
19. manure; compost
21. area drained by river/lake
25. provide nutrients
26. acid rain
27. hydrosphere
28. carbon dioxide
29. (a) economic benefit
    (b) other benefit
    (c) economic benefit
    (d) cultural benefit
    (e) other benefit
    (f) other benefit
    (g) other benefit
36. light, nutrients, acidity, temperature, salinity
37. introduce species; physical barriers
45. increase carbon storage

Unit C

Section 5.2, p. 179

1. sweet, brown, melts easily
2. (a) hard, brittle
   (b) 25.5 cm
3. (a) absorbs moisture, will not dissolve in water
   (b) soft enough to mark on a surface
4. (a) tan, rough, square, flexible
   (b) roughness
5. (a) to hold things together
   (b) flexible, sticky

Section 5.3, p. 181

1. (a) chemical
   (b) physical
   (c) chemical
   (d) physical

Section 5.4, p. 185

1. (a) physical
   (b) physical
   (c) chemical
   (d) physical
   (e) chemical
2. (a) no new substance forms
   (b) no new substance forms
   (c) rust is a new substance
   (d) no new substance forms
   (e) ashes are a new substance
4. (a) A gas is produced.
   (b) The odour changes.
   (c) It gives off light and thermal energy.
   (d) The colour changes.

Section 5.7, p. 192

2. freezing point of the water–salt mixture is lower
7. its freezing point

Chapter 5 Self-Quiz, pp. 202–203

1. (b)
2. (d)
3. (d)
4. (a)
5. (b)
7. T
9. T
10. density
11. physical
12. characteristic
13. (a) (iv)
    (b) (v)
    (c) (ii)
    (d) (iii)
    (e) (i)
15. (a) flexible, soft
    (b) not brittle, hard, low density
    (c) flexible, hard, low density
16. hard, not brittle
17. (a) flame test
22. They are the same temperature.
23. (a) physical
    (b) chemical
    (c) physical
    (d) physical

Chapter 5 Review, pp. 204–205

3. chemical
5. (a) Glass is transparent.
6. (a) 64 °C
7. (a) qualitative
   (b) quantitative
   (c) qualitative
   (d) quantitative

Section 6.1, p. 217

1. protons, neutrons, electrons
2. 12
4. (a) They both contain potassium.
5. (a) Al
   (b) F
   (c) P
   (d) Ti
   (e) C
   (f) Au
   (g) K
   (h) Zn

6. (a) arsenic
   (b) boron
   (c) copper
   (d) lead
   (e) silver
   (f) calcium
   (g) iron
   (h) tungsten
9. (a) 7
   (b) 7
   (c) 5
10. one electron in the outer orbit

Section 6.4, p. 227

1. brittle; have a dull lustre; poor conductors

Section 6.5, p. 229

2. along the "staircase" between metals and non-metals
7. Zn, Ge, Br

Section 6.6, p. 232

1. conducts electricity
2. charcoal is brittle and burns easily; graphite is soft; diamonds are hard

Chapter 6 Self-Quiz, pp. 238–239

1. (c)
2. (c)
3. (b)
4. (b)
5. (d)
6. T
7. F
8. F
9. T
10. F
11. non-metals; metals
12. ductility; malleability
13. protons; neutrons
14. non-metals
15. atom; atomic number
16. charcoal; graphite; diamond
17. (a) (iii)
    (b) (i)
    (c) (iv)
    (d) (v)
    (e) (ii)

Chapter 6 Review, pp. 240–241

1. (a) Ba
   (b) Cl
   (c) Na
   (d) O
   (e) Hg
   (f) H
2. (a) carbon
   (b) nitrogen
   (c) phosphorus
   (d) titanium
   (e) silicon
   (f) magnesium
3. boundary between metals and non-metals
4. same number of electrons in their outer orbit
5. arranged in numerical order by increasing atomic number (number of protons)
13. (a) 9; 9
    (b) electrons; same number of electrons as protons
    (c) Group 17
14. (a) carbon
    (b) oxygen
    (c) nitrogen
    (d) different
    (e) carbon is Group 14; oxygen is Group 16; and nitrogen is Group 15

Section 7.2, p. 251

2. (a) element
   (b) compound
   (c) compound
   (d) compound
   (e) element
   (f) element
3. (a) (iii)
   (b) (ii)
   (c) (i)

Chapter 7 Self-Quiz, pp. 270–271

1. (b)
2. (b)
3. (c)
4. (a)
5. (b)
6. (b)
7. (a)
8. F
9. T
10. F
11. compounds
12. atom
13. subscripts
14. (a) (i)
    (b) (ii)
    (c) (ii)
    (d) (ii)
15. (a) (i)
    (b) (ii)
    (c) (i)
    (d) (i)

Chapter 7 Review, pp. 272–273

1. (a)
2. (d)
3. (a)
4. (b)
5. (c)
6. (a)
7. They can be a diatomic molecule such as oxygen.
8. Water breaks up into hydrogen and oxygen.
10. (a) atom
    (b) molecule
    (c) molecule
    (d) atom
11. (a) element
    (b) compound
    (c) element
    (d) element
12. ozone
13. catalyst
14. carbon dioxide
15. recycled

Unit C Self-Quiz, pp. 278–279

1. (b)
2. (d)
3. (b)
4. (a)
5. (d)
6. (c)
7. T
8. T
9. F
10. T
11. subscripts
12. lustre
13. chemical
14. metalloids
15. (a) (i)
    (b) (ii)
    (c) (i)
    (d) (ii)

16. (a) physical
    (b) chemical
    (c) chemical
    (d) physical
    (e) physical
22. (a) non-metal
    (b) metal
    (c) non-metal
    (d) metal
    (e) non-metal
    (f) metalloid
26. (a) 19
    (b) 19
    (c) electron is given away

Unit C Review, pp. 280–283

1. (c)
2. (c)
3. (b)
4. (c)
5. (d)
6. (b)
7. (d)
8. (b)
9. (c)
10. (d)
11. (b)
12. (c)
13. (b)
14. (b)
15. (b)
16. solubility
17. chemical
19. to prevent rusting
22. (a) physical
    (b) physical
    (c) chemical
    (d) physical
    (e) chemical
    (f) physical
    (g) physical
25. both are unreactive; both are gases
30. the staircase
35. (a) 2
    (b) 3
36. it decomposed

Unit D

Section 8.1, p. 296

2. Universe, galaxy, solar system, star, planet, Moon
3. Earth's rotation on its axis
5. terrestrial planets; gas giants

Section 8.2, p. 299

2. reflecting use mirrors; refracting use lenses
3. (a) size limit of lenses; Earth's atmosphere
   (b) replace lenses with mirrors; place telescopes in space

Section 8.3, p. 304

2. vibration experiments

Section 8.4, p. 307

1. Mercury, Venus, Mars, Jupiter, and Saturn

Section 8.5, p. 311

1. distance is too great; light year
2. absolute magnitude; compares using same distance from Earth
3. dimmer

Section 8.6, p. 315

1. Sun's core

Section 8.8, p. 319

1. spiral and lenticular
2. takes light a long time to travel to Earth
4. galaxies are too distant

Chapter 8 Self-Quiz, pp. 322–323

1. (d)
2. (b)
3. (d)
4. (b)
5. T
6. F
7. F
8. F
9. T
10. T
11. spiral
12. solar winds
13. nuclear reactions
14. absolute magnitude
15. light pollution
16. spiral
17. luminous
18. sun
19. reflecting
20. dwarf planet
21. (a) (iv)
    (b) (ii)
    (c) (i)
    (d) (iii)
22. Telescopes were put on high mountaintops or into space.
23. light pollution

Chapter 8 Review, pp. 324–325

1. lenticular
3. (a) Milky Way
   (b) looks like milk spilled on a road
6. flat and circular with curving arms
8. unit of measure based on the distance light travels in a year
9. nuclear reactions
20. Deneb; lowest absolute magnitude

Section 9.1, p. 335

1. sun will rise and set; day is much longer on Mercury
4. 365 times
5. one rotation takes 24 h

Section 9.2, p. 339

1. tilt of Earth on its axis
5. no seasons

Section 9.3, p. 342

2. Earth rotates west to east
4. never sets below the horizon

Section 9.6, p. 350

1. gravitational forces

Section 9.7, p. 352

1. The corona is visible.

Section 9.8, p. 355

1. the greater the shift, the faster the motion
3. no, took billions of years to form
4. study the Universe and its origins

Chapter 9 Self-Quiz, pp. 360–361

1. (b)
2. (d)
3. (a)
4. (d)
5. F
6. F
7. T
8. F
11. summer
12. rotate
13. Big Bang
14. axis
15. (a) (iii)
    (b) (i)
    (c) (iv)
    (d) (ii)
16. gravitational forces
19. galaxy is moving away

Chapter 9 Review, pp. 362–363

1. rotation of Earth
8. (c)
9. (b)
10. other galaxies; that they were moving
12. (a) summer
    (b) winter
14. located directly above the Earth over the North Pole
15. Sun's corona
20. Light rays bend as they enter Earth's atmosphere.
21. Jupiter has no tilted axis.

Section 10.1, p. 373

2. travel at a speed of 7.5 km/s

Section 10.3, p. 386

1. dizziness and nausea; bone loss; muscle weakness
2. lack of oxygen and air pressure; small meteorites

Section 10.4, p. 389

1. (a) training in microgravity environments
   (b) This environment does not exist on Earth.

Chapter 10 Self-Quiz, pp. 398–399

1. (c)
2. (d)
3. (c)
4. (d)
5. T
6. F
7. F
8. T
9. low-Earth orbit
10. microgravity
11. thrust
12. spacecraft
13. (a) (iii)
    (b) (ii)
    (c) (iv)
    (d) (i)

Chapter 10 Review, pp. 400–401

2. (a) global positioning system
   (b) finds locations on Earth
3. (a) land humans on the Moon
   (b) landed six spacecrafts with astronauts
4. provides oxygen and pressure; protect against radiation and small meteorites
5. (a) robotic arm that performs tasks
   (b) assists astronauts on spacewalks outside the ISS
10. must maintain a balance between the pull of Earth's gravity and its forward speed

Unit D Self-Quiz, pp. 406–407

1. (d)
2. (c)
3. (b)
4. (c)
5. (a)
6. (b)
7. (b)
8. (b)
9. T
10. T
11. F
12. F
13. T
14. T
15. F
16. F
17. T
18. F
19. F
20. asterism
21. solar eclipse
22. microgravity
23. refracting
24. galaxy
25. sun spots

Unit D Review, pp. 408–411

1. (b)
2. (b)
3. (a)
4. (c)
5. (a)
6. (a)
7. (c)
8. F
9. T
10. T
11. F
12. T
13. T
14. F
15. T
16. F
17. T
18. T
19. F
20. F
21. F
22. T
23. F
24. T
25. circumpolar
26. azimuth
27. waxing
28. gravity
29. Moon, Sun
30. apparent magnitude

31. inner planets
32. (a) (iv)
    (b) (iii)
    (c) (ii)
    (d) (i)
34. (a) east to west
    (b) Earth's rotation

Unit E

Section 11.1, p. 423

1. radiant energy (light) and thermal energy

Section 11.2, p. 427

2. natural gas, coal, oil

Section 11.3, p. 431

3. solar and wind power

Section 11.5, p. 438

5. $12.80

Section 11.7, p. 445

2. purchase price and cost to operate

Chapter 11 Self-Quiz, pp. 452–453

1. (d)
2. (b)
3. (c)
4. (c)
5. (a)
6. (d)
7. T
8. F
9. F
10. T
11. uranium
12. steam
13. nuclear
14. light
15. (a) (ii)
    (b) (iii)
    (c) (i)
    (d) (iv)
16. $4.20
17. $90.00

Chapter 11 Review, pp. 454–455

10. (a) $65.70
    (b) $98.50
11. (a) $0.60
    (b) $0.43
12. (a) stove: $360.00
    microwave: $12.00
    TV: $21.00
    computer: $17.50
    washer: $7.80
    dryer: $39.20

Section 12.1, p. 463

2. (a) electrons
5. positively charged
6. positively charged

Section 12.2, p. 467

1. electric discharge

Section 12.3, p. 471

1. (a) charging by friction
   (b) glass rod becomes negative
3. moisture in air will drain charge from decal
5. (a) sphere, rod
   (b) positive

Section 12.5, p. 479

1. (b) and (c)
2. (a) temporary positive charge
   (b) positive charge
   (c) positive charge
   (d) plate charged by friction; lid charged by induction
5. negative; it induced positive charge

Section 12.8, p. 488

1. (a) conductor
   (b) insulator
   (c) insulator
   (d) conductor
   (e) conductor
   (f) insulator
4. Electric current must flow back to its source.

Chapter 12 Self-Quiz, pp. 492–493

1. (d)
2. (b)
3. (d)
4. (c)
5. (c)
6. F
7. T
8. F
9. F
10. T
11. F
12. F
13. induction
14. repel
15. insulator
16. electrostatic discharge
17. opposite
18. (a) (iii)
    (b) (iv)
    (c) (v)
    (d) (ii)
    (e) (vi)
    (f) (i)
22. (a) attraction between positive and negative charges
    (b) electrostatic discharge

Chapter 12 Review, pp. 494–495

1. (a) not charged; equal number of protons and neutrons
   (b) charging over a distance without direct contact
   (c) a material that resists or blocks the flow of electric current
   (d) a material's relative ability to conduct electrons
   (e) sudden release of electric charge; produces a spark
   (f) movement of electrons through a conductor in a closed path
3. vinyl cling advertising, photocopiers and printers
8. insulators resist flow of electrons

Section 13.1, p. 503

1. bulb
4. parallel
5. electric fan, electric motor

Section 13.4, p. 513

2. potential difference: 120 V; current: 12 A; resistance: 10 Ω
3. longer, thinner cord; greater resistance, more thermal energy

Section 13.7, p. 521

1. resistance
2. resistance increases, current decreases
3. use different metal with lower resistance
6. current increases

Section 13.8, p. 524

1. too much current, wires overheat causing fires
2. in series all current in circuit passes through device
3. circuit could carry more current than designed for
5. GFCI protects against water hazards; AFCI protects against damage to wiring

Section 13.9, p. 527

1. more electricity changes to thermal energy; wire becomes hotter
3. sealed heating elements
4. Ceramics heat and cool down quickly for safer use.
5. danger of severe burns

Chapter 13 Self-Quiz, pp. 530–531

1. (a)
2. (c)
3. (a)
4. (d)
5. (c)
6. (b)
7. T
8. F
9. F
10. source
11. series
12. parallel
13. decreases
14. (a) (iv)
    (b) (i)
    (c) (ii)
    (d) (iii)
15. (a) series
    (b) parallel
    (c) parallel
    (d) series
18. decrease
21. insulation on wires; sealed heating elements
22. resistance decreases
24. (a) resistance increases
    (b) current decreases
25. prevents heating element from touching food

Chapter 13 Review, pp. 532–533

1. current flow stops
2. parallel
3. turn off current
4. current increases
6. ohm
7. increases
8. fuse melts
9. voltage
10. voltmeter
11. fuse, circuit breaker, GFCI outlets
13. high internal resistance
15. (a) parallel
    (b) series, one path
    (c) parallel, four paths
    (d) series, one path
21. flashlight will not light

Unit E Self-Quiz, pp. 538–539

1. (c)
2. (d)
3. (c)
4. (c)
5. (d)
6. (a)
7. resistance
8. discharge
9. fossil
10. electroscope
11. F
12. F
13. T
14. F
15. T
16. (a) (vi)
    (b) (iv)
    (c) (iii)
    (d) (v)
    (e) (ii)
    (f) (vii)
    (g) (i)

Unit E Review, pp. 540–543

1. (c)
2. (a)
3. (a)
4. (b)
5. (d)
6. (c)
7. (c)
8. (a)
9. (c)
10. (a)
11. (d)
12. (d)
13. (d)
14. (a)
15. (b)
16. (c)
17. (d)
18. (b)
19. (c)
20. (b)
21. (c)
22. (d)
23. voltage
24. 1000
25. ammeter
26. ground the object
27. electrons
28. neutral
35. parallel; more than one complete path

Glossary

A

abiotic factors [AY-bye-aw-tik FAK-turz]
the non-living characteristics of an ecosystem (p. 30)

absolute magnitude [AB-suh-loot MAG-nuh-tyood]
a measurement of how bright an object would be if it were 33 ly from Earth (p. 309)

acid precipitation [A-sid pruh-si-puh-TAY-shuhn]
precipitation that has been made more acidic than normal by combining certain chemicals in the air with water vapour (p. 105)

agroecosystem [AG-ro-EE-ko-sis-tuhm]
an agricultural ecosystem designed and maintained by humans to meet the human need for food (p. 128)

alloy [AL-oy]
a solid solution of two or more metals (p. 220)

alternating current (AC)
an electric current that repeatedly reverses direction (p. 487)

altitude [AL-tuh-tyood]
the angular height of a celestial object above the horizon; measured vertically (in °) from the horizon (p. 344)

ammeter [AM-eet-ur]
a device used to measure current (p. 508)

ampere (A) [AM-peer]
the unit of electric current (p. 508)

apparent magnitude [uh-PEHR-ent MAG-nuh-tyood]
a measurement of how bright an object is when viewed from Earth; it depends on luminosity and distance (p. 309)

asterism [AST-uh-riz-uhm]
a pattern of stars in Earth's night sky (p. 340)

asteroids [AS-tuh-roydz]
small, rocky, irregularly shaped celestial objects that orbit the Sun (p. 301)

astronomical [AS-truh-NAW-mik-uhl] unit (AU)
the average distance between Earth and the Sun; about 150 million km (p. 331)

atmosphere [AT-muhs-feer]
the thin layer of gases that surrounds Earth (p. 28)

atomic number
the number of protons in an atom of an element; each element has a different atomic number (p. 212)

aurora [uh-RAW-ruh]
coloured lights in the sky produced when the solar wind comes in contact with Earth's magnetic field and the atmosphere (p. 314)

axis
an imaginary line around which an object rotates (p. 333)

azimuth [A-zi-muhth]
the horizontal angular distance from north measured (in °) eastward along the horizon to a point directly below a celestial object (p. 344)

B

Big Bang theory
the theory that the Universe began in an incredibly hot, dense expansion about 14 billion years ago (p. 354)

bioaccumulation [BYE-o-uh-KYOO-myoo-LAY-shuhn]
the concentration of a substance, such as a pesticide, in the body of an organism (p. 142)

bioamplification [BYE-o-AM-pluh-fuh-KAY-shuhn]
the increase in concentration of a substance, such as a pesticide, as it moves higher up the food chain or food web (p. 143)

biodiversity [BYE-o-duh-VUR-suh-tee]
the variety of species in an ecosystem (p. 88)

biogeochemical [BYE-o-JEE-o-KEM-uh-kuhl] cycles
movement of matter through the biotic and abiotic parts of the environment (p. 48)

biomes [BYE-oms]
large geographical regions defined by climate with a specific set of biotic and abiotic features (p. 60)

biosphere [BYE-os-feer]
the zone in, on, and around Earth where life can exist (p. 29)

biotic factors [bye-AW-tik FAK-turz]
living things, their remains, and the features, such as nests, associated with their activities (p. 30)

C

carbon cycle
the series of processes that moves carbon compounds throughout the lithosphere, atmosphere, hydrosphere, and biosphere (p. 50)

carrying capacity [KER-ee-ing kuh-PA-suh-tee]
the maximum population size of a species that a given ecosystem can sustain (p. 57)

catalyst [KAT-uh-list]
a substance that speeds up a chemical change without being consumed or chemically changed itself (p. 258)

celestial [suh-LES-tee-uhl] object
any object that exists in space (p. 292)

cellular respiration [SEL-yoo-lur res-puh-RAY-shuhn]
the process by which sugar and oxygen are converted into carbon dioxide, water, and energy to provide energy for the cell (p. 40)

characteristic property
a property that is unique to a substance and that can be used to identify the substance (p. 190)

charging by conduction
charging an object by contact with a charged object (p. 470)

charging by friction
the charging of two neutral objects made of different materials by rubbing them together (p. 468)

charging by induction
charging a neutral object by bringing another charged object close to, but not touching, the neutral object (p. 474)

chemical change
a change in which one or more new substances is formed (p. 183)

chemical property
a description of what a substance does when it reacts to produce new substances (p. 180)

circuit [SUR-kuht] diagram
a way of drawing an electric circuit using standard symbols (p. 502)

circumpolar [sur-kuhm-POL-ur]
a star or constellation that never sets below the horizon (p. 341)

combustibility [kuhm-bus-tuh-BIL-uh-tee]
the ability of a substance to burn (p. 180)

combustion [kuhm-BUS-chuhn]
a rapid chemical change involving oxygen and a fuel that releases large amounts of energy (p. 264)

comet
a large chunk of ice, rock, and dust that orbits the Sun (p. 332)

command module
the section of a spacecraft in which astronauts live, communicate with Earth, and operate controls during a flight (p. 375)

community
all the living organisms, from all the populations, that live in the same ecosystem (p. 31)

compound
a pure substance made up of two or more elements that are chemically joined (p. 249)

conductivity [kuhn-duk-TI-vuh-tee]
a measure of a material's ability to conduct thermal energy or electricity (pp. 220, 485)

conductor
a material that transmits thermal energy or electrical energy easily (p. 484)

constellation [kawn-stuh-LAY-shuhn]
a region of stars in the night sky (p. 340)

consumer
an organism that obtains its energy by consuming other organisms, or the products of other organisms (p. 40)

controlled experiment
an experiment in which the independent variable is purposely changed to find out what change, if any, occurs in the dependent variable (p. 8)

cosmologists [kaws-MAW-luh-jist]
scientists who study the makeup and origins of the Universe (p. 354)

craters
circular or bowl-shaped holes in surfaces caused by the impact of a celestial object (p. 301)

D

decomposition
the breaking down of a compound into simpler substances (p. 256)

density
a measure of how much mass is contained in a given volume of a substance; you calculate density by dividing the mass of a sample by its volume (p. 191)

dependent variable
a variable that changes in response to the change in the independent variable (p. 7)

direct current (DC)
an electric current in which the flow of electrons travels in one direction only (p. 486)

ductile [DUHK-tye-uhl]
a material's ability to be pulled into a wire (p. 220)

dwarf planet
a celestial object that orbits the Sun and has a spherical or round shape, but does not clear its orbit (p. 305)

E

ecological footprint
a measure of the land area required to sustain a given population of humans (p. 150)

ecological niche [EE-kuh-LAW-ji-kuhl NEESH]
the function a species serves in its ecosystem, including where it lives, what it eats, what eats it, and how it behaves (p. 42)

ecosystem [EE-ko-sis-tuhm]
all the living organisms in an area and their non-living environment (p. 30)

efficiency [e-FISH-en-see]
a measure of how much useful energy an electrical device produces compared with the amount of energy that was supplied to the device (p. 436)

electrical energy
a type of energy that comes from the flow of charged particles; it allows electrical devices to function (p. 420)

electrical energy consumption
the amount of electrical energy used, often measured in kilowatt hours (p. 434)

electrical energy distribution grid
a network that carries electricity from the energy generating station to the consumer (p. 422)

electrical resistance (R)
the ability of a material to oppose the flow of electrons through it (p. 510)

electric charge
positive or negative charge on objects that exert an electric force (p. 460)

electric circuit
a continuous path along which electrons flow that includes an energy source, a load, conductors, and usually a switch (p. 500)

electric current
the flow of electricity through a conductor in a closed path (p. 486)

electric current (*I*)
the rate of flow of electrons past a certain point in a circuit (p. 508)

electric discharge
the rapid transfer of electrons from one object to another (p. 464)

electric force
force of attraction or repulsion between charged objects (p. 460)

electrolysis [el-ek-TRAW-luh-sis]
the process during which electricity is used to break down a compound into simpler substances (p. 256)

electron
a negatively charged particle found in an orbit outside the nucleus of an atom (p. 210)

element
a pure substance that cannot be broken down into simpler substances, either physically or chemically (p. 211)

energy source
a device that provides electrical energy in a circuit, such as a battery or an electrical outlet (p. 500)

equilibrium [EE-kwuh-LIB-ree-uhm]
the state of an ecosystem in which the biotic and abiotic features remain relatively constant over a long period of time (p. 82)

equinoxes [EK-wuh-nawk-sez]
days when the hours of daylight equal the hours of darkness (p. 338)

experimental design
a brief description of the procedure by which a hypothesis or prediction is tested (p. 10)

extinct [eks-TINKT]
no longer existing on Earth (p. 93)

F

fertilizer
a substance applied to an area to provide the nutrients plants need for growth (p. 131)

food web
a representation of the feeding relationships within a community (p. 45)

G

galaxies [GA-luhk-sees]
huge, rotating collection of gas, dust, stars, planets, and other celestial objects (p. 294)

generator
device that produces electricity by rotating a wire coil near a magnet (p. 422)

geocentric [jee-o-SEN-trik] model
model of the Universe in which Earth is at the centre (p. 330)

gravitational [GRA-vuh-TAY-shuhn-uhl] force
the force of attraction between two massive objects; gravity (p. 348)

grounding
connecting an object to a large body, like Earth, that can remove an electric charge from the object (p. 476)

group or chemical family
a vertical column in the periodic table of elements (p. 212)

H

habitat
the environment where an organism lives (p. 28)

heliocentric [HEE-lee-o-SEN-trik] model
model of the Universe in which the Sun is at the centre (p. 330)

hypothesis [hye-PAW-thuh-sis]
a possible answer or untested explanation that relates to the initial question in an experiment (p. 10)

I

independent variable
a variable that is changed by the investigator (p. 7)

induced charge separation
a shift in the position of electrons in a neutral object that occurs when a charged object is brought near it (p. 474)

insulator [IN-suh-lay-tur]
a material that resists or blocks the flow of electrons through it (p. 485)

International Space Station (ISS)
a research laboratory in low orbit around Earth (p. 369)

invasive species [in-VAY-siv SPEE-seez]
a non-native species that has a negative impact on the natural environment (p. 101)

K

kilowatt
1000 watts (p. 423)

kilowatt hour
a unit of energy equal to using 1000 W for one hour (p. 423)

L

Law of Electric Charges
objects that have opposite charges attract each other; objects that have like charges repel each other (p. 462)

light year
the distance that light travels in a year; the unit is ly (p. 308)

limiting factor
any factor that restricts the size of a population or where it can live (p. 54)

load
an electrical device that converts electrical energy into another form of energy (p. 500)

luminosity [loo-muh-NAW-suh-tee]
the amount of energy released by a star each second (p. 308)

luminous [LOO-muh-nuhs]
producing and giving off light; shining (p. 292)

lunar cycle
the sequence of phases made by the Moon during one orbit around Earth (p. 349)

lunar eclipse
when the Moon passes through Earth's shadow and does not reflect as much sunlight as usual (p. 352)

M

malleable [MA-lee-uh-bul]
a material's ability to be hammered into different shapes (p. 220)

metal
a shiny substance that usually is malleable (bendable) and is a good conductor of electricity (p. 220)

meteor
a meteoroid that burns up as it travels through Earth's atmosphere; it forms a bright streak of light in the night sky (p. 332)

meteoroid
a small celestial object made of rock or metal that is smaller than an asteroid (p. 332)

microgravity [MYE-kro-gra-vuh-tee] environment
an environment in which objects behave as though there is very little gravity affecting them (p. 387)

molecule [MAW-luh-kyool]
two or more atoms that are chemically joined (p. 248)

monoculture [MAWN-o-kul-chur]
an agroecosystem (a farm) consisting of a single crop in an area (p. 128)

multimeter
a device used to measure different circuit quantities (p. 513)

N

natural fertilizers
plant nutrients that have been obtained from natural sources and have not been chemically changed by humans (p. 131)

natural satellite
a celestial object that orbits a larger celestial object (p. 294)

negatively charged object
an object that has more electrons than protons (p. 461)

neutral [NYOO-truhl]
having no charge or an equal number of positive and negative charges (p. 460)

neutral object
an object that has an equal number of protons and electrons (p. 461)

neutron [NYOO-trawn]
a particle in an atom that has no electrical charge (p. 210)

non-metal
an element that is usually a gas at room temperature or a dull, brittle solid, and a poor conductor of thermal energy and electricity (p. 224)

non-renewable energy source
an energy source that either cannot be renewed, or that takes millions of years to renew naturally (p. 424)

nuclear reactions
reactions in which the nuclei of atoms change; these types of reactions produce large amounts of energy (p. 312)

O

observational study
the careful watching and recording of a subject or event to gather scientific information to answer a question (p. 8)

ohm (Ω) [O-uhm]
the unit of electrical resistance (p. 510)

ohmmeter [O-uhm-mee-tur]
a device used to measure electrical resistance (p. 510)

Ohm's law
the relationship between resistance, current, and potential difference; for a given potential difference, as the resistance increases, the current decreases (p. 518)

orbit
the closed path that a celestial object takes around another object; elliptical or circular in shape (p. 293)

P

parallel circuit
a circuit in which two or more loads are connected so that electrons can follow more than one path (p. 503)

period
a horizontal row in the periodic table of elements (p. 212)

periodic table
the table that lists all the known elements in rows and columns based on patterns of similar properties (p. 211)

pest
any plant, animal, or other living organism that causes illness or harm and is an annoyance to humans (p. 128)

pesticide [PEST-uh-syed]
a substance used to kill a pest (p. 139)

phases of the Moon
the changing amount of illuminated Moon surface visible from Earth (p. 349)

photosynthesis [FO-to-SIN-thuh-sis]
the process in which the Sun's energy is converted into chemical energy (p. 39)

photovoltaic [fo-to-vol-TAY-ik] cell
a device that converts light directly into electrical energy (p. 429)

physical change
a change in a substance that does not involve the formation of a new substance (p. 182)

physical property
a characteristic of a substance that you can observe or measure without forming a new substance (p. 178)

planet
a large, round celestial object that travels around a star (p. 293)

population
all the members of the same species that live in the same ecosystem (p. 31)

positively charged object
an object that has fewer electrons than protons (p. 461)

potential difference or voltage (*V*)
the difference in electric potential energy per unit of charge measured between two points (p. 509)

precipitate [pruh-SIP-uh-tayt]
a solid that forms when two liquids are mixed (p. 185)

prediction
a statement that predicts the outcome of a controlled experiment (p. 10)

producer
an organism that makes its own energy-rich food using the Sun's energy (p. 39)

proton [PRO-tawn]
a positively charged particle in an atom (p. 210)

Q

qualitative [KWA-luh-tay-tiv] observation
a non-numerical observation that describes the qualities of objects or events (p. 12)

quantitative [KWAN-tuh-tay-tiv] observation
a numerical observation based on measurements or counting (p. 11)

R

radiant [RAY-dee-ent] energy
energy that travels through empty space (p. 38)

red shift
light from a galaxy shifting toward the red end of the visible spectrum (p. 354)

reflecting telescope
a telescope that uses mirrors to gather and focus light (p. 298)

refracting telescope
a telescope that uses lenses to gather and focus light (p. 298)

renewable energy source
an energy source that can be replaced in a relatively short period of time (p. 428)

revolve
to orbit a central point or object (p. 333)

rocket booster
a powerful rocket engine that contains fuel and attaches to a spacecraft for use during launch (p. 370)

rotates
moves, or turns, around a centre point (p. 333)

rusting
a slow chemical change of iron reacting with oxygen and water (p. 265)

S

series circuit
a circuit in which two or more loads are connected one after another so that electrons follow only one path (p. 502)

solar eclipse
when the Moon passes between the Sun and Earth and blocks some or all of the Sun's light (p. 351)

solar flare
gases and charged particles that are blasted above an active region of the photosphere (p. 314)

Solar System
the Sun and all the celestial objects that travel around it (p. 293)

solar wind
electrically charged particles that flow from the Sun's corona out into space (p. 314)

solstice [SOL-stis]
a day when Earth's axis is most inclined toward, or away from, the Sun (p. 338)

solubility
the ability to dissolve in another substance (p. 178)

spacecraft
a robotic or human-occupied vehicle used to explore space and celestial objects (p. 368)

space junk
wreckage from artificial objects orbiting Earth (p. 382)

space technology spinoffs
technologies originally designed for space exploration that have made their way into everyday use (p. 391)

star
a massive collection of gases in space that emits large amounts of energy (p. 292)

star cluster
a dense group of stars held together by gravity (p. 319)

static electricity
an imbalance of electric charge at rest on the surface of an object (p. 464)

stewardship [STOO-uhrd-ship]
taking responsibility for managing and protecting the environment (p. 113)

subscript
the small number that follows the symbol for an element in a chemical formula and shows how many atoms of the element are present (p. 248)

succession [suk-SESH-uhn]
the observed changes over time in an ecosystem, following a disturbance (p. 83)

sunspot
a dark area of the Sun's surface that is cooler than the area surrounding it (p. 313)

sustainable [suhs-TAY-nuh-bul] ecosystem
an ecosystem that maintains relatively constant biotic and abiotic features forever (p. 34)

switch
a control device that allows you to safely open or close a circuit (p. 500)

synthetic fertilizers
plant nutrients that are produced by subjecting natural materials to chemical processes (p. 131)

T

thermal generation
a method for generating electricity in which water is heated to produce steam, which spins a turbine connected to a generator (p. 424)

thrust
the force of a rocket engine (p. 370)

time zone
a vertical section of Earth's surface that has a uniform time (p. 334)

tolerance [TAWL-uh-rens] range
the range of abiotic conditions within which a species can survive (p. 54)

trophic [TRO-fik] level
the level of an organism in an ecosystem, depending on its feeding position along a food chain (p. 44)

U

Universe
everything that exists, including all matter, energy, and space (p. 294)

V

variable [VEHR-ee-uh-buhl]
any condition that changes or varies the outcome of a scientific inquiry (p. 7)

volt (V)
the unit of potential difference (p. 509)

voltmeter
a device used to measure potential difference (p. 509)

W

waning
decreasing in size; a period during which the visible surface of the Moon decreases (p. 349)

water cycle
the series of processes that moves water through the environment (p. 49)

watershed
the land area drained by a particular river or lake; also called a drainage basin (p. 63)

waxing
increasing in size; a period during which the visible surface of the Moon increases (p. 349)

Index

F

G

H

I

J

K

L

M

N

O

P

Q

R

S

Credits

This page constitutes an extension of the copyright page. We have made every effort to trace the ownership of all copyrighted material and to secure permission from copyright holders. In the event of any question arising as to the use of any material, we will be pleased to make the necessary corrections in future printings. Thanks are due to the following authors, publishers, and agents for permission to use the material indicated.

Table of Contents. v © LouLouPhotos/Shutterstock; vii © Chip Simons/Getty Images; viii © Stocktrek Images, Inc./Alamy; x © Gaertner/Alamy

Unit A

2 AP Photo/Paul Sancya/CP IMAGES; 3 top to bottom © LouLouPhotos/ Shutterstock, © Chip Simons /Getty Images, © Stocktrek Images, Inc./Alamy, LouLouPhotos/Shutterstock

Chapter 1. 4 David Cooper/Toronto Star; 5 left to right © Greenpeace/Ron Scheffler, CP PICTURE ARCHIVE/AP, Heribert Proepper; 7 Jim Richardson/ Getty; 9 photos.com; 10 © NASA Images/Alamy; 11 top to bottom © Ron Niebrugge/Alamy, bottom Daniel Cox/Photolibrary; 12 © Image Source/Alamy; 14 Mark C. Burnett/Photo Researchers; 16 Philippe Henry/First Light; 17 top left to right photos.com, John Zoiner/Getty, © Ron Niebrugge/Alamy; bottom © Tim Pannell/Corbis; 18 Huguette Roe/Shutterstock; 19 top left to right Jim Richardson/Getty, Daniel Cox/Photolibrary, bottom left to right Huguette Roe/ Shutterstock ©Tim Pannell/Corbis

Unit B

20 © LouLouPhotos/Shutterstock; 21 top to bottom © Bruce MacQueen/ Shutterstock; © John Cancalosi/Photolibrary; © Robert Foeger/Shutterstock; 22 left to right © Randolph Images/Alamy; © Francois Gohier/Photo Researchers, Inc.; Graça Victoria/Shutterstock; 23 © Bruce MacQueen/ Shutterstock.

Chapter 2. 24 © Randolph Images/Alamy; 25 top to bottom © blickwinkel/ Alamy; © Artostock.com/Alamy; 26 left top to bottom wave/First Light; © Zach Holmes/Alamy; James Steid/Shutterstock; right top to bottom © esemelwe/ iStockphoto; Oxana Zubof/Shutterstock; Ragnarock/Shutterstock; 27 © First Light/Alamy; 28 left to right Paul B. Moore/Shutterstock; © Poelzer Wolfgang/ Alamy; 30 left top to bottom © Rob Crandall/Alamy; © imagebroker/Alamy; right top to bottom © Keith Levit/Photolibrary; © photosbygar/Alamy; 32 left Courtesy of Verena Tuniclife; right top to bottom © Gunter Marx/Alamy; © Brandon Cole Marine Photography/Alamy; © imac/Alamy; Michael P. Gadomski/Photo Researchers, Inc.; 34 © First Light/Alamy; 35 © Jodiejohnson/ Dreamstime; 36 Boreal Laboratories; © M.Brodie/Alamy; 38 Gail Jankus/ Photo Researchers, Inc.; 39 top right © Monkey Business Images/Shutterstock bottom left to right © Logray-2008/iStockphoto; 42 © Bill Brooks/Alamy; 43 © moodboard/Alamy; 48 © Digital Vision/Alamy; 55 left to right © Jan Baks/ Alamy; © tbkmedia.de/Alamy; 56 left to right © David M Dennis/Photolibrary; Michael Poliza/Getty Images; © Andrew Syred/Photo Researchers, Inc.; 57 left to right © Bill Bachman/Alamy; © Photoshot Holdings Ltd/Alamy; 58 © Radius Images/Alamy; 59 © Angelera/iStockphoto; 60 © Visual&Written SL/Alamy; 61 top to bottom Garry Black/Masterfile; © imagebroker/Alamy; 4loops/iStockphoto; 63 © Bill Brooks/Alamy; 64 © Bob Halstead/Lonely Planet Images; © Hemis/Alamy; © Visuals&Written SL/Alamy; 65 top right Paul Chesley/Getty Images; bottom left to right Courtesy of Scott Baker; ©Stephen Frink Collection/Alamy; 66 © aaaah/Shutterstock; © Pjorg/Shutterstock; © TerryM/Shutterstock; © David J. Green/Alamy; © Spencer Grant/PhotoEdit; © WhiskyBottle/Alamy; © Katmay/Shutterstock; © Christian Delbert/ Shutterstock; 67 iStockphoto; 68 © Visuals&Written SL/Alamy; 69 left top to bottom wave/First Light; © Zach Holmes/Alamy; James SteidlShutterstock; right top to bottom © esemelwe/iStockphoto; Oxana Zubof/Shutterstock; Ragnarock/Shutterstock.

Chapter 3. 74 © Alaska Stock LLC/Alamy; © Francois Gohier/Photo Researchers, Inc.; 76 left top to bottom © First Light/Alamy; © Brent Linton/ Canadian Press Images; © KONRAD WOTHE/MINDEN PICTURES/National Geographic Stock; right top to bottom © All Canada Photos/Alamy; © blickwinkel/Alamy; ©Photos.com; 78 © Jim West/Alamy; 79 left to right © Glowimages RM/Alamy; © ying /Shutterstock; © Jim Jurica/iStockphoto; 80 © Dr. Morley Read/Shutterstock; © Roger Ressmeyer/CORBIS;

81 © Photos.com; 82 © All Canada Photos/Alamy; 83 left to right © Thomas & Pat Leeson/Photo Researchers, Inc.; © Science Source/USGS/Photo Researchers, Inc.; © Tony Craddock/Photo Researchers, Inc.; 85 © Tom Uhlman/Alamy; 86 Dufferin Aggregates, a division of of Holcim (Canada) Inc.; 87 © Gunter Marx/Alamy; 88 left to right © Radius Images/Alamy; Steve Allen/ Brand X Pictures/Getty Images; 89 © Photos.com; 90 © Photos.com; 92 © Arco Images GmbH/Alamy; 94 Richard Ellis/Photo Researchers, Inc.; 96 top © Gaertner/Alamy; bottom Doug Fraser; 97 © Ibpics/Alamy; 98 WWF 2006, Global Environment Outlook 4, United Nations Environment Programme Figure 5.1; 99 top © NORBERT WU/MINDEN PICTURES/National Geographic Stock; bottom left to right © james woodhead/iStockphoto; © Michael Stubblefield/ Alamy; 100; ©Ken and Linda Parker/www.SweetGrassGardens.com; 101 left to right © John A. Anderson/Shutterstock; © blickwinkel/Alamy; 102 top to bottom © tonisalado/Shutterstock; © john t. fowler/Alamy; 103 Patrick Hirlehey; 104 left to right © Maryann Frazier/Photo Researchers, Inc.; © Ted Kinsman/ Photo Researchers, Inc.; 105 top right © gary corbett/Alamy; bottom left to right © Keith Levit/Shutterstock; © foment/Shutterstock; Donald Neusbaum/ Getty Images; 107 © National Geographic/Getty Images; 108 top © Photos. com; © Luminis/Shutterstock; middle © Photos.com; © Glowimages RM/ Alamy; bottom © vita khorzhevska/Shutterstock; © Jamesmcq24/iStockphoto; © 4736202690/Shutterstock; 110 left © Michael Macor/San Francisco Chronicle/Corbis; right © Lisa Craig Gautier, MatterOfTrust.org http:// MatterOfTrust.org; 112 top © Realimage/Alamy; bottom National Geographic/ Getty Images; 116 top left to right © Jim West/Alamy; © Tony Craddock/Photo Researchers, Inc.; © Photos.com; bottom left to right ©Doug Fraser © Keith Levit/Shutterstock; National Geographic/Getty Images; 117 left top to bottom © First Light/Alamy; ©Brent Linton/Canadian Press Images; © KONRAD WOTHE/MINDEN PICTURES/National Geographic Stock; right top to bottom © All Canada Photos/Alamy; © blickwinkel/Alamy; © Photos.com.

Chapter 4. 122 © Graça Victoria/Shutterstock; 123 © PR NEWSWIRE/ Associated Press; 124 left top to bottom © nolie/Shutterstock; © ABN Stock Images/Alamy; © Photos.com; right top to bottom © Photos.com; © Photos. com; © Mishkaki/Dreamstime; 126 © Photos.com; 127 top © All Canada Photos/Alamy; bottom iStockphoto; 129 © Dave Bevan/Alamy; 132 top © BSIP MARTIN PL./Science Photo Library/Photo Researchers, Inc.; bottom © National Geographic Image Collection/Alamy; 133 top © AGStockUSA, Inc./ Alamy; bottom © Joseph Gareri/Alamy; 134 © Barbro Bergfeldt/Shutterstock; 135 Courtesy BEST Energies, Inc.; 136 left to right © Khorzhevska/Dreamstime; © PhotoAlto/Alamy; © intraprese/iStockphoto; 139 top © Nigel Cattlin/Alamy; bottom © Images-USA/Alamy; 140 top Paul Grebliunas/Getty Images; bottom © JTB Photo Communications, Inc./Alamy; 142 © BRUCE COLEMAN INC./ Alamy; 144 top © roy cano/iStockphoto; bottom left to right © angelo gilardelli/ Shutterstock; ©Photos.com; © yamix/Shutterstock; © justin maresch/ Shutterstock; © Photos.com; 146 left to right © Jay Crihfield/Shutterstock; © Jaren Jai Wicklund/Shutterstock; 148 left to right © Oleksiy Maksymenko/ Alamy; © Luke Buchan/Canadian Press Images; © Bill Brooks/Alamy; 150 top © Ronald Greer/Alamy; bottom © Alison Wright/Corbis; 152 top © Photos.com; bottom left to right © Images-USA/Alamy; © BRUCE COLEMAN INC./Alamy; © Oleksiy Maksymenko/Alamy; 153 left top to bottom © nolie/Shutterstock; © ABN Stock Images/Alamy; © Photos.com; right top to bottom © Photos.com; © Photos.com; © Mishkaki/Dreamstime.

Unit B Looking Back. 158 left © Visuals&Written SL/Alamy; centre top to bottom © Jim West/Alamy; © Tony Craddock/Photo Researchers, Inc.; Photos. com; Doug Fraser; © Keith Levit/Shutterstock; National Geographic/Getty Images; right top to bottom © Photos.com; © Images-USA/Alamy; © BRUCE COLEMAN INC./Alamy; © Oleksiy Maksymenko/Alamy; © BRUCE COLEMAN INC./Alamy; © Oleksiy Maksymenko/Alamy; 159 Barrett & MacKay/All Canada Photos; 160 U.S. Fish and Wildlife Service; PARTSCH/Science Photo Library; 167 © sébastien Baussais/Alamy.

Unit C

168 © Chip Simons/Getty Images; 169 top to bottom © SCIENCE PHOTO LIBRARY/Photolibrary; © Macduff Everton/CORBIS; © Blend Images/ Alamy; © D. Hurst/Alamy; 170 left to right © murboy/iStockphoto; © Oleksiy Maksymenko/Alamy; ©Gunter Marx/Alamy; 171 top left © Comstock/ Jupiterimages; bottom left © Iurii Konoval/Shutterstock; top right © Hemera Technologies/Getty Images/Jupiterimages; bottom right © Imagesource/ Photolibrary.

Chapter 5. 172 background © murboy/iStockphoto; top © Dennis Kunkel Microscopy, Inc./Visuals Unlimited/Corbis; bottom © National Geographic/ Getty Images; 173 top to bottom © PoodlesRock/Corbis; ©Hugh Threlfall/ Alamy; © The Hope Diamond/Smithsonian Institution, Washington DC, USA/ The Bridgeman Art Library; 174 left top to bottom picamaniac/Shutterstock; © Arcticphoto/Alamy; © Charles D. Winters/Photo Researchers, Inc.; right top to bottom © Corbis/Jupiterimages; © David Young-Wolff/PhotoEdit; ©Danita Delimont/Alamy; 176 top left to right © Vadim Kozlovsky/Dreamstime; © Danny Smythe/Dreamstime; bottom left to right © Hemera Technologies/

Getty Images/Jupiterimages; © Varyaphoto/Dreamstime.com; 178 top to bottom © JoeFox/Alamy; © Mike Powell/Getty Images; 180 © Stephen Harrison/Alamy; 181 © Ebby May/Jupiterimages; 182 top left to right © Envision/Corbis; © Design Pics/Jupiterimages; 183 © Mark Fairey/Alamy; 184 top left to right letty17/Shutterstock; bottom left to right Mark Boulton/Photo Researchers, Inc.; © Bon Appetit/Alamy; 185 © Charles D. Winters/Photo Researchers, Inc.; 188 top left to right © Georgios Alexandris/Dreamstime; © Ron Kloberdanz/Shutterstock; middle left to right © Daver0316/Dreamstime; © Brian Weed/Dreamstime; © drflet /iStockphoto; bottom left to right © Yoav Levy/Phototake; © Ron Chapple Studios/Dreamstime; 190 © Yoav Levy/PHOTOTAKE/Alamy; 191 top to bottom © Vincon/Klein/plainpicture/Corbis; © 81a/Alamy; 193 top T-Pool/Getty Images; bottom © blickwinkel/Alamy; 194 top left to right © csheezio/iStockphoto; © Marek Uliasz/Dreamstime; © Aweech/Dreamstime; © Viktorfischer/Dreamstime; bottom left to right © Helen Sessions/Alamy; 196 Julien/Shutterstock; 198 top left to right © Science First; middle left to right © Ggw1962/Dreamstime; © Svlumagraphica/Dreamstime; © eNASCO; bottom left to right © Andrew Lambert Photography/ Photo Researchers, Inc.; © Benjamin Mills; © Helen Sessions/Alamy; © Martyn F. Chillmaid/Photo Researchers, Inc.; 200 left top to bottom © Ebby May/Jupiterimages; © Mike Powell/Getty Images; right top to bottom © Vincon/Klein/plainpicture/Corbis; © Design Pics/Jupiterimages; 201 left top to bottom picamaniac/Shutterstock; ©Arcticphoto/Alamy; © Charles D. Winters/Photo Researchers, Inc.; right top to bottom © Corbis/Jupiterimages; © David Young-Wolff/PhotoEdit; © Danita Delimont/Alamy; 205 Scientifca/Visuals Unlimited/Getty Images.

Chapter 6. 206 © Oleksiy Maksymenko/Alamy; 208 left top to bottom © Stock Connection Blue/Alamy; © Hornbil Images/Alamy; © Margaret S/Alamy; right top to bottom © Paul Rapson/Photo Researchers, Inc.; © Bon Appetit/Alamy; 211 © Image Source/Alamy; 216 © Andrew Lambert Photography/Photo Researchers, Inc.; © Arno Enzerink/Alamy; 218 left to right © Judith Collins/Alamy; © Les Cunliffe/Dreamstime; © Fabrizio Zanier/Dreamstime; © Ra3rn/Dreamstime; © Knorre/Dreamstime; © Fabrizio Troiani/Dreamstime; © Brian Weed/Dreamstime; 220 left to right © Michael Pearcy/Alamy; © Didier Carpentier; © Phil Degginger/Alamy; 221 right © Bernard O'Kane/Alamy; left © Steve Nutt/Handout/Reuters/Corbis; 222 © The Starry Night, June 1889 (oil on canvas), Gogh, Vincent van (1853–90)/Museum of Modern Art, New York, USA/The Bridgeman Art Library; 223 top Theodore Gray/Visuals Unlimited, Inc; bottom iStock photo; 224 © Picture Contact/Alamy; 225 © Natural Visions/Alamy; 226 Andrew Lambert Photography/Science Photo Library; 227 © RTimages/Alamy; 228 left to right © Russell Lappa/Photo Researchers, Inc.; © Joyce Photographics/Photo Researchers, Inc.; © Charles D. Winters/Photo Researchers, Inc.; © Charles D. Winters/Photo Researchers, Inc.; 229 © Adam Hart-Davis/Photo Researchers, Inc.; 230 top to bottom © Hugh Threlfall/Alamy; robophobic/Shutterstock; South 12th Photography/Shutterstock; 231 CP PHOTO/Adrian Wyld; 232 © Reuters/CORBIS; 233 top right © Dellnesco/Alamy; top left © Ben Johnson/Photo Researchers, Inc.; bottom © Antonia Reeve/Photo Researchers, Inc.; 234 © Charles D. Winters/Photo Researchers, Inc.; 236 top left to right Charles D. Winters/Photo Researchers, Inc., bottom left to right © Joyce Photographics/Photo Researchers, Inc.; South 12th Photography/Shutterstock; Andrew Lambert Photography/Science Photo Library; 237 left top to bottom © Stock Connection Blue/Alamy; © Hornbil Images/Alamy; © Margaret S/Alamy; right top to bottom © Paul Rapson/Photo Researchers, Inc.; © Bon Appetit/Alamy.

Chapter 7. 242 © Gunter Marx/Alamy; 243 © J.F.T.L IMAGES/Alamy; 244 left top to bottom © Vince Clements/Shutterstock; ©sciencephotos/Alamy; © wolv/iStockphoto; right top to bottom © Dan Barba/Photolibrary; © Jamdotsi/Dreamstime; 248 top © Kenneth Eward/Photo Researchers, Inc.; 249 top ©Stocktrek Images, Inc./Alamy; 252 right © Serg_Velusceac/iStockphoto; 254 left to right Charles D. Winters/Photo Researchers, Inc.; © Bruce Macqueen/Dreamstime; © Aweech/Dreamstime; © Home Science Tools; 256 © Lester V. Bergman/CORBIS; 257 left to right © Westend61 GmbH/Alamy; © Elly Godfroy/Alamy; © rmarnold/iStockphoto; 258 left © Image Source Pink/Alamy; right © Helen Sessions/Alamy; 260 right © Vince Clements/Shutterstock; 261 left to right © Kaupang/iStockphoto; ©Image Source/Alamy; © Vince Clements/Shutterstock; 262 © sébastien Baussais/Alamy; 264 © A. T. Willett/Alamy; 265 top © Xu Xiaolin/Corbis; bottom © Rannoch/Dreamstime.com; 266 © Mitch1921/Dreamstime; 268 top left to right © Kenneth Eward/Photo Researchers, Inc.; © Alex Staroseltsev/Shutterstock; bottom left to right © Lester V. Bergman/CORBIS; © A. T. Willett/Alamy; © Image Source Pink/Alamy; 269 left top to bottom © Vince Clements/Shutterstock; © sciencephotos/Alamy; © wolv/iStockphoto; right top to bottom © Dan Barba/Photolibrary; © Jamdotsi/Dreamstime.

Unit C Looking Back. 274 Left top to bottom © Ebby May/Jupiterimages; © Arcticphoto/Alamy; © Mike Powell/Getty Images; © Design Pics/Jupiterimages; centre top to bottom Charles D. Winters/Photo Researchers, Inc., © Joyce Photographics/Photo Researchers, Inc.; South 12th Photography/Shutterstock; Andrew Lambert Photography/Science Photo Library; right top to bottom Kenneth Eward/Photo Researchers, Inc.; © Alex Staroseltsev/Shutterstock; © Lester V. Bergman/CORBIS; © A. T. Willett/Alamy; ©Image Source Pink/Alamy; 275 © Crystal Craig/Dreamstime; 276 left to right ©Photos.com; © Floortje/iStockphoto; © billnoll/iStockphoto; 281 iStockphoto; 283 top All Canada Photos/Alamy; bottom left to right © Lisa F. Young /Shutterstock; © Bloomberg/Getty Images.

Unit D

284 © Stocktrek Images, Inc./Alamy; 285 top to bottom © Blend Images/Alamy © titelio/shutterstock, © Tomas Bercic/istock photo, © Me issa King/istockphoto; 286 left to right © The Print Collector/Alamy, NASA/courtesy of nasaimages.org, NASA/SCIENCE PHOTO LIBRARY; 287 top left to bottom right Ragnarock/Shutterstock, Stephen Firmender/shutterstock, © Dennis MacDonald/Alamy

Chapter 8. 288 © The Print Collector/Alamy; 289 Top to bottom Mount Stromlo And Siding Springs Observatories, ANU/Photo Researchers, Inc., Paul Kalas (UCB), Michael Fitzgerald (LLNL/UCLA), Franck Marchis (SETI Institute/UCB), James Graham (UCB); 290 left top to bottom Eckhard Slawick/Photo Researchers, Inc., Na Gen Imaging/Getty Images, Ron Erwin/All Canada Photos; 290 right top to bottom Earth Imaging/Getty Images, patrimonio designs limited/shutterstock, European Southern Observatory/Photo Researchers, Inc.; 292 PeKKa Parviainen/Science Photo Library; 293 Diego Barucco/Alamy; 295 © PCN Photography/Alamy; 298 © J Marshal —Tribaleye Images/Alamy; 299 © Philippe Widling/istock photo; 300 NASA/courtesy of nasaimages.org; 301 © Diego Barucco/istockphoto; 302 margin © Alaska Stock LLC/Alamy; 303 NASA/David Scott; 304 NASA/Photo Researchers, Inc.; 306 Friedrich Saurer/Photo Researchers, Inc.; 308 © Lee C. Coombs/PHOTOTAKE/Alamy; 309 John Chumack/Photo Researchers, Inc.; 313 left NASA/SCIENCE PHOTO LIBRARY, right © david gregs/Alamy; 314 Eric Gevaert/shutterstock; 315 © Kevin Schafer/Alamy; 316 Dr. Seth Shostak/Photo Researchers, Inc.; 318 left to right Science Source/Photo Researchers, Inc., John Chumack/Photo Researchers, Inc., W. Keel/U. Alabama/NASA/ESA/STSCI/ Photo Researchers, Inc., Science Source/Photo Researchers, Inc.; 319 top Baback Tafreshi/Photo Researchers, Inc., bottom ROYAL OBSERVATORY, EDINBURGH/AAO/SCIENCE PHOTO LIBRARY; 320 top left to right Friedrich Saurer/Photo Researchers, Inc., PeKKa Parviainen/Science Photo Library, © Alaska Stock LLC/Alamy; bottom left to right © Lee C. Coombs/PHOTOTAKE/Alamy, © david gregs/Alamy, John Chumack/Photo Researchers, Inc.; 321 left top to bottom Eckhard Slawick/Photo Researchers, Inc., Na Gen Imaging/Getty Images, Ron Erwin/All Canada Photos; right top to bottom Earth Imaging/Getty Images, patrimonio designs limited/shutterstock, European Southern Observatory/Photo Researchers, Inc.; 324 W. Keel/U. Alabama/NASA/ESA/STSCI/Photo Researchers, Inc.

Chapter 9. 326 NASA; 327 top R.Creation/Getty Images; bottom Image Source/Veer; 328 left top to bottom Dan Suzio/Photo Researchers, Inc., © Rudolph Pizzonia/istock photo, © NASA/Alamy; right top to bottom Baback Tafreshi/Photo Researchers, Inc., © Laura Coelho/Alamy, John Szenke/Shutterstock; 332 margin PLANETOBSERVER/Science Photo Library, main © blickwinkle/Alamy; 333 © PSL Images/Alamy;; 336 David A. Hardy/Photo Researchers, Inc., 338 top Photoshot Holdings Ltd/Alamy, bottom © Jason Gallier/Alamy; 340 margin David A. Hardy/Photo Researchers, Inc.; 343 main © Cosmo Condina Mexico/Alamy, margin Eckhard Slawik/Photo Researchers, Inc.; 344 © Jack Fields/CORBIS; 345 background left to right © Steve Bloom/Alamy; 349 Eckhard Slawick/Science Photo Library; 351 Babak Tafreshi/Photo Researchers, Inc.; 352 Jerry Schad/Photo Researchers, Inc.; 353 top Babk Tafreshi/Photo Researchers, Inc.; bottom left Emilio SegrÈ Visual Archives/American Institute of Physics/Photo Researchers, Inc.; bottom right Robert Gendler; 356 © fStop/Alamy; 358 top middle © PSL Images/Alamy; bottom left to right © Cosmo Condina Mexico/Alamy, Babak Tafreshi/Photo Researchers, Inc., Robert Gendler; 359 left top to bottom Gerard Lodriguss/Photo Researchers, Inc., © Rudolph Pizzonia/istock photo, © NASA/Alamy; right top to bottom Baback Tafreshi/Photo Researchers, Inc., ©Laura Coelho/Alamy, John Szenke/Shutterstock

Chapter 10. 364–365 © NASA/Science Photo Library; 365 top © Roger Ressmeyer/CORBIS bottom © Corbis/Superstock; 366 left top to bottom © Grant Heilman Photography/Alamy, NASA/courtesy of nasaimages.org, AFP/Getty Images; right top to bottom © NASA/Roger Ressmeyer/CORBIS, John Lee/Aurora Photos, © ANATOLY MALTSEV/epa/Corbis; 368 TRW and Ball Aerospace/European Space Agency; 369 clockwise Stocktrek Images/Getty Images, NASA/courtesy of nasaimages.org, Science and Society/SuperStock,

David Ducros/Photo Researchers, Inc.; 371 main © Space Photos/Alamy; margin ©NASA/Courtesy of Apercu/CP Images; 373 AFP/Getty Images; 374 Getty Images; 375 left to right NASA/courtesy of nasaimages.org, © NASA/Roger Ressmeyer/CORBIS, NASA/courtesy of nasaimages.org; 376 margin © NASA TV/Reuters/Corbis, main NASA/SPL/Photo Researchers, Inc.; 378 © NASA/Reuters/Corbis; 380 AFP/Getty Images; 381 top right Getty Images bottom right Dennis Hallinan/Alamy; 382 LACROIX/Photo Researchers, Inc.; 383 DETLEV VAN RAVENSWAAY/SCIENCE PHOTO LIBRARY; NASA/SPL/Photo Researchers, Inc.; 387 NASA/SPL/Photo Researchers, Inc.; 388 Stocktrek Images; 389 DEXTRE robotic arm, http://www.asc-csa.gc.ca/eng/missions/sts-123/dextre.asp, CSA 2008; 390 left NASA, right corner background © Stocktrek Images, Inc./Alamy ; 391 main © VO TRUNG DUNG/CORBIS SYGMA, margin NASA/Photo Researchers, Inc.; 392 left © D. Hurst/Alamy; right © roberta casaliggi/istock photos; 394 NASA/JPL/UA/Lockheed Martin/courtesy of nasaimages.org; 395 Bristol Aerospace Limited; 396 top left to right ©NASA/Courtesy of Apercu/CP Images, NASA/courtesy of nasaimages.org, AFP/Getty Images; bottom left to right NASA/SPL/Photo Researchers, Inc., DEXTRE robotic arm, http://www.asc-csa.gc.ca/eng/missions/sts-123/dextre.asp, CSA 2008, © D. Hurst/Alamy; 397 left top to bottom © Grant Heilman Photography/Alamy, NASA/courtesy of nasaimages.org, AFP/Getty Images; right top to bottom © NASA/Roger Ressmeyer/CORBIS, John Lee/Aurora Photos, © ANATOLY MALTSEV/epa/Corbis; 401 Stocktrek Images, Inc./Alamy

Unit D Looking Back. 402 left top to bottom PeKKa Parviainen/Science Photo Library, © Alaska Stock LLC/Alamy, © david gregs/Alamy, John Chumack/Photo Researchers, Inc.; middle top to, © Cosmo Condina Mexico/Alamy, Robert Gendler; right top to bottom ©NASA/Courtesy of Apercu/CP Images, NASA//courtesy of nasaimages.org, AFP/Getty Images, NASA/SPL/Photo Researchers, Inc., DEXTRE robotic arm, http://www.asc-csa.gc.ca/eng/missions/sts-123/dextre.asp, CSA 2008, © D. Hurst/Alamy; 407 NASA Archive/Alamy ; 410 Stocktrek Images, Inc./Alamy.

Unit E

412 © Gaertner/Alamy; 413 top to bottom © Kingjon/Dreamstime; © Jose Gil/Shutterstock; © Andersen Ross/Jupiterimages; Christoforos Giatzakis/Alamy; 414 left to right © All Canada Photos/Alamy; © Sam Javanrouh—ddoi.ca /Getty Images; © Rick Gayle Studio/CORBIS; 415 left top left to right © Grant Heilman Photography/Alamy; © K-PHOTOS/Alamy; left middle © Joe Belanger/Alamy; bottom right © Gunter Marx/Alamy.

Chapter 11. 416 © All Canada Photos/Alamy; 417 top © Reuters/CORBIS; bottom left to right © Studiotouch/Shutterstock; © Karen Andersen /Canadian Press Images; 418 left top to bottom © Jose Fuste Raga/Zeta/CORBIS; Chantal Cameron/Niagara Falls (Ontario) Public Library; © NASA/Corbis; © Abemossop/Dreamstime.com; © Andrew Buckin/Shutterstock; © David Stuckel/Alamy; 420 © Erik Isakson/Corbis; 421 © Jonathan Nourok/PhotoEdit; 424 CP PHOTO/Dave Chidley; 425 © John La Gette/Alamy; 426 top Kevin Frayer/CP PHOTO; bottom Francois Mori/AP Photo/CP PHOTO; 428 top © TOM MARESCHAL/Alamy; bottom © Karen Tweedy-Holmes/CORBIS; 429 © Photos.com; bottom Bill Brooks/Alamy; 430 © Arctic-Images/Corbis; 434 both © Photos.com; 435 top © Michael Willis/Alamy; © KAKIMAGE/Alamy; 436 top Alexandra McLeod; bottom left to right Jonathan Vasata/Shutterstock; © Photographer's Choice/Getty Images; FloridaStock/Shutterstock; 439 © Kevin Foy/Alamy; 440 © Alfonso de Tomás /Shutterstock; 442 © Imagemore Co., Ltd./Corbis; 443 © Reproduced with permission of the Minister of Natural Resources, Canada 2010; 444 top © Reproduced with permission of the Minister of Natural Resources, Canada 2010; bottom © greenwales/Alamy; 446 © Alfonso de Tomás/Shutterstock; 450 top left to right ©Photos.com; Bill Brooks/Alamy; bottom left to right Francois Mori/AP Photo/CP PHOTO; ©Reproduced with permission of the Minister of Natural Resources, Canada 2010;. 451 left top to bottom ©Jose Fuste Raga/Zeta/CORBIS; Chantal Cameron/Niagara Falls (Ontario) Public Library; © NASA/Corbis; right top to bottom © Abemossop/Dreamstime.com; © Andrew Buckin/Shutterstock; © David Stuckel/Alamy; 454 © Reproduced with permission of the Minister of Natural Resources, Canada 2010.

Chapter 12. 456 © Sam Javanrouh—ddoi.ca /Getty Images; 457 © Luscious Frames/Alamy; 458 left top to bottom © Dana Hoff/Getty Images; ©Michael Newman/Photo Edit; © David R. Frazier Photolibrary, Inc./Alamy; right top to bottom Charles D. Winters/Photo Researchers, Inc.; © Frank Vetere/Alamy; 464 © Mark Burnett/Photo Researchers, Inc.; 465 © José Elias—Technology series/Alamy; 466 top to bottom left © WILDLIFE GmbH/Alamy; © INSADCO Photography/Alamy; right © Ali Kabas/Alamy; 480 Science Kit and Boreal Laboratories; 482 Science Kit and Boreal Laboratories; 484 left ©Phillip Duf/Alamy; right © Ken Kaminesky/Take 2 Productions/Corbis; 485 © alistair siddons/Alamy; 489 left top to bottom © Takeshi Takahara/Photo Researchers, Inc.; © Mike Goldwater/Alamy; right © Mike Hill/Alamy; 490 top left to right © Ali Kabas/Alamy; © INSADCO Photography/Alamy; bottom left to right © Ken Kaminesky/Take 2 Productions/Corbis; © alistair siddons/Alamy; 491 left top to bottom © Dana Hoff/Getty Images; © Michael Newman/Photo Edit; © David R. Frazier Photolibrary, Inc./Alamy; right top to bottom Charles D. Winters/Photo Researchers, Inc.; © Frank Vetere/Alamy; 493 Science Kit and Boreal Laboratories; 494 Andrew Lambert Photography/Photo Researchers, Inc.

Chapter 13. 496 © Rick Gayle Studio/CORBIS; background © da-kuk/iStockphoto; 497 top to bottom © Mark Boulton/Alamy; © mr. Brightside/Shutterstock; background © timsa/iStockphoto; © zcw26/iStockphoto; 498 left top to bottom © KAKIMAGE/Alamy; © Patrick Ryan/Getty Images; © Michael Newman/PhotoEdit; right top to bottom © amana images inc./Alamy; © petpics/Alamy; © Doug Martin/Photo Researchers, Inc.; 504 © cloki/Shutterstock; © David J. Green/Alamy; 506 © cloki/Shutterstock; © David J. Green/Alamy; 508 left top © NRT-Sports/Alamy; 510 © Andrew Scheck/Shutterstock; 511 © Stock Connection/Alamy; 512 left to right © c./Shutterstock; © David Zimmerman/CORBIS; 513 © Photos.com; 514 left to right ©Andrew Scheck/Shutterstock; © Photos.com; © David J. Green/Alamy; 516 left to right © Andrew Scheck/Shutterstock; © Photos.com; © David J. Green/Alamy; 518 top to bottom © DK Limited/CORBIS; © Jurgen Reisch/Getty Images; 519 © david pearson/Alamy; 520 left to right © Getty Images; © Purestock/Alamy; 522 top to bottom © St. Stephen's Hospital, London/Photo Researchers, Inc.; © Andrew Lambert Photography/Photo Researchers, Inc.; © Photos.com; 523 top to bottom © Leland Bobbe/Getty Images; © Photos.com; © Tony Freeman/PhotoEdit; 525 left © ICP-DE/Alamy; right top to bottom © Spencer Grant/PhotoEdit; © Najlah Feanny/Corbis; 526 left © Philip Lewis/Alamy; bottom right to left © Photos.com; © Sean MacD/Shutterstock; 527 top to bottom © Adrov Andriy/Shutterstock; © Aleksandar Jocic/Dreamstime; ©Martyn F. Chillmaid/Photo Researchers, Inc.; 528 bottom left to right © Philip Lewis/Alamy; © DK Limited/CORBIS; 529 left top to bottom © KAKIMAGE/Alamy; © Patrick Ryan/Getty Images; © Michael Newman/PhotoEdit; right top to bottom © amana images inc./Alamy; © petpics/Alamy; © Doug Martin/Photo Researchers, Inc.

Unit E Looking Back. 534 left top to bottom photos.com; Bill Brooks/Alamy; Francois Mori/AP Photo/CP PHOTO; © Reproduced with permission of the Minister of Natural Resources, Canada 2010; middle top to bottom © Ali Kabas/Alamy; © INSADCO Photography/Alamy; © Ken Kaminesky/Take 2 Productions/Corbis; © alistair siddons/Alamy; right top to bottom © Philip Lewis/Alamy; © DK Limited/CORBIS; 535 © Lisa F. Young/Alamy; 536 left to right © cloki/Shutterstock; © David J. Green/Alamy.

Skills Handbook

544 Veer Royalty Free; 545 top left to right photos.com, © Nikreates/Alamy; bottom left to right © Peter Arnold, Inc./Alamy, Jose Gil /Shutterstock; 557: top right © George Steinmetz, bottom left to right © Arnulf Husmo/Getty, Charles D. Winters/Photo Researchers, Inc., © Yoav Levy/PHOTOTAKE/Alamy; 558 Charles D. Winters/Photo Researchers, Inc.; 561 © sciencephoto/ Alamy; 563 Corbis Premium RF/Alamy; 566 Ian Shaw/Alamy; 567 left to right David J.Green/Alamy, Corbis Premium RF/Alamy; 571 David Grossman/Alamy; 591 © Eraxion/Dreamstime, Wim van Egmond/Visuals Unlimited/Getty Images, © 2009 David B. Fleetham/Oxford Scientific/Jupiterimages Corporation, © All Canada Photos.

Studio Photography: Dave Starrett

Nelson Science Connections

Key

atomic number	26
most common ion charge	3+
other ion charge	2+
symbol of element (solids in black, gases in red, liquids in blue)	Fe
name of element	iron
atomic mass (u)—based on C-12	55.85

Period	1	2	3	4	5	6	7	8	9
1	1 1+ 1− **H** hydrogen 1.01								
2	3 1+ **Li** lithium 6.94	4 2+ **Be** beryllium 9.01							
3	11 1+ **Na** sodium 22.99	12 2+ **Mg** magnesium 24.31							
4	19 1+ **K** potassium 39.10	20 2+ **Ca** calcium 40.08	21 3+ **Sc** scandium 44.96	22 4+ 3+ **Ti** titanium 47.87	23 5+ 4+ **V** vanadium 50.94	24 3+ 2+ **Cr** chromium 52.00	25 2+ 4+ **Mn** manganese 54.94	26 3+ 2+ **Fe** iron 55.85	27 2+ 3+ **Co** cobalt 58.93
5	37 1+ **Rb** rubidium 85.47	38 2+ **Sr** strontium 87.62	39 3+ **Y** yttrium 88.91	40 4+ **Zr** zirconium 91.22	41 5+ 3+ **Nb** niobium 92.91	42 6+ **Mo** molybdenum 95.94	43 7+ **Tc** technetium (98)	44 3+ 4+ **Ru** ruthenium 101.07	45 3+ **Rh** rhodium 102.91
6	55 1+ **Cs** cesium 132.91	56 2+ **Ba** barium 137.33	57 3+ 2+ **La** lanthanum 138.91	72 4+ **Hf** hafnium 178.49	73 5+ **Ta** tantalum 180.95	74 6+ **W** tungsten 183.84	75 7+ **Re** rhenium 186.21	76 4+ **Os** osmium 190.23	77 4+ **Ir** iridium 192.22
7	87 1+ **Fr** francium (223)	88 2+ **Ra** radium (226)	89 3+ 2+ **Ac** actinium (227)	104 **Rf** rutherfordium (261)	105 **Db** dubnium (262)	106 **Sg** seaborgium (266)	107 **Bh** bohrium (264)	108 **Hs** hassium (277)	109 **Mt** meitnerium (268)
	Alkali metals	Alkaline earth metals							

Period					
6	58 3+ **Ce** cerium 140.12	59 3+ **Pr** praseodymium 140.91	60 3+ **Nd** neodymium 144.24	61 3+ **Pm** promethium (145)	62 3+ 2+ **Sm** samarium 150.36
7	90 4+ **Th** thorium 232.04	91 5+ 4+ **Pa** protactinium 231.04	92 6+ 4+ **U** uranium 238.03	93 5+ **Np** neptunium (237)	94 4+ 6+ **Pu** plutonium (244)

- Metals
- Metalloids
- Nonmetals
- Hydrogen